Advances in Organometallic Chemistry

EDITED BY

F. G. A. STONE

DEPARTMENT OF INORGANIC CHEMISTRY

THE UNIVERSITY

BRISTOL, ENGLAND

ROBERT WEST

DEPARTMENT OF CHEMISTRY

UNIVERSITY OF WISCONSIN

MADISON, WISCONSIN

VOLUME 28

ACADEMIC PRESS, INC.

Harcourt Brace Jovanovich, Publishers

San Diego   New York   Berkeley   Boston
London   Sydney   Tokyo   Toronto

Advances in
ORGANOMETALLIC CHEMISTRY

VOLUME 28

Academic Press, Inc.
1250 Sixth Avenue
San Diego, California 92101

United Kingdom Edition published by
ACADEMIC PRESS INC. (LONDON) LTD.
24-28 Oval Road, London NW1 7DX

LIBRARY OF CONGRESS CATALOG CARD NUMBER: 64-16030

ISBN 0-12-031128-3   (alk. paper)

PRINTED IN THE UNITED STATES OF AMERICA
88 89 90 91   9 8 7 6 5 4 3 2 1

Contents

Interaction of Ketenes with Organometallic Compounds: Ketene, Ketenyl, and Ketenylidene Complexes

GREGORY L. GEOFFROY AND SHERRI L. BASSNER

Graphite–Metal Compounds

RENÉ CSUK, BRIGITTE I. GLÄNZER, AND ALOIS FÜRSTNER

Nucleophilic Activation of Carbon Monoxide: Applications to Homogeneous Catalysis by Metal Carbonyls of the Water Gas Shift and Related Reactions

PETER C. FORD AND ANDRZEJ ROKICKI

Organopalladium and Platinum Compounds with Pentahalophenyl Ligands

RAFAEL USÓN AND JUAN FORNIÉS

H–H, C–H, and Related Sigma-Bonded Groups as Ligands

ROBERT H. CRABTREE AND DOUGLAS G. HAMILTON

Organometallic Compounds Containing Oxygen Atoms

FRANK BOTTOMLEY AND LORI SUTIN

Recent Developments in NMR Spectroscopy of Organometallic Compounds

BRIAN E. MANN

ADVANCES IN ORGANOMETALLIC CHEMISTRY, VOL. 28

Interaction of Ketenes with Organometallic Compounds: Ketene, Ketenyl, and Ketenylidene Complexes[1]

GREGORY L. GEOFFROY and SHERRI L. BASSNER

Department of Chemistry
The Pennsylvania State University
University Park, Pennsylvania 16802

I

INTRODUCTION

Ketenes are reactive organic molecules whose chemistry has been investigated since the early 1900s, when Staudinger in Germany and Wilsmore in England conducted many of their pioneering studies. The characteristic organic chemistry of ketenes is now well defined, and excellent reviews on various aspects of the synthesis and properties of ketenes are available (*1–3*). In contrast, the *organometallic* chemistry of ketenes is only now beginning to be seriously developed, although a number of interesting metal–ketene reactions were described in the early organometallic literature. There have been no reviews of the interaction of metals with ketenes, and the purpose of this chapter is to collect and interpret the available data so as to provide a foundation for future studies.

Four separate but related aspects of the interaction of ketenes with metals are reviewed here. These are

(a) The reactions of ketenes with organometallic complexes.
(b) The preparation, structures, and reactions of stable ketene complexes.
(c) The chemistry of *ketenyl* complexes that have a metal as one of the ketene substituents.
(d) The chemistry of *ketenylidene* complexes in which the ketene carbon has only metals as substituents.

[1] *Abbreviations*: Cp, η-C_5H_5; Cp*, η-C_5Me_5; Cp′, η-C_5H_4Me; Tf, OSO_2CF_3; Tp′, $HB\{Me_2C_3HN_2\}_3$; dppm, $Ph_2PCH_2PPh_2$; DMAP, *p*-$NC_5H_4NMe_2$.

Also discussed are reactions in which ketene complexes are suspected as intermediates, but proof of their structure and formulations is lacking. Coverage of the literature is through 1986.

A brief review of the characteristic chemistry of ketenes will help to place in perspective reactivity studies of ketene complexes. Much of the reactivity of ketenes can be rationalized by the polarization of the molecule shown below (*2b*). Frontier orbital arguments give essentially the same

$$\overset{\delta^-}{R_2C}=\overset{\delta^+}{\underset{\alpha}{C}}=\overset{\delta^-}{\underset{\beta}{O}}$$

conclusions in regard to reactivity since the lowest unoccupied molecular orbital (LUMO) is mainly localized on the β carbon and the highest occupied molecular orbital (HOMO) is localized on the α carbon and the oxygen atom (*4*). Nucleophiles tend to add to the β carbon, whereas electrophiles add to the terminal atoms. Characteristic reactions of ketenes are addition of polar H–X bonds across the ketene C=C bond [Eq. (1)] (*2a*). Ketenes, particularly those with electron-withdrawing substituents,

$$
\begin{array}{c}
\overset{\delta^-}{R_2C}=\overset{\delta^+}{C}=\overset{\delta^-}{O} \\
+ \\
\overset{\delta^+}{H}-\overset{\delta^-}{X}
\end{array}
\longrightarrow
R_2\underset{H}{C}-C\overset{O}{\underset{X}{\diagup}}
\tag{1}
$$

$$X = OR, \ NR_2, \ SR, \ etc.$$

undergo cycloaddition reactions with electron-rich olefins, alkynes, and other unsaturated organics, e.g., Eq. (2) (*2a*). These reactions are thought

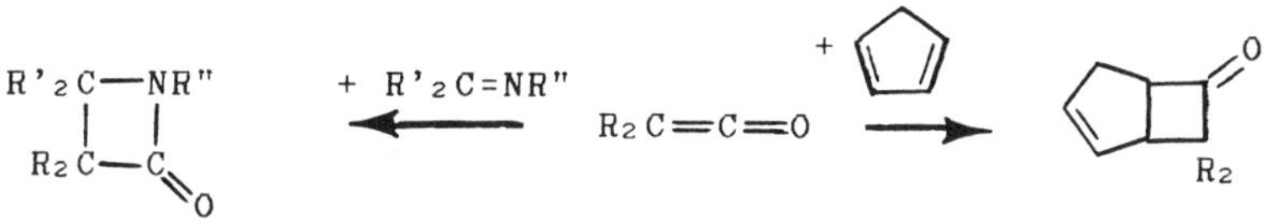

$$
\begin{array}{c}
R'_2C-NR'' \\
| \quad\quad | \\
R_2C-C \\
\quad\quad\ \diagdown O
\end{array}
\underset{\longleftarrow}{\overset{+\ R'_2C=NR''}{}}
R_2C=C=O
\longrightarrow
\tag{2}
$$

to proceed via an allowed antarafacial $[\pi 2s + \pi 2a]$ concerted process (*4*). Such a reaction accounts for the facile dimerization that many ketenes undergo and accounts for their instability. Dimerization can occur in either head-to-head or head-to-tail fashion, depending on the ketene substituents [Eq. (3)] (*1,2a*).

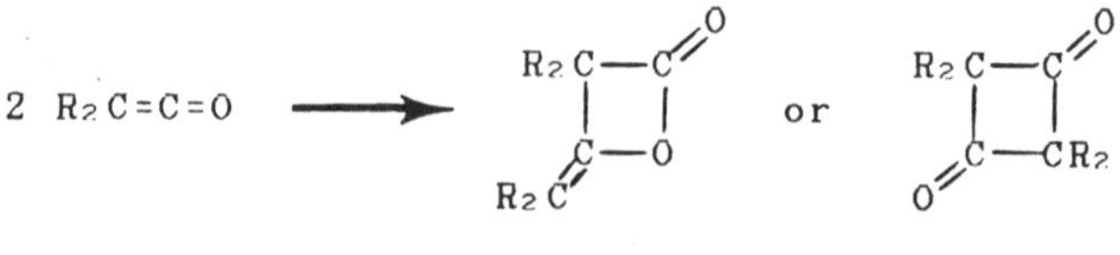

$$
2\ R_2C=C=O \longrightarrow
\begin{array}{c}
R_2C-C\overset{O}{\diagup} \\
| \quad\quad | \\
\underset{R_2C}{\diagup}C-O
\end{array}
\quad or \quad
\begin{array}{c}
R_2C-C\overset{O}{\diagup} \\
| \quad\quad | \\
\underset{O}{\diagdown}C-CR_2
\end{array}
\tag{3}
$$

II

REACTIONS OF KETENES WITH ORGANOMETALLIC COMPLEXES

Ketenes have been observed to react with many metal complexes to give products arising from (a) simple addition of the ketene to the metal, (b) insertion into metal–ligand bonds, (c) reactions at coordinated ligands, (d) transformation of the ketene into new ligands, and (e) metal-induced conversion of the ketene into other organics. Each of these classes of reactions is discussed in detail below.

A. Addition of Ketenes to Metals to Form Stable Ketene Complexes

This is one of the more important ways to prepare ketene complexes and a number of such reactions are known. However, discussion of these reactions is deffered until Section III, when all the stable ketene complexes are discussed together.

B. Insertion of Ketenes into Metal–Ligand Bonds

1. Insertion into Metal–Hydride Bonds

Many unsaturated organic molecules insert into metal–hydride bonds, but surprisingly only two examples of such reactions with ketenes are known [Eqs. (4) and (5)] (5,6). Such a reaction would seem to offer a

$$\text{H-Mn(CO)}_5 \quad + \quad \text{CH}_2=\text{C}=\text{O} \quad \xrightarrow{-30^\circ \text{C}} \quad \text{CH}_3\overset{\text{O}}{\overset{\|}{\text{C}}}\text{-Mn(CO)}_5 \qquad (4)$$

$$\text{H-Co(CO)}_4 \quad + \quad \text{RR'C}=\text{C}=\text{O} \quad \xrightarrow{-78^\circ \text{C}} \quad \overset{\text{R}}{\underset{\text{R'}}{\diagdown}}\text{CH}\overset{\text{O}}{\overset{\|}{\text{C}}}\text{-Co(CO)}_4 \qquad (5)$$

$$\text{R}=\text{R'}=\text{CH}_3 \quad (90\%)$$
$$\text{R}=\text{Et}, \ \text{R'}=\text{H} \quad (95\%)$$

convenient way to prepare metal acyl complexes if the high yields of Eq. (5) are typical. The mechanisms of these reactions are unknown, but it is important to note that the addition of these acidic H–M reagents follows the same pattern as addition of many other H–X reagents to ketenes [Eq. (1)] (1).

The reverse of such an insertion reaction has recently been demonstrated. Convincing evidence was presented for the evolution of ketene during the reaction of Eq. (6), and it was suggested that the intermediate acyl complex (**2**) lost ketene by a β-elimination process (*7,8*).

$$
Ru(CO)_3(triphos) + CH_3-\overset{O}{\overset{\|}{C}}-Cl \longrightarrow
$$
(**1**)

$$
[(triphos)(CO)_2Ru-\overset{O}{\overset{\|}{C}}-CH_3]^+Cl^- + CO
$$
(**2**)

$\downarrow$ $-$ CH$_2$=C=O

$$
[(triphos)(CO)_2Ru-H]^+Cl^-
$$

(6)

Ketenes have also been observed to insert into μ-H bonds [Eq. (7)] (*9*).

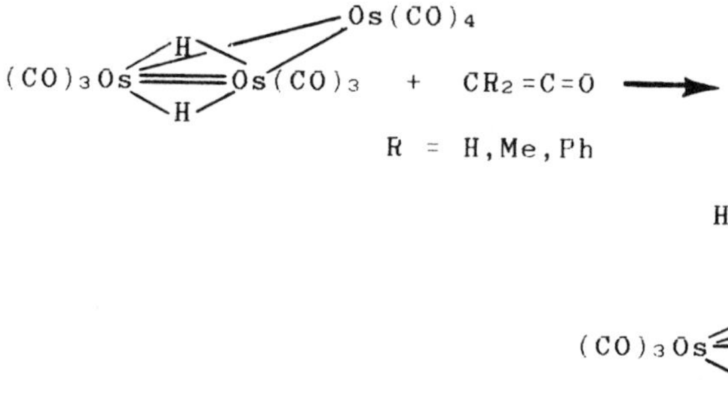

$$
(CO)_3Os\overset{H}{\underset{H}{=\!=\!=}}Os(CO)_3 \quad + \quad CR_2=C=O \longrightarrow
$$
with Os(CO)$_4$ bridging

R = H, Me, Ph

(7)

This reaction may proceed via initial coordination of the ketene to the metal framework followed by insertion into the M–H bond. Note, however, that addition of M–H across the ketene O=C bond has occurred, in contrast to the additions across the C=C bond in Eqs. (4) and (5).

2. Insertion into Metal–Alkyl Bonds

Only two reports have described the insertion of ketenes into metal–alkyl bonds. The complex Cp$_2$ZrMe$_2$ reacts with diphenylketene to give the insertion products **3** and **4** [Eq. (8)] (*10*). The first insertion occurs much faster than the second, allowing isolation of both complexes and their subsequent characterization by X-ray diffraction. In contrast to the metal–hydride reactions described above, insertion has occurred across the C=O bond. Although the detailed mechanisms of these reactions are not known, the insertion is that expected given the oxophilicity of Zr^{4+} and the polarity of the $\overset{\delta+}{Zr}-\overset{\delta-}{C}H_3$ bond. The reaction can be viewed as addition of nucleophilic CH$_3^-$ to the electropositive ketene carbon, in line with the established ketene reactivity pattern noted in the Introduction.

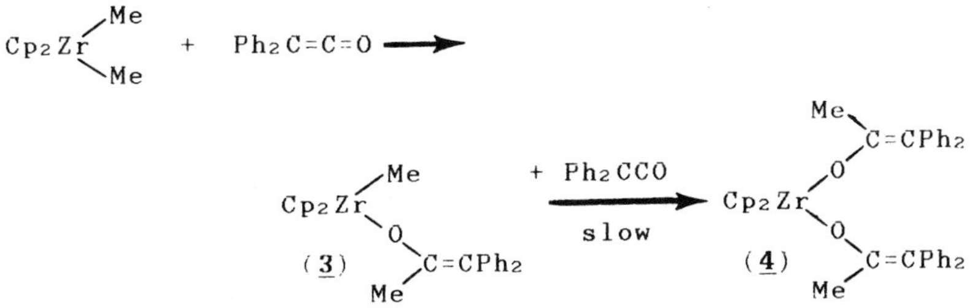

$$Cp_2 Zr \begin{matrix} Me \\ \\ Me \end{matrix} \quad + \quad Ph_2 C=C=O \longrightarrow$$

$$Cp_2 Zr \begin{matrix} Me \\ \\ O \\ \\ \underline{(3)} \quad C=CPh_2 \\ Me \end{matrix} \quad \xrightarrow[slow]{+ \ Ph_2 CCO} \quad Cp_2 Zr \begin{matrix} Me \\ O \\ \\ O \\ \underline{(4)} \quad C=CPh_2 \\ Me \end{matrix} \quad \begin{matrix} C=CPh_2 \end{matrix} \quad (8)$$

A similar reaction apparently occurs with the nickel complex shown in Eq. (9), although the structural assignment for the product was not firmly established (*11*).

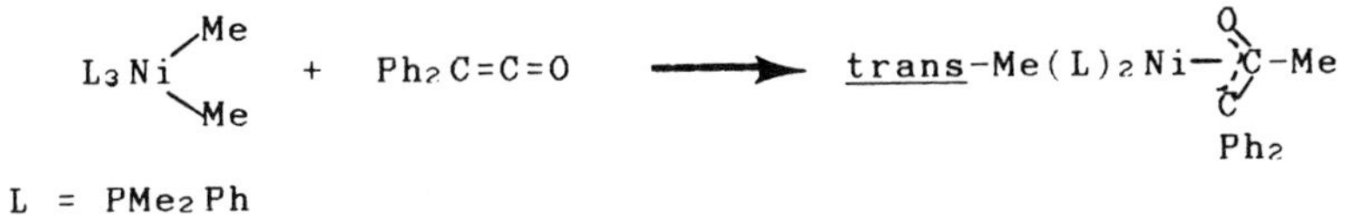

$$L_3 Ni \begin{matrix} Me \\ \\ Me \end{matrix} \quad + \quad Ph_2 C=C=O \quad \longrightarrow \quad \underline{trans}\text{-}Me(L)_2 Ni\text{-}\overset{O}{\underset{Ph_2}{\overset{\|}{C}}}\text{-}Me \quad (9)$$

$$L = PMe_2 Ph$$

3. Insertion into Metal–OR and Metal–NR₂ Bonds

Ketenes readily insert into Ti–OR bonds as illustrated in Eq. (10) (*12*).

$$Ti(OR)_4 \quad + \quad n \ Ph_2 C=C=O \quad \longrightarrow \quad (RO)_{4-n} Ti[CPh_2 - \overset{O}{\overset{\|}{C}} - OR]_n \quad (10)$$

$$R = Et, {}^i Pr, Ph \qquad\qquad n = 1,2$$

A later study demonstrated a similar insertion for $Zr(OBu)_4$ and also determined a preference for ketene insertion into metal–amido over metal–alkoxy bonds [Eq. (11)] (*13*).

$$Ti(OEt)_2 (NMe_2)_2 \quad + \quad Ph_2 C=C=O \longrightarrow (EtO)_2 Ti(CPh_2 - \overset{O}{\overset{\|}{C}} - NMe_2)_2$$
$$(11)$$

Ketones and aldehydes undergo subsequent insertion into the metal alkyl bonds generated from ketene insertion into M–OR bonds and on hydrolysis give a novel preparation of β-hydroxyesters [Eq. (12)] (*14*).

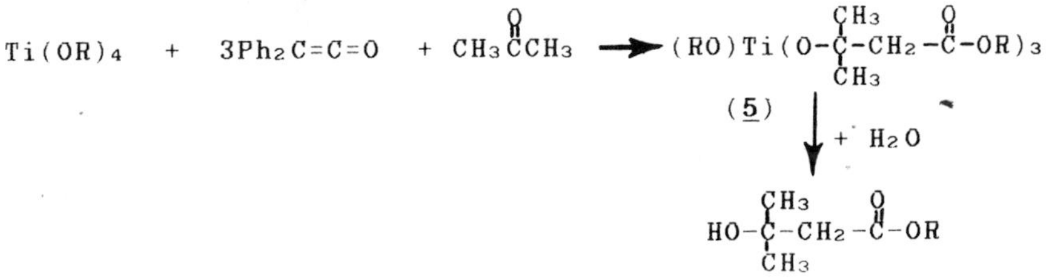

$$Ti(OR)_4 \quad + \quad 3Ph_2 C=C=O \quad + \quad CH_3 \overset{O}{\overset{\|}{C}} CH_3 \longrightarrow (RO)Ti(O-\overset{CH_3}{\underset{CH_3}{\overset{|}{C}}}-CH_2 -\overset{O}{\overset{\|}{C}}-OR)_3$$

$$\underline{(5)} \quad \downarrow + \ H_2 O$$

$$HO-\overset{CH_3}{\underset{CH_3}{\overset{|}{C}}}-CH_2 -\overset{O}{\overset{\|}{C}}-OR$$

$$(12)$$

The initially formed complexes (5) in these reactions belong to the $Ti(OR)_{4-n}(OR')_n$ family and as such can undergo further insertion of ketene into the metal–oxygen bonds. Repetition leads to alternating insertion of up to four ketene and ketone units, and hydrolysis of the resultant ligands from the metal gives a series of *poly*esters (15,16). Several industrial patents have described the $Ti(OR)_4$ and other metal-assisted formation of such polyesters from ketene and crotonaldehyde followed by hydrolysis to give sorbic acid [Eq. (13)] (17).

$$CH_2=C=O + CH_3CH=CH-CHO \xrightarrow{Ti(OR)_4} (RO)_3Ti-(-O-CH-CH_2-\overset{O}{\overset{\|}{C}}-)-O-CH-CH_2-\overset{O}{\overset{\|}{C}}-OR$$

$$\underset{CH=CHCH_3}{\qquad} \times \underset{CH=CHCH_3}{\qquad}$$

$$\downarrow + HCl$$

$$n\ CH_3CH=CH-CH=CH-CO_2H$$

$$(13)$$

Another reaction that likely involves insertion of ketene into a metal–OR bond is that shown in Eq. (14) (18). This reaction may proceed by

$$(py)_2ClPd-CH_2\overset{O}{\overset{\|}{C}}CH_2\overset{O}{\overset{\|}{C}}OCH_2Ph + CH_2=C=O + ROH \longrightarrow$$

$$(py)_2ClPd-CH_2-\overset{O}{\overset{\|}{C}}-OR \qquad (14)$$

initial addition of the alcohol across the Pd–C bond to form the free β-ketoester and a new Pd–OR complex. Insertion of ketene into the latter would give the observed organometallic product.

One interesting aspect of these various insertions into M–OR bonds is the addition across the ketene C=C bond. Addition across the C=O bond, as with the metal–alkyl complexes discussed above, might have been expected with the oxophilic Ti and Zr complexes but such addition did not occur.

4. Insertion into M=O and M–O₂ Bonds

The only reported examples of reaction of ketenes with metal–oxide complexes involves the [2 + 3] and [2 + 2] cycloadditions across Re=O bonds to give complexes 6 and 7 [Eqs. (15) and (16)] (19). Both complexes

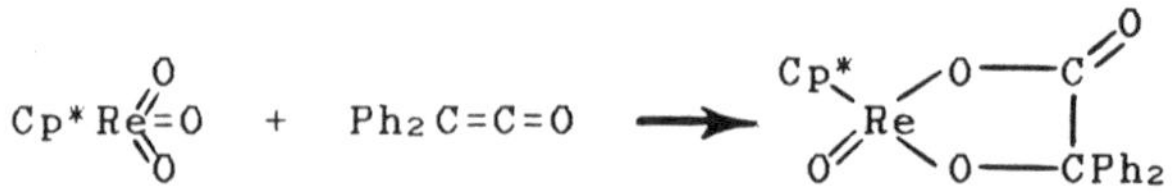

$$Cp^*R\overset{O}{\underset{O}{\overset{\diagup}{Re}}}=O + Ph_2C=C=O \longrightarrow \underset{(\underline{6})}{Cp^*\overset{\diagup O-C\overset{\diagup O}{\diagdown}}{\underset{O\diagdown O-CPh_2}{Re}}} \qquad (15)$$

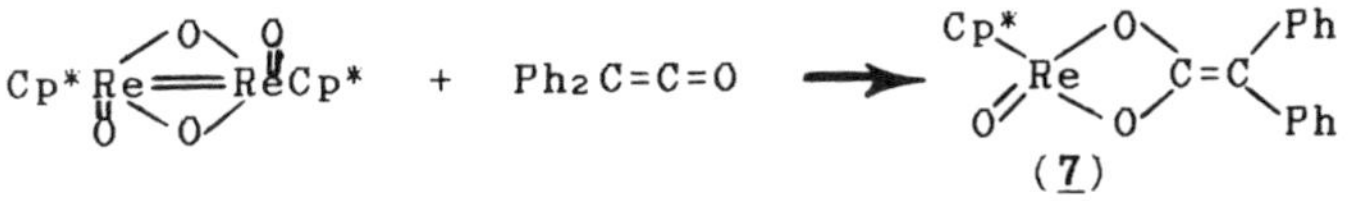

$$Cp^*\,\overset{O}{\underset{O}{Re}}\!\!=\!\!\overset{O}{\underset{O}{Re}}Cp^* \;+\; Ph_2C{=}C{=}O \;\longrightarrow\; Cp^*\,\overset{O}{\underset{O}{Re}}\overset{O}{\underset{O}{\diagdown}}C{=}C\!\!\overset{Ph}{\underset{Ph}{\diagup}} \tag{16}$$

$$(\underline{7})$$

were structurally characterized and shown to possess nearly planar metallacycles.

The dioxygen complex **8** reacts with $Ph_2C{=}C{=}O$ to give a product which was proposed to be **9** on the basis of preliminary spectroscopic evidence [Eq. (17)] (*20*). A similar reaction with the corresponding

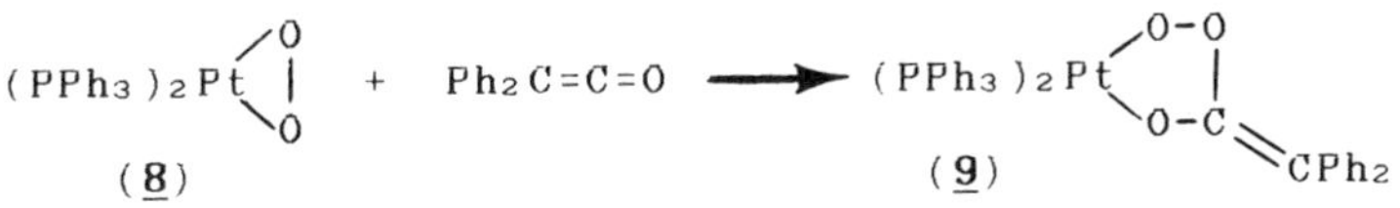

$$(PPh_3)_2Pt\!\!\overset{O}{\underset{O}{\diagup}} \;+\; Ph_2C{=}C{=}O \;\longrightarrow\; (PPh_3)_2Pt\!\!\overset{O-O}{\underset{O-C}{\diagup}}_{\!\!\diagdown CPh_2} \tag{17}$$

$$(\underline{8}) \qquad\qquad\qquad\qquad (\underline{9})$$

palladium complex was suggested to form complex **10** [Eq. (18)], which was subsequently converted into enolate complexes by reaction of substrates with acidic hydrogens (*18*).

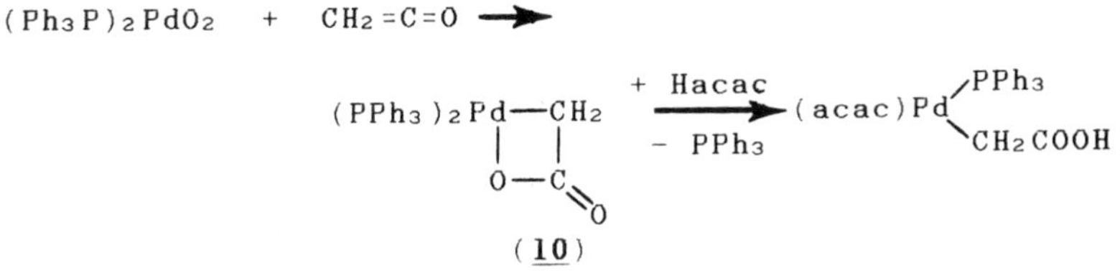

$$(Ph_3P)_2PdO_2 \;+\; CH_2{=}C{=}O \;\longrightarrow$$

$$(PPh_3)_2Pd\!-\!CH_2 \;\underset{-\;PPh_3}{\overset{+\;Hacac}{\longrightarrow}}\; (acac)Pd\!\!\overset{PPh_3}{\underset{CH_2COOH}{\diagup}} \tag{18}$$

$$\underset{(\underline{10})}{O\!-\!C\!\!\underset{\diagdown O}{}}$$

C. Reactions of Ketenes with Coordinated Ligands

There are numerous examples of reactions of ketenes with coordinated ligands with little or no direct interaction of the ketene with the metal. Additions of ketenes to allyl, diene, alkynyl, and an alkynyl-substituted alkyl ligand are shown in Eqs. (19)–(22) [Fp = $Cp(CO)_2Fe$] (*21–24*). The

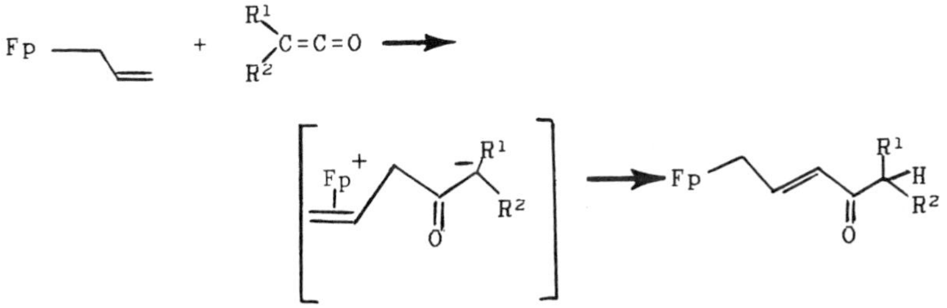

$$Fp\!-\!\diagup\!\!\diagdown \;+\; \overset{R^1}{\underset{R^2}{\diagup}}C{=}C{=}O \;\longrightarrow$$

$$\left[\;\overset{+}{Fp}\diagup\!\!\diagup\!\!\overset{R^1}{\underset{\overset{\|}{O}}{\diagdown}}R^2\;\right]^{\!-} \;\longrightarrow\; Fp\diagup\!\!\diagup\!\!\overset{R^1}{\underset{\overset{\|}{O}}{\diagdown}}\!\!\overset{H}{\underset{R^2}{}} \tag{19}$$

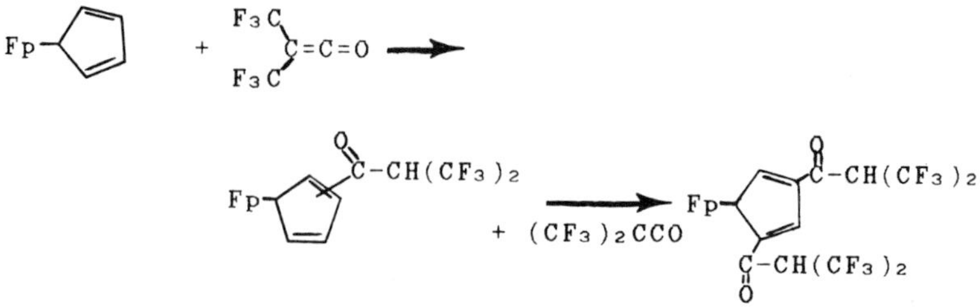

$$Fp-\hspace{-1mm}\bigcirc \quad + \quad \begin{matrix} F_3C \\ \\ F_3C \end{matrix}\hspace{-2mm}C=C=O \longrightarrow$$

(20)

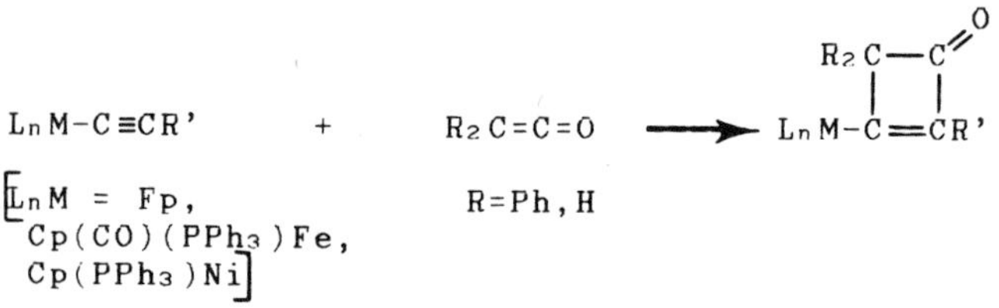

$$LnM-C\equiv CR' \quad + \quad R_2C=C=O \longrightarrow \begin{matrix} R_2C-C \nwarrow^O \\ | \quad\quad | \\ LnM-C=CR' \end{matrix} \quad (21)$$

$$\left[\begin{matrix} LnM = Fp, \\ Cp(CO)(PPh_3)Fe, \\ Cp(PPh_3)Ni \end{matrix} \right] \quad\quad R = Ph, H$$

$$Fp-CH_2C\equiv CR \quad + \quad \begin{matrix} R^1 \\ \\ R^2 \end{matrix}\hspace{-2mm}C=C=O \longrightarrow$$

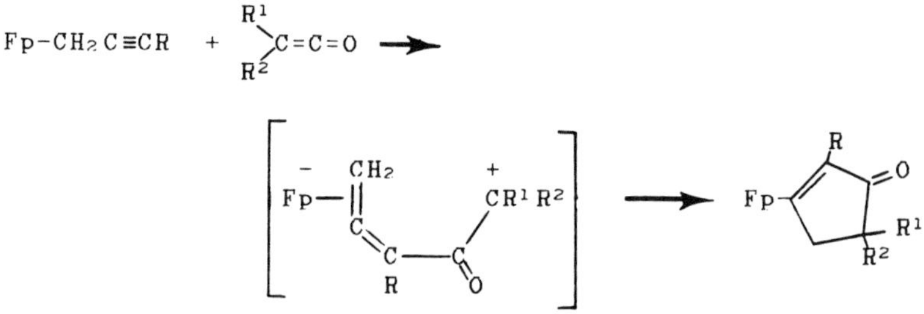

(22)

latter reaction is an example of a 1,3-dipolar cycloaddition, and it also occurs for the corresponding $Cp(CO)_3Mo-CH_2C\equiv CR$ complex (24).

A reaction similar to the addition of aryl lithium reagents to free ketenes occurs with an aryl lithium complex [Eq. (23)] (25). Ketenes have also

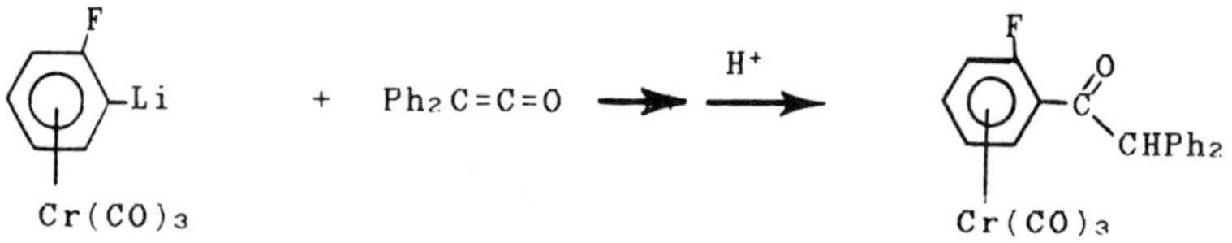

$$+ \quad Ph_2C=C=O \xrightarrow{\quad H^+ \quad}$$

(23)

been shown to undergo [2 + 2] cycloaddition reactions with coordinated polyenes [Eqs. (24) and (25)] (26–28), typical of reactions of ketenes with

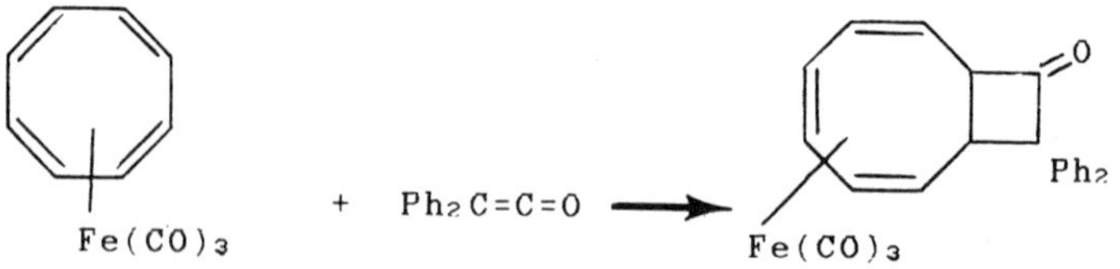

$$+ \quad Ph_2C=C=O \longrightarrow$$

(24)

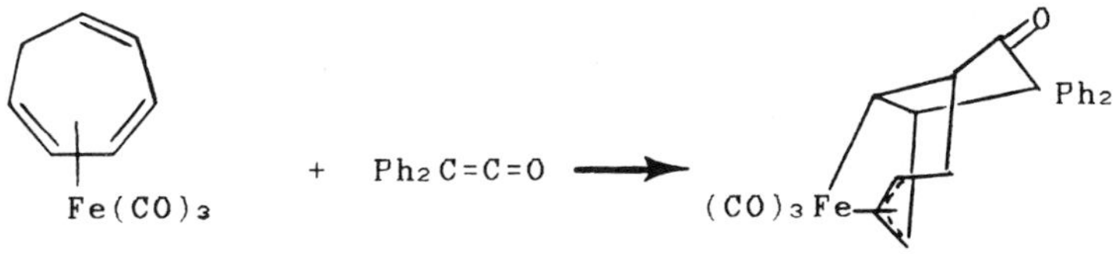

Fe(CO)3
+ Ph2C=C=O
(CO)3Fe
Ph2
O

(25)

free olefins. However, a Wittig-type reaction occurs with coordinated tropone [Eq. (26)], possibly via an intermediate such as **11** (*29*). Such

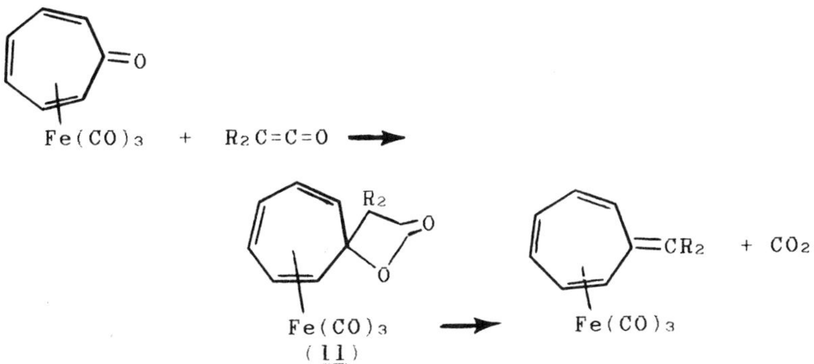

O
Fe(CO)3
+ R2C=C=O
R2
O
O
Fe(CO)3
(11)
CR2
+ CO2
Fe(CO)3

(26)

extrusions of CO_2 commonly occur from the initial cycloaddition products of the reactions of ketenes with ketones (*2a*), and the metal seems to have little effect on this transformation.

A similar cycloaddition reaction with an η^5-Cp ligand has been described [Eq. (27)], but the overall reaction is complicated by the incorporation of two ketenes into the final product **12** (*30–32*). It should be noted that the

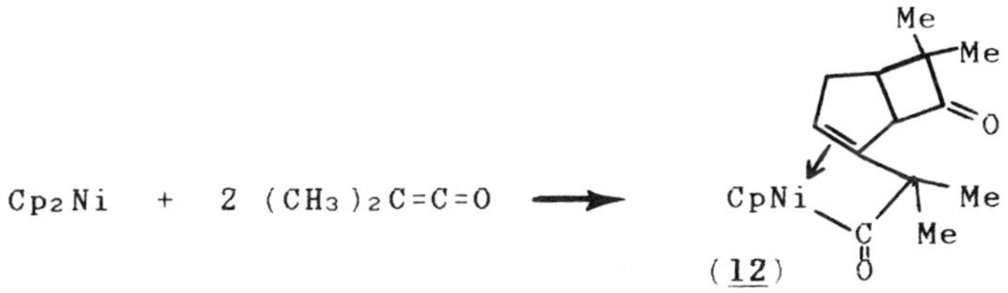

Cp2Ni + 2 (CH3)2C=C=O
Me
Me
O
CpNi
Me
Me
C
O
(12)

(27)

structure assigned to **12** in the original synthesis report (*30*) was later shown to be incorrect by two simultaneous but independent X-ray diffraction studies (*31*,*32*).

Addition of $Ph_2C{=}C{=}O$ to a coordinated azacyclopentadiene ligand leads to the conversion of **13** to **14** by the proposed complex series of transformations shown in Eq. (28) (*33*). A key step is the addition of the H–O bond of the presumed but unobserved **15** across the ketene C=C bond.

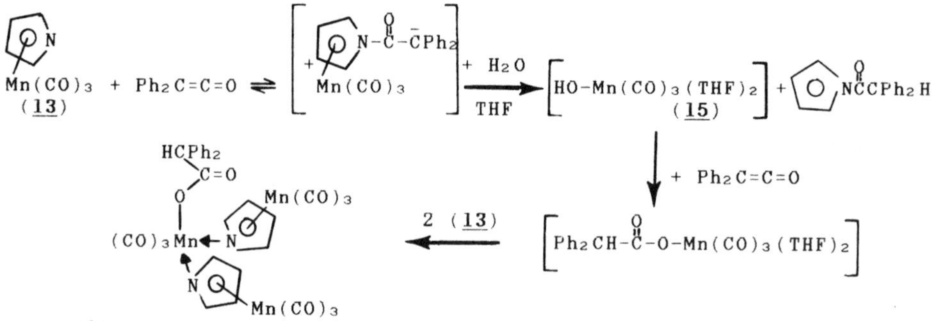

$$Mn(CO)_3 \ (\underline{13}) + Ph_2C=C=O \ \rightleftharpoons \left[\substack{N-\overset{O}{\overset{\|}{C}}-\bar{C}Ph_2 \\ Mn(CO)_3} \right]^+ \xrightarrow[\text{THF}]{+ H_2O} \left[HO-Mn(CO)_3(THF)_2 \ (\underline{15}) \right] + \substack{O \\ N\overset{\|}{C}CPh_2H}$$

$$\substack{HCPh_2 \\ C=O \\ O \\ (CO)_3Mn \quad Mn(CO)_3 \\ N \\ Mn(CO)_3} \ (\underline{14}) \quad \xleftarrow{2 \ (\underline{13})} \quad \left[Ph_2CH-\overset{O}{\overset{\|}{C}}-O-Mn(CO)_3(THF)_2 \right] \quad \xleftarrow{+ Ph_2C=C=O}$$

$$(28)$$

A key step is the addition of the H–O bond of the presumed but unobserved **15** across the ketene C=C bond.

A similar addition of a coordinated S–H bond across the ketene C=C bond has been demonstrated [Eq. (29)] (*34*). The binuclear complex **16** also reacts with diphenylketene to give complex **17** [Eq. (30)] (*34*). Both of

$$[(CO)_5W-SH]^- + Ph_2C=C=O \longrightarrow [(CO)_5W-S-\overset{O}{\overset{\|}{C}}-CHPh_2]^- \quad (29)$$

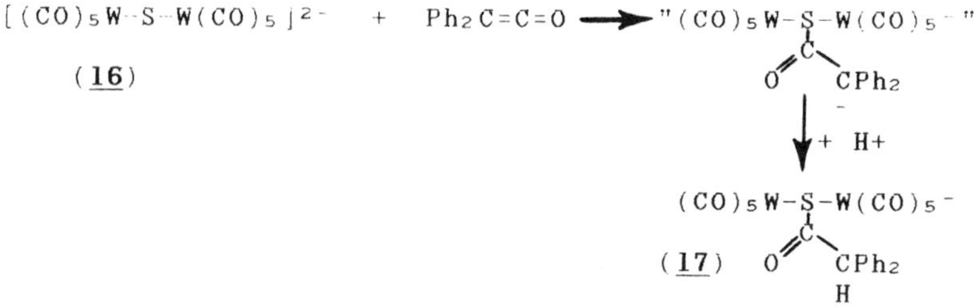

$$[(CO)_5W-S-W(CO)_5]^{2-} + Ph_2C=C=O \longrightarrow {}^{\prime\prime}(CO)_5W-\underset{\underset{O}{\overset{\diagup}{C}}\diagdown CPh_2}{S}-W(CO)_5 {}^{\prime\prime}$$

$$(\underline{16})$$

$$\downarrow + H^+ \quad (30)$$

$$(CO)_5W-\underset{\underset{O}{\overset{\diagup}{C}}\diagdown CPh_2}{S}-W(CO)_5^- \quad (\underline{17}) \qquad H$$

these reactions follow the typical ketene reactivity pattern of nucleophile ($-S^-$) addition to the electropositive ketene β carbon.

Ketenes have also been shown to add between two bridging sulfur atoms in binuclear **18** to give **19** [Eq. (31)] (*35*). Complex **18** is also known to

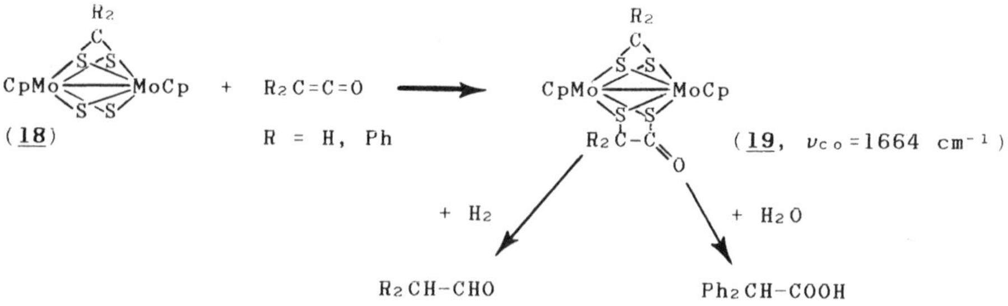

$$\underset{(\underline{18})}{CpMo\underset{S\diagdown S}{\overset{\overset{R_2}{\overset{|}{C}}\diagdown S}{\diagup}}MoCp} + R_2C=C=O \longrightarrow \underset{R_2C-C\diagdown O}{CpMo\underset{S\diagdown S}{\overset{\overset{R_2}{\overset{|}{C}}\diagdown S}{\diagup}}MoCp} \ (\underline{19}, \ \nu_{co}=1664 \ cm^{-1})$$

$$R = H, \ Ph$$

$$\swarrow + H_2 \qquad \qquad + H_2O \searrow$$

$$R_2CH-CHO \qquad \qquad Ph_2CH-COOH$$

$$(31)$$

similarly add olefins and alkynes (*35*). Interestingly, the "coordinated" ketene in **19** reacts with H_2O to form diphenylacetic acid and with H_2 to form acetaldehyde [Eq. (31)]. The latter is especially interesting since ketene, formed via the combination of methylene ligands with CO, is a likely intermediate in the catalytic reduction of CO by H_2 (Fischer–Tropsch chemistry). Molybdenum sulfides are catalysts for this reaction (*36*), and the above chemistry provides a good model for this reaction step.

Equation (32) illustrates a novel nitrogen–carbon bond-forming reaction between a coordinated diimine ligand and added ketene (*37*). Note that

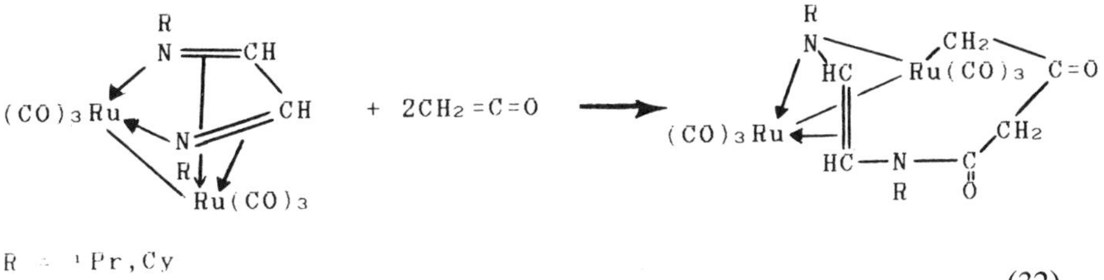

$$(32)$$

the two ketenes which are incorporated in the new ligand have undergone head-to-tail dimerization, which is unprecedented for free ketene.

D. *Transformation of Ketenes into Other Types of Ligands*

1. *Into Carbene Ligands*

A number of reports have described the formation of carbene ligands from the reaction of ketenes with metal complexes. Carbenes can be produced from ketenes either by metal-induced decarbonylation of the ketene or by transfer of the ketene oxygen to a CO ligand to form CO_2 and a vinylidene ligand. An example of the latter is the reaction of $Fe(CO)_5$ with diphenylketene to give the μ-vinylidene complex **20** (*38,39*) in low yield along with the ketene complex **21** (*40,41*) [Eq. (33)]. The latter complex is discussed more fully in Section III.

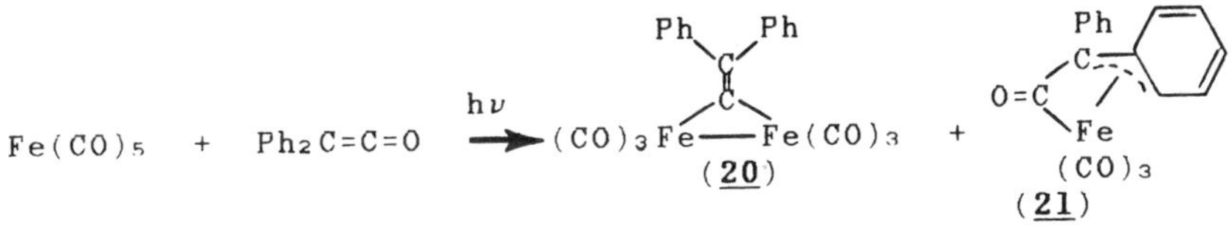

$$(33)$$

A similar but undetected μ-vinylidene complex (22) was proposed to account for the formation of 23 and 24 from the reaction of dimethylketene with $Co_2(CO)_8$ [Eq. (34)] (42). The use of $Co_2(CO)_6(PPh_3)_2$, $Fe(CO)_5$,

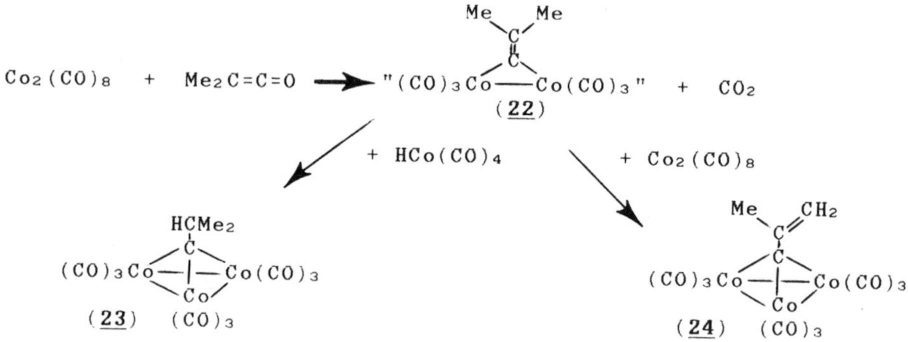

$$Co_2(CO)_8 \ + \ Me_2C=C=O \ \longrightarrow \ "(CO)_3Co\!-\!\!-\!Co(CO)_3" \ + \ CO_2$$

with μ-vinylidene bridge labeled Me, Me, C, C above (22)

(23): HCMe₂ bridged structure $(CO)_3Co\!-\!|\!-\!Co(CO)_3$ over Co(CO)₃

(24): Me, CH₂ bridged structure $(CO)_3Co\!-\!|\!-\!Co(CO)_3$ over Co(CO)₃

+ HCo(CO)₄ → (23)

+ Co₂(CO)₈ → (24)

(34)

or $Fe_2(CO)_9$ in place of $Co_2(CO)_8$ in this reaction gave only high-yield formation of the polyester 25 [Eq. (35)] (42), which is known to form from dimethylketene in the presence of base catalysts (2a).

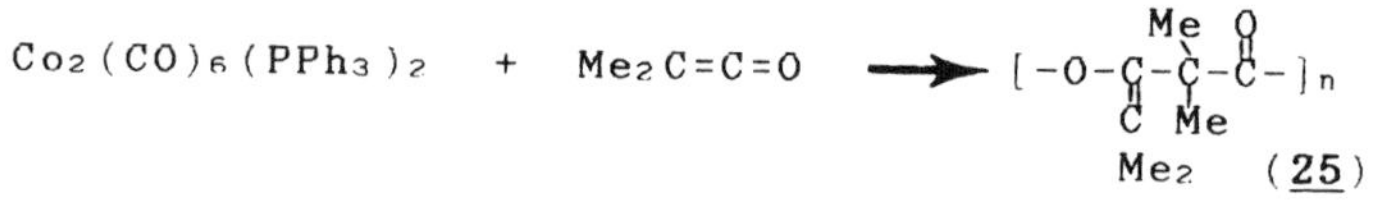

$$Co_2(CO)_6(PPh_3)_2 \ + \ Me_2C=C=O \ \longrightarrow \ [-O-\underset{Me_2}{\overset{Me}{C}}-\overset{O}{\underset{Me}{C}}-\overset{O}{C}-]_n \quad (35)$$

(25)

If $Ph_2C=C=O$ instead of $Me_2C=C=O$ is treated with $Co_2(CO)_8$, only decarbonylation of the ketene occurs to yield $Ph_2C=CPh_2$ [see

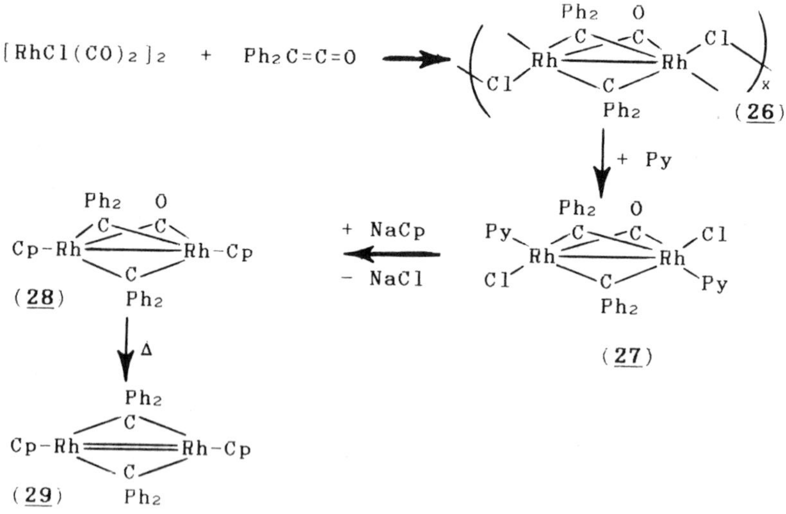

$$[RhCl(CO)_2]_2 \ + \ Ph_2C=C=O \ \longrightarrow \ (26)$$

(26): Ph₂, O bridged Rh–Rh dimer with Cl bridges, subscript x

(26) → + Py → (27)

(27): Py, Ph₂, O bridged Rh–Rh with Cl, Py ligands

(27) → + NaCp / − NaCl → (28)

(28): Cp–Rh bridged with Ph₂, O, C units, Rh–Cp

(28) → Δ → (29)

(29): Cp–Rh≡≡Rh–Cp bridged with Ph₂, C units

(36)

Eq. (39)] (48). Decarbonylation of diphenylketene also occurred when it was treated with [RhCl(CO)$_2$]$_2$, and relatively stable diphenylcarbene complexes were formed [Eq. (36)] (43,44). The oligomeric complex 26 was isolated as an air-stable but insoluble product. The authors proposed a structure with terminal carbene ligands, but the structure drawn in Eq. (36) seems more likely on the basis of the crystallographically established structure of 27 (44) and the tendency of carbene ligands to occupy bridging positions. Complexes 28 and 29 have also been structurally characterized (45), although initially a trinuclear structure was proposed for 29 (43).

Decarbonylation of ketene and formation of a μ-methylene ligand occurred when it was allowed to react with the solvated Os$_3$ cluster 30 [Eq. (37)] (46).

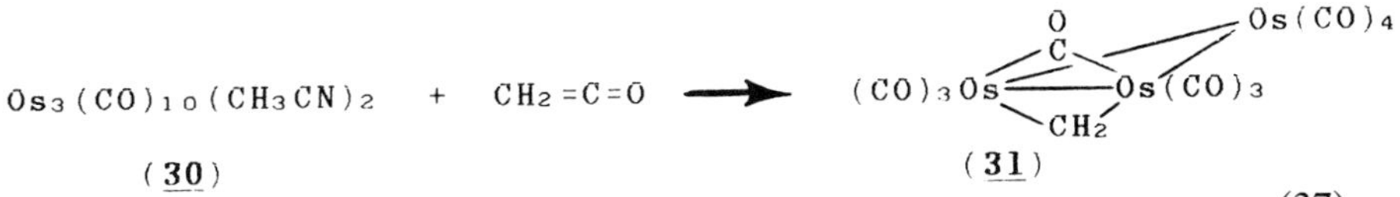

$$Os_3(CO)_{10}(CH_3CN)_2 \ + \ CH_2{=}C{=}O \ \longrightarrow \ (CO)_3Os{\Longleftarrow}Os(CO)_3$$

(**30**)   (**31**)

(37)

2. Into Other Ligands

Reaction of diphenylketene with Ni(COD)$_2$ in the presence of bidentate amine ligands gives a product 33 with a ligand formed by a head-to-tail dimerization of the ketene (47). This was deduced from analysis of the organics produced on reaction of 33 with acid and CO [Eq. (38)], but the exact structure of 33 is unknown. The two proposed structures are shown.

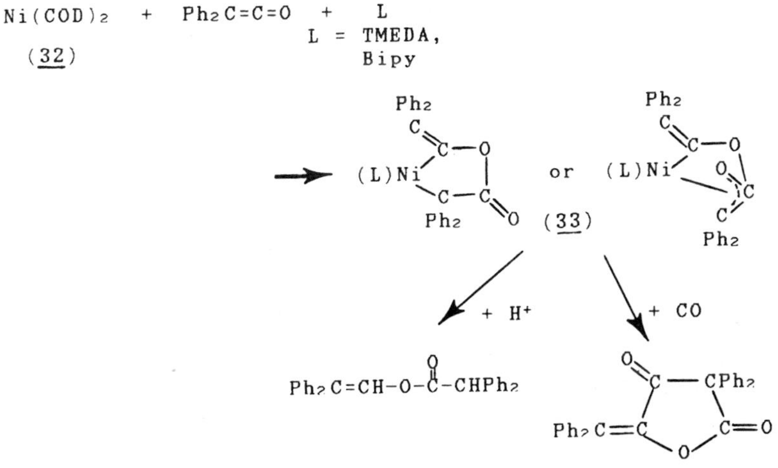

$$Ni(COD)_2 \ + \ Ph_2C{=}C{=}O \ + \ L$$

L = TMEDA,

(**32**)   Bipy

(38)

$$Ph_2C{=}CH{-}O{-}\overset{O}{\overset{\|}{C}}{-}CHPh_2$$

E. Metal-Induced Conversion of Ketenes into Other Organics

The reactions of ketenes with metals do not always lead to new organometallic complexes but often lead to transformations of the ketene. For example, $Co_2(CO)_8$, $CpCo(CO)_2$, $Co_4(CO)_{12}$, and $RhCl(PPh_3)_3$ have all been reported to catalyze decarbonylation of $Ph_2C{=}C{=}O$ [Eqs. (39) and (40)] (48). Both reactions likely involve the initial formation of unstable diphenylcarbene complexes. Tetraphenylethylene can

$$Ph_2C{=}C{=}O \quad \xrightarrow[\substack{10 \text{ hrs,} \\ \text{Toluene}}]{Co_2(CO)_8} \quad Ph_2C{=}CPh_2 \tag{39}$$

$$RhCl(PPh_3)_3 \quad + \quad Ph_2C{=}C{=}O \quad \longrightarrow$$
$$RhCl(CO)(PPh_3)_2 \quad + \quad Ph_2C{=}C{=}CPh_2 \quad + \quad Ph_3PO \tag{40}$$

arise from coupling of two such carbene ligands. Combination of a diphenylcarbene ligand with PPh_3 would lead to the ylide $Ph_2C{=}PPh_3$, which could subsequently react with $Ph_2C{=}C{=}O$ to account for the formation of tetraphenylallene and Ph_3PO in Eq. (40). Dimethylketene is also decarbonylated by the rhodium complex of Eq. (40), but the organic products are those shown in Eq. (41) (42). That carbene complexes are

$$RhCl(PPh_3)_3 \quad + \quad Me_2C{=}C{=}O \quad \longrightarrow RhCl(CO)(PPh_3)_2 \quad + \quad Me_2C{=}CMe_2$$
$$+ \quad Me_2CH{-}CHMe_2 \tag{41}$$

likely intermediates in the formation of the olefin products of Eqs. (40) and (41) is indicated by the formation of stable carbene complexes **26** and **27** from the reaction of $Ph_2C{=}C{=}O$ with $[RhCl(CO)_2]_2$ [Eq. (36)] and by the observation that $Ph_2C{=}CPh_2$ is formed on treatment of **26** with PPh_3 (43).

It was noted in Eq. (39) above that $Co_2(CO)_8$ catalyzes the decarbonylation of $Ph_2C{=}C{=}O$. However, if this reaction is conducted under CO pressure, deoxygenation of diphenylketene occurs and the organic **34** is formed in which a total of three ketenes have been assembled (49). It was suggested that a vinylidene complex such as **35** may be an intermediate in this transformation, as a similar complex **22** was in Eq. (34), with the final product forming from ketene addition to **36** [Eq. (42)]. The major uncertainty in this process is how the **35**-to-**36** conversion occurs.

$Co_2(CO)_8$ has been reported to assist the cycloaddition of diphenylketene to styrene [Eq. (43)] (48). If diarylacetylenes and $Fe(CO)_5$ are used in a similar reaction, CO incorporation occurs to give **37–39** [Eq. (44)]

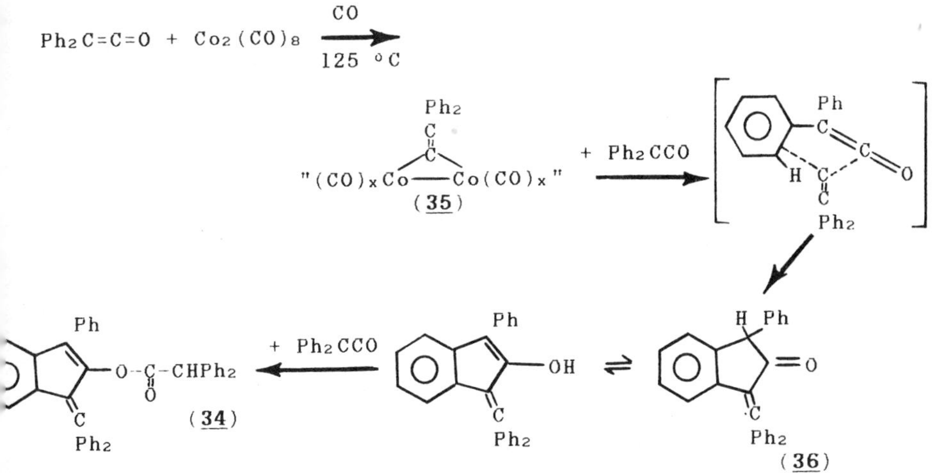

$$Ph_2C=C=O \ + \ Co_2(CO)_8 \xrightarrow[125\ ^\circ C]{CO}$$

$$\text{"}(CO)_xCo\!-\!Co(CO)_x\text{"} \quad (35) \quad + \ Ph_2CCO \longrightarrow$$

(42)

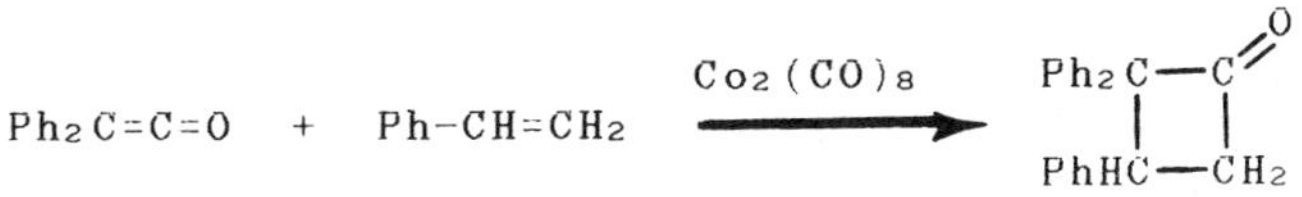

$$+ \ Ph_2CCO \quad (34) \longleftarrow \quad (36) \quad \rightleftharpoons$$

$$Ph_2C=C=O \ + \ Ph-CH=CH_2 \xrightarrow{Co_2(CO)_8} \quad (43)$$

$$Ph_2C=C=O \ + \ Ar-C\equiv C-Ar \xrightarrow[150^\circ C]{Fe(CO)_5}$$

$$Ar = Ph, Tol$$

(37) (62%) $\qquad$ Ph Ph (trace) (38)

(40) (CO)$_3$

$+$

(39) (trace)

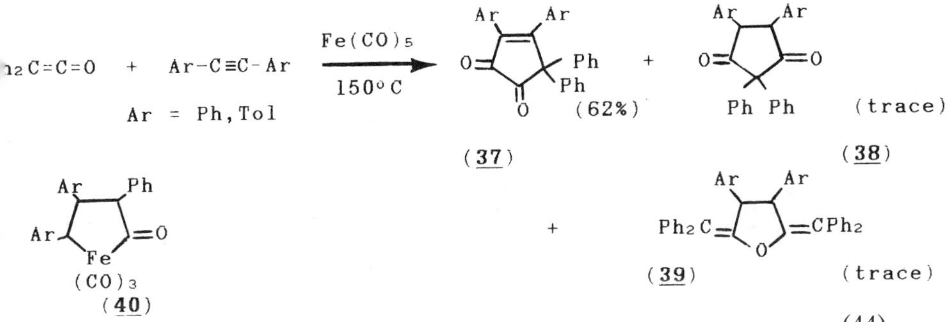

(44)

(50). The organics **37** and **38** were suggested to form via carbonylation of **40**, which would result from coupling of the alkyne and ketene reagents. Complex **39** could arise from cycloaddition of alkyne and ketene with a vinylidene complex such as **20** that is known to form in the reaction of $Fe(CO)_5$ with $Ph_2C=C=O$ (27,38). If $Ni(CO)_4$ is used in the reaction, the product mixture is dependent on solvent but **38** is formed in modest yield.

The reaction of diphenylcyclopropenone with ketenes in the presence of $Ni(CO)_4$ has been reported to give products similar to those above [Eq. (45)] (51). The first step in the reaction is likely to be insertion of Ni into a C–C single bond of cyclopropenone to give **41**, which then inserts ketene into the Ni–C bonds to give the observed products.

$$\begin{array}{ccc}
\text{Ph} \diagdown \;\;\;\; \diagup \text{Ph} \\
\text{C} = \text{C} \\
\diagdown \;\; \diagup \\
\text{C} \\
\text{O}
\end{array}
\quad
\xrightarrow{\text{Ni(CO)}_4}
\quad
"\;
\begin{array}{c}
\text{PhC} = \text{CPh} \\
|\;\;\;\;\; | \\
\text{C} - \text{Ni(CO)}_3 \\
\text{O}\diagup
\end{array}
\;"
\quad + \; \text{Ph}_2\text{C}=\text{C}=\text{O}
\xrightarrow{}
\underline{\mathbf{37}} \; + \; \underline{\mathbf{38}}$$

(low yield)

$$\downarrow (\underline{\mathbf{41}})$$

$$\text{PhC} \equiv \text{CPh}$$

(45)

A remarkable catalytic reaction is the $Rh_4(CO)_{12}$-assisted addition of benzene to diphenylketene [Eq. (46)] (52,53). This reaction was suggested

$$\text{Ph}_2\text{C}=\text{C}=\text{O} \; + \; \text{Ph-H} \quad \xrightarrow[\text{CO (30 atm), 5hr}]{\text{Rh}_4(\text{CO})_{12}, \; 200°\text{C}} \quad
\begin{array}{c}
\quad\;\; \text{O} \\
\quad\;\; \| \\
\text{Ph}_2\text{C}-\text{C}-\text{Ph} \\
\text{H}
\end{array}
\quad
\begin{array}{c}
(68\%, \;\; 144 \\
\text{turnovers})
\end{array}$$

(46)

to proceed by insertion of Rh into an Ar–H bond and then insertion of ketene into the resultant Rh–H bond followed by reductive elimination of the aryl and acyl groups. The order of arene reactivity is toluene < benzene < anisole < fluorobenzene.

Several patents have described the formation of malonic acid diesters from CO, ketene, and organonitrites using soluble or heterogeneous Ni, Pd, and Pt catalysts [Eq. (47)] (54–56), but no mechanistic details were presented.

$$\text{CO} \; + \; \text{CH}_2=\text{C}=\text{O} \; + \; \text{RONO} \quad \xrightarrow{\text{MX}_2\text{L}_2} \quad
\begin{array}{c}
\;\;\;\text{O}\;\;\;\;\;\;\;\;\;\text{O} \\
\;\;\;\|\;\;\;\;\;\;\;\;\;\| \\
\text{RO-C-CH}_2\text{-C-OR}
\end{array}$$

(47)

III

COMPLEXES POSSESSING KETENE LIGANDS

A number of organometallic complexes have been prepared that contain ketene and substituted ketene ligands, but these have a variety of different ketene coordination modes as illustrated in **I–IX**. Type I is an η^2-(C,C)

$\underline{\mathbf{I}}(57-71)$ $\underline{\mathbf{II}}(74-79)$ $\underline{\mathbf{III}}(80)$ $\underline{\mathbf{IV}}(81-89)$ $\underline{\mathbf{V}}(82,83,87,\;91-97)$

$\underline{\mathbf{VI}}(85)$ $\underline{\mathbf{VII}}(98-108)$ $\underline{\mathbf{VIII}}(109-115)$ $\underline{\mathbf{IX}}(116-118)$

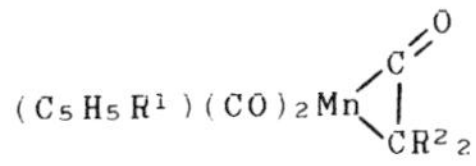

$(C_5H_5R^1)(CO)_2Mn$ — C=O, CR^2_2 (**42^a**) (*57-61*)

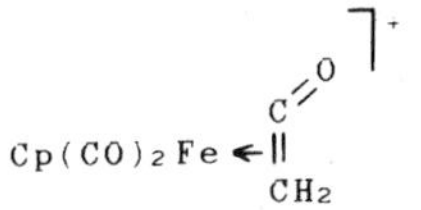

$Cp(CO)_2Fe$ ⟵ C=O, CH_2]$^+$ (**43**) (*62-64*)

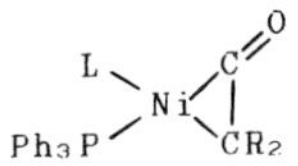

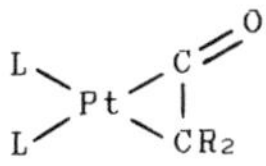

L, Ph_3P — Ni — C=O, CR_2]$^+$ (**44^b**) (*65-67*)

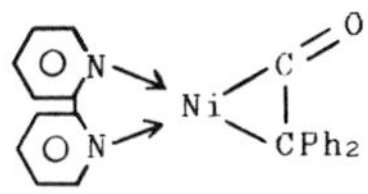

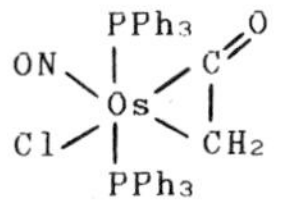

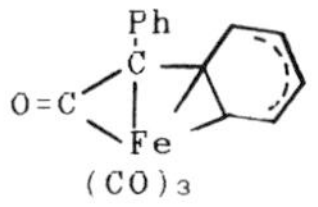

Ni — C=O, CPh_2 (**45**) (*65*)

L, L — Pt — C=O, CR_2 (**46^c**) (*65,68,69*)

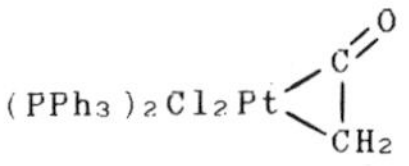

$(PPh_3)_2Cl_2Pt$ — C=O, CH_2 (**47**) (*69*)

ON, PPh_3, Cl — Os — C=O, CH_2, PPh_3 (**48**) (*70*)

$O=C$ — Ph, C, Fe, $(CO)_3$ (**49**) (*40,41,71*)

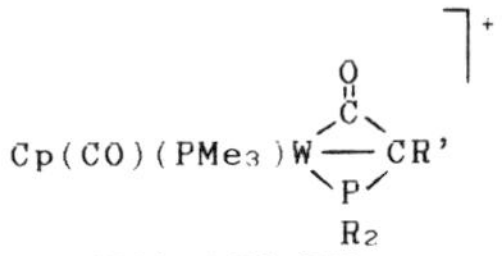

$Cp(CO)(PMe_3)W$ — C=O, CR', P, R_2]$^+$ (**50**) (*72,73*)

FIG. 1.   Type I ketene complexes. ^{a}42a, R^1 = H, R^2 = Ph.; 42b, R^1 = Me, R^2 = Ph; 42c, R^1 = H, R^2_2 = $C_{13}H_8O$; 42d, R^1 = Me, R^2_2 = $C_{13}H_8O$; 42e, R^1 = H, R^2_2 = $C_{14}H_{12}$; 42f, R^1 = Me, R^2_2 = $C_{14}H_{12}$; 42g, R^1 = H, R^2_2 = $C_{14}H_{10}$; 42h, R^1 = Me, R^2_2 = $C_{14}H_{10}$, ^{b}44a, R = Ph, L = PPh_3; 44b, R = Ph, L = py; 44c, R = H, L = PPh_3. ^{c}46a, R = Ph, L = PPh_3; 46b, R = H, L = PPh_3; 46c, R = H, L = Cl.

ligand bound to a single metal center typically found for low-valent middle to late transition metals. The known examples of these are illustrated in Fig.1. In type II complexes, a μ_2,η_2-(C,C) ligand bridges adjacent metals in binuclear and cluster compounds, and these are shown in Fig. 2. The only known example of a type III compound in which the carbonyl oxygen also interacts with a third metal atom is 57 (*80*). Oxophilic early transition

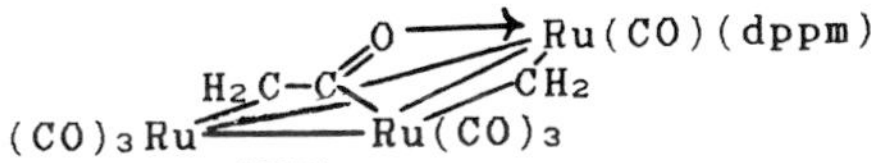

$(CO)_3Ru$ — H_2C-C — O → Ru(CO)(dppm), CH_2, Ru(CO)$_3$ (**57**)

metals generally yield η^2-(C,O) ketene complexes with this ligand bound to a single metal as in IV (Fig. 3) or to two metals as in V (Fig. 4) and VI, the known examples of which are 73 aand 74 (*85*). Several vinylketene

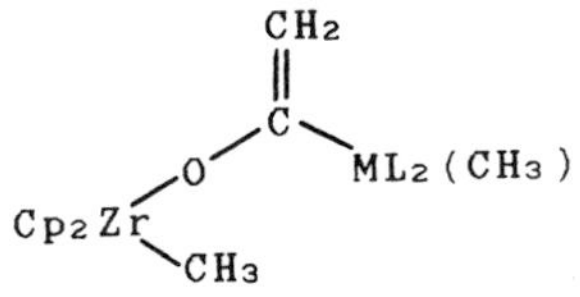

Cp_2Zr — O — C(=CH$_2$) — $ML_2(CH_3)$, CH_3

(**73**, M=Zr, L=Cp)
(**74**, M=Pt, L=PMe$_3$)

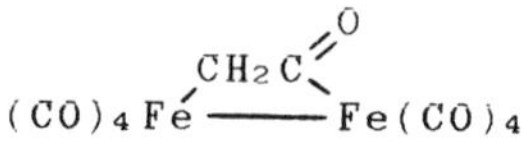

$$(CO)_4Fe\underset{(\underline{51})\ (74)}{\overset{CH_2C{\overset{O}{\diagdown}}}{\rule{3em}{0.4pt}}}Fe(CO)_4$$

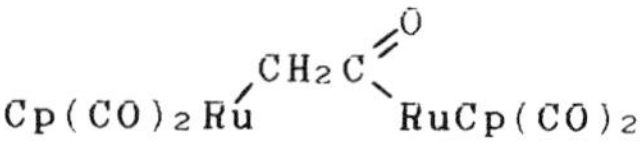

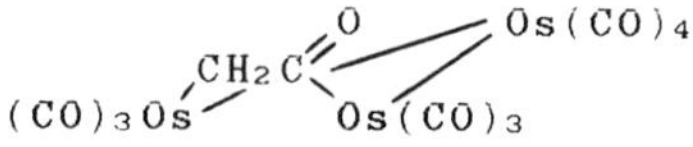

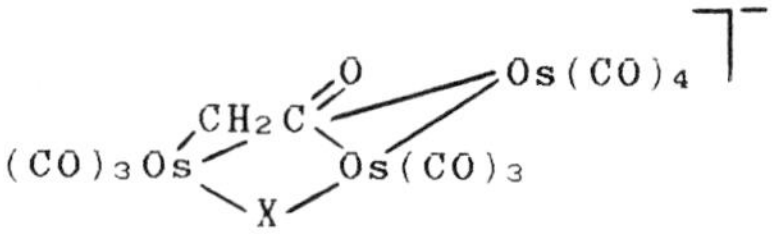

$$Cp(CO)_2Ru\underset{(\underline{52^a})\ (75)}{\overset{CH_2C{\overset{O}{\diagdown}}}{\diagdown}}RuCp(CO)_2$$

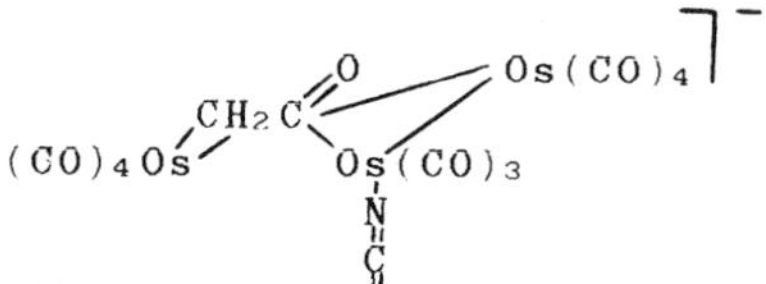

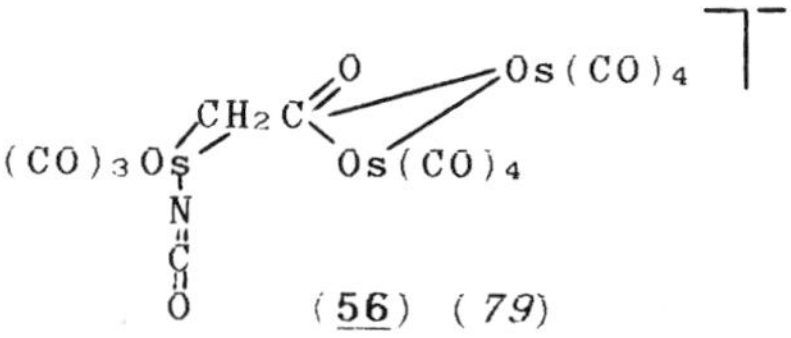

$$(CO)_3Os\underset{(\underline{53})\ (76,77)}{\overset{CH_2C}{\diagdown}}\overset{Os(CO)_4}{Os(CO)_3}$$

$$(CO)_3Os\underset{(\underline{54^b})\ (78,79)}{\overset{CH_2C}{\diagdown}\underset{X}{\diagdown}}\overset{Os(CO)_4}{Os(CO)_3}\Bigg]^-$$

$$(CO)_4Os\underset{\substack{(\underline{55})\ (79)}}{\overset{CH_2C}{\diagdown}}\overset{Os(CO)_4}{\underset{\substack{N\\ \overset{C}{}\\ O}}{Os(CO)_3}}\Bigg]^-$$

$$(CO)_3Os\underset{\substack{(\underline{56})\ (79)}}{\overset{CH_2C}{\diagdown}}\overset{Os(CO)_4}{\underset{\substack{N\\ \overset{C}{}\\ O}}{Os(CO)_4}}\Bigg]^-$$

FIG. 2. Type II ketene complexes. [a]52a, Cp = C_5H_5, L = CO; 52b, Cp = C_5H_5, L = PMe_3; 52c, Cp = C_5Me_5, L = CO. [b]54a,X = Cl; 54b, X = Br; 54c, X = I; 54d, X = NCO. [c]MLn = $Cp(CO)_2Fe$, $(C_5H_4Me)Fe(CO)_2$, $Cp(CO)_2(PPh_3)Mo$, Cp(CO)Ni, $Mn(CO)_5$, $Co(CO)_3(PMe_2Ph)$.

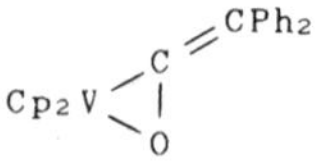

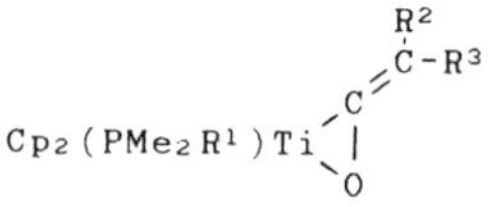

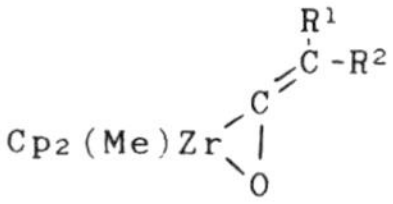

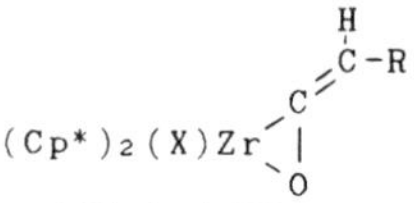

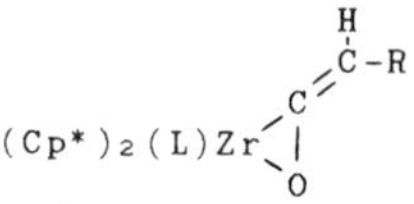

$$Cp_2V\underset{(\underline{58})\ (81,82)}{\overset{C}{\diagdown}}\overset{CPh_2}{\underset{O}{\diagdown}}$$

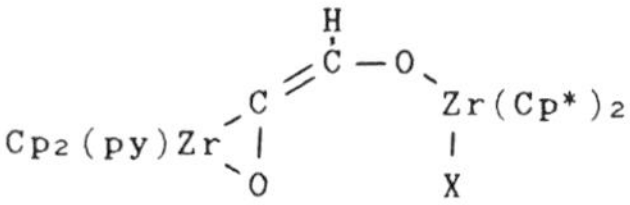

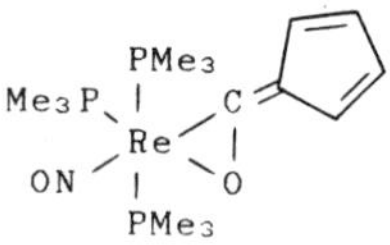

$$Cp_2(PMe_2R^1)Ti\underset{(\underline{59})\ (83,84)}{\overset{C}{\diagdown}}\overset{\overset{R^2}{C}-R^3}{\underset{O}{\diagdown}}$$

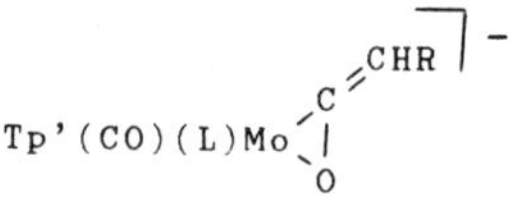

$$Cp_2(Me)Zr\underset{(\underline{60^b})\ (83,85)}{\overset{C}{\diagdown}}\overset{\overset{R^1}{C}-R^2}{\underset{O}{\diagdown}}$$

$$(Cp^*)_2(X)Zr\underset{(\underline{61^c})\ (86)}{\overset{C}{\diagdown}}\overset{\overset{H}{C}-R}{\underset{O}{\diagdown}}$$

$$(Cp^*)_2(L)Zr\underset{(\underline{62^d})\ (86,87)}{\overset{C}{\diagdown}}\overset{\overset{H}{C}-R}{\underset{O}{\diagdown}}$$

$$Cp_2(py)Zr\underset{(\underline{63^e})\ (88)}{\overset{C}{\diagdown}}\overset{\overset{H}{C}-O}{\underset{O}{\diagdown}}\underset{X}{\overset{Zr(Cp^*)_2}{|}}$$

$$\underset{ON}{\overset{Me_3P}{\diagdown}}\underset{\underset{PMe_3}{|}}{\overset{|}{Re}}\underset{(\underline{64})\ (89)}{\overset{PMe_3}{\overset{|}{C}}\underset{O}{\overset{C}{\diagdown}}}$$

$$Tp'(CO)(L)Mo\underset{(\underline{65^f})\ (90)}{\overset{C}{\diagdown}}\overset{CHR}{\underset{O}{\diagdown}}\Bigg]^-$$

FIG. 3. Type IV ketene complexes. [a]59a, R^1 = Ph, R^2 = R^3 = H; 59b, R^1 = Me, R^2 = Ph, R^3 = (Ph)C=CH_2; 59c, R^1 = Me, R_2 = Ph, R^3 = (Me_3Si)C=CH_2; 59d, R^1 = Me, R^2 = Ph, R^3 = (Me)C=CH_2; 59e, R^1 = R^2 = Me, R^3 = (Me)C=CH; 59f, R^1 = Me, R_2 = Et, R_3 = (Et)C = CH. [b]60a, R^1 = R^2 = H, 60b, R^1 = H, R^2 = Me; 60c, R^1 = R^2 = Me. [c]61a, X = Br, R = H; 61b, X = Cl, R = Bu^t. [d]62a, L = py, R = H; 62b, L = CO, R = Bu^t; 62c, L = CH_2PMe_3, R = Bu_t; 62d, L = DMAP; R = CH_2Bu^t (Cp^* = Cp). [e]63a, X = H; 63b, X = I. [f]65a, L = CO, R = H; 65b, L = CO, R = CH_3; 65c, L = $P(OPh)_3$, R = CH_3.

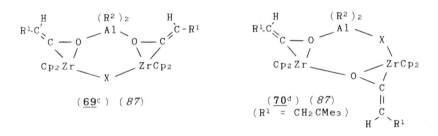

$$[Cp_2M(OCCH_2)]_n$$

$$(\underline{68a}, \ M=Ti) \ (83,87)$$
$$(\underline{68b}, \ M=Zr) \ (87)$$

$$(\underline{66}^a, \ M=Ti) \ (82,91,92)$$
$$(\underline{67}^b, \ M=Zr) \ (83,87,93)$$

$$(\underline{69}^c) \ (87)$$

$$(\underline{70}^d) \ (87)$$
$$(R^1 = CH_2CMe_3)$$

$$(\underline{71}^e) \ (94-97)$$

$$(\underline{72}^f) \ (87)$$

Fig. 4. Type V ketene complexes. [a]$R^1 = R^2 = $ Ph. [b]**67a**, $R^1 = R^2 = $ Ph; **67b**, $R^1 = $ H, $R^2 = CH_2Bu^t$. [c]**69a**, $X = $ Me, $R^1 = CH_2Bu^t$, $R^2 = $ Me; **69b**, $X = $ Cl, $R^1 = CH_2Bu^t$, $R^2 = $ Me; **69c**, $X = $ H, $R^1 = CH_2Bu^t$, $R^2 = $ Et; **69d**, $X = $ Cl, $R^1 = $ H, $R^2 = $ Me. [d]**70a**, $X = $ Me, $R^2 = $ Me; **70b**, $X = $ Cl, $R^2 = $ Me; **70c**, $X = $ H, $R^2 = $ Et. [e]**71a**, $R = $ Me; **71b**, $R = $ Ph. [f]**72a**, $X = $ Me; **72b**, $X = $ Cl.

complexes of type VII are known in which the ketene ligand binds in a fashion similar to the bonding of butadiene to metals, and these are shown in Fig. 5. Type VIII complexes have a metal as a substituent in one of the groups attached to the ketene α carbon. The only examples of these are **87** (109–114) and **88a** (115), although complexes **88b** have been proposed as important reaction intermediates (119,120). The single example of a type IX ketene complex (**89**) is perhaps better described as a metal enolate, but

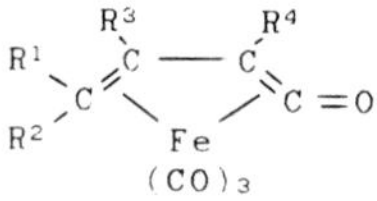

(**75a**) (*98–104* 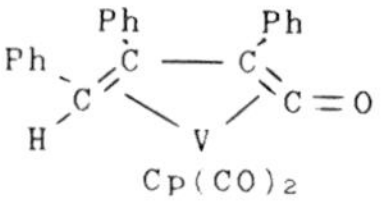)   (**76**) (*105*)   (**77b**) (*103,105*)

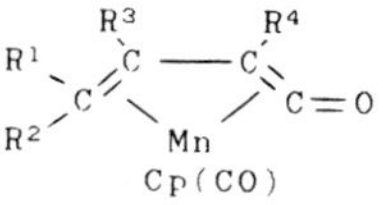

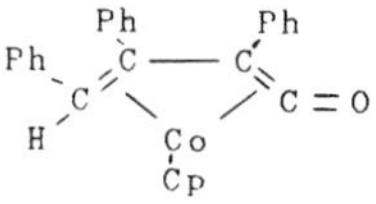

(**78**) (*105* 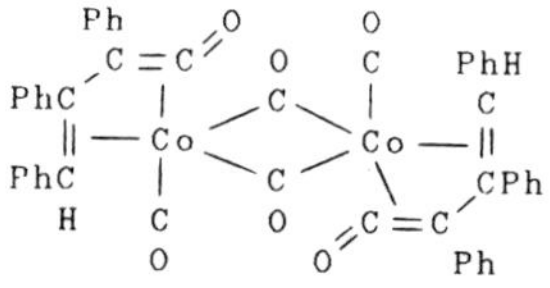)   (**79**) (*105*)   (**80c** , M=Mo) (*106*)
(**81d** , M=W) (*106*)

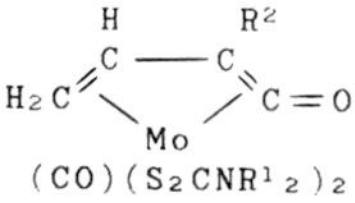

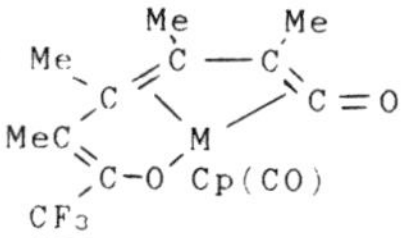

(**82** , M=Mo) (*107*)   (**84** , M=Mo) (*107*)   (**86**) (*108*)
(**83** , M=W) (*107*)   (**85** , M=W) (*107*)

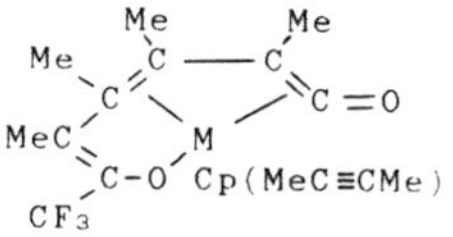

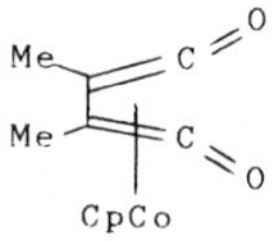

FIG. 5. Type VII vinylketene complexes. [a]**75a**, $R^1 = R^2 = R^3 = Me$, $R^4 = H$; **75b**, $R^1 = R^2 = R^4 = H$, $R^3 = OCH_3$; **75c**, $R^1 = CO_2Me$, $R^2 = R^3 = R^4 = Ph$; **75d**, $R^1 = CO_2Et$, $R^2 = R^3 = R^4 = Ph$; **75e**, $R^1 = R^2 = H$, $R^3 = CO_2Me$, $R^4 = OMe$; **75f**, $R^1 = R^2 = Me$, $R^3 = R^4 = H$; **75g**, $R^1 = R^2 = PhC=C(Ph)C(O)O$, $R_3 = R_4 = Ph$. [b]**77a**, $R^1 = H$, $R^2 = R^3 = R^4 = Ph$; **77b**, $R^1 = R^2 = Me$; $R^3 = R^4 = H$. [c]**80a**, $R^1 = Et$, $R^2 = H$; **80b**, $R^1 = Pr^i$, $R^2 = H$. [d]**81a**, $R^1 = Me$, $R^2 = H$; **81b**, $R^1 = Et$, $R^2 = H$; **81c**, $R^1 = Ph$; $R^2 = H$; **81d**, $R^1 = Et$, $R^2 = Me$; **81e**, $R^1 = Pr^i$, $R^2 = Me$; **81f**, $R^1_2 = C_2H_4$, $R^2 = Me$.

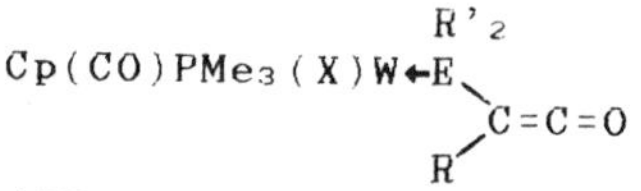

(**87** , E=P,As; R=Me,Tol;   (**88**)
R'=Me,Ph; X=Cl,I)

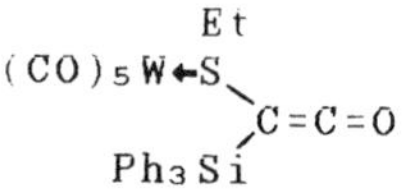

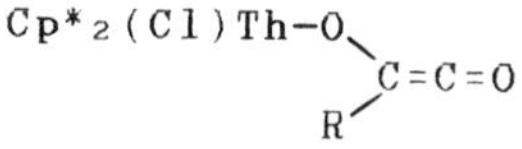

(**88b** , R=Ph,Me)

since it possesses the elements of the ketene ligand it is included here for completeness (*116–118*). The various synthetic methods that have been used for these complexes are described below, followed by sections summarizing their spectroscopic and structural data and their reactivity characteristics.

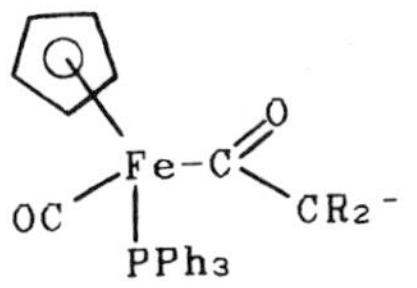

$$\underset{(\underline{89})}{\overset{\displaystyle \bigodot}{\underset{\underset{\displaystyle PPh_3}{\displaystyle |}}{OC \diagdown} \overset{\displaystyle Fe-C}{\diagup} \overset{\displaystyle \diagup O}{\diagdown} CR_2{}^-}}$$

A. Synthetic Methods for Ketene Complexes

A variety of synthetic procedures have been used to prepare ketene complexes, with the most useful involving direct addition of ketenes to unsaturated organometallic complexes, carbonylation of carbene complexes, and deprotonation of acyl complexes.

1. Addition of Ketenes to Unsaturated Organometallic Complexes

This is the most straightforward route to organometallic ketene complexes. However, it has been successfully applied in only a few cases, mainly because of the instability of ketene complexes under the conditions necessary to open coordination sites in their precursor compounds. If an open coordination site is not inherently present, as in the reactant of Eq. (48) (81,82), it is best formed by dissociation of weakly coordinated ligands such as ethylene [Eq. (49)] (65,68,69) or THF solvent [Eq. (50)] (57) or by photodissociation of a CO ligand [Eq. (51)] (40,41,71).

$$Cp_2V \ + \ Ph_2C=C=O \ \xrightarrow[15 \ min]{22^\circ C} \ Cp_2V(OCCPh_2) \qquad (48)$$
$$(\underline{58}, \ 64\%)$$

$$(PPh_3)_2M(CH_2=CH_2) \ + \ Ph_2C=C=O \ \xrightarrow{22^\circ C} \ (PPh_3)_2M(Ph_2CCO) \quad (49)$$
$$(\underline{44}, \ M = Ni)$$
$$(\underline{46}, \ M = Pt)$$

$$Cp(CO)_2Mn(THF) \ + \ Ph_2C=C=O \ \xrightarrow[2 \ h]{22^\circ C} \ Cp(CO)_2Mn(Ph_2CCO) \quad (50)$$
$$(\underline{42a})$$

$$Fe(CO)_5 \ + \ Ph_2C=C=O \ \xrightarrow[6.5 \ h]{15^\circ C, \ h\nu} \ (CO)_3Fe(Ph_2CCO) \quad (51)$$
$$(\underline{49}, \ 38\%)$$

Note that all of the above reactions occur under relatively mild conditions. Under more severe conditions, such reactions of ketenes with organometallics often transform the ketene into other types of ligands or into organic molecules, as discussed above in Section II.

2. Carbonylation of Metal Carbenes

A number of ketene complexes have been prepared by the addition of carbon monoxide to preformed metal–carbene complexes. Terminal carbene complexes typically yield type I or type IV ketene complexes with the ligand attached to a single metal, whereas carbonylation of bridging carbene complexes leads to μ-ketene complexes of types II and III. The first example of the former reaction is that shown in Eq. (52) (59,60). Note the high CO pressure required for this reaction.

$$Cp(CO)_2Mn=CPh_2 \ + \ CO \ (650 \ atm) \xrightarrow[\substack{25 \ h}]{30-50 \ ^\circ C} Cp(CO)_2Mn\{Ph_2CCO\}$$

$$(\underline{42a}, \ 45\%)$$

$$(52)$$

It was later shown that the anthronylketene complex **42c** formed from the addition of 9-diazoanthrone to **90** [Eq. (53)], a reaction that seemingly proceeds via a similar carbonylation of an initially formed anthronylcarbene complex (61). However, a careful mechanistic study showed that the

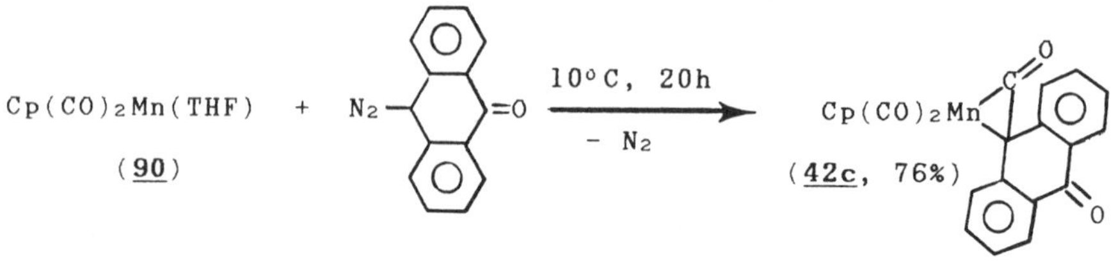

$$(53)$$

reaction actually proceeds by direct formation of the ketene ligand via addition of carbene to a coordinated CO (61). This was mainly indicated by the sole formation of $(MeC_5H_4)(CO)_2Mn(ketene)$ on addition of the diazo carbene precursor to an equimolar mixture of $(MeC_5H_4)(CO)_3Mn$ and $Cp(CO)_2Mn(THF)$.

The carbonylation of $[Fp=CH_2]^+$ [Eq. (54) (62) proceeds under much milder conditions than those needed to prepare **42a** [Eq. (52)], suggesting that the facility of these carbonylation reactions is a sensitive function of the carbene substituents. A labeling study showed that the CO incorporated into the ketene ligand in **43** came from the added CO but not from

$$[Cp(CO)_2Fe=CH_2]^+ \ + \ CO \ (85 \ psi) \ \xrightarrow[\substack{0.5 \ h}]{25°C} \ [Cp(CO)_2Fe\{CH_2CO\}]^+$$

$$(\underline{43}, \ 48\%)$$

$$(54)$$

one of the original complex CO's, implying direct addition of CO to the methylene ligand (*62*).

A brief report has mentioned the similar carbonylation of an osmium–methylene complex [Eq. (55)] (*70*). The ketene ligand was suggested to

$$(PPh_3)_2Cl(NO)Os=CH_2 \ + \ CO \ \longrightarrow \ (PPh_3)_2Cl(NO)Os\{CH_2CO\}$$

$$(\underline{48})$$

$$(55)$$

form by initial coordination of CO to the metal, which was facilitated by the nitrosyl ligand changing from a linear 3 e⁻ donor to a bent 1 e⁻ donor. The reverse of the latter would then accompany the presumed migration of CO to the methylene ligand.

A type IV ketene complex has also been formed by a carbene carbonylation reaction. Pyridine addition to the zirconoxy–carbene complex **91**, which was generated *in situ* from the addition of $Cp^*_2ZrH_2$ ($Cp^* = \eta$-C_5Me_5) to $Cp_2Zr(CO)_2$, induced CO insertion into the Zr–carbon bond to give ketene complex **63a** [Eq. 56] (*88*).

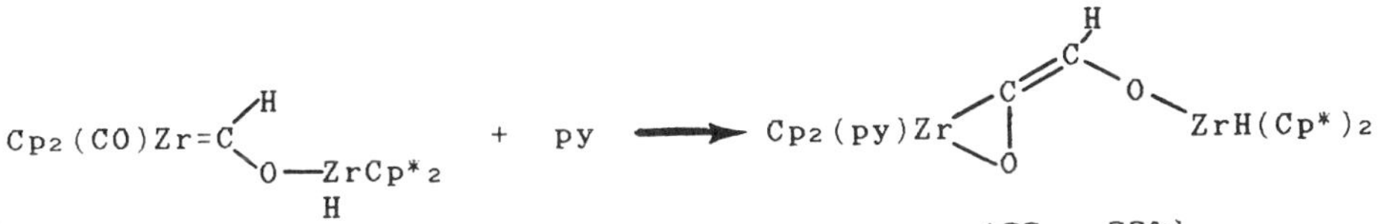

$$(56)$$

There are other examples of preparations of ketene complexes that apparently proceed via intermediate, but undetected, carbene complexes. Transient methylene ligands are apparently formed in Eqs. (57) (*69*) and (58) (*66,67*) by halide abstraction from CH_2Br_2 and by $CH_2{=}CMe_2$ elimination, respectively. A cyclopentadienyl–carbene complex is presumably formed in Eq. (59) by CH_4 loss (*89*).

$$PtL_4 \ + \ CH_2Br_2 \ + \ \underset{(3 \ atm)}{CO} \ \xrightarrow{20°C, \ 10h} \ L_2Pt\{CH_2CO\} \quad (57)$$

$$L = PPh_3 \qquad\qquad (\underline{46b}, \ 23\%)$$

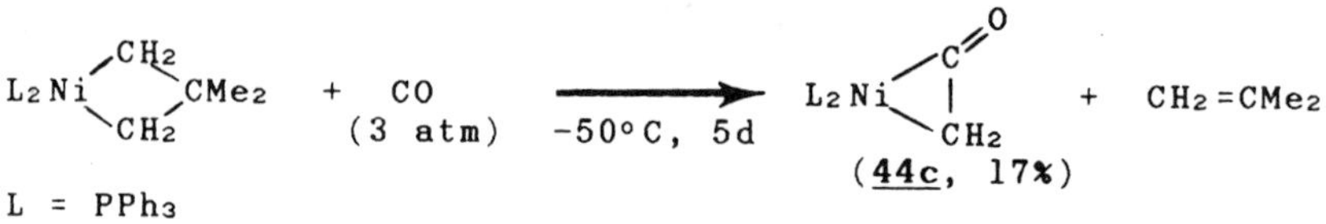

$$L_2Ni\overset{\displaystyle CH_2}{\underset{\displaystyle CH_2}{\diagdown}}CMe_2 \; + \; \underset{(3\ atm)}{CO} \quad \xrightarrow{-50°C,\ 5d} \quad L_2Ni\overset{\displaystyle C{\nwarrow}^O}{\underset{\displaystyle CH_2}{\diagdown}}\Big| \quad + \quad CH_2=CMe_2$$

L = PPh₃   (**44c**, 17%)

$$(58)$$

$$L_2(CO)(NO)(CH_3)Re{\diagup}\overset{H}{\diagdown}\text{(cyclopentene)} \; + \; PMe_3 \quad \xrightarrow[\displaystyle 17h]{\displaystyle 72°C} \quad CH_4 \; + \; L_3(NO)Re{\diagup}\overset{C(cyclopentadienyl)}{\diagdown}O$$

L = PMe₃   (**64**)

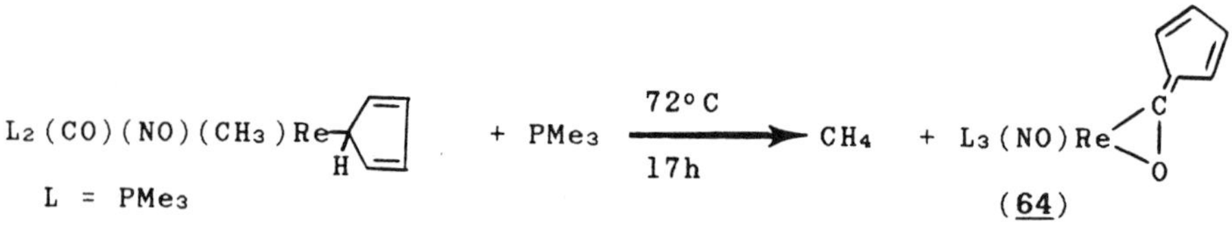

$$(59)$$

Bridging ketene ligands have also been prepared by carbonylation of μ-carbenes, although nearly all of the known examples involve μ-CH₂ ligands. This reaction was first suggested to account for the formation of alkyl acetates in Eq. (60) (*74*). The ketene complex **51** was not observed

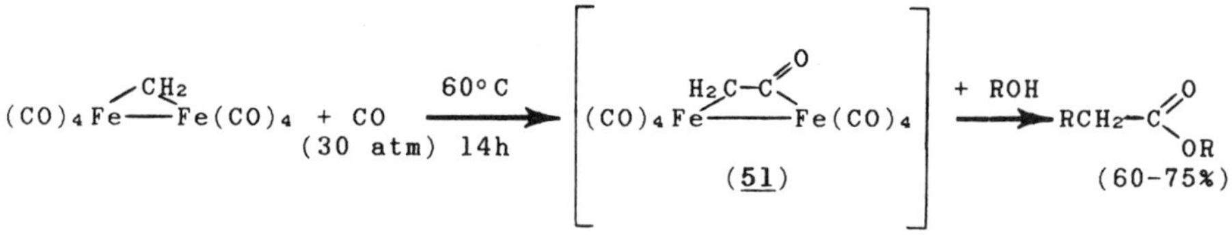

$$(CO)_4Fe\overset{\displaystyle CH_2}{\diagup\diagdown}Fe(CO)_4 \; + \; \underset{(30\ atm)}{CO} \quad \xrightarrow{\displaystyle 60°C} \quad \left[\; (CO)_4Fe\overset{\displaystyle H_2C-C{\nwarrow}^O}{\diagup\diagdown}Fe(CO)_4 \;\right] \quad \xrightarrow{+\ ROH} \quad RCH_2-C{\overset{\displaystyle O}{\underset{\displaystyle OR}{\diagup}}}$$

14h   (**51**)   (60–75%)

$$(60)$$

spectroscopically, but its presence was inferred from the color change that occurred on addition of CO and from the formation of a solution similar color by the reaction of $[Fe_2(CO)_8]^{2-}$ with $ClCH_2C(O)Cl$ (*74*). Another synthesis of a μ-ketene complex that likely proceeds via carbonylation of an unstable μ-CH₂ intermediate is that shown in Eq. (61) (*80*).

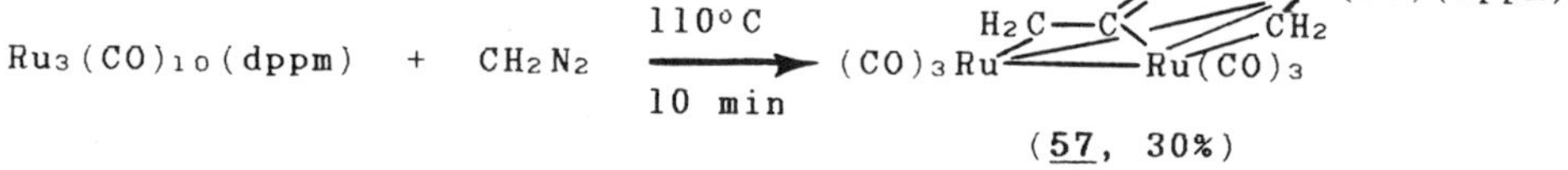

$$Ru_3(CO)_{10}(dppm) \; + \; CH_2N_2 \quad \xrightarrow[\displaystyle 10\ min]{\displaystyle 110°C} \quad (CO)_3Ru\overset{\displaystyle H_2C-C{\overset{O}{\diagup}}{\diagdown}Ru(CO)(dppm)}{\diagdown Ru(CO)_3}CH_2$$

(**57**, 30%)

$$(61)$$

Definitive examples of the formation of ketenes via the carbonylation of μ-CH₂ complexes are illustrated in Eqs. (62)–(64) (*75,76–78*). The latter

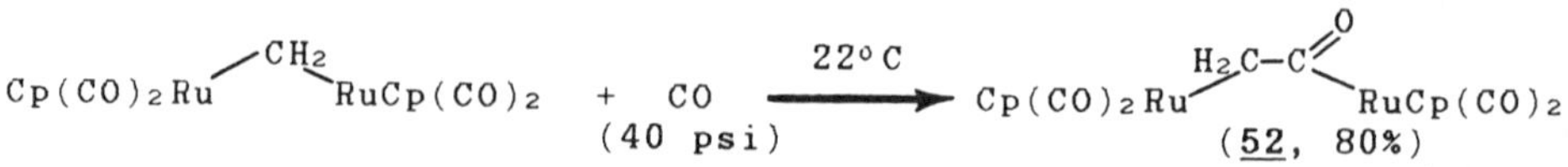

$$Cp(CO)_2Ru\overset{\displaystyle CH_2}{\diagup\diagdown}RuCp(CO)_2 \; + \; \underset{(40\ psi)}{CO} \quad \xrightarrow{\displaystyle 22°C} \quad Cp(CO)_2Ru\overset{\displaystyle H_2C-C{\nwarrow}^O}{\diagup\diagdown}RuCp(CO)_2$$

(**52**, 80%)

$$(62)$$

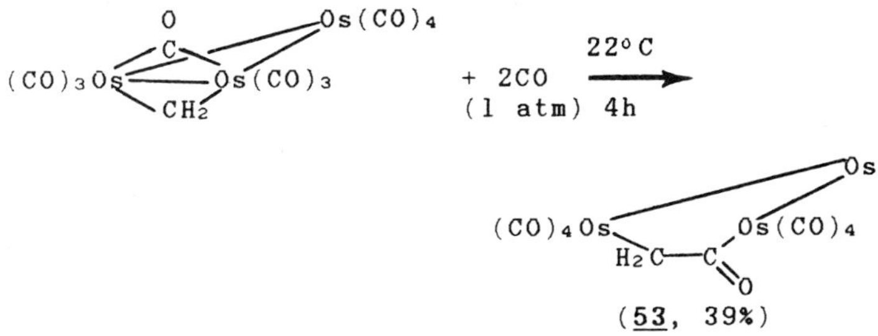

$$(CO)_3Os \underset{CH_2}{\overset{O}{\underset{\diagdown}{\overset{\diagup}{C}}}} Os(CO)_3 \quad \overset{Os(CO)_4}{} \quad + \ 2CO \quad \overset{22^\circ C}{\underset{(1 \ atm) \ 4h}{\longrightarrow}}$$

$$(CO)_4Os \overset{Os(CO)_4}{\underset{H_2C-C}{\diagdown}} Os(CO)_4 \quad (63)$$

$$\underset{(\underline{53}, \ 39\%)}{}$$

$$\left[(CO)_3Os \underset{CH_2}{\overset{I}{\underset{\diagdown}{\overset{\diagup}{}}}} Os(CO)_3 \right]^- \overset{Os(CO)_4}{} \quad + \ CO \quad \underset{(1 \ atm) \ 1 \ min}{\overset{22^\circ C}{\rightleftharpoons}} \quad \left[(CO)_3Os \underset{H_2C-C}{\overset{I}{\underset{\diagdown}{\overset{\diagup}{}}}} Os(CO)_3 \right]^- \overset{Os(CO)_4}{}$$

$$\underset{(\underline{54c}, \ >95\%)}{}$$

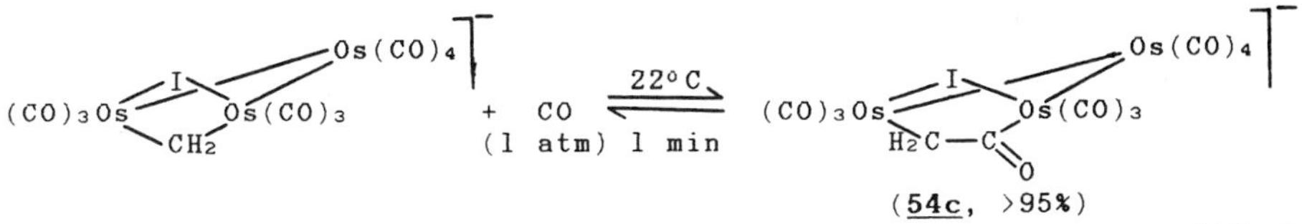

$$(64)$$

two reactions illustrate an important example of a halide promoting effect on the carbonylation reaction as the rate of reaction (64) was ~100 times greater than that of reaction (63) (76–78). Also, reaction (64) completely reverses on removing the CO atmosphere, whereas reaction (63) does not. Labeling studies showed that the CO incorporated into the ketene ligands in both 53 and 54c came from one of the original cluster COs and not from the added CO (76–78). This implies that the μ-ketene ligands were formed by a CO insertion process rather than from direct addition of CO to the methylene ligand as was found for $[Cp(CO)_2Fe=CH_2]^+$ (62). Binuclear $Os_2(CO)_8(\mu\text{-}CH_2)$ also undergoes carbonylation to form a ketene complex, similarly to reaction (63), but details have not yet been published (121).

There are two examples of the formation of *free* ketenes or ketene-derived products on carbonylation of μ-carbene complexes, and these reactions likely proceed via the initial formation of ketene complexes. That is surely the case in reaction (65), in which free $Ph_2C=C=O$ was produced on carbonylation of 92 (122). Similarly, carbonylation of a series of

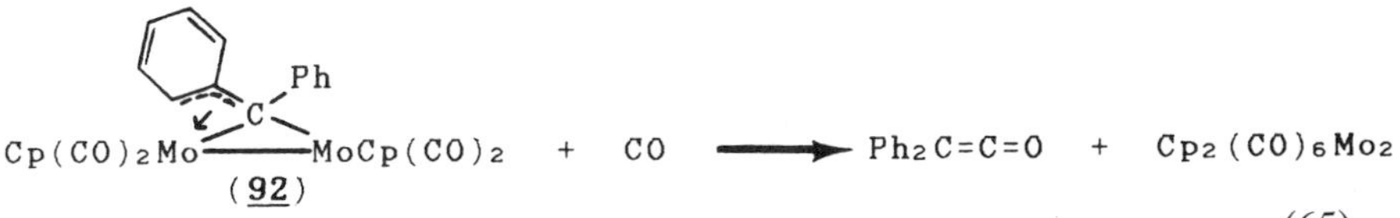

$$Cp(CO)_2Mo \overset{\overset{Ph}{\underset{\diagup}{C}}}{\rule{1cm}{0.4pt}} MoCp(CO)_2 \quad + \quad CO \quad \longrightarrow \quad Ph_2C=C=O \quad + \quad Cp_2(CO)_6Mo_2$$

$$\underset{(\underline{92})}{}$$

$$(65)$$

μ-CRR' complexes 93 in the presence of CH_3OH gave formation of substituted methyl acetates and complex 94 [Eq. (66)] (123a). This reaction likely proceeds via initial formation of a transient ketene complex,

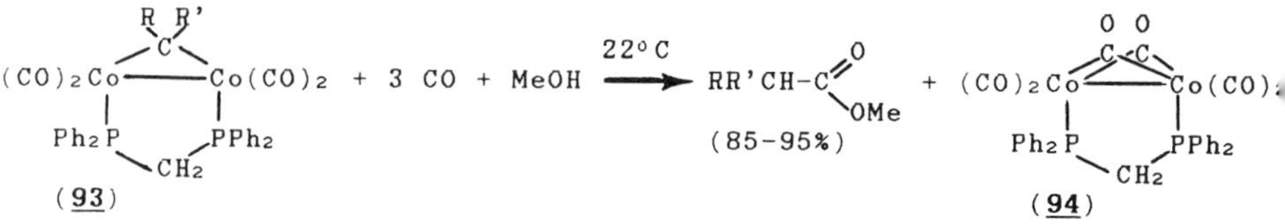

$$(66)$$

since ketene ligands are known to rapidly react with alcohols in this fashion (77,79). Free ketene was also observed by gas chromatography when the reaction was carried out in the absence of MeOH (123a). A variety of R and R' groups were used in this study, which gave some indication of the relative tendencies of substituted μ-carbene groups to undergo carbonylation. For example, the μ-CH_2 ligand reacted much faster than did the μ–$CHCO_2Et$ ligand.

There also exist a number of reactions that appear to proceed via transient but undetected ketene complexes formed by addition of CO to carbene ligands. One of the earliest involved the formation of $Ph_2C{=}C{=}O$ on addition of Ph_2CN_2 to $Ni(CO)_4$, a reaction that likely generates a transient diphenylcarbene complex which is rapidly carbonylated (123b). Thermal decomposition of $(CO)_5W{=}CPh_2$ gave 20–70% yields of diphenylketene, which was suggested to form on the tungsten center by carbonylation of the carbene ligand (123c). Similar carbonylation of $(CO)_5W{=}C(OEt)SiPh_3$ to form a ketene intermediate that subsequently undergoes hydrolysis was invoked to explain the formation of the anhydride $Ph_3Si(OEt)CHC\{O\}OC\{O\}CH(OEt)SiPh_3$ (123d). The syntheses of a number of interesting organic products using transition metal carbene complexes appear to involve the formation of vinylketene complexes by carbonylation of the carbene ligand (123e–m).

3. Deprotonation of Metal Acyl Ligands

Ketene ligands have also been formed by deprotonation of acyl ligands [Eqs. (69) and (70)] (83,87). Mononuclear type IV ketene complexes are apparently the initial products of both of the above reactions,

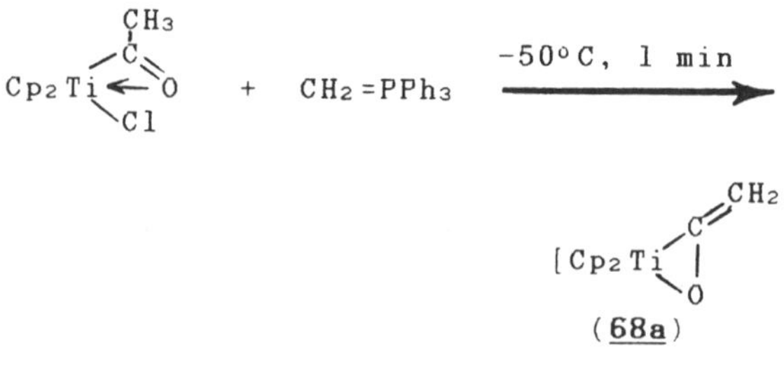

$$(67)$$

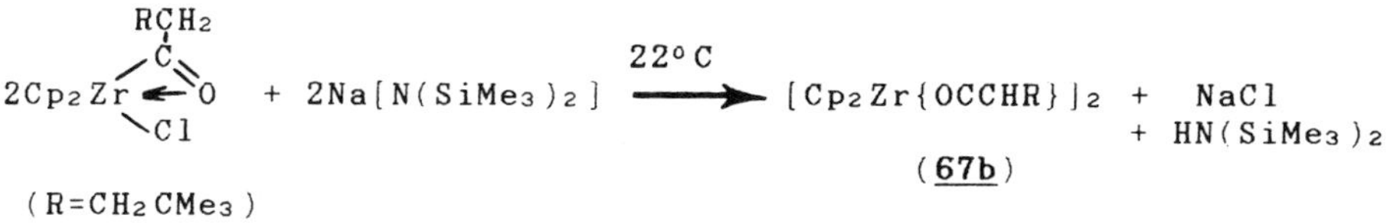

$$2 \, Cp_2 Zr \overset{RCH_2}{\underset{Cl}{\overset{C}{\diagdown}}} O \; + \; 2 \, Na[N(SiMe_3)_2] \; \xrightarrow{22^\circ C} \; [Cp_2 Zr\{OCCHR\}]_2 \; + \; NaCl \; + \; HN(SiMe_3)_2$$

$$(R = CH_2 CMe_3)$$

$$(\mathbf{67b})$$

(68)

but they must then oligomerize and dimerize to give the observed products.

Acyl protons have also been removed from the acyl complexes **95** and **96** [Eqs. (69) and (70)], but the ketene products **89** (*116–118*) and **65** (*90*) have proved too unstable to isolate, although **65** was spectroscopically characterized. Although these complexes formally contain a ketene

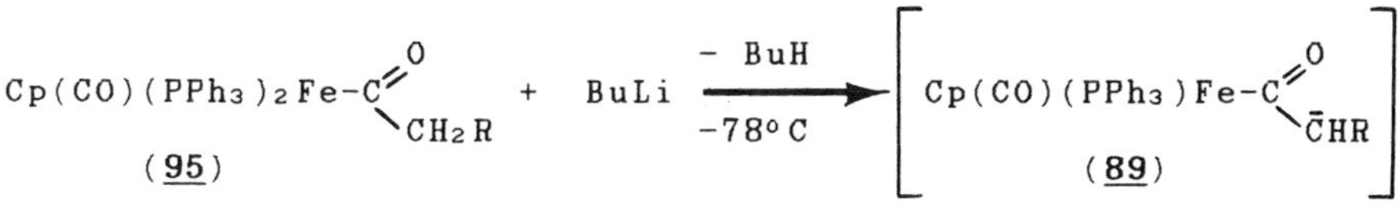

$$Cp(CO)(PPh_3)_2 Fe-C\overset{O}{\underset{CH_2R}{\diagdown}} \; + \; BuLi \; \xrightarrow[-78^\circ C]{-\, BuH} \; \left[Cp(CO)(PPh_3)Fe-C\overset{O}{\underset{\bar{C}HR}{\diagdown}} \right]$$

$$(\mathbf{95}) \qquad\qquad\qquad\qquad\qquad (\mathbf{89})$$

(69)

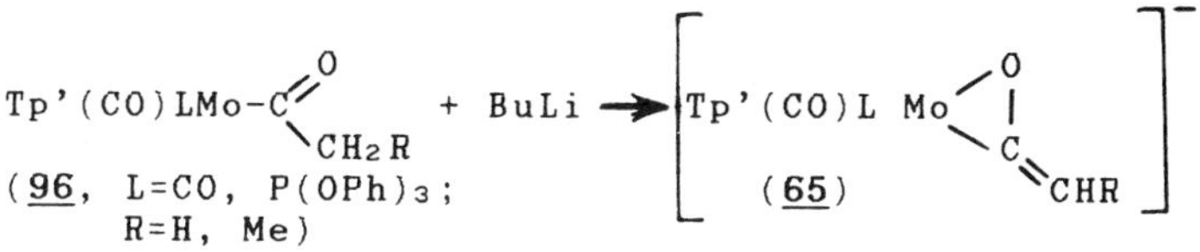

$$Tp'(CO)LMo-C\overset{O}{\underset{CH_2R}{\diagdown}} \; + \; BuLi \; \longrightarrow \; \left[Tp'(CO)L \, Mo\overset{O}{\underset{\overset{|}{C}}{\diagup}} \right]^{-} \qquad (70)$$

$$(\mathbf{96}, \; L=CO, \; P(OPh)_3; \qquad\qquad (\mathbf{65}) \qquad \overset{C}{\underset{CHR}{\diagdown}}$$
$$R=H, \; Me)$$

ligand, their regiospecific alkylation at the carbon atom to form a new acyl ligand has led them to be viewed instead as enolate complexes (*90,116–118*). We found that similar metallation of **89** with BrM(CO)$_5$ led to the isolable but unstable μ-ketene complex **97** [Eq. (71)] (*124*).

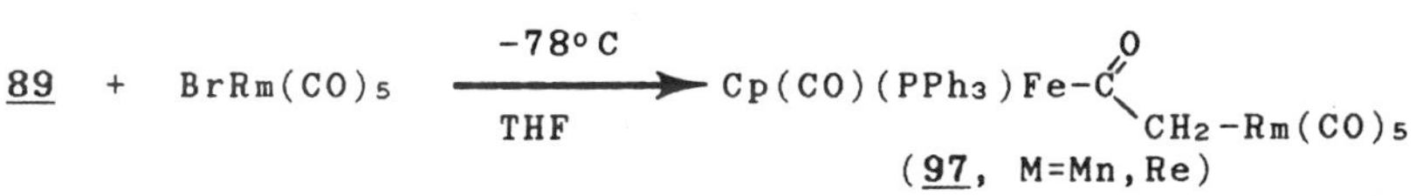

$$\mathbf{89} \; + \; BrRm(CO)_5 \; \xrightarrow[THF]{-78^\circ C} \; Cp(CO)(PPh_3)Fe-C\overset{O}{\underset{CH_2-Rm(CO)_5}{\diagup}}$$

$$(\mathbf{97}, \; M=Mn, Re)$$

(71)

The generality of this synthetic method involving deprotonation of acyl ligands will be limited by the susceptibility of other ligands, particularly cyclopentadienyl ligands, to undergo deprotonation in preference to the acyl ligand. Nevertheless, it remains a fertile area of investigation, particularly with early transition metal complexes, where it seems to work best. It should be noted that similar proton abstraction from bridging acetyl ligands would give μ-ketene complexes [Eq. (72)]. However, no examples of such chemistry have yet been demonstrated even though a number of suitable μ-acyl complexes are available for study.

$$M\underset{O=C}{\overset{M}{\diagdown}} + B^- \longrightarrow M\underset{O-C}{\overset{M}{\diagdown}} + BH \qquad (72)$$

4. Addition of Metal Carbonyl Anions to Chloroacetyl Chloride

Displacement of both chlorides from $ClCH_2C(O)Cl$ has been twice used to prepare μ-ketene complexes [Eq. (73) and (74a)] (75,76). The unstable ketene complex **51** was not isolated but evidence for its formation came

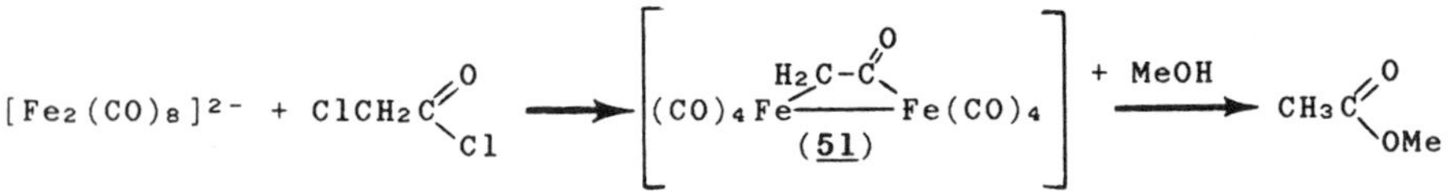

$$[Fe_2(CO)_8]^{2-} + ClCH_2C\overset{O}{\underset{Cl}{\diagdown}} \longrightarrow \left[(CO)_4Fe\overset{H_2C-C\overset{O}{\diagdown}}{\underset{(51)}{\rule{1.5cm}{0.4pt}}}Fe(CO)_4 \right] \xrightarrow{+ MeOH} CH_3C\overset{O}{\underset{OMe}{\diagdown}}$$

$$(73)$$

$$2 [Cp(CO)_2Ru]^- + ClCH_2C\overset{O}{\underset{Cl}{\diagdown}} \longrightarrow Cp(CO)_2Ru\overset{H_2C-C\overset{O}{\diagdown}}{\underset{(52a)}{\diagdown}}RuCp(CO)_2$$

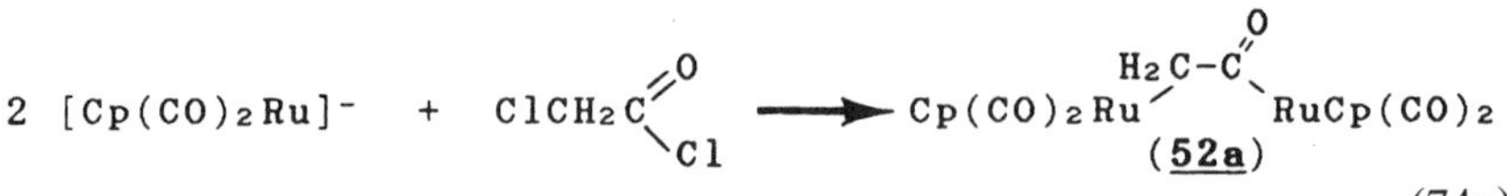

$$(74a)$$

from the observed production of methyl acetate on reaction with methanol (74). Complex **52a** was isolated in pure form, although it was noted that its yield (unspecified) was less than in the direct carbonylation route of Eq. (62).

With judicious choice of reagents and reaction conditions the overall utility of this synthetic method would seem to be far greater than these two examples. However, the reaction will likely be complicated by electron transfer reactions with the strongly reducing metal carbonyl anions that would typically be employed. Also, the resultant ketene complex may not be stable, as apparently indicated by the recently reported reaction of $[Mn(CO)_5]^-$ with $BrCH_2C\{O\}Cl$ to give free ketene rather than a μ-ketene complex (79a).

An important variation of this method which promises to have considerable generality involves the reaction of metal anions with $Cp(CO)_2Fe\!-\!CH_2C\{O\}Cl$, which was itself generated in situ [Eq. (74b)]

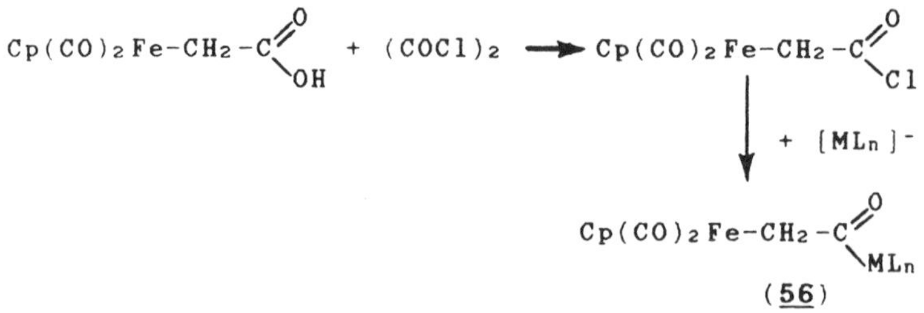

$$Cp(CO)_2Fe\!-\!CH_2\!-\!C\overset{O}{\underset{OH}{\diagdown}} + (COCl)_2 \longrightarrow Cp(CO)_2Fe\!-\!CH_2\!-\!C\overset{O}{\underset{Cl}{\diagdown}}$$

$$\Big\downarrow + [ML_n]^- \qquad (74b)$$

$$Cp(CO)_2Fe\!-\!CH_2\!-\!C\overset{O}{\underset{ML_n}{\diagdown}}$$

$$(56)$$

(*79b*). A variety of metal anions have been used—$[Cp(CO)_2Fe]^-$, $[(C_5H_4Me)(CO)_2Fe]^-$, $[Cp(CO)_2(PPh_3)Mo]^-$, $[Cp(CO)Ni]^-$, $[(CO)_5Mn]^-$, $[(CO)_3(PMe_2Ph)Co]^-$—and the μ-ketene products **56** are sufficiently stable to survive chromatography (*79b*).

5. *Modification of Existing Ketene Complexes*

A synthetic method that has often been used to prepare new ketene complexes involves modification of the coordination sphere of existing ketene complexes. For example, the sole examples of type VI ketene complexes **73** and **74** were easily formed from the monometallic ketene complex **60a** [Eq. (75)] (*85*). Trinuclear type V complexes resulted from similar reactions with alkyl aluminum reagents [Eq. (76)] (*87*).

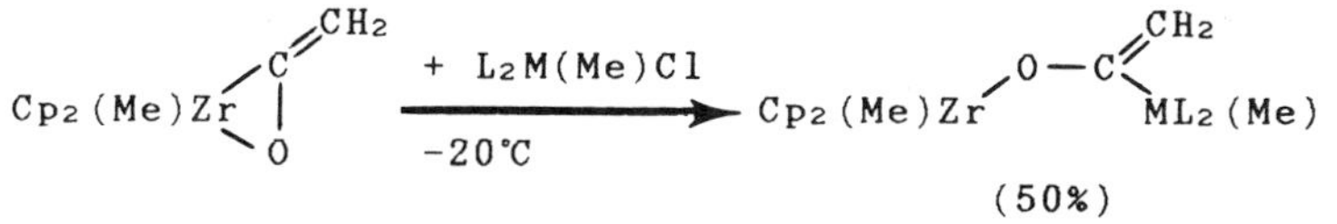

$$
Cp_2(Me)Zr\underset{O}{\overset{CH_2}{\Big|}}C \quad \xrightarrow[-20°C]{+\ L_2M(Me)Cl} \quad Cp_2(Me)Zr\overset{CH_2}{\underset{}{O-C}}ML_2(Me)
$$

$$(50\%)$$

60a

($\underline{73}$, M=Zr, L=Cp, ~50%)
($\underline{74}$, M=Pt, L=PMe$_3$)

$$(75)$$

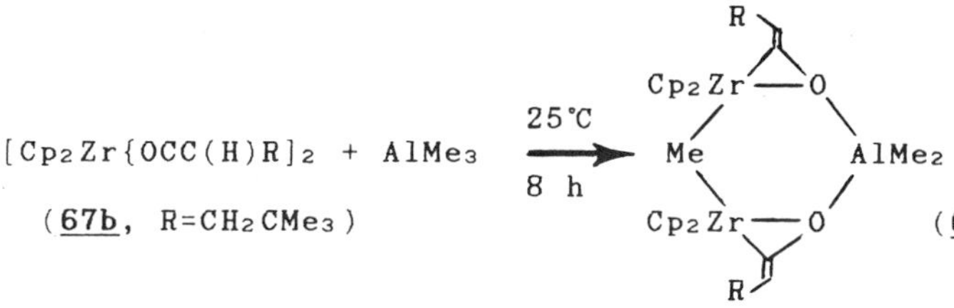

$$
[Cp_2Zr\{OCC(H)R\}]_2 \ + \ AlMe_3 \quad \xrightarrow[8\ h]{25°C} \quad
\begin{array}{c}
R \\
Cp_2Zr\!\!-\!\!O \\
Me \qquad AlMe_2 \\
Cp_2Zr\!\!-\!\!O \quad (\underline{69a}) \\
R
\end{array}
$$

(**67b**, R=CH$_2$CMe$_3$)

$$(76)$$

Other ketene complexes have been prepared by ligand substitution in existing ketene complexes [Eqs. (77) and (78)] (*65,86*), by phosphine displacement of the bridging ketene oxygen in oligomeric type V ketene complexes [Eq. (79)] (*83*), and by ligand addition reactions [Eq. (80)] (*69*).

$$
[(Cp^*)_2(X)Zr\{OCCHR\}]^- \ + \ L \quad \longrightarrow \quad (Cp^*)_2(L)Zr\{OCCHR\} \ + \ X^-
$$

(**61**, X=Cl,Br)

(**62**, L = py, CO, CH$_2$PMe$_3$)

$$(77)$$

$$
(PPh_3)_2Ni\{Ph_2CCO\} \ + \ py \quad \longrightarrow \quad (PPh_3)(py)Ni\{Ph_2CCO\} \ + \ PPh_3
$$

(**44b**)

$$(78)$$

$$[Cp_2 Ti\{OCCH_2\}]_n \quad + \quad PMe_2 Ph \longrightarrow n \left[Cp_2 (PMe_2 Ph) Ti\{OCCH_2\} \right]$$

$$(\underline{68a}) \qquad\qquad\qquad\qquad\qquad (\underline{59a})$$

$$(79)$$

$$Cl_2 Pt\{CH_2 CO\} \quad + \quad 2\ PMe_2 Ph \longrightarrow Cl_2(PMe_2 Ph)_2 Pt\{CH_2 CCO\}$$

$$(\underline{47},\ 48\%)$$

$$(80)$$

6. Other Synthetic Routes

An excellent route to vinylketene complexes is the metal-induced ring opening of cyclopropenes with concomitant insertion of CO [Eq. (81)] (98,99). This reaction has been shown to tolerate a variety of substituents on the cyclopropene ring and has been used with several metal

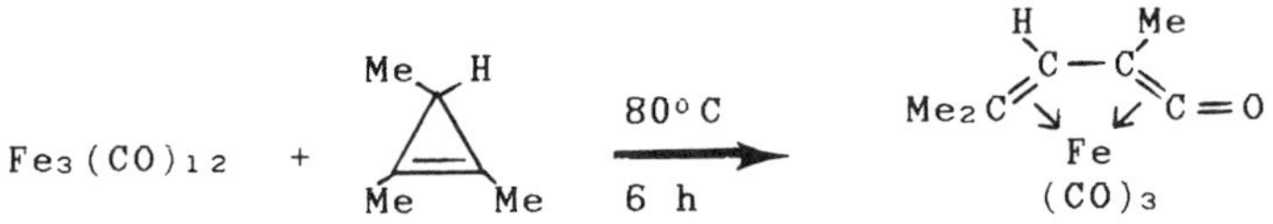

$$Fe_3(CO)_{12} \quad + \quad \underset{Me \quad Me}{\overset{Me \quad H}{\triangle}} \quad \xrightarrow[\text{6 h}]{80^\circ C} \quad \underset{Fe}{\overset{H \quad Me}{\underset{(CO)_3}{Me_2 C}}} C=0 \quad (81)$$

$$(\underline{75a},\ 5\%)$$

complexes including $Fe_2(CO)_9$, $CpV(CO)_4$, $CpCo(CO)_2$, and $Co_2(CO)_8$ (101,103–106).

Vinylketene complexes have also been formed by more serendipitous routes, for example, the type IV vinylketene complex **59** formed on carbonylation of the titanacyclobutene **98** [Eq. (82)] (84). The mechanism

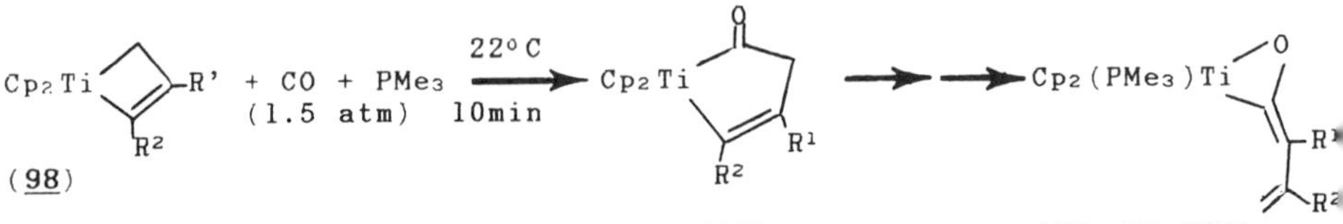

$$Cp_2 Ti\underset{R^2}{\overset{}{<}}\!\!-R' + CO + PMe_3 \xrightarrow[\text{10min}]{\underset{(1.5\ atm)}{22^\circ C}} Cp_2 Ti\overset{}{<}\underset{R^2}{\overset{}{\searrow}}R^1 \longrightarrow \longrightarrow Cp_2(PMe_3)Ti\overset{}{<}\!\!-R^1_{R^2}$$

$$(\underline{98}) \qquad\qquad\qquad\qquad (\underline{99}) \qquad\qquad (\underline{59},\ 70\text{–}92\%)$$

$$(82)$$

by which this reaction occurs is obviously complicated. An acyl intermediate (**99**) was isolated from the reaction but the path by which it converts into **59** was not elucidated.

Reaction of the vinyl ether **100** with $Fe_2(CO)_9$ led to the formation of the vinylketene complex **75b**, apparently via chloride abstraction by Fe followed by insertion of CO into the resultant metal–carbon bond and then proton abstraction [Eq. (83)] (*100*). Molybdenum and tungsten vinyl-ketene complexes were isolated from the thermal and photolytic interactions of 2-butyne with $M-C\{O\}CF_3$ complexes [Eq. (84)] (*107*).

$$CH_2 = C(OMe)CH_2Cl + Fe_2(CO)_9 \xrightarrow[1 \text{ h}]{40\,°C} (CO)_3Fe\{H_2C=C(OMe)HC=C=O\}$$

$(\underline{100})$ $(\underline{75b},\ 10\%)$

$$(83)$$

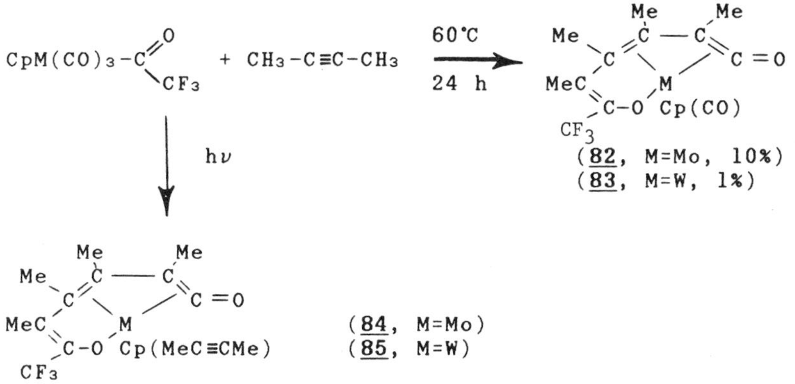

$(\underline{82},\ M=Mo,\ 10\%)$
$(\underline{83},\ M=W,\ 1\%)$

$(\underline{84},\ M=Mo)$
$(\underline{85},\ M=W)$

$$(84)$$

The unusual bis(ketene) complex **86** was prepared by photoinduced CO loss from **101** [Eq. (85)] (*108*). Complex **86** is a valuable precursor for the synthesis of substituted quinones via its reaction with alkynes (*108*)).

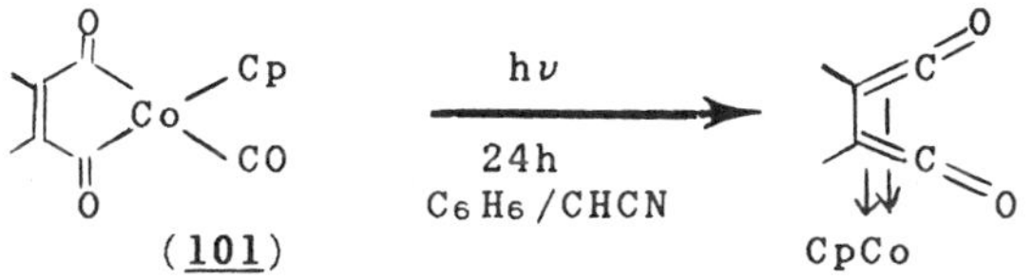

$(\underline{101})$

$(\underline{86},\ 66\%)$

$$(85)$$

Arsenic- and phosphorus-substituted ketenes that are metal coordinated via the As or P atoms have been prepared by addition of the appropriate R_2XE reagents to η^2-*ketenyl* complexes (see Section IV) [Eq. (86)] (*109,111,114*). When the reaction was run at low temperature, the type I ketene complex **50** [Eq. (85)], formed by addition of PMe_2 across a $W-C$ bond, was isolated as an intermediate and structurally characterized (*111*). When warmed, this species converted into **87**, implying its importance in the overall reaction.

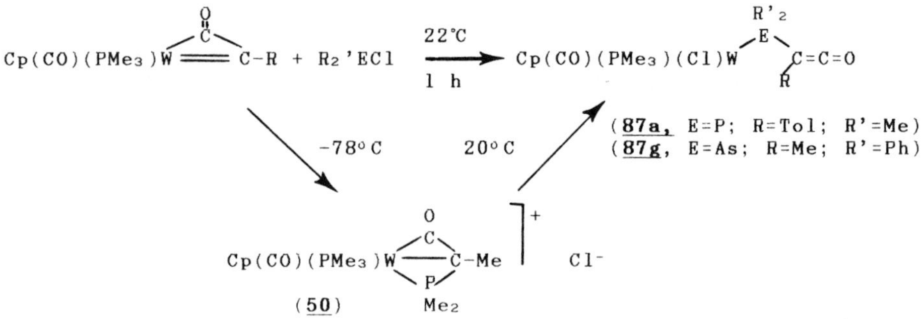

$$\text{Cp(CO)(PMe}_3)\text{W} \equiv\!\equiv\!\equiv \text{C-R} \ + \ \text{R}_2\text{'ECl} \ \xrightarrow[\text{1 h}]{22°C} \ \text{Cp(CO)(PMe}_3)(\text{Cl})\text{W}\underset{\text{R}}{\overset{\text{R'}_2}{\diagup\text{E}}}\text{C=C=O}$$

$$(\underline{87a}, \ \text{E=P}; \ \text{R=Tol}; \ \text{R'=Me})$$
$$(\underline{87g}, \ \text{E=As}; \ \text{R=Me}; \ \text{R'=Ph})$$

$$-78°\text{C} \qquad 20°\text{C}$$

$$\left[\text{Cp(CO)(PMe}_3)\text{W}\underset{\text{Me}_2}{\overset{\text{O}}{\diagup\text{P}}}\text{C-Me} \right]^+ \ \text{Cl}^-$$

$$(\underline{50})$$

(86)

A recently reported serendipitous synthesis of the type II ketene complex **52d** from a vinylidene species is that outlined in Eq. (86a) (*75b*). It

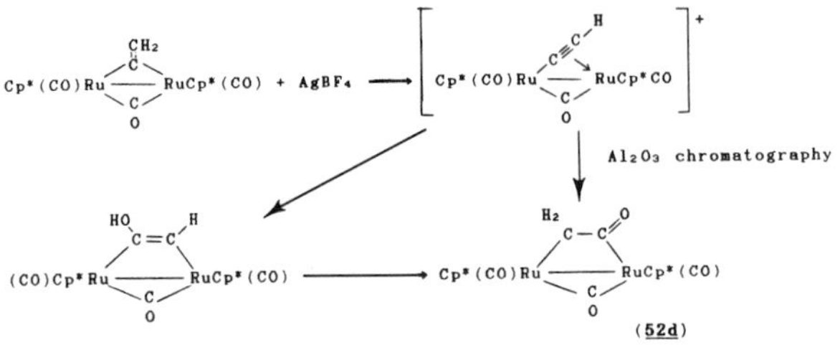

(86a)

was suggested that this species formed via Al_2O_3-induced hydroxylation of the initially formed alkynyl complex followed by tautomerization to give the product **52d**.

B. Spectroscopic Characterization of Ketene Complexes

The most useful spectroscopic data for characterizing ketene complexes are the infrared CO stretching frequency of the ketene ligand [$\nu(\text{cco})$], the ^{13}C NMR chemical shifts of the ketene carbon resonances, and the ^1H NMR chemical shifts and $^{13}\text{C}-^1\text{H}$ coupling constants of hydrogens attached to the ketene ligand. Table I summarizes the data ranges that have been found for these spectroscopic parameters for the different types of ketene complexes.

Free ketenes generally have their carbonyl vibrations in the 2100–2200 cm^{-1} range (*1*), but when coordinated to metals this band shifts to 1520–1787 cm^{-1}. The only exceptions are the ketene complexes **82**, in which the "ketene" part of the molecule is basically a free ketene, and complex **43** [$\nu(\text{cco}) = 2179$ cm^{-1}], which is perhaps best described as having a ketene ligand weakly coordinated to a $[\text{Cp(CO)}_2\text{Fe}]^+$ fragment (*62*). Types I and II ketene complexes typically have their ν_{cco} stretch at

TABLE I

REPRESENTATIVE SPECTROSCOPIC DATA FOR KETENE COMPLEXES

Complex type	v(cco) (cm^{-1})	δ(CHR)	J(CH) (Hz)	δ(CCO)	δ(CCO)
I	1600–1787	1.6–2.4	159–160	−6–75	169–256
II	1550–1622	2.5–4.1	125–137	21–51	219–258
III	1569	1.8–2.6	148	37	203
IV	1518–1620	3.4–6.3	148–160	73–124	166–228
V	1595–1620	3.9–6.2	146–148	87–103	178–197
VI	1538	4.0–5.0	148–159	93–110	202–209
VII	1520–1830	3.5–6.7	—	38–103	186–242
VIII	2090–2112	—	—	10–63	183–257

$$\underset{\displaystyle CH_2}{\overset{\displaystyle \overset{O}{\underset{\displaystyle C}{\|}}}{Cp(CO)_2\overset{+}{Fe}\longleftarrow \|}}$$

higher energies than do types III–VI complexes, principally because the oxygen atom is bound to the metal in the latter but not in the former. However, these ranges overlap significantly. Consequently, IR data are of limited use in determining which type of ketene complex has been made.

The most useful spectroscopic data for determining the ketene bonding mode are the ^{13}C NMR resonances for the ketene α and β carbons. In ketene complexes of all types except VIII, the β carbon is coordinated to the metal and shows a downfield chemical shift in the $\delta165 \rightarrow \delta258$ range typical of metal acyls (125) with little dependence on type of complex. However, the α carbon in ketene complexes of types I–III is metal coordinated, whereas it is not in types IV–VI. Accordingly, the ^{13}C NMR resonance for the α carbon in the latter complexes is typically found in the $\delta73 \rightarrow \delta125$ range characteristic of sp^2 carbon atoms (126). Complexes of types I–III with metal-coordinated α carbons generally have ^{13}C NMR resonances for this carbon in the $\delta6 \rightarrow \delta75$ range more typical of sp^3-hybridized metal alkyl ligands (125,126).

Proton NMR data can be useful for ketene compounds that have hydrogens attached to the α carbons. The protons of the $CH_2{=}C{=}O$ ligand are typically inequivalent and appear as a pair of doublets, but their position is a function of the type of ketene complex. For the type I–III complexes that have metal-bound α carbons, the proton resonances are generally in the $\delta1.5 \rightarrow \delta4.0$ region, upfield of the normal sp^2 vinyl proton region ($\delta4.5 \rightarrow \delta7.5$, 126). More downfield resonances in the $\delta3.5 \rightarrow \delta7.0$

region are typically found for complexes of types IV–VI with the α carbon not metal bound.

C. Structural Characterization

Important structural data for ketene complexes that have been characterized by X-ray diffraction studies are set out in Table II, organized by complex type. The one property that structurally characterized ketene complexes of all types have in common is a significant bending of the linear

TABLE II

STRUCTURAL DATA FOR KETENE COMPLEXES

Complex	C–C (Å)	C–O (Å)	C–C–O (deg)	Ref.
Type I				
Cp(CO)$_2$Mn(Ph$_2$CCO), **42a**	1.35(2)	1.21(2)	145	58
(C$_5$H$_4$Me)(CO)$_2$Mn (C$_{15}$H$_8$O$_2$), **42d**	1.448(8)	1.194(8)	139.8(6)	61
(C$_5$H$_4$Me)(CO)$_2$Mn (C$_{16}$H$_{12}$O), **42f**	1.401(9)	1.224(8)	142.5(7)	60
(CO)$_3$Fe(Ph$_2$CCO), **49**	1.40(2)	1.23(1)	a	40
	1.470(3)^b	1.204(3)^b	135.8(2)^b	41
Cp(CO)(PMe$_3$)W {Me$_2$PC(Me)CO}, **50**	1.44(2)	1.21(1)	a	111
Type II				
Os$_3$(CO)$_{12}$(μ-CH$_2$CO), **53**	1.470(14)	1.26(1)	117.9(9)	77
Type III				
Ru$_3$(CO)$_7$(dppm)(μ-CH$_2$) (μ-CH$_2$CO), **57**	1.43(2)	1.30(2)	119(2)	80
Type IV				
Cp$_2$V(OCCPh$_2$), **58**	1.340(7)	1.290(6)	135.9(5)	81
[Cp$_2$(Me)Zr(OCCH$_2$)]$^-$, **60a**	1.32(1)	1.339(9)	124.1(8)	85
(Cp*)$_2$(py)Zr(OCCH$_2$), **62a**	1.333(3)	1.338(2)	126.3(2)	86
Cp$_2$(PMe$_3$)Ti{OCC(Ph)C (SiMe$_3$)CH$_2$}, **59c**	1.349(6)	1.298(6)	130.0(4)	84
Type V				
[Cp$_2$Ti(μ-OCCPh$_2$)]$_2$, **66**	1.357(4)	1.311(4)	128.8(3)	92
[Cp$_2$Zr(μ-OCCPh$_2$)]$_2$, **67a**	1.333(5)	1.371(4)	123.6(3)	93
[Cp$_2$Zr{OCC(H)CH$_2$CMe$_3$}]$_2$ (AlMe$_2$)(μ-Me), **69a**	1.323(9)	1.398(7)	124.7(6)	87
	1.330(9)	1.413(7)	122.2(5)	
[Cp$_2$Zr{OCC(H)CH$_2$CMe$_3$}]$_2$ (AlMe$_2$)(μ-Cl), **69b**	1.319(9)	1.390(7)	a	87
	1.339(9)	1.393(7)	a	
[Cp$_2$Zr{OCC(H)CH$_2$CMe$_3$}]$_2$ (AlMe$_2$)(μ-H), **69c**	1.315(5)	1.392(4)	a	87
[Cp$_2$Zr{OCC(H)CH$_2$CMe$_3$}]$_2$ (AlMe$_2$H), **70c**	1.308(8)	1.406(7)	a	87
	1.318(8)	1.372(6)	a	

TABLE II (*continued*)

Complex	C–C (Å)	C–O (Å)	C–C–O (deg)	Ref.
[Cp$_2$ZrCl]$_2${OCC(H)PPh$_2$],	1.31(1)	1.33(2)	125	*94a*
73a	1.327(8)	1.31(1)	*a*	*94b*
	1.35(2)	1.35(2)	123(2)	*94c*
[Cp$_2$ZrCl]$_2${OCC(H)PMe$_2$},				
73b				
Type VII				
(CO)$_3$Fe{Me$_2$CC(H)C(Me)	1.439(5)	*a*	137	*99*
CO}, **75a**				
(CO)$_3$Fe{H$_2$CC(CO$_2$Me)	1.48(1)	1.21(1)	*a*	*102*
C(OMe)CO}, **75e**				
(CO)$_3$Fe{Me$_2$CC(H)C(H)	1.442(9)	1.188(8)	136.5(6)	*103*
CO}, **75f**				
Cp(CO)Mo{CF$_3$C(O)C(Me)	1.47(2)	1.22(2)	131(1)	*107*
C(Me)C(Me)C(Me)CO}, **82**	1.48(2)	1.21(2)	131(1)	
Cp(CO)Mn{PhHCC(Ph)C(Ph)	1.48(1)	1.20(1)	134.7(5)	*105*
CO}, **77a**	1.47(1)	1.23(1)	132.7(7)	
CpCo{PhHCC(Ph)C(Ph)CO},	1.439(8)	1.208(7)	139.5(6)	*105*
78	1.455(5)	*a*	137.8(3)	*108*
CpCo{OCC(Me)(Me)CCO},	1.455(5)	*a*	138.5(4)	
86				

a Not available.
b Neutron diffraction data.

ketene ligand on coordination, with the C–C–O angles ranging from 122 to 145°. Other structural features are dependent on the particular type of ketene complex.

Two extreme descriptions of the bonding in type 1 complexes are the metallacyclopropanone (**102**) and coordinated ketene (**103**) structures drawn below. The observed bending of the ketene ligand suggests the

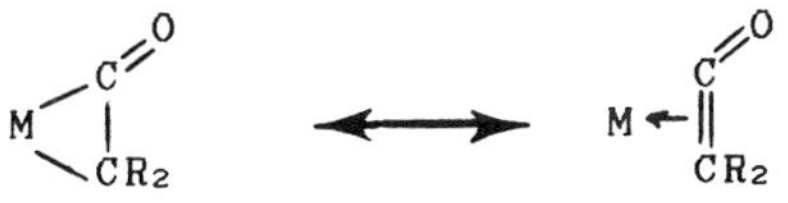

(**102**)　　　　　　(**103**)

relative unimportance of **103** since carbon–carbon π bonding cannot be substantial in a bent structure, if the C=O π bond is maintained. The structural data show that the C—O bond length does not significantly change on coordination, with the 1.19–1.23-Å range in the five structurally characterized type I complexes being only slightly longer than the 1.179-Å C–O bond length in free dimethylketene (*127*). In contrast, the C–C bond

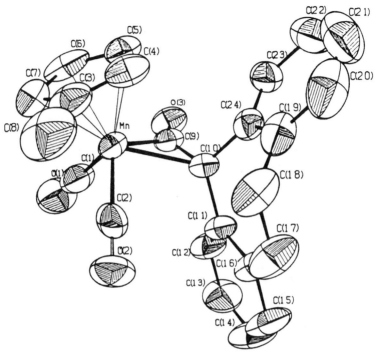

Fig. 6. Molecular structure of $(C_5H_4Me)(CO)_2Mn(C_{14}H_{12}CCO)$ (**42f**). (Reprinted with permission from Ref. *60*.)

of the coordinated ketene ligand is significantly elongated from the 1.271-Å length in free $Me_2C=C=O$ (*127*) to the 1.42-Å average value for the five type I complexes. This value is intermediate between typical C–C single (1.54 Å) and double (1.33 Å) bond values (*128*). Figure 6 shows an ORTEP drawing of complex **42f**, which is representative of this family of type I complexes.

Complexes $(CO)_3Fe(Ph_2CCO)$ (**49**) and $Cp(CO)(PMe_3)W\{Me_2PC(Me)CCO\}$ (**50**) are special examples of type 1 complexes since one of the ketene substituents is also coordinated to the metal. In **49**, one of the Ph substituents coordinates to iron with two phenyl carbons, and the ketene α carbon forms a coordinated π-allyl ligand system, Fig. 7 (*40,41*). The

$$O=C\begin{array}{c}Ph\\|\\C\\\diagdown\\Fe\\(CO)_3\end{array}$$

(**49**)

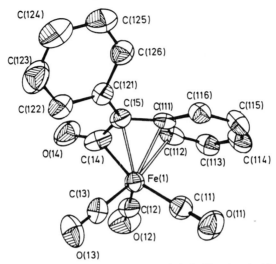

FIG. 7. Molecular structure of (CO)₃Fe(Ph₂CCO) (**49**). (Reprinted with permission from Ref. *40*.).

relatively short iron–carbon bond [1.914(2) Å] to the ketene carbonyl suggests that this portion of the structure should be described as an acyl complex. The bond distances from Fe to the π-bonded hydrocarbon unit are substantially longer (2.133–2.260 Å). The ketene C–C bond [1.470(3) Å] is also much longer than the corresponding bond in free dimethylketene (1.271 Å, *120*). Overall, the ketene ligand behaves as a 4-e⁻ donor in this complex, as required to give the metal the necessary 18-e⁻ count.

In **50**, the PMe₂ substituent of the ketene ligand is also coordinated to the metal. However, this has little effect on the coordination of the ketene ligand since the structural parameters within this ligand are typical of the remainder of the type I complexes given in Table II. The W–C distance to the ketene α carbon is 2.36(1) Å, slightly longer than the 2.13(1)-Å distance to the ketene β carbon, but still within bonding distance (*111*).

An ORTEP drawing of Os₃(CO)₁₂{μ-CH₂CO} (**53**), the only type II compound to be structurally characterized, is shown in Fig. 8 (*77*). The structural data, particularly the ketene Os–C–C bond angle of 112.6(6)°, indicate that the α carbon is nearly fully *sp³* hybridized and the ligand retains little, if any, of its original ketene character. The C–C bond distance of 1.47(1) Å is close to typical C–C single bond distances for carbons adjacent to carbonyls (1.52 Å, *128*), and the bond angles about the ketene carbonyl average 120.0°. The molecule would appear to be

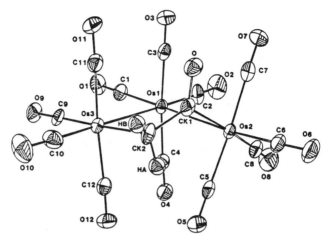

FIG. 8.   Molecular structure of $Os_3(CO)_{12}(\mu\text{-}CH_2CO)$ (53). (Reprinted with permission from Ref. 77.)

better described as a metallated acyl complex rather than a ketene complex, although it does contain the elements of the ketene ligand.

The structural parameters for the closely related type III compound $Ru_3(CO)_7(dppm)(\mu\text{-}CH_2)(\mu\text{-}CH_2CO)$ (57) are similar to those of 53, with little effect on the ketene part of the molecule by coordination of the oxygen to the third metal atom (80). The Ru-O distance is 2.11(1) Å, well within normal bonding values. Figure 9 shows an ORTEP drawing of the molecule, which, when compared to the structure of 53 in Fig. 6, illustrates the effect of this Ru-O coordination on the overall cluster geometry.

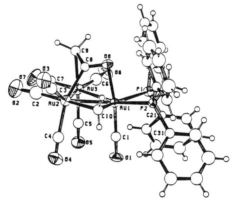

FIG. 9.   Molecular structure of $Ru_3(CO)_7(dppm)(\mu\text{-}CH_2)(\mu\text{-}CH_2CO)$ (57). (Reprinted with permission from Ref. 80.).

In type IV, V, and VI ketene complexes, the bonding of the ketene ligand to the metal is markedly different from that in types I–III complexes, and this difference is reflected in the structural parameters. ORTEP drawings of Cp$_2$V(OCCPh$_2$) (**58**, *81*), (Cp*)$_2$(py)Zr(OCCH$_2$) (**62a**, *86*),[Cp$_2$TiOCCPh$_2$)]$_2$ (**66**, *92*), and [Cp$_2$Zr{OCC(H)CH$_2$CMe$_3$}]$_2$-(AlMe$_2$) (μ-Me) (**69a**, *87*), which are representative of these types of complexes, are shown in Fig. 10–13. Coordination of the ketene ligand through the carbon–oxygen bond leads to a significant elongation of this bond, with an average of 1.32 Å for the three structurally characterized type IV complexes. This compares to the 1.23 Å value for the seven type I–III complexes in Table II and the 1.179 Å value for free dimethylketene (*127*). Coordination of the ketene oxygen to a second metal as in the type V complexes leads to further lengthening of this bond to an average value of 1.37 Å for the nine structurally characterized compounds of this class.

Since the ketene C=C bond is not metal coordinated in the type IV–VI complexes, this bond should retain more of its original double bond character and be relatively short. Indeed, it averages 1.33 Å for both the type IV and V complexes, which is a typical carbon–carbon double bond value (1.33 Å, *128*), compared to the 1.42 Å average for the type I–III complexes, in which the C=C bond is metal coordinated. However, the ketene ligand C=C bond in the type IV and V complexes is lengthened from the 1.271 Å value for free dimethylketene, indicating a long-range effect of coordination to the carbon–oxygen bond. The overall structural parameters of these type IV–VI complexes suggest that they be described

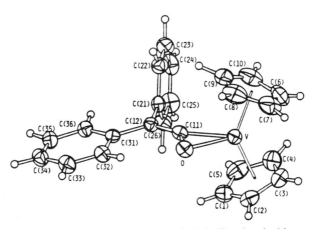

FIG. 10.   Molecular structure of Cp$_2$V(OCCPh$_2$) (**58**). (Reprinted with permission from Ref. *81*.).

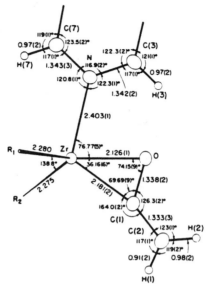

FIG. 11.   Molecular structure of $(Cp^*)_2(py)Zr(OCCH_2)$ (**62a**). (Reprinted with permission from Ref. *86.*)

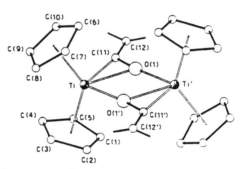

FIG. 12.   Molecular structure of $[Cp_2Ti(OCCPh_2)]_2$ (**66**). (Reprinted with permission from Ref. *92.*)

as either enolate complexes, particularly for anionic **60a** (*85*), or metallaoxirane complexes (*87*). Complex **69a**, Fig. 13, is significant in that it possessses the first example of a symmetrically bridging methyl ligand with the structural parameters indicating a distorted trigonal bipyramidal arrangement of the metals and hydrogens about the μ-carbon atom (*87*).

Several vinylketene complexes of type VIII have been structurally characterized (Table II), and an ORTEP drawing of $(CO)_3Fe\{Me_2CC(H)$

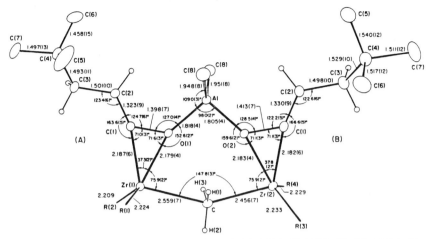

FIG. 13.   Molecular structure of [Cp$_2$Zr{OCC(H)CH$_2$CMe$_3$}]$_2$(AlMe$_2$)(μ-Me) **(69a)**. (Reprinted with permission from Ref. *87.*).

C(Me)CO}, **75a** (*99*), which is representative of this family, is shown in Fig. 14. The structural data for all the complexes imply contributions from three resonance forms, **104–106**. The marked bending of the C–C–O bond

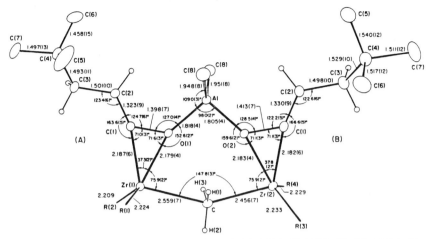

in all of the complexes, together with relatively short M–C{O} bonds to the ketene carbonyl carbon (1.918 Å in **75a**, *99*), signals the importance of the acyl canonical structures **104** and **105**. These differ only in the bonding description of the remaining three carbon atoms, which simply represent the two canonical forms that can be drawn for η^3-allyl complexes. For example, in complex **75e** the C–C distances within this three-carbon unit are nearly equal, causing the authors to favor resonance form **104** (*102*). However, in **75a** these C–C distances are quite different, 1.414(4) and 1.397(4) Å, **105** (*99*).

An ORTEP drawing showing the structure of the unusual bis(ketene) complex **86** is shown in Fig. 15 (*108*). The near equality of the C–C distances within this ligand suggests some dienelike character (**107**) in the bonding but the bent C–C–O angles and slightly shorter Co–C{O} bonds

η^1-Ketenyl Complexes

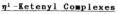

Cp(CO)L(PR'$_3$)M-C=C=O with R

(**142**, M=Mo) (*136,138*)
(**143**[a], M=W) (*132-140*)
ν(CCO)=1996→2029 cm^{-1}
^{13}C: δ 6→δ-15 (C̲CO)
       δ156→δ164 (CC̲O)

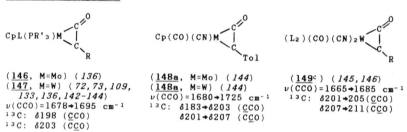

(**144a**) (*141a*)
ν(CCO)=2068cm^{-1}

(**144b**) (*141a*)
ν(CCO)=2080 cm^{-1}

Cp(PMe$_3$)LFe-C=C=O with SiMe$_3$

(**145**[b]) (*141b*)
ν(CCO)=2025s

η^2-Ketenyl Complexes

CpL(PR'$_3$)M with R

(**146**, M=Mo) (*136*)
(**147**, M=W) (*72,73,109,*
       133,136,142-144)
ν(CCO)=1678→1695 cm^{-1}
^{13}C: δ198 (C̲CO)
^{13}C: δ203 (CC̲O)

Cp(CO)(CN)M with Tol

(**148a**, M=Mo) (*144*)
(**148a**, M=W) (*144*)
ν(CCO)=1680→1725 cm^{-1}
^{13}C: δ183→δ203 (C̲CO)
       δ201→δ207 (CC̲O)

(L$_2$)(CO)(CN)$_2$W with R

(**149**[c]) (*145,146*)
ν(CCO)=1665→1685 cm^{-1}
^{13}C: δ201→205(C̲CO)
       δ207→211(CC̲O)

μ_2-η^1-Ketenyl Complexes

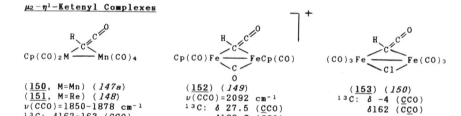

Cp(CO)$_2$M——Mn(CO)$_4$ with R

(**150**, M=Mn) (*147a*)
(**151**, M=Re) (*148*)
ν(CCO)=1850-1878 cm^{-1}
^{13}C: δ162-163 (C̲CO)

Cp(CO)Fe——FeCp(CO) with H and O

(**152**) (*149*)
ν(CCO)=2092 cm^{-1}
^{13}C: δ 27.5 (C̲CO)
       δ162.6 (CC̲O)

(CO)$_3$Fe——Fe(CO)$_3$ with H and Cl

(**153**) (*150*)
^{13}C: δ -4 (C̲CO)
       δ162 (CC̲O)

Fig. 14. Ketenyl complexes. [a]L = CO, PMe$_3$, PPh$_3$, MeC≡CNEt$_2$; [b]L = CO, PMe$_3$;
[c]L$_2$ = 1,10-phenanthroline, 2,2'-bipyridine.

to the ketene carbonyl carbons argue for a significant contribution from the bis(acyl) resonance form **108**.

(**107**)   (**108**)

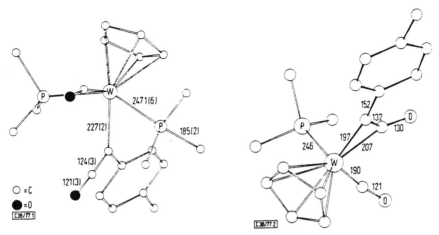

FIG. 15. Comparison of the molecular structures of Cp(CO)(PMe₃)₂W–C(Tol)CO (**143**, left) and Cp(CO)(PMe₃)W{TolCCO} (**147**, right). (Reprinted with permission from Ref. *133a*.)

D. Reactivity of Coordinated Ketene Ligands

Free ketenes are very reactive molecules, and the relatively few reactivity studies of coordinated ketenes also indicate high reactivity. These ligands are susceptible to both nucleophilic and electrophilic attack, they undergo insertion reactions, and they can be hydrogenated to produce aldehydes and alcohols. However, it must be emphasized that the *reactivity* of ketene complexes is a neglected area, with relatively few studies having been conducted. The sections below summarize the more important reaction classes, beginning with the simple displacement of the ketene ligand by other ligands.

1. Displacement of the Ketene Ligand

In several complexes the ketene ligand has been found to be easily displaced by other ligands, as illustrated by the substitution of $Ph_2C{=}C{=}O$ in **42a** by PPh_3 and ethylene [Eq. (87)], although the fate of

$$
\begin{array}{c}
\xrightarrow[\text{60°C, 10h}]{+\ PPh_3} Cp(CO)_2Mn-PPh_3 \\
Cp(CO)_2Mn(Ph_2C{=}C{=}O) \\
\textbf{42a} \\
\xrightarrow[\text{55 atm}]{+\ C_2H_4} Cp(CO)_2Mn(CH_2{=}CH_2) \\
(21\%)
\end{array}
$$

(87)

the released ketene was not determined (57). In this study it was noted that although the carbonyl ligands in complex **42a** are not equivalent, as one is trans to the ketene α carbon and the other is trans to the β carbon, only one ^{13}C NMR CO signal was detected as low as $-55°C$ (58). This suggests either rapid rotation of the ketene ligand on the NMR time scale or reversible dissociation and recoordination of the ketene. The latter was ruled out by the demonstration of no exchange between coordinated $Ph_2C{=}C{=}O$ and free $Me(Ph)C{=}C{=}O$ under the conditions of the NMR experiment. However, it was not stated whether exchange of ketenes occurs on more typical laboratory time scales.

The reaction of the arsanylketene complex **87** with PMe_3 in refluxing toluene gave near-quantitative yield of the free arsanylketene **109** [Eq. (88)], a substituted ketene that had previously been unattainable with traditional organic methodology (*110*). It was noted that under milder

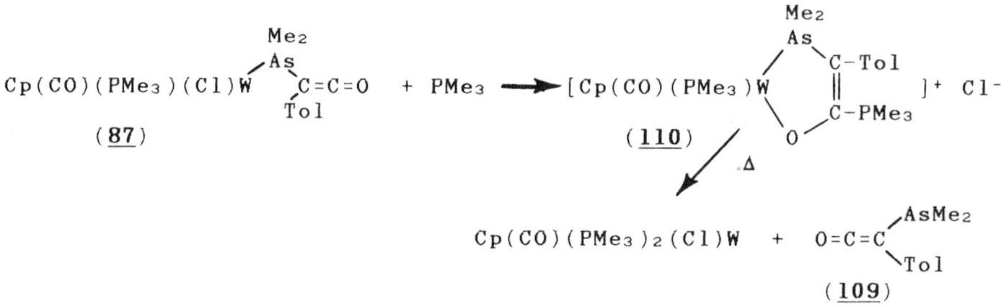

$$Cp(CO)(PMe_3)(Cl)W \overset{Me_2}{\underset{Tol}{\overset{\diagup As}{\diagdown}}} C{=}C{=}O \ + \ PMe_3 \ \longrightarrow \ [Cp(CO)(PMe_3)W \overset{Me_2}{\underset{}{\overset{As}{\diagdown}}} \overset{C-Tol}{\underset{C-PMe_3}{\overset{\|}{\diagdown}}}]^+ \ Cl^-$$

$$(\underline{87}) \qquad\qquad\qquad\qquad (\underline{110}) \qquad \downarrow \Delta$$

$$Cp(CO)(PMe_3)_2(Cl)W \ + \ O{=}C{=}C \overset{\diagup AsMe_2}{\diagdown Tol}$$

$$(\underline{109})$$

$$\tag{88}$$

reaction conditions the added PMe_3 ligand adds to the ketene β carbon to give an arsatungstaheterocycle (**110**), and this species is an intermediate in the formation of **109**. Arsanylketene is also formed along with much decomposition when its displacement is accomplished with CO (100°C, 60 atm) (*110*).

Release of the ketene ligand from the type IV complex **58** was accomplished by oxidation with I_2 [Eq. (89)] or O_2 (81) (79):

$$Cp_2V(O{=}C{=}CPh_2) \ + \ I_2 \ \longrightarrow \ Ph_2C{=}C{=}O \ + \ Cp_2VI_2 \tag{89}$$

$$(\underline{58})$$

The organometallic product of the O_2-induced oxidation was not identified.

2. Insertion Reactions

An insertion reaction is one in which another ligand or potential ligand inserts into a metal–ketene bond. These reactions could prove useful as they provide a way to incorporate the ketene synthon into desirable organic reaction products. An excellent example is shown in Eq. (90), in which addition of alkynes to the unusual bis(ketene) complex **86** led to the formation of substituted quinones (*108*). The mechanism of this

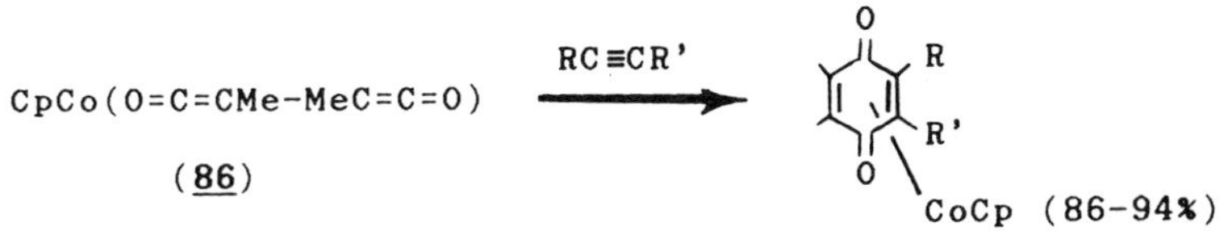

$$CpCo(O=C=CMe-MeC=C=O) \xrightarrow{\quad RC\equiv CR' \quad} \text{[quinone–CoCp]} \quad (90)$$

(**86**)   CoCp (86-94%)

reaction is not known, but it may involve initial coordination of alkyne to cobalt.

Insertion of a second ketene molecule into the Ti–ketene bond of **63** has been shown to give the novel complex **111** [Eq. (91)] (*91*). This reaction

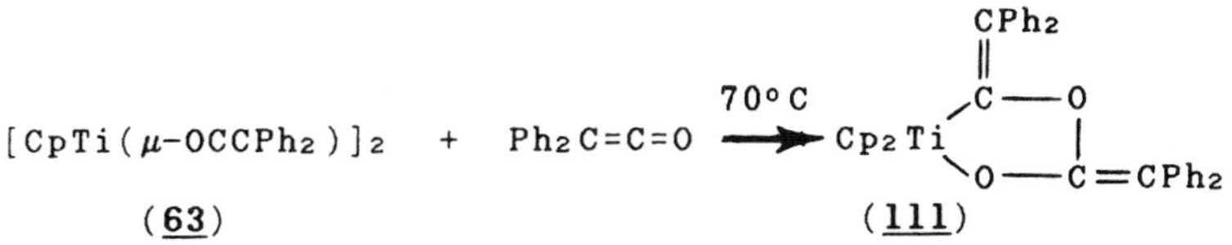

$$[CpTi(\mu\text{-}OCCPh_2)]_2 \; + \; Ph_2C=C=O \xrightarrow{\;70^\circ C\;} \text{Cp}_2\text{Ti} \begin{bmatrix} C-O \\ O-C=CPh_2 \end{bmatrix} \quad (91)$$

(**63**)   (**111**)

could proceed via insertion of the C=O bond of free ketene into either the Ti–O or Ti–C bond of the coordinated ketene ligand, although the authors favored the former. Surprisingly, this reaction is reversible since **63** was regenerated along with free Ph$_2$C=C=O when **111** was heated in solution for 2 h.

An analogous insertion of ketone solvents into the Re–O bond of **64** results in a structurally similar product **112** [Eq. (92)] (*89*). This reaction is

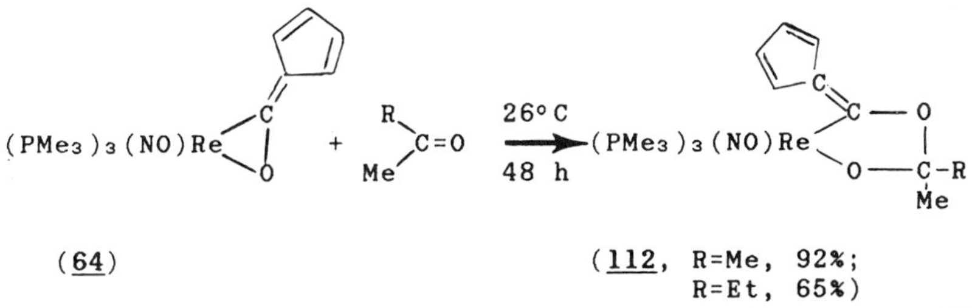

$$(PMe_3)_3(NO)Re\begin{bmatrix} C \\ O \end{bmatrix} \; + \; \begin{matrix} R \\ Me \end{matrix}C=O \xrightarrow[48\ h]{26^\circ C} (PMe_3)_3(NO)Re\begin{bmatrix} C & C-O \\ O-C-R \\ Me \end{bmatrix}$$

(**64**)   (**112**, R=Me, 92%; R=Et, 65%)

$$(92)$$

also reversible, as shown by the clean conversion back into **64** when **112** (R = Me) was heated at 80°C ($t_{1/2} \sim 1$ h). When compound **113**, the precursor to ketene complex **64**, was heated in the presence of CO, the unusual binuclear complex **114** was formed (*90*). The mechanism of this obviously complicated reaction is unknown, but it is tempting to suggest

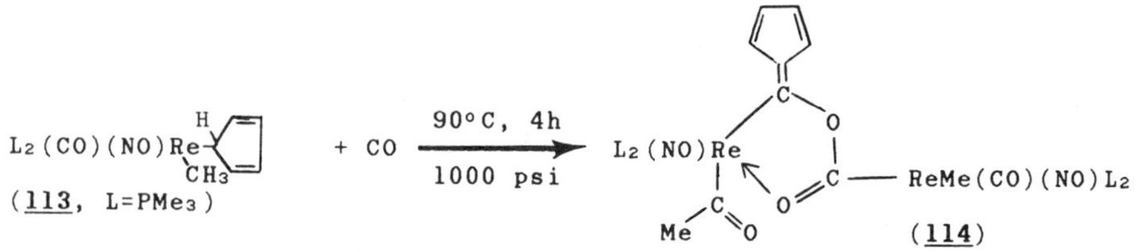

$$L_2(CO)(NO)Re{\overset{H}{\underset{CH_3}{\diagup}}} \quad + \quad CO \quad \xrightarrow[\text{1000 psi}]{90°C, \ 4h} \quad L_2(NO)Re \cdots\text{(114)} \cdots ReMe(CO)(NO)L_2$$

$(\underline{113}, \ L=PMe_3)$

$(\underline{114})$

$$(93)$$

the initial formation of ketene complex **64**, which then inserts a *coordinated CO* of an $Re(CO)_2L_2(Me)(NO)$ complex.

Very promising reactions, because of what they signal for the use of ketene complexes in synthetic organic chemistry, are the demonstrated insertions of ethylene and acetylene into the Ti–carbon bond of the type V ketene complex **68a** to give the metallacycles **112** and **113** [Eq. (94)] (*83*).

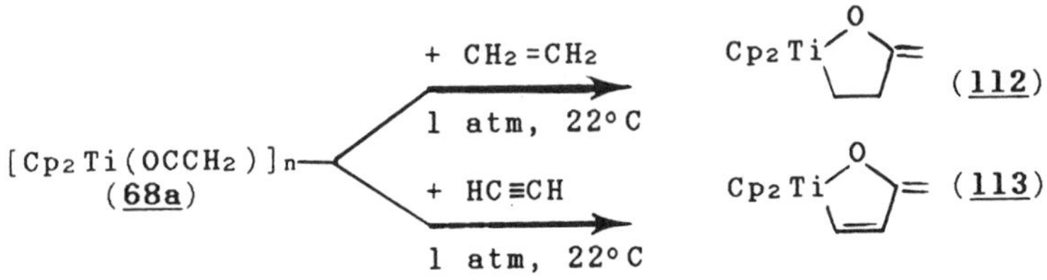

$$[Cp_2Ti(OCCH_2)]_n \quad (\underline{68a})$$

$$+ \ CH_2=CH_2 \xrightarrow{1 \ atm, \ 22°C} Cp_2Ti\cdots (\underline{112})$$

$$+ \ HC\equiv CH \xrightarrow{1 \ atm, \ 22°C} Cp_2Ti\cdots (\underline{113})$$

$$(94)$$

The same product **113** forms on reaction of $HC\equiv CH$ with **59a** (*83*). However, ethylene and acetylene did not react with the type V ketene complexes $[Cp_2Zr\{\mu\text{-}OCC(H)CH_2CMe_3\}]$ (**67b**) (*83*).

Similar insertion of acetylene into the metal–ketene bond occurs with ketene complex **69a** (see Fig. 4) to give near-quantitative yield of the metallacycles **114** and **115** [Eq. (95)] (*87*). Metallacycle **115** becomes the

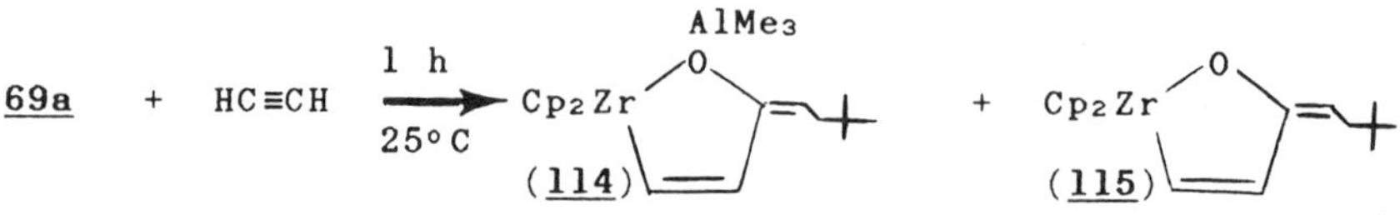

$$\underline{69a} \quad + \quad HC\equiv CH \xrightarrow[25°C]{1 \ h} Cp_2Zr\cdots(\underline{114}) \quad + \quad Cp_2Zr\cdots(\underline{115})$$

$$(95)$$

sole product when the reaction is run in THF solution, where the $AlMe_3$ is complexed by solvent molecules. However, insertion of substituted alkynes or ethylene into the Zr–ketene bond of **69a** did not occur, and acetylene did not react with the chloro analog of **69a** (**69b**) presumably due to less facile opening of a coordination site by cleavage of the Zr—X—Zr bridge (X = Me, Cl) (*87*). Insertion of CO into the Zr–carbon bond of ketene complex **69a** led to the acyl complex **116** [Eq. (96)] (*87*).

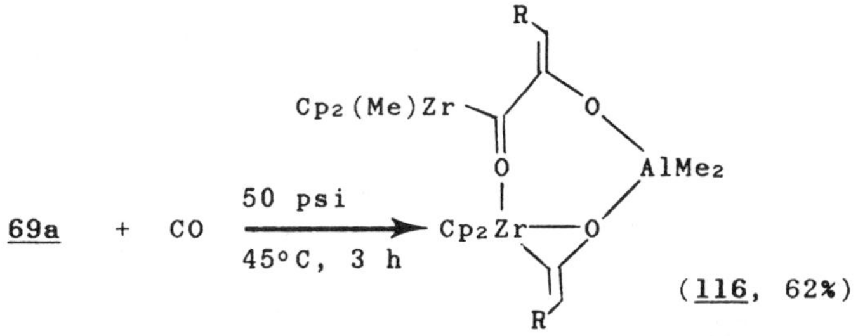

$$\mathbf{69a} \ + \ CO \ \xrightarrow[\substack{45^{\circ}C, \ 3 \ h}]{50 \ psi} \ \text{(116, 62\%)} \tag{96}$$

Although strictly not an insertion reaction, that shown in Eq. (97) is an interesting example of the construction of an η^4-oxaallyl ligand by combination of coordinated ketene and methylene ligands (80). Complex 117 is

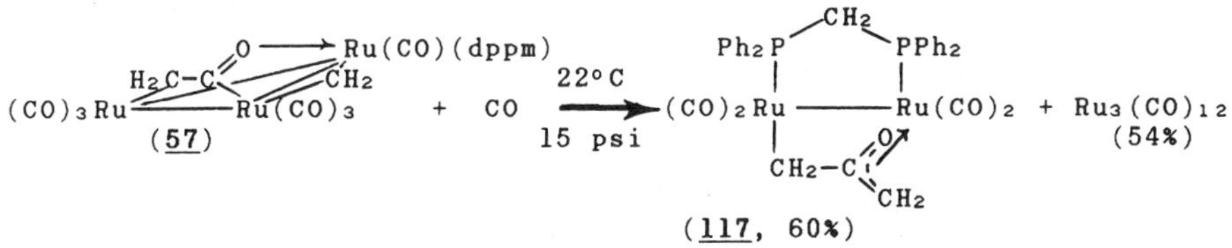

$$\text{(57)} \ + \ CO \ \xrightarrow[\substack{15 \ psi}]{22^{\circ}C} \ \text{(117, 60\%)} \ + \ Ru_3(CO)_{12} \ \text{(54\%)} \tag{97}$$

the first example of an isolable η^3-enolate complex (part of the η^4-oxaallyl ligand) and reaction (97) is the first example of the construction of a three-carbon unit from two methylene fragments and a carbonyl ligand via an intermediate ketene complex.

3. Decarbonylation of Coordinated Ketenes

There are several reported reactions in which a coordinated ketene ligand undergoes a decarbonylation process prior to forming other products. A clear indication that the ketene carbonyl can migrate to the metal comes from a ^{13}CO exchange study with $(CO)_3Fe\{Ph_2C=C=O\}$ (49) in which it was found that the ketene carbonyl and the metal carbonyls were equally enriched with ^{13}CO after stirring under a ^{13}CO atmosphere at 22°C for 3 days (40,41). A double labeling experiment clearly showed the exchange to be strictly intramolecular. Further evidence for decarbonylation of the ketene ligand in this complex came from its reaction with $Fe_2(CO)_9$ to give the novel binuclear complex 118, in which H migration from Ph to the ketene β carbon has occurred in addition to CO loss [Eq. (98)] (40,41). The same workers showed that the ketene carbonyl in 49 can be replaced by an ethylidene group, the latter formed via metal-induced rearrangement of added ethylene [Eq. (99)] (71).

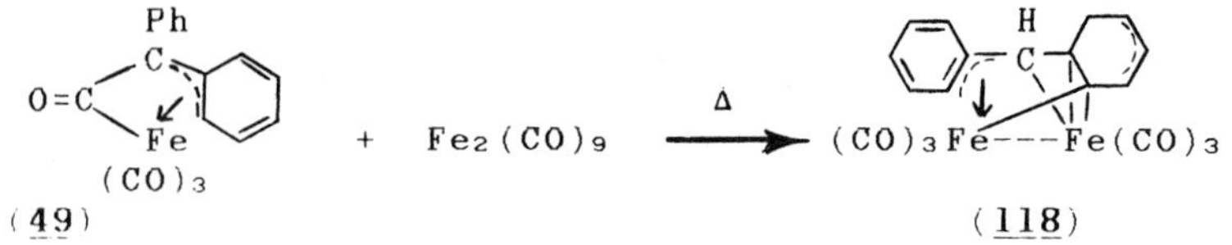

$$(98)$$

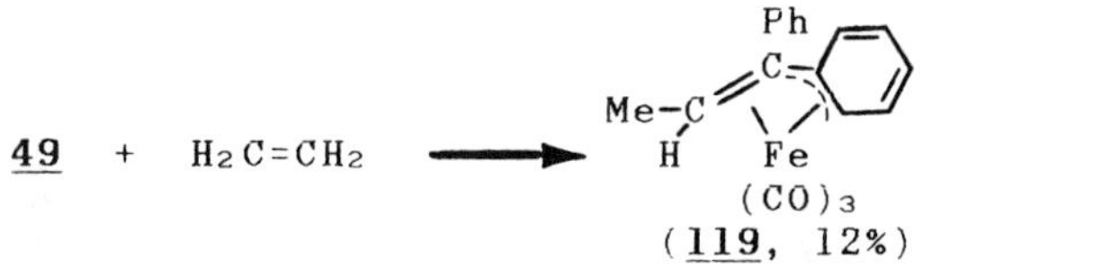

$$(99)$$

The vinylketene complex **75g** was also shown to undergo *reversible* loss of CO to form vinylcarbene complex **120** [Eq. (100)] (*104*). This

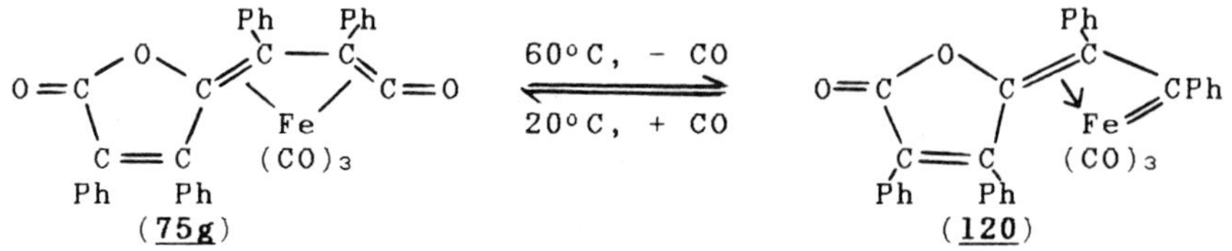

$$(100)$$

work has precedent in an earlier demonstration of the PPh$_3$ and CO-induced carbonylation of the vinylcarbene complex $(CO)_3Fe=C(OMe)$ $C(CO_2Me)=CH_2$ to form the vinylketene complex **75e** (*102*).

An example of decarbonylation of a type IV ketene complex is that which occurs on removal of pyridine from ketene complex **63a** [Eq. (101)] (*88*). This ligand loss opens a coordination site on Zr, which is then

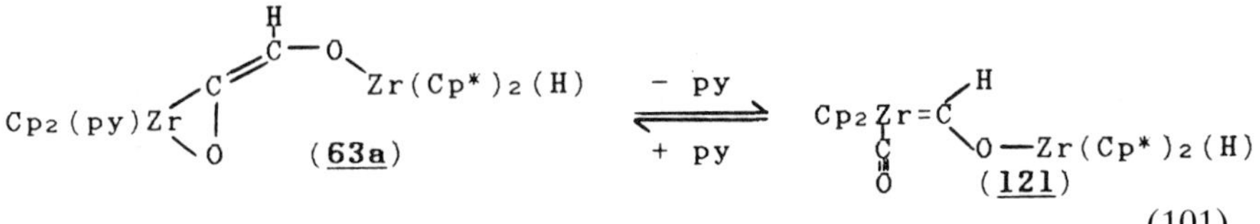

$$(101)$$

occupied by the ketene carbonyl following deinsertion. This reaction is reversible and addition of py to **121** was the method by which **63a** was prepared.

Reaction (102) illustrates the decarbonylation of the novel phosphine-substituted ketene ligand in **50**, which yields a rare example of an η^2-phosphinocarbene ligand (*73*).

There are also several examples of the formation of hydrocarbon product on thermal decomposition of ketene complexes, which presumably proceeds via initial decarbonylation of the coordinated ketene to form metal carbenes, which then couple or scavenge hydrogen to give the observed products (*e.g.* Eq. (103)] (*65*). Similarly, ethylene was produced in

$$[Cp(CO)(PMe_3)W \overset{\overset{\displaystyle O}{\underset{\displaystyle C}{\diagup}}\diagdown}{\underset{\overset{\displaystyle P}{\diagup}}{}\underset{Ph_2}{}}CMe]^+ \xrightarrow{\Delta} [Cp(CO)(PMe_3)W \overset{\overset{\displaystyle Me}{\underset{\displaystyle C}{\diagdown}}}{\underset{\overset{\displaystyle P}{|}}{}\underset{Ph_2}{}}]^+ + CO$$

$$(\underline{50}) \qquad\qquad\qquad\qquad (102)$$

$$(PPh_3)_2Ni\{Ph_2C=C=O\} \xrightarrow{80°C} Ph_2CH_2 + Ph_2CH-CHPh_2 + Ph_2C=CPh_2$$
$$\qquad\qquad\qquad\qquad\qquad\qquad (\sim 2\%) \qquad\quad (\sim 5\%) \qquad\qquad (47\%)$$

$$+ (PPh_3)_2Ni(CO)$$

$$(103)$$

76% yield on thermal decomposition of $(PPh_3)_2Ni\{CH_2=C=O\}$ along with a poorly characterized $[Ni(PPh_3)(CO)]_n$ complex (66). Likewise, thermolysis of $(PPh_3)_2Pt\{CD_2=C=O\}$ at 55°C gave C_2D_4 (17%), C_3D_6 (16%), C_4D_8 (7%), and $Pt(CO)(PPh_3)_2$ (36%) (69).

The type II ketene complexes $[Cp_2(CO)Ru]_2(\mu\text{-}CO)$ (52d) and $[Os_3(CO)_{10\mu-x}(\mu\text{-}CH_2CO)]^-$ (54) have also been shown to readily decarbonylate to yield the corresponding μ-methylene complexes (75b,78).

4. Nucleophilic Attack on Coordinated Ketenes

One of the characteristic reactions of free ketenes is their reaction with nucleophiles at the electrophilic β carbon [Eq. (1)]. An important question concerns how coordination of ketene, and particularly the type of coordination, affects the reactivity of this ligand. While only a few studies have been conducted, the results indicate that the β carbon remains susceptible to nucleophilic attack by reagents such as OMe^-, but the overall reactivity of the ligand is attenuated by metal coordination.

The only type I complex for which nucleophilic reactions have examined is $[Fe\{CH_2CO\}]^+$. As expected, the positive charge of the complex makes this species highly electrophilic and reactions with I^-, OMe^-, $MeOH$, and water have been reported [Eq. (104)] (62,63). The iodoacyl species 124 is

$$[Cp(CO)_2Fe\{CH_2CO\}]^+ \quad\begin{cases} \xrightarrow[\text{or } OMe^-]{+ MeOH} Cp(CO)_2Fe-CH_2-C\overset{\displaystyle O}{\underset{\displaystyle OMe}{\diagup\diagdown}} + H^+ \\ \\ \xrightarrow{+ H_2O} Cp(CO)_2Fe-CH_2-C\overset{\displaystyle O}{\underset{\displaystyle OH}{\diagup\diagdown}} + H^+ \\ \\ \xrightarrow{+ I^-} Cp(CO)_2Fe-CH_2-C\overset{\displaystyle O}{\underset{\displaystyle I}{\diagup\diagdown}} \end{cases}$$

$$(\underline{43}) \qquad\qquad (\underline{122}) \qquad\qquad (\underline{123}) \qquad\qquad (\underline{124})$$

$$(104)$$

unstable and has not been well characterized; however, its ready solvolysis to **123** and **124** supports its formulation. Nucleophilic addition to the μ-ketene ligand in the type II complex **54c** gives similar products [Eq. (105)] (*129,130*). The latter are rare examples of alkyl-substituted cluster compounds.

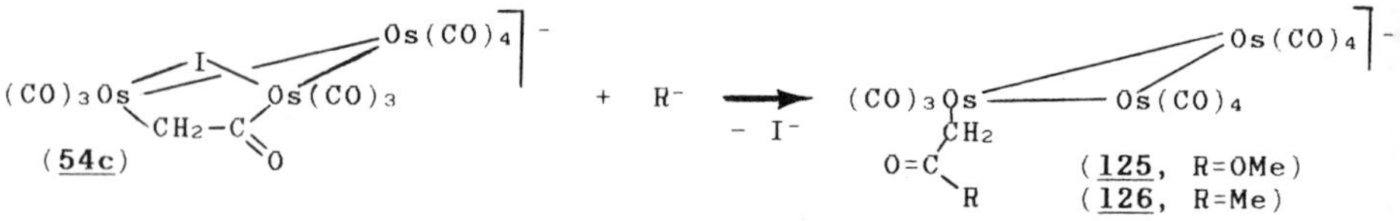

$$(105)$$

Type VIII ketene complexes also react with nucleophiles at the β carbon to give products that depend on the nature of the nucleophile [Eq. (106)] (*109,112,113*). Compound **127** formed as two diastereomers due to the

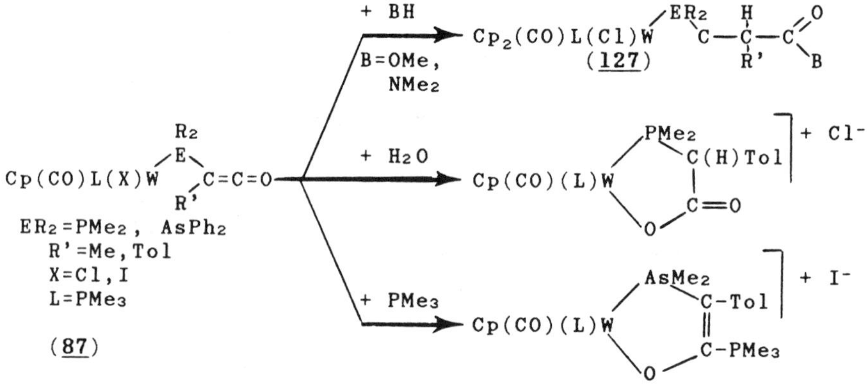

$$(106)$$

chirality of the tungsten center as well as the alkyl ligand. Similar addition of PMe_3 to the presumed intermediate **88b** was proposed to account for the formation of **128** [Eq. (107)] (*119*).

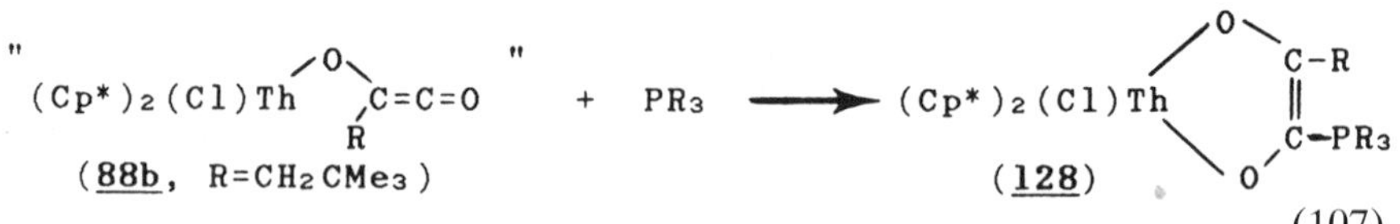

$$(107)$$

A number of ketene complexes have been shown to react with alcohols and water to give free alkyl acetates and acetic acid, reactions that presumably proceed via initial nucleophilic addition to the ketene β carbon similar to the above reactions but followed by protonolysis of the resultant metal–alkyl bond. Such reactions have been described with $(PPh_3)_2Ni(Ph_2CCO)$ (65), $Fe_2(CO)_8(\mu\text{-}CH_2CO)$ (74), $[Cp(CO)_2Ru]_2(\mu\text{-}CH_2CO)$ under photochemical activation (75), $Os_3(CO)_{12}(\mu\text{-}CH_2CO)$ (77), and the vinylketene complex $(CO)_3Fe\{Me_2CC(H)C(H)CCO\}$ (103). Indeed, the reaction of ketene complexes with methanol and water to form coordinated alkyl (enolate) ligands or free methyl acetate and acetic acid appears to be a general and diagnostic reaction for ketene complexes of types I–III and VII and VIII. The lack of examples of such reactions with the (C,O)-bound ketenes of types IV–VI may be due to their low electrophilicity or perhaps to the fact that such reactions have not been examined.

Although the reaction mechanism is not known, the reported $BH_3 \cdot THF$ reduction of the μ-ketene ligand in $[Cp^*(CO)Ru]_2(\mu\text{-}CH_2CO)(\mu\text{-}CO)$ (52d) to yield the ethylene and μ-ethylidene complexes $Cp^*_2Ru_2\text{-}(CH_2{=}CH_2)(CO)(\mu\text{-}CO)_2$ and $[Cp^*(CO)Ru]_2(\mu\text{-}CHMe))\mu\text{-}CO)$ presumably proceeds via initial hydride addition to the μ-ketene carbonyl carbon (75b).

5. Electrophilic Attack on Coordinated Ketenes

Whereas the (C,O)-bound ketene complexes do not appear to be susceptible to nucleophilic attack, they do readily undergo electrophilic addition either to the ketene carbonyl oxygen [Eq. (97)] (89) or to the ketene α carbon for the more oxophilic Ti and Zr complexes [Eqs. (108)–(110)] (83,85). Since 60a is formed by abstraction of a proton from a Zr–acetyl complex similar to 129, the formation of 60a followed by reaction (109) represents a useful elaboration of the acetyl ligand.

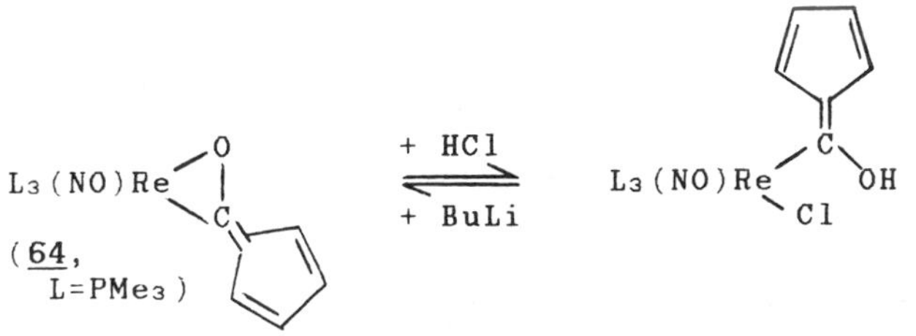

$$L_3(NO)Re \overset{O}{\underset{C}{\diagup}} \quad \underset{+\ BuLi}{\overset{+\ HCl}{\rightleftharpoons}} \quad L_3(NO)Re \overset{C}{\underset{Cl}{\diagup}} OH \qquad (108)$$

(64, L=PMe₃)

$$Cp_2(Me)Zr \overset{CH_2}{\underset{O}{\diagup}} \overset{-}{\bigg|} \quad +\ RX \quad \longrightarrow \quad Cp_2(Me)Zr \overset{CH_2R}{\underset{O}{\diagdown}} \qquad (109)$$

(60a)   R=Me, CH₂Ph, (129)
     SiMe₃

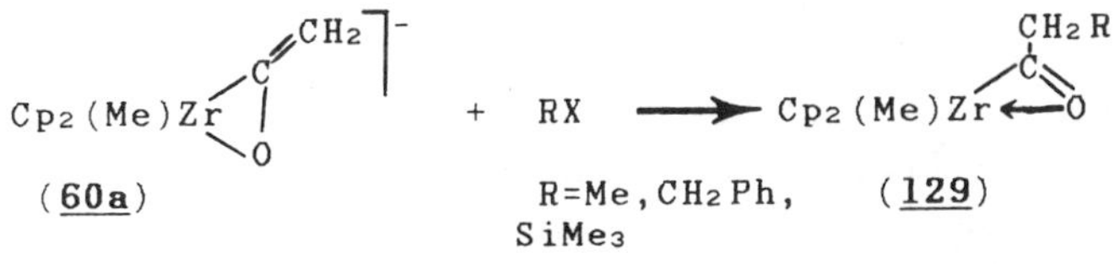

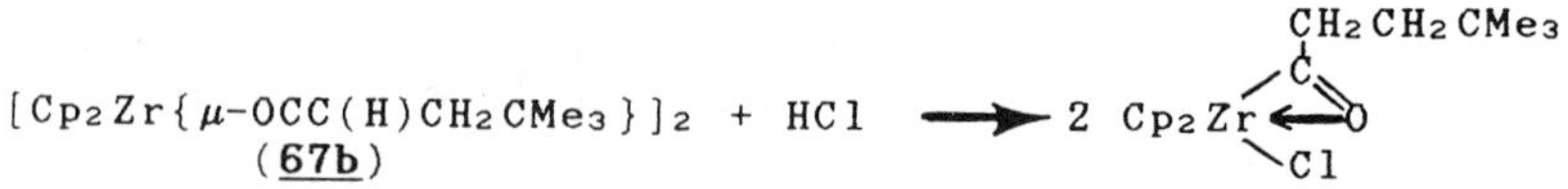

$$[Cp_2Zr\{\mu\text{-}OCC(H)CH_2CMe_3\}]_2 \; + \; HCl \; \longrightarrow \; 2 \; Cp_2Zr \underset{Cl}{\overset{C}{\longleftarrow}} O$$
$$\text{(67b)}$$

(110)

An impressive synthetic application of the alkylation of ketene complexes is found in the several studies with $[Cp(CO)(PPh_3)Fe—C(O)CHR]^-$ (89) and its derivatives (116–118). Recall that this species was generated by proton abstraction from the corresponding acetyl complex [Eq. (69)]. Alkylation of 89 with RI [Eq. (111)] generates a new acyl complex, which subsequently can be deprotonated and alkylated to give elaboration of the acyl ligand. Significantly, it was recently found that these

$$Cp(CO)(PPh_3)Fe—C\overset{O}{\underset{CH_2^-}{\diagup}} \; + \; RI \; \longrightarrow \; Cp(CO)(PPh_3)Fe—C\overset{O}{\underset{CH_2R}{\diagup}} \; + \; I^-$$

(111)

reactions are highly stereospecific due to the chirality of the metal complex, signaling potential synthetic organic applications for this chemistry (117,118). Similar stereospecific alkylations were recently reported for the related Mo complex $[Tp'(CO)(L)Mo\{OCCHR\}]^-$ (65) (90).

Protonation of the ketene complex 58 with carboxylic acids was reported to give complex 130 along with $Ph_2CH—CHPh_2$ and CO and the mechanistic route of Eq. (112) was proposed (81). A key element in this proposal

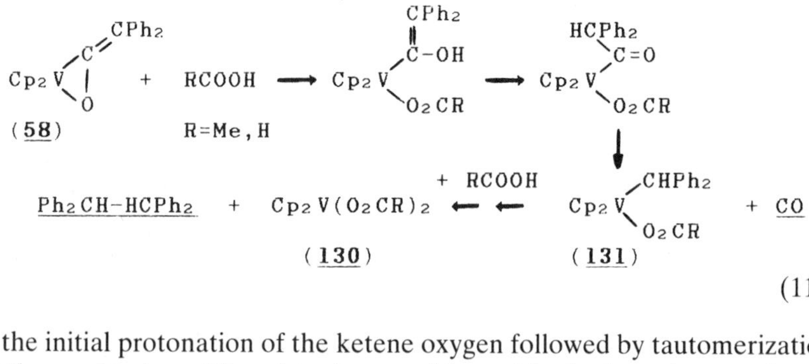

$$Cp_2V\overset{CPh_2}{\underset{O}{\diagdown}} \; + \; RCOOH \; \longrightarrow \; Cp_2V\overset{\overset{CPh_2}{\parallel}{\underset{O_2CR}{C\text{-}OH}}} \; \longrightarrow \; Cp_2V\overset{HCPh_2}{\underset{O_2CR}{C=O}}$$
$$\text{(58)} \qquad\qquad R=Me,H$$

$$Ph_2CH\text{-}HCPh_2 \; + \; Cp_2V(O_2CR)_2 \; \longleftarrow \overset{+\;RCOOH}{\longleftarrow} \; Cp_2V\overset{CHPh_2}{\underset{O_2CR}{\diagup}} \; + \; CO$$
$$\text{(130)} \qquad\qquad\qquad\qquad \text{(131)}$$

(112)

is the initial protonation of the ketene oxygen followed by tautomerization of the resultant enol ligand to the acetyl form. The hydrocarbon product is presumably formed by decomposition of unstable 131. A similar mechanism would explain the formation of Ph_2CH_2 and $Ph_2CH—CHPh_2$ on protonation of $[Cp_2Ti\{Ph_2CCO\}_2]_2$ (66) (91,92).

The only (C,C)-bound ketene ligand that has been reported to undergo electrophilic attack is that found in the cluster compound 54c in which the ketene ligand is converted into vinyl and acetyl ligands [Eq. (113)]

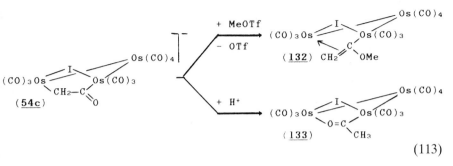

(113)

(*129,130*). These reactions suggest the importance of oxycarbene (**134**) and oxyvinyl (**135**) resonance forms in which the negative charge is localized on the ketene oxygen. The oxycarbene resonance form is known to be

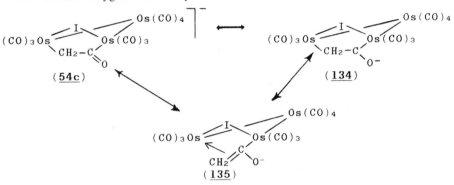

important for anionic metal acyl complexes and contributes to the ease with which these are alkylated at the carbonyl oxygen by R^+ reagents to yield metal carbenes. The conversion of **54c** into the observed methoxyvinyl complex **132** occurs by similar addition of CH_3^+ to the carbonyl oxygen. Reasonable paths for the protonation-induced conversion of **2** into the μ-acetyl complex **5** are outlined in Eq. (114) (*129,130*). These

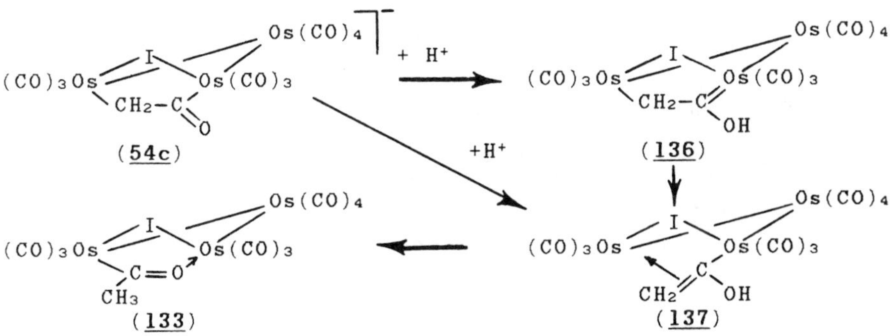

(114)

involve protonation of the ketene carbonyl oxygen either to produce a hydroxycarbene complex **136** which then rearranges to the hydroxyvinyl complex **137** or to **137** from **54c** directly. Tautomerization of **137** would yield **133**, similar to organic keto–enol tautomerization. Infrared and ^{1}H NMR spectra clearly showed the presence of an intermediate in the **54c** → **133** conversion, which was proposed to be the hydroxyvinyl cluster **136** on the basis of the similarity of its IR spectrum to that of the methoxyvinyl cluster **132** (*129,130*).

6. Cycloaddition Reactions

A characteristic reaction of free ketenes is their cycloaddition with unsaturated organic substrates [Eq. (2)] (*2a*). Surprisingly, there is only one report of a similar reaction with a coordinated ketene ligand [Eq. (115)] (*57*). It was noted that reaction (104) was far slower than the

$$\text{Cp(CO)}_2\text{Mn} \overset{\displaystyle \nearrow^O}{\underset{\displaystyle \searrow \text{CPh}_2}{\text{C}}} \quad + \quad \square \quad \xrightarrow[81^\circ \text{C}]{\text{slow}} \quad \boxed{} \overset{\displaystyle \nearrow^O}{\underset{\displaystyle \text{CPh}_2}{\text{C}}} \quad (115)$$

(**42a**)

reaction of free $Ph_2C{=}C{=}O$ with cyclopentadiene under comparable conditions, indicating deactivation of the ketene on coordination or a rate-limiting dissociation of ketene from the metal.

7. Hydrogenation

Ketenes have often been proposed to be important intermediates in the metal-catalyzed reduction of CO to hydrocarbons (*131*), and thus the hydrogenation of coordinated ketenes has received attention in order to model this proposal. Several examples of such reaction are now known with aldehydes and alcohols being the typical products [Eqs. (116)–(119)] (*59,69,77*). However, not all coordinated ketene complexes undergo such

$$\text{Cp(CO)}_2\text{Mn(Ph}_2\text{C}{=}\text{C}{=}\text{O)} \xrightarrow[50^\circ\text{C},20\text{h}]{\text{H}_2\,(750\ \text{Bar})} \underset{\substack{1:3 \\ (51\%\ \text{overall})}}{\text{PhCH}_2\text{CHO} + \text{Ph}_2\text{CHCH}_2\text{OH}} \quad (116)$$

(**42**)

$$(\text{PPh}_3)_3\text{Pt(CH}_2{=}\text{C}{=}\text{O)} + \underset{(3\ \text{atm})}{\text{H}_2} \xrightarrow{26^\circ\text{C}} \underset{(20\%)}{\text{C}_2\text{H}_5\text{OH}} + \underset{(36\%)}{\text{CH}_3\text{CHO}} + \underset{(27\%)}{\text{CH}_3\text{OH}}$$

(**46b**)

$$(117)$$

$$Cl_2Pt(CH_2=C=O) + H_2 \xrightarrow{26°C} C_2H_5OH + CH_3CHO + CH_3OH$$
$$(\underline{46c}) \qquad (3 \text{ atm}) \qquad (55\%) \qquad (21\%) \qquad (2\%)$$
$$+ CH_4 + C_3H_8$$
$$(3\%) \quad (9\%)$$

$$(118)$$

$$Os_3(CO)_{12}(CH_2CO) + H_2 \xrightarrow[5 \text{ d}]{22°C} CH_3CHO$$
$$(\underline{53}) \qquad\qquad\qquad\qquad\qquad (20\%)$$
overall 65%

$$(119)$$

reduction, as indicated by the inertness of $Ru_3(dppm)(CO)_7(\mu\text{-}CH_2)(\mu\text{-}CH_2CO)$ (57) when treated with H_2 (80 psi, 110°C) (80).

There is only one example of partial hydrogenation of a ketene ligand, which in this case gives the enolate complex 138 [Eq. (120)] (86).

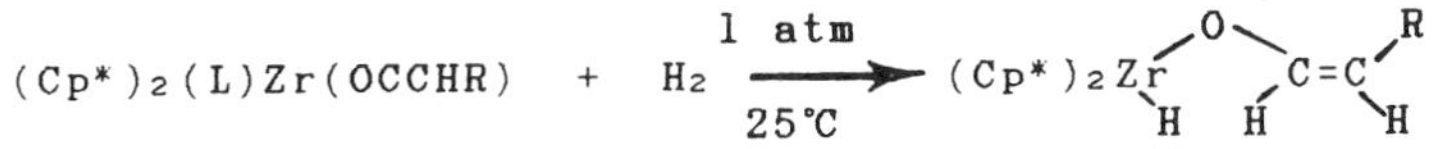

$$(Cp^*)_2(L)Zr(OCCHR) + H_2 \xrightarrow[25°C]{1 \text{ atm}} (Cp^*)_2Zr\overset{O}{\underset{H}{\diagdown}}\overset{R}{\underset{H}{C=C}}\overset{}{\underset{H}{}} \quad (120)$$
$$\underline{62} \qquad\qquad\qquad\qquad\qquad\qquad (\underline{138})$$

8. Isomerization

A remarkable reaction is the recently reported isomerization of a type II ketene ligand to a formylmethylene ligand [Eq. (120a)] (75b). However, no mechanism was suggested for this process.

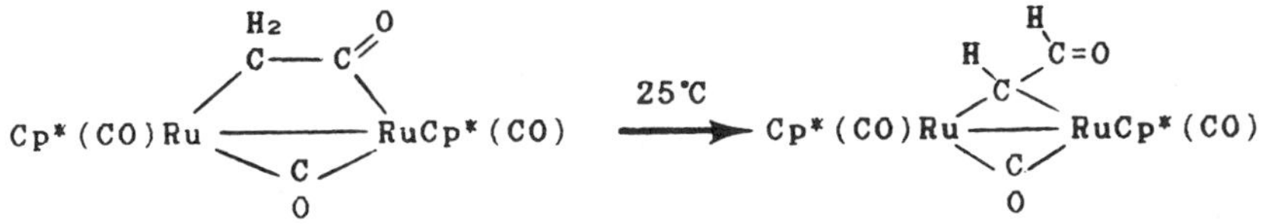

$$\text{Cp}^*(\text{CO})\text{Ru} \overset{H_2}{\underset{O}{\diagup}}\overset{O}{\underset{}{C}} \xrightarrow{25°C} \text{Cp}^*(\text{CO})\text{Ru} \cdots \text{RuCp}^*(\text{CO})$$

$$(\underline{52d})$$

$$(120a)$$

IV

KETENYL COMPLEXES

Ketenyl complexes have a metal as one of the ketene substituents, and the ligand is bound in either η^1 (139) or η^2 (140) mode to a single metal or less commonly as a $\mu_2\text{-}\eta^1$ ligand (141) bridging two metals. The known classes of ketenyl complexes are illustrated in Fig. 14 along with representative IR and ^{13}C NMR data. Both of these spectroscopic methods are

TABLE III

Representative IR and ^{13}C NMR Data for Ketenyl Complexes

Coordination mode	$\nu(CC{=}O)$ (cm^{-1})	$\delta(C{=}C{=}O)$	$\delta(C{=}C{=}O)$
η^1-Ketenyl	1950–2100	10 → −30	150 → 170
η^2-Ketenyl	1600–1700	190 → 215	190 → 215
μ_2-η^1Ketenyl	1850–2100	10 → −30	150 → 170

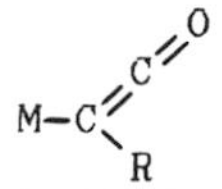

$$\eta^1 \quad (\underline{139}) \qquad (\underline{140}) \quad \eta^2 \qquad (\underline{141}) \quad \mu_2-\eta^1$$

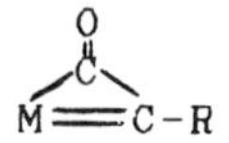

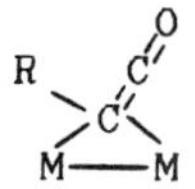

sufficient to distinguish between the η^1 and η^2 coordination modes, as illustrated by the correlations of Table III.

The discovery of this class of compounds and many of the subsequent synthetic and reactivity studies were accomplished by F. Kreissl and his co-workers in Munich (132). Their studies have focused on Mo and W complexes and accordingly there are few ketenyl complexes known with other metals.

A. Synthesis of Ketenyl Complexes

Ketenyl complexes have typically been made by ligand-induced CO migration to a carbyne ligand [Eq. (121)] (132,133,135–137). Addition of

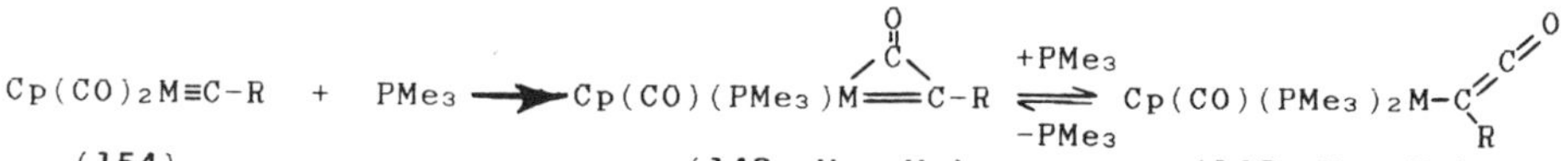

$$\text{(121)}$$

one equivalent of PMe$_3$ to the stable carbyne complex 154 gives rapid and quantitative formation of the η^2-ketenyl complexes 142 and 143. In the presence of excess PMe$_3$, the η^1-ketenyl complexes 146 and 147 form and can be isolated in good yield. Most studies have focused on the tungsten complexes, and various carbyne substituents have been used in these reactions. Although mechanistic details have not been published, the much greater rate of the reaction with PMe$_3$ compared to PPh$_3$ (see below) suggests an associative process with the phosphine inducing migration of a coordinated CO ligand to the carbyne.

When PPh_3 is used in place of PMe_3, a similar but markedly slower transformation of **154** to an η^1-ketenyl complex has been claimed (*133a*), although we have found that the latter reaction proceeds efficiently only under photochemical conditions [Eq. (122)] (*151*). We believe that migration of CO to the carbyne ligand is photoinduced and that this opens a

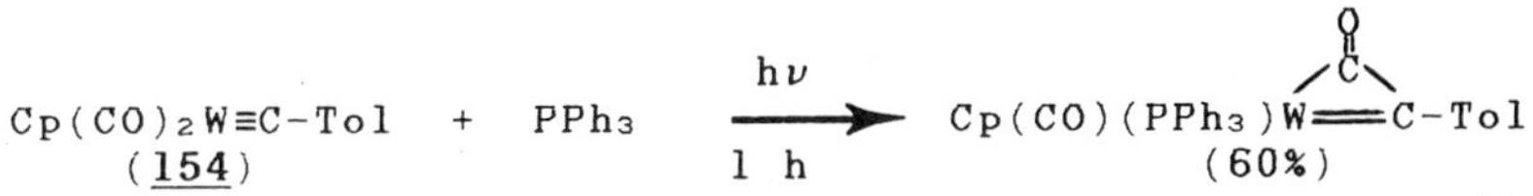

$$Cp(CO)_2W{\equiv}C\text{-}Tol \; + \; PPh_3 \; \xrightarrow[\;1\text{ h}\;]{h\nu} \; Cp(CO)(PPh_3)W{=}C\text{-}Tol$$
$$(\underline{154}) \qquad\qquad\qquad\qquad\qquad\qquad (60\%)$$

(122)

coordination site for PPh_3 addition. In the absence of added phosphine, starting complex **154** adds to the presumed coordinatively unsaturated ketenyl complex (**155**) to give binuclear **156** [Eq. (123)] (*151*).

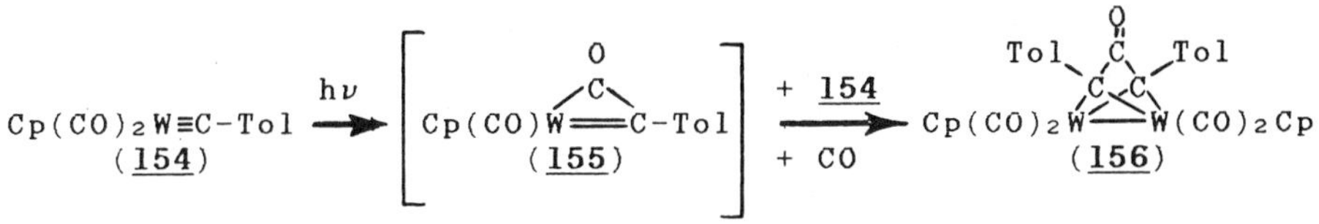

$$Cp(CO)_2W{\equiv}C\text{-}Tol \xrightarrow{h\nu} \left[Cp(CO)W{=}C\text{-}Tol \right] \xrightarrow[+\ CO]{+\ \underline{154}} Cp(CO)_2W{=}W(CO)_2Cp$$
$$(\underline{154}) \qquad\qquad (\underline{155}) \qquad\qquad\qquad (\underline{156})$$

(123)

Cyanide ion is also an effective nucleophile for inducing carbyne carbonylation to yield η^2-ketenyl complexes [Eqs. (124) (*144*) and (125) (*145,146*)]. Reaction (125) also represents a significant extension of the ketenyl-forming reaction to carbyne complexes other than **154**.

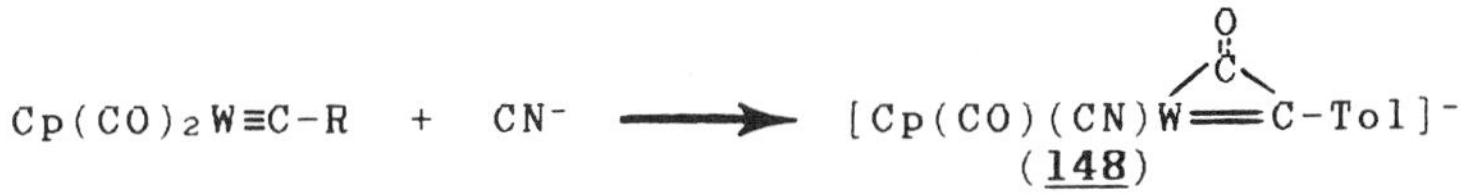

$$Cp(CO)_2W{\equiv}C\text{-}R \; + \; CN^- \; \longrightarrow \; [Cp(CO)(CN)W{=}C\text{-}Tol]^- \quad (124)$$
$$(\underline{148})$$

$$Br(CO)_2(L_2)W{\equiv}C\text{-}R \; + \; CN^- \longrightarrow [(CO)(L_2)(CN)_2W{=}C\text{-}R]^- \quad (125)$$
$$(\underline{149})$$

A related reaction involving the carbonylation of a high-oxidation-state tungsten–carbyne complex leads to a binuclear species possessing two metallated η^2-ketenyl ligands [Eq. (126a)] (*147b*). However, the X-ray

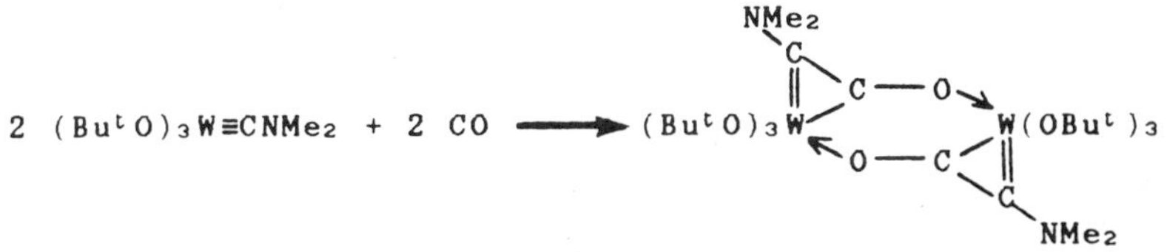

$$2\ (Bu^tO)_3W{\equiv}CNMe_2 \; + \; 2\ CO \; \longrightarrow \; (Bu^tO)_3W \cdots W(OBu^t)_3$$

(126a)

structural data for the product imply that the ketenyl ligands are better described as coordinated alkynes, $Me_2N—C≡C—OWL_3$. The overall reaction is thus similar to the reported addition of electrophiles to the η^2-ketene oxygen in $Cp(CO)(PMe_3)W(\eta^2\text{-TolCCO})$ to form alkyne ligands (*143*) [see Eq. (136)].

Carbonylation of a *bridging* carbyne complex yields a μ_2-η^1-ketenyl complex [Eq. (126b)] (*149*). As shown in Eq. (*126b*), a ^{13}CO labeling study

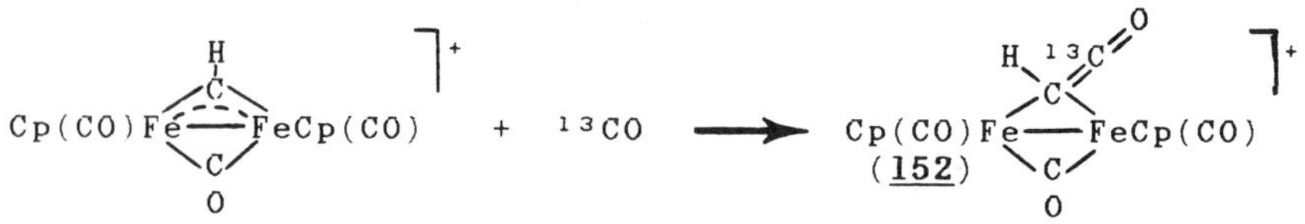

$$(126b)$$

showed that exogenous CO adds to the carbyne ligand and that this reaction does not involve migration of a metal-bound CO to the coordinated carbyne. Similar cabonylation of an intermediate carbyne complex is presumably involved in the reactions of Eq. (127) (*150*).

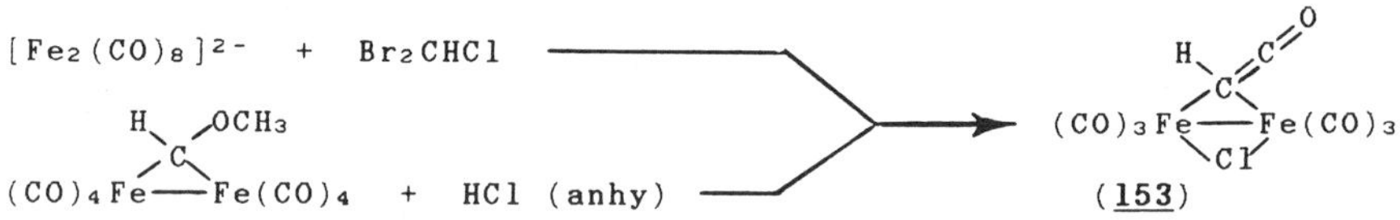

$$(127)$$

Bridging ketenyl complexes **150** and **151** form on addition of $[Mn(CO)_5]^-$ to $[Cp(CO)_2M≡C—Tol]^+$ [Eq. (128)] (*147a,148*), but the

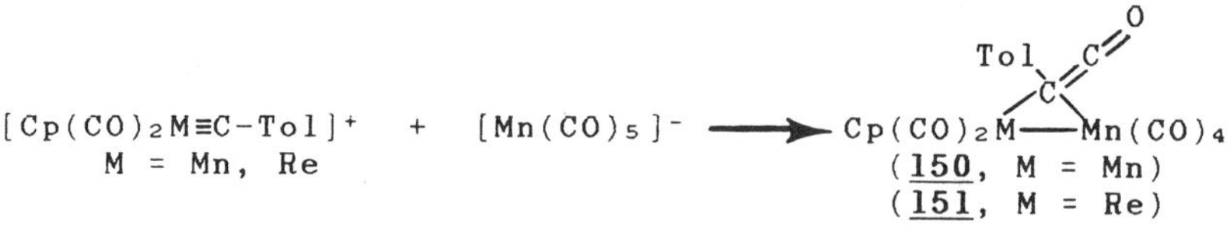

$$(128)$$

mechanisms of these reactions are not clear. They may proceed via initial formation of a reactive μ-carbyne complex which is then carbonylated by released CO. Altenatively, the added $[Mn(CO)_5]^-$ may induce migration of CO to the carbyne ligand through an associative process.

The only other syntheses of ketenyl complexes involve insertion of carbon suboxide into metal hydride bonds [Eqs. (129) and (130)] (*141a*), modification of the bridging *carbene* ligand in **155** [Eq. (131a)] (*150*), and base-induced rearrangement of a cationic ylide complex to give **145** [Eq. (131b)] (*141b*).

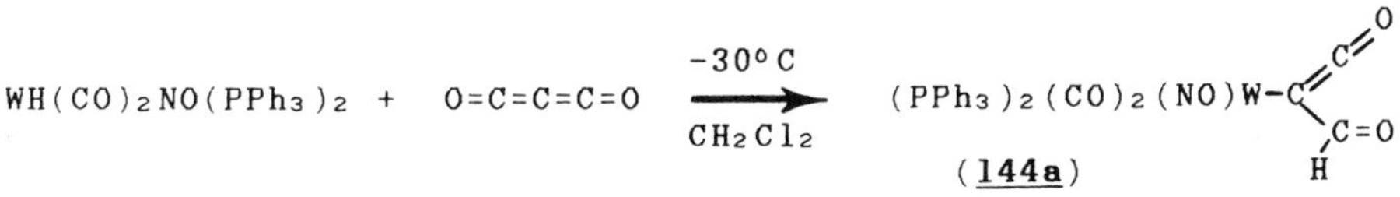

$$WH(CO)_2NO(PPh_3)_2 \ + \ O=C=C=C=O \ \xrightarrow[CH_2Cl_2]{-30°C} \ (PPh_3)_2(CO)_2(NO)W-C\begin{smallmatrix} C=O \\ \\ C=O \\ H \end{smallmatrix}$$

(**144a**)

(129)

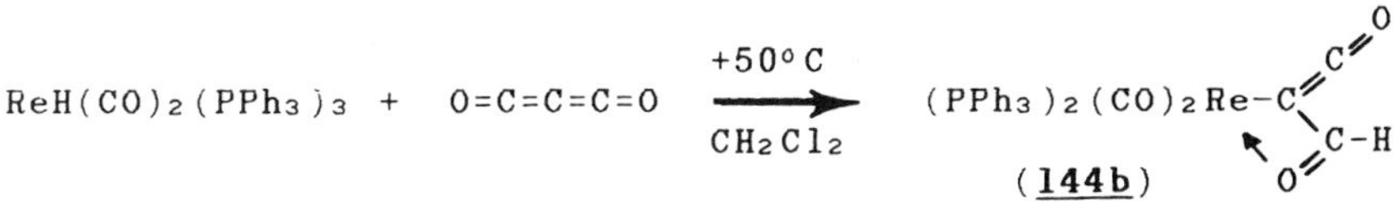

$$ReH(CO)_2(PPh_3)_3 \ + \ O=C=C=C=O \ \xrightarrow[CH_2Cl_2]{+50°C} \ (PPh_3)_2(CO)_2Re-C\begin{smallmatrix} C=O \\ \\ C-H \\ O \end{smallmatrix}$$

(**144b**)

(130)

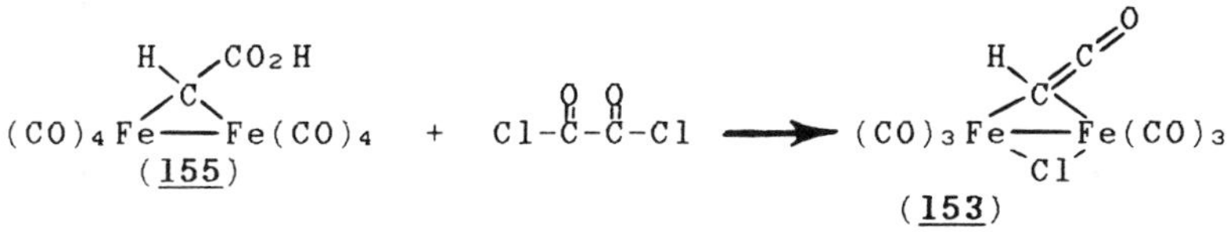

$$(CO)_4Fe\underset{(155)}{\overset{H\diagdown C\diagup CO_2H}{-\!\!\!-}}Fe(CO)_4 \ + \ Cl-\overset{O}{\underset{}{C}}-\overset{O}{\underset{}{C}}-Cl \ \longrightarrow \ (CO)_3Fe\underset{Cl}{\overset{H\diagdown C\diagup C=O}{-\!\!\!-}}Fe(CO)_3$$

(**153**)

(131a)

$$\left[Cp(CO)_2Fe-C\begin{smallmatrix} PMe_3 \\ \\ SiMe_3 \\ H \end{smallmatrix} \right]^+ \ + \ OMe^- \ \longrightarrow \ \left[Cp(CO)_2Fe=C\begin{smallmatrix} PMe_3 \\ \\ SiMe_3 \end{smallmatrix} \right]$$

$$Cp(CO)(PMe_3)Fe-C\begin{smallmatrix} C=O \\ \\ SiMe_3 \end{smallmatrix} \ \longleftarrow \ \left[Cp(CO)Fe\begin{smallmatrix} C=O \\ | \\ C \\ PMe_3 \ SiMe_3 \end{smallmatrix} \right]$$

(**145**)

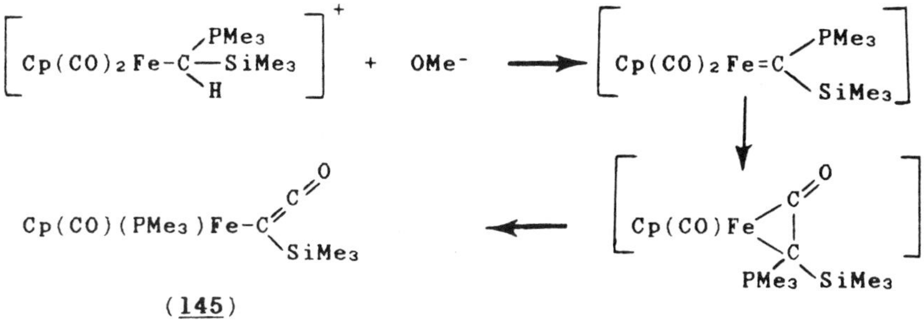

(131b)

If ketenyl complexes are to be developed as useful synthetic reagents it will be necessary to develop better synthetic routes to them. The present methodology involving carbonylation of carbyne ligands suffers from the multistep syntheses necessary for most carbyne complexes and their general lack of availability. Furthermore, only **154** has been shown to readily undergo the carbyne carbonylation reaction with some generality. A potentially viable but presently unexplored route to ketenyl complexes is the displacement of halide from haloketenes by carbonylmetallates [Eq. (132)]. Assessment of the utility of this reaction will have to await the results of suitably designed experiments.

$$[(CO)_xL_yM]^- \ + \ \underset{R}{\overset{Cl}{\diagdown}}C=C=O \ \longrightarrow \ Cl^- \ + \ \underset{R}{\overset{(CO)_xL_yM}{\diagdown}}C=C=O$$

(132)

B. *Structural Characterization*

Only six ketenyl complexes have been structurally characterized, but these have been informative in regard to the bonding within this family of complexes. Relevant structural data are summarized in Table IV, where the ketene carbonyl carbon is designated the β carbon. Representative ORTEP drawings that illustrate the structures are shown in Fig. 15–17.

Four possible resonance structures for an η^1-ketenyl ligand are the nonpolar (156), acylium (157), and dipolar (158 and 159) structures drawn below. The high-ν(CCO) IR bands observed for η^1-ketenyl complexes

$$
\underset{M}{\overset{R}{\diagdown}}C=C=O \quad\longleftrightarrow\quad \underset{M}{\overset{R}{\diagdown}}\bar{C}-C\equiv O^{+} \quad\longleftrightarrow\quad \underset{M}{\overset{R}{\diagdown}}C=C-\bar{O}^{+} \quad\longleftrightarrow\quad \underset{M}{\overset{R}{\diagdown}}\bar{C}-C=O^{+}
$$
$$
\quad(\underline{156})\qquad\qquad(\underline{157})\qquad\qquad(\underline{158})\qquad\qquad(\underline{159})
$$

(Table III) suggest the importance of the acylium resonance form (157). However, the structural data for the two structurally characterized complexes of this class do not allow an adequate assessment to be made of the relative importance of 156–159. The C_β–O distances of 1.21 and 1.173 Å for 143a and 143b are significantly longer than the corresponding distance

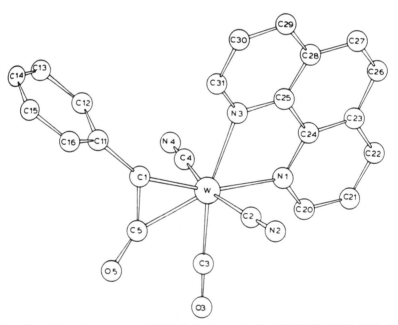

Fig. 16. Molecular structure of $(CN)_2(1,10\text{-phenanthroline})(CO)W\{PhCCO\}$ (149). (Reprinted with permission from Ref. *145.*)

TABLE IV

Structural Data for Ketenyl Complexes

Complex	M–C_A (Å)	C_A–C_B (Å)	C_B–O (Å)	C_A–C_B–O (deg)	M–C_A–C_B (deg)	Ref.
η^1-*Ketenyl complexes*						
Cp(CO)L$_2$W–C(Tol)CO (143a, L = PMe$_3$)	2.27(2)	1.24(3)	1.21(3)	a	a	132
η^2-*Ketenyl complexes*						
Cp(CO)LW{TolCCO} (147, L = PMe$_3$)	1.97 M–C_β = 2.07 Å	1.32	1.30	a	a	133b
(CN)$_2$L$_2$(CO)W{PhCCO} (149, L$_2$ = 1,10-phenanthroline)	1.97(1) M–C_β = 2.14(2) Å	1.41(2)	1.25(2)	145(2)	77(1)	145
μ_2-η^1-*Ketenyl complexes*						
Cp$_2$(CO)$_3$Fe$_2$(μ-HCCO) (152)	1.994(4)	1.338(8)	1.135(7)	174.9	110.6	149
Cp(CO)$_6$Mn$_2$(μ-TolCCO) (150)	2.128(4)	1.326(6)	1.167(5)	a	a	147
Cp(CO)$_6$MnRe(μ-PhCCO) (151)	2.21(3) (M = Mn) 2.24(3) (M = Re)	1.32(4)	1.18(4)	169(3)	96(2) (M = Mn) 111(2) (M = Re)	148

^{a} Not published.

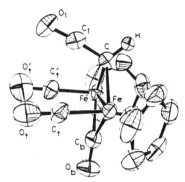

FIG. 17.   Molecular structure of $Cp_2(CO)_3Fe_2(\mu\text{-HCCO})$ (**152**). (Reprinted with permission from Ref. *149*.)

of 1.11(1) Å in the acylium ion $CH_3C\equiv O^+$ (*155*), but they are also longer than those in free ketene [1.16(2)Å *154*], $Me_2C=C=O$ (1.179 Å *127*), and $MeHC=C=O$ [1.171(2) Å *157*]. This suggests greater importance of resonance form **158**. The C_α–C_β distance of 1.24 Å for **143** is shorter than the corresponding distances in $MeC\equiv O^+$ [1.39(2) Å *155*], $Me_2C=C=O$ (1.271 Å *127*) and $CH_2=C=O$ (1.31 Å *154*), implying significant multiple-bond character in this bond. The W–C_α distances are typical of W–C single bonds.

The structural data for all three μ_2-η^1-ketenyl complexes are similar and imply contributions from both acylium (**160**) and ketene (**161**) resonance forms (*149*). For example, the C_β–O bond length within the ketenyl ligand

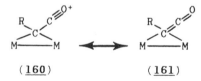

(**160**)   (**161**)

of **152** is 0.03 Å longer than the C–O bond of the acylium salt $CH_3C\equiv O^+SbF_6^-$ (*155*) but 0.03 Å shorter than the C–O bond of $CH_2=C=O$ (*154*). Likewise, the C_α–C_β distance is 0.05 Å shorter than the C–C bond of the acylium salt and 0.03 Å longer than the C–C bond of ketene. The structural data also imply that the acylium resonance form **160** is more important for cationic **152** than it is for neutral **150** and **151**. This is indicated by the longer C_β–O and shorter C_α–C_β distances for the latter two compounds compared to **152**. Even so, the acylium description appears more important for the three μ_2-η^1 complexes than for any of the other ketene complexes since the C_β–O distances for **150–152** are shorter than those found in any of the other structurally characterized compounds

(Table IV). Note that the C_α–C_β–O angles for the three μ_2-η^1 complexes are all close to 180°C, as expected for a metal-substituted ketene. Data for **147** and **149**, the only two η^2-ketenyl complexes to be structurally characterized, indicate that **162** is the major resonance form for this class of compounds rather than the alternative descriptions **163** and **164**. This is indicated by the inequality in the W–C_α and W–C_β distances

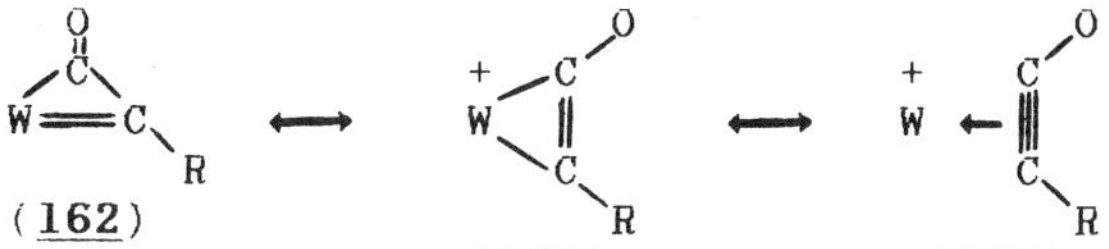

and the fact that the W–C_α distance of 1.97 Å for both compounds is significantly shorter than typical W–C single bonds (*cf* 2.27 Å in **143**), implying a high degree of multiple bond character. In contrast, the W–C_β distances are more characteristic of W–C single bonds.

C. Reactions of Ketenyl Complexes

The chemistry of the ketenyl ligand has only begun to be developed, with nearly all of the existing studies being performed by Kreissl and co-workers. The η^1-ketenyl ligand is essentially a metal-substituted ketene, and thus it can be expected to display ketenelike behavior. A characteristic reaction of free ketenes is addition of nucleophiles to the β carbon [Eq. (1)], and similar reactions occur for both terminal and bridging η^1-ketenyl complexes [Eqs. (133) (*139,140*) and (134) (*149*)]. The intermediacy of a μ_2-η^1-ketenyl complex that undergoes similar nucleophilic addition of methoxide has also been invoked to explain formation of **165** [Eq. (135)] (*152*). η^2-Ketenyl complexes are apparently not susceptible to similar nucleophilic additions, since no examples of such reactions have been reported and η^2-ketenyl complexes have actually been prepared in methanol solvent (*146*).

Whereas nucleophiles add to η^1-ketenyl complexes, η^2-ketenyl ligands undergo electrophilic attack at the oxygen atom to form alkyne complexes [Eq. (136)] (*143*). A similar reaction may also be involved in an early

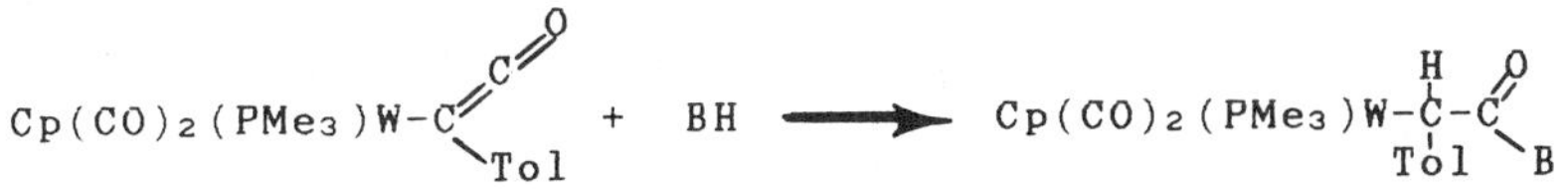

$$B = OCH_3, \ OH, \ NMe_2$$

(133)

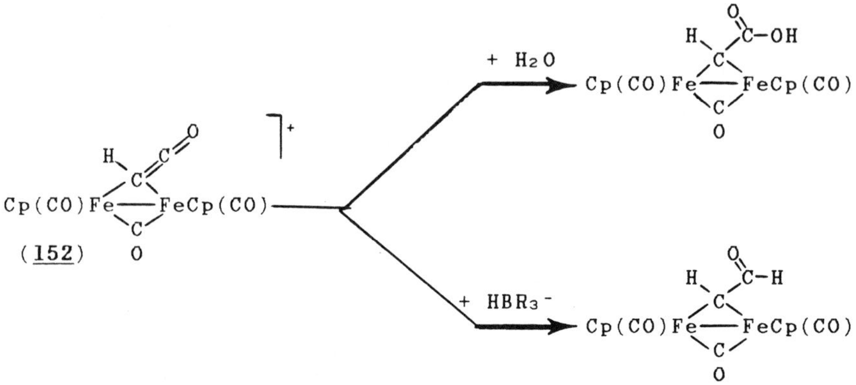

(134)

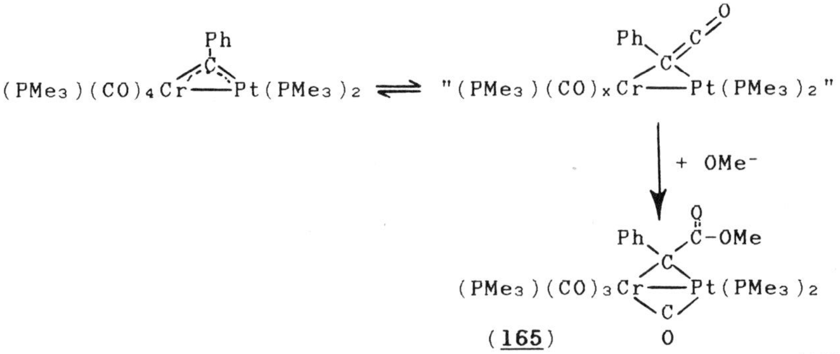

(135)

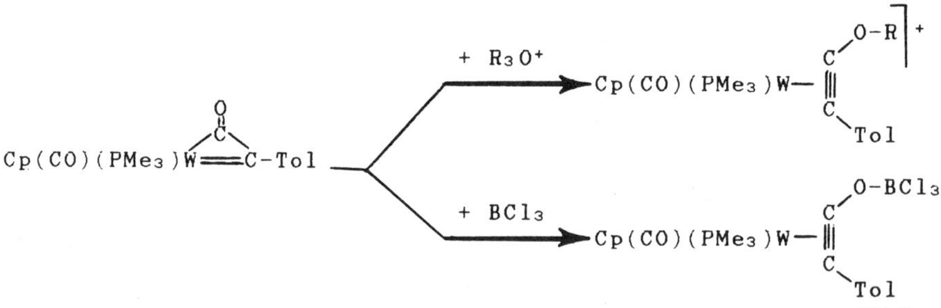

(136)

report of the first characterization of a hydroxyalkyne ligand (166) formed by photolysis of $Cl(CO)_4 W\equiv CPh$ in the presence of acetylacetone (Hacac) (153). We suggest that this reaction proceeds via photoinduced migration of CO to the carbyne ligand to generate an unsaturated η^2-ketenyl complex, which is then protonated by Hacac to give 166 [Eq. (137)] (151).

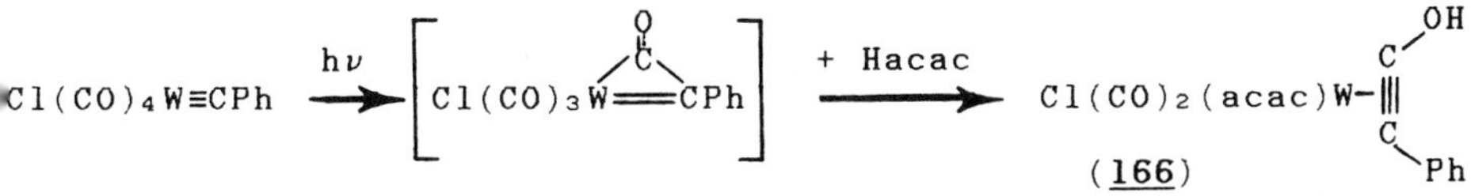

$$Cl(CO)_4W\equiv CPh \xrightarrow{h\nu} \left[Cl(CO)_3\overset{\overset{O}{\parallel}}{W}\!\!=\!\!\!=CPh \right] \xrightarrow{+\ Hacac} Cl(CO)_2(acac)W\!\!-\!\!\overset{C\diagup OH}{\underset{C\diagdown Ph}{\parallel}}$$

$$(\underline{166})$$

(137)

A remarkable reaction is the CO-induced conversion of η^1-ketenyl ligands into alkynyl ligands by a deoxygenation reaction [Eq. (138)] (*134,138*). This reaction tolerates a wide variation in ketenyl substituents

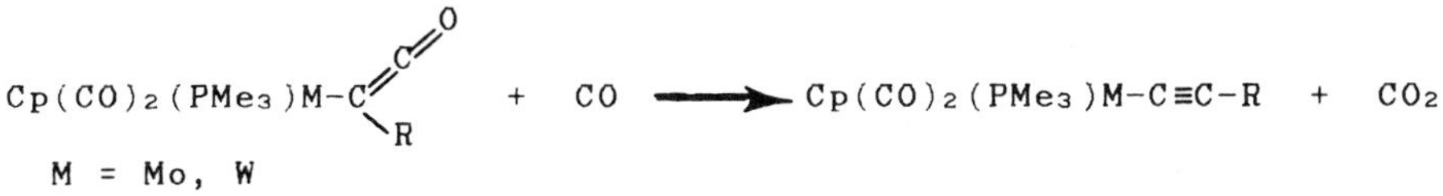

$$Cp(CO)_2(PMe_3)M\!-\!\overset{\overset{O}{\parallel}}{\underset{R}{C\diagdown}}C \quad + \quad CO \quad \longrightarrow \quad Cp(CO)_2(PMe_3)M\!-\!C\equiv C\!-\!R \quad + \quad CO_2$$

M = Mo, W

(138)

R, but its mechanism is unknown. It was suggested that the first step involves loss of CO to generate an η^2-ketenyl complex, which in an undefined manner interacts with CO to yield the observed alkynyl products (*134*).

The insertion of XPR_2 and $XAsR_2$ (X = Cl, I; R = Ph, Me) into the metal–carbon bond of η^2-ketenyl complexes was shown to yield the first syntheses of arsino- and phosphinoketenes (**87**) [Eq. (139)] (*110,111,114*).

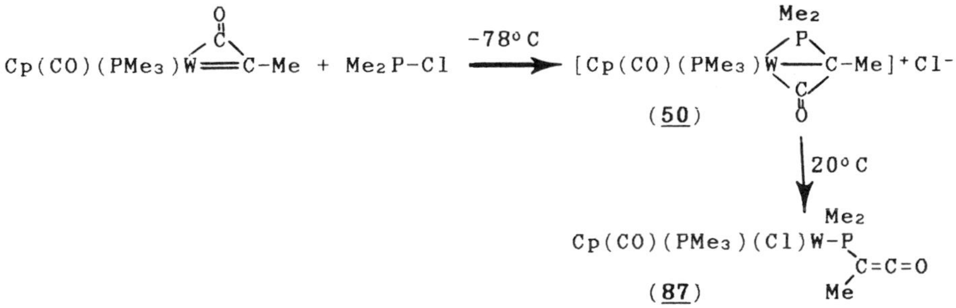

$$Cp(CO)(PMe_3)\overset{\overset{O}{\parallel}}{W}\!\!=\!\!\!=C\!-\!Me \ + \ Me_2P\!-\!Cl \ \xrightarrow{-78^\circ C} \ [Cp(CO)(PMe_3)\overset{\diagup P\diagdown Me_2}{W\!\!-\!\!\underset{\underset{O}{\parallel}}{C}}\!\!C\!-\!Me]^+Cl^-$$

$$(\underline{50})$$

$$\Big\downarrow 20^\circ C$$

$$Cp(CO)(PMe_3)(Cl)W\!-\!\overset{Me_2}{\underset{Me}{P}}\diagup \!\!\underset{}{\overset{}{}}C\!=\!C\!=\!O$$

$$(\underline{87})$$

(139)

The initial formation of the salt **50** occurs at low temperature, and this species has been isolated and structurally characterized (see Section III). The free arsinoketenes have been liberated from their complexes by displacement with CO at high pressures or by addition of PMe₃ [Eq. (88)] (*110*).

The formation of a novel sulfine (**167**) from an η^2-ketenyl complex has also been accomplished, although the isolated yield was only 9% [Eq. (140)] (*142*).

$$
Cp(CO)(PMe_3)W \equiv\!\!\equiv C-Tol \quad + \quad SOCl_2 \quad \longrightarrow \quad O=S=C \begin{matrix} Cl \\ \diagdown C=O \\ \diagup \\ Tol \end{matrix} \quad (140)
$$

$$(167)$$

A characteristic reaction of free ketenes is their cycloaddition to unsaturated organic molecules, but to our knowledge such reactions have never been examined with coordinated ketenyl ligands. This could be a profitable area in view of the potential stereospecificity of such reactions due to the inherent chirality of the metal center in several such compounds (e.g., **143**).

V

KETENYLIDENE COMPLEXES

Ketenylidene complexes have only metal atoms attached to the ketene α carbon. These complexes have also been termed acylium and carbonyl methylidyne complexes. The first such species, $[Co_3(CO)_9\{CCO\}]^+$ (**168**), was reported in 1972 (*157,158*), and its chemistry has been extensively examined by Seyferth and co-workers. These studies were reviewed in 1976 (*159*), and only a brief account will be included here. Several ketenylidene complexes have been made in the ensuing years, and their chemistry is described in more detail. All of the characterized ketenylidene complexes have three metals as substituents, although a tetrametallic ketenylidene has been invoked as a reaction intermediate (*160,161*). The known ketenylidene complexes are illustrated in Fig. 18. Also known are Cu, Ag, and Au ketenylidenes of the general formula M_2CCO (*171–173*). However, their total insolubility has precluded their definitive characterization and they will not be included here.

Both "tilted" (**177**) and "perpendicular" (**178**) structures are possible for ketenylidene complexes with little apparent energy difference between them. Although not all ketenylidene complexes have been structurally

$$(177)$$ $$(178)$$

characterized, it appears that those with first-row metals tend to have the tilted structure while those of third-row metals have the perpendicular

$$(CO)_3Co \overset{\overset{\displaystyle O}{\overset{\|}{C}}}{\underset{\underset{Co(CO)_3}{|}}{\diagdown}} Co(CO)_3$$

(**168**) (*157,159*)

$$\left[(CO)_3Fe \overset{\overset{\displaystyle O}{\overset{\|}{C}=C}}{\underset{\underset{Fe(CO)_3}{|}}{\diagdown}} Fe(CO)_3\right]^{2-}$$

(**169**) (*162–165*)
^{13}C: δ 90.1 ($\underline{C}CO$)
δ182.2 ($C\underline{C}O$)
$J_{CC}=73.2Hz$

$$\left[(CO)_3Co \overset{\overset{\displaystyle O}{\overset{\|}{C}=C}}{\underset{\underset{Fe(CO)_3}{|}}{\diagdown}} Fe(CO)_3\right]^{-}$$

(**170**) (*164*)
^{13}C: δ 82.8 ($\underline{C}CO$)
δ172.5 ($C\underline{C}O$)
$J_{CC}=82.8Hz$

$$(CO)_3Os \overset{\overset{\displaystyle O}{\overset{\|}{C}}}{\underset{\underset{(CO)_3}{|}}{\diagdown}} Os(CO)_3 \atop H-Os-H$$

(**171**) (*166,167*)
^{13}C: δ8.6 ($\underline{C}CO$)
δ160.3 ($C\underline{C}O$)
$_{CC}=86$ Hz

$$(CO)_3Ru \overset{\overset{\displaystyle O}{\overset{\|}{C}}}{\underset{\underset{(CO)_3}{|}}{\diagdown}} Ru(CO)_3 \atop H-Ru-H$$

(**172**) (*168*)

$$\left[(CO)_2Ru \overset{\overset{\displaystyle O}{\overset{OC}{\diagup}}\overset{\|}{C}=C}{\underset{\underset{(CO)_2}{|}}{\diagdown}} Ru(CO)_2 \atop OC-Ru-CO\right]^{2-}$$

(**173**) (*169a*)
^{13}C: δ–28.3 ($\underline{C}CO$)
δ159.1 ($C\underline{C}O$)
$J_{CC}=96Hz$

$$\left[(CO)_3Ru \overset{\overset{\displaystyle O}{\overset{\|}{C}}}{\underset{\underset{(CO)_3}{|}}{\diagdown}} Ru(CO)_3 \atop H-Ru-H\right]^{+}$$

(**174**) (*168*)

$$\left[(CO)_3Os \overset{\overset{\displaystyle O}{\overset{\|}{C}}}{\underset{\underset{(CO)_3}{|}}{\diagdown}} Os(CO)_3 \atop H-Os-H\right]^{+}$$

(**175**) (*167*)

$$\left[(CO)_3Co \overset{\overset{\displaystyle O}{\overset{\|}{C}}}{\underset{\underset{(CO)_3}{|}}{\diagdown}} Mo(CO)_2Cp \atop Co\right]^{+}$$

(**176a**) (*170*)

$$\left[(CO)_3Co \overset{\overset{\displaystyle O}{\overset{\|}{C}}}{\underset{\underset{(CO)_3}{|}}{\diagdown}} Mn(CO)_3 \atop Co\right]^{-}$$

(**176b**) (*169b*)

Fɪɢ. 18. Ketenylidene complexes.

structure. There are three equivalent positions in the tilted structure of a homometallic cluster, and the ketenylidene ligand can shift from one to another. A recent variable-temperature ^{13}C NMR study demonstrated this process for $[Co_3(CO)_9(CCO)]^{+}$ and gave an activation barrier of only 9 kcal/mol (*174b*). Calculations for a similar process for the related vinyl-idene cluster $[Co_3(CO)_9(C{=\!=}CH_2)]^{+}$ have given an activation barrier of 16 kcal/mol for such a migration of the μ_3-$C{=\!=}CH_2$ ligand around the metal triangle (174a), but this barrier is somewhat higher than that indicated by NMR (175). A similar calculation for $[Co_3(CO)_9(CCO)]^+$ indicated the tilted structure to be more favorable than the perpendicular, but less so than for $[(Co_3(CO)_9(CCH_2)]^+$ (174b). The average metal environment of a rapidly interconverting tilted structure is the same as that of the perpendicular structure, and as such it can be difficult to distinguish between these two possiblities by NMR analysis. We have arbitrarily chosen to drawn the perpendicular structure for those ketenylidene complexes not characterized by X-ray diffraction.

Spectroscopic characterization of ketenylidene complexes is usually accomplished through ^{13}C NMR studies. The metal-bound ketenylidene α carbon generally appears in the $\delta 28 \rightarrow \delta 90$ range, whereas the carbonyl β carbon is further upfield in the $\delta 160 \rightarrow \delta 182$ range. However, the latter overlaps with the metal carbonyl region, leading to possible difficulties in spectral assignments. In no case has a $\nu(CCO)$ stretch for the coordinated ketenylidene been identified, perhaps because it also overlaps with the metal carbonyl vibrations or because it may be unusually weak. Details of the synthesis, structure, and reactivity of the known ketenylidene clusters are outlined below.

A. Synthetic Methods

All but one of the reported synthetic methods for the characterized ketenylidene clusters involve the modification of a ligand or a metal vertex in a trinuclear precursor cluster. The synthesis of the first member of this family of compounds was accomplished by protonation of **179** in concentrated H_2SO_4 [Eq. (141)] (157). This procedure was developed

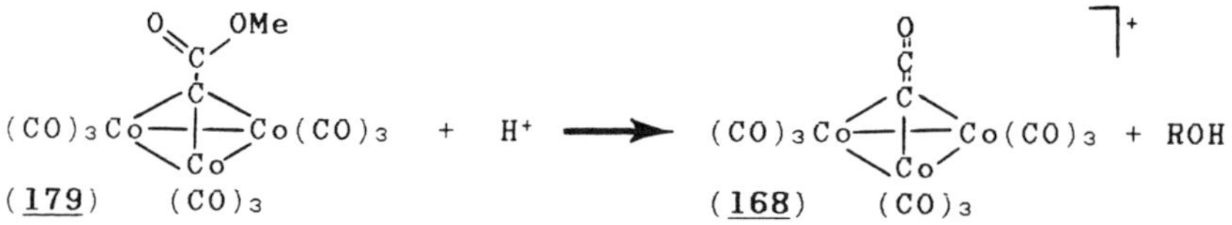

$$(141)$$

by analogy to a similar procedure for effecting the esterification of sterically hindered carboxylic acids (176). It has been used more recently to prepare the MoCo$_2$ ketenylidene cluster **176** by treating CpMoCo$_2$-$(CO)_8CC\{O\}OCHMe_2$ with HPF$_6$ (170).

An alternative procedure was later developed for preparing **168** by treating **180** with AlCl$_3$ [Eq. (142)] (177). The mechanism of this reaction is unknown, but it probably involves AlCl$_3$-assisted extraction of

$$\begin{array}{c}\text{Cl}\\\text{(CO)}_3\text{Co}\underset{\underset{\text{(180)}}{\overset{\text{C}}{\underset{\text{Co}}{\diagdown}}}}{\overset{\diagup}{\overset{\text{C}}{\diagdown}}}\text{Co(CO)}_3\end{array}\begin{array}{c}\text{(CO)}_3\end{array} + \text{AlCl}_3 \xrightarrow[30\ \text{min}]{22°C} \begin{array}{c}\overset{\overset{O}{\|}}{\overset{C}{}}\\\text{(CO)}_3\text{Co}\underset{\underset{\text{(168)}}{\overset{\text{C}}{\underset{\text{Co}}{\diagdown}}}}{\overset{\diagup}{\overset{\|}{\diagdown}}}\text{Co(CO)}_3\end{array}\begin{array}{c}\end{array}^{+}\ \text{AlCl}_4^{-}$$

$$(142)$$

chloride from the methylidyne carbon simultaneous with CO migration from cobalt to the exposed carbon. This method was also used to generate $[\text{H}_3\text{Ru}_3(\text{CO})_9\{\text{CCO}\}]^{+}$ **(174)** by treatment of $\text{H}_3\text{Ru}_3(\text{CO})_9(\mu_3\text{-}$ CBr) with AlCl_3 *(178)*.

A related method involves the reductive cleavage of an activated ligand C–O bond to generate an "exposed" carbon atom to which CO migrates subsequent to or concomitant with cleavage of the C–O bond [Eq. (143)] *(162,163)*. The same method was also used to prepare **173**, the Ru analog of **169** *(169)*.

$$\begin{array}{c}\text{R}\\\text{O}\\\text{CO)}_3\text{Fe}\underset{\underset{\text{O}}{\overset{\text{C}}{\diagdown}}}{\overset{\overset{\text{C}}{\diagup}}{=}}\text{Fe(CO)}_3\end{array}\begin{array}{c}-\\\text{Fe(CO)}_3\end{array} + \text{Na/Ph}_2\text{CO} \longrightarrow \begin{array}{c}\overset{\overset{O}{\diagdown}}{\overset{\text{C}}{\diagup}}\overset{O}{}\\\text{(CO)}_3\text{Fe}\underset{\underset{\text{(169, 68\%)}}{\overset{\text{C}}{\underset{\text{Fe(CO)}_3}{\diagdown}}}}{\overset{\diagup}{\diagdown}}\text{Fe(CO)}_3\end{array}^{2-}$$

181, R = Me, C{O}Me)

$$(143)$$

Oxidative cleavage of metal–metal and metal–carbon bonds to expose a carbon atom has been proposed to give a tetranuclear ketenylidene intermediate (**182**), which rapidly undergoes solvolysis to give the observed product **183** [Eq. (144)] *(160,161)*. The ketenylidene cluster, however, was not detected.

$$\text{Fe}_6\text{C(CO)}_{16}]^{2-} + \text{C}_7\text{H}_7^{+} \longrightarrow \begin{array}{c}\overset{\overset{O}{\diagdown}}{\overset{\text{C}}{\diagup}}\overset{O}{}\\\text{(CO)}_3\text{Fe}\underset{\underset{\underset{\text{(182)}}{\text{Fe}}}{\text{(CO)}_3\text{Fe}}}{\overset{\diagup}{\diagdown}}\underset{\underset{\text{(CO)}_3}{\text{(CO)}_3}}{\text{Fe}}\end{array}^{2-} \xrightarrow{+\ \text{MeOH}} [\text{Fe}_4(\text{CO})_{12}\text{CCO}_2\text{Me}]^{-}$$

(183)

$$(144)$$

The Os_3 ketenylidene cluster **171** forms on thermolysis of both the methylene cluster **184** and the methylidyne cluster **185**, probably via the reaction path of Eq. (145) *(166,167)*. The same reactions occur at much lower temperature for the Ru_3 analogs to ultimately yield Ru_3H_2-$(\text{CO})_9\{\text{CCO}\}$ **(172)** *(168)*.

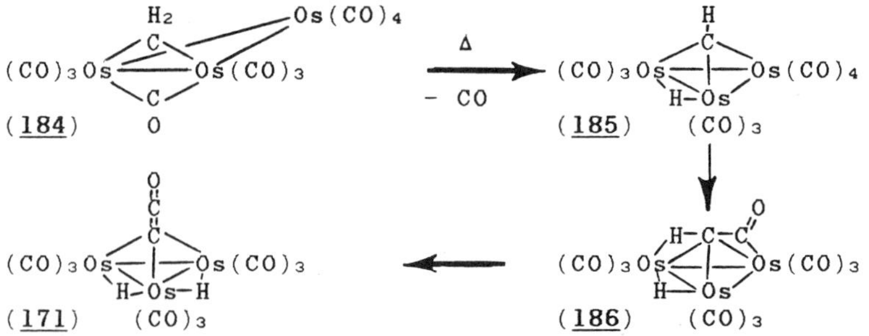

$$
\begin{array}{c}
\text{(184)} \quad \text{(185)} \\
\end{array}
$$

(145)

Many trinuclear clusters that possess μ_3 capping ligands are known to undergo metal exchange reactions (179), and this has also been shown to occur for μ_3-ketenylidene clusters [Eq. (146a)] (164). This type of reaction

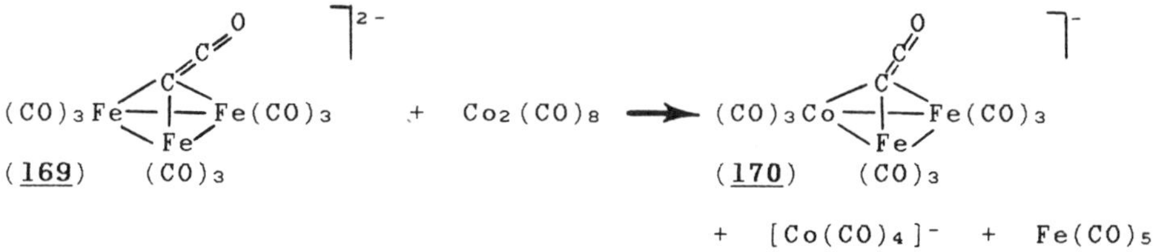

$$
(CO)_3Fe\!-\!\overset{|}{\underset{Fe}{}}\!-\!Fe(CO)_3 \quad + \quad Co_2(CO)_8 \longrightarrow (CO)_3Co\!-\!\overset{|}{\underset{Fe}{}}\!-\!Fe(CO)_3
$$

$$
\text{(169)} \quad (CO)_3 \qquad\qquad \text{(170)} \quad (CO)_3
$$

$$
+ \quad [Co(CO)_4]^- \quad + \quad Fe(CO)_5
$$

(146a)

may have utility beyond the preparation of **170**, but its use will be complicated by cluster expansion reactions to yield tetranuclear carbide clusters; see below (164,165).

The only synthesis of a ketenylidene cluster beginning with mononuclear reagents is that shown in Eq. (146b) (169b). The mechanism of this

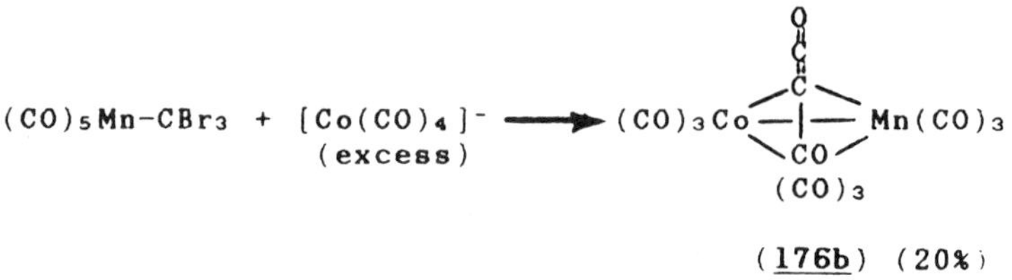

$$
(CO)_5Mn\!-\!CBr_3 \quad + \quad [Co(CO)_4]^- \longrightarrow (CO)_3Co\!-\!\overset{|}{\underset{CO}{}}\!-\!Mn(CO)_3
$$
$$
\text{(excess)}
$$
$$
(CO)_3
$$

(**176b**) (20%)   (146b)

reaction is unknown, but it was anticipated that a tetranuclear carbide cluster would form from this combination of reagents.

B. Structural Characterization

Structural data for the four ketenylidene clusters that have been characterized by X-ray diffraction studies are presented in Table V and representative ORTEP drawings of tilted (**169**) and perpendicular (**171**)

TABLE V

STRUCTURAL DATA FOR KETENYLIDENE COMPLEXES

Complex	C_A-C_B (Å)	C_B-O (Å)	C_A-C_B-O (deg)	Tilt (deg)	Reg.
Fe$_3$(CO)$_9${CCO}]$^{2-}$, 169	1.28(3)	1.18(3)	172.8	33.5	163
Fe$_2$Co(CO)$_9${CCO}]$^-$, 170	a	a	a	24	164
Ru$_3$(CO)$_9${CCO}]$^{2-}$, 173	1.30(1)	1.17(8)	178.8	11.3	169a
H$_2$Os$_3$(CO)$_9${CCO}, 171	1.26(4)	1.15(3)	178.6	a,b	167

^{a} Not reported.
^{b} Assumed to be near zero.

structures are shown in Fig. 19 and 20. The ketenylidene ligand is essentially linear and has a very short $C_\beta-O$ distance in all the compounds, indicating a significant contribution from the acylium resonance form **187**. The average $C_\beta-O$ distance for the four compounds with data

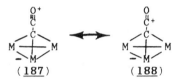

(**187**)   (**188**)

in Table V is 1.17 Å, only slightly longer than the corresponding distance of 1.11(1) Å in the acylium ion $CH_3C{\equiv}O^+$ (*155*). However, the facile addition of nucleophiles to the β carbon of ketenylidene clusters (see below) indicates that resonance form **188** cannot be ignored.

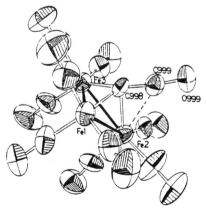

FIG. 19.   Molecular structure of [Fe$_3$(CO)$_9${CCO}]$^{2-}$ (**169**). (Reprinted with permission from Ref. *163*.)

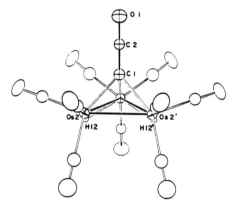

FIG. 20.   Molecular structure of $H_2Os_3(CO)_9\{CCO\}$ (**171**). (Reprinted with permission from Ref. *167*.)

The most interesting structural feature of ketenylidene complexes is the variable degree of "tilt" of the ketenylidene ligand toward the plane of the metal triangle. The degree of tilt varies from perpendicular for $H_2Os_3(CO)_9\{CCO\}$ (**171**) to slightly tilted (11.3°) for $[Ru_3(CO)_9\{CCO\}]^{2-}$ (**173**) to significantly tilted for $[Fe_3(CO)_9\{CCO\}]^{2-}$ (**169**, 33.5°) and $[Fe_2Co(CO)_9\{CCO\}]^-$ (**170**, 24°). As noted above, the closely related $[Co_3(CO)_9\{CCH_2\}]^+$ cluster (**189**) has been theoretically analyzed and an energy minimum was found as a result of a preference for an empty orbital on the β carbon to interact with a set of filled metal orbitals. However, the ketenylidene ligand, in contrast to the vinylidene ligand in **189**, has its p-π^* orbitals in degenerate pairs, and there is no obvious need for it to tilt away from perpendicular (*167*). Nevertheless, it does in some of the compounds, and the reasons for this will have to await a more detailed theoretical analysis of this family of compounds.

C. Reactions of Ketenylidene Complexes

The reactivity characteristics of ketenylidene clusters appear to be a sensitive function of the cluster charge and the metals involved. Cationic and neutral ketenylidene clusters are more electrophilic, whereas the anionic clusters show pronounced nucleophilic character. The effect of metals is best illustrated by the observed protonation of the Fe_3 cluster **169** at the ketenylidene α carbon (*162,163*), but in contrast protonation of the analogous Ru_3 cluster **173** occurs on the metal framework (*169*).

For simplicity, the following discussion is divided into electrophilic, nucleophilic, and cluster expansion reactions.

1. *Electrophilic Behavior*

The most representative example of electrophilic ketenylidene clusters is $[Co_3(CO)_9\{CCO\}]^+$ (**168**), whose chemistry has been reviewed by Seyferth (*159*). This cluster behaves much as an organic acylium ion, but stabilized by the metal framework, and adds a variety of nucleophiles as summarized in Fig. 21. The MoCo$_2$ ketenylidene cluster **176** shows similar behavior [Eq. (147)] (*170*). Similar addition of methanol to a ketenylidene ligand in a proposed Fe$_4$ cluster was suggested to account for the

$$
[CpMoCo_2(CO)_8\{CCO\}]^+ \quad (\textbf{176})
$$

$$
\xrightarrow{+ \ HNEt_2} \quad CpMoCo_2(CO)_8\,(\mu_3-C-C\overset{O}{\underset{NEt_2}{\diagdown}})
$$

$$
\xrightarrow{+ \ ROH} \quad CpMoCo_2(CO)_8\,(\mu_3-C-C\overset{O}{\underset{OR}{\diagdown}})
$$

$$
\xrightarrow{+ \ Ph-NMe_2} \quad CpMoCo_2(CO)_8\,(\mu_3-C-C\overset{O}{\diagdown})
$$

(147)

production of **183** [Eq. (132)] (*160,161*). Methanol also adds to the ketenylidene β carbon in **174** and **175** [Eq. (148)] (*167,168*).

$$
\begin{array}{ll}
[H_3M_3(CO)_9\{CCO\}]^+ \ + \ MeOH \ \longrightarrow \ H_3M_3(CO)_9\,(\mu_3-C-C\overset{O}{\underset{OMe}{\diagdown}}) \\
(\textbf{174},\ M = Ru) \qquad\qquad\qquad\qquad\qquad (\textbf{190},\ M = Ru) \\
(\textbf{175},\ M = Os) \qquad\qquad\qquad\qquad\qquad (\textbf{191},\ M = Os)
\end{array}
$$

(148)

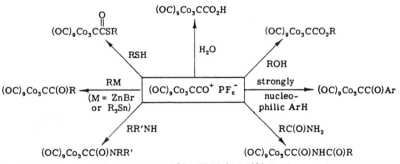

Fig. 21. Representative reactions of $[Co_3(CO)_9\{CCO\}]^+$ (**168**). (Reprinted with permission from Ref. *159*.)

Protonation of the neutral ketenylidene clusters **171** and **172** occurs at the metal framework to give the trihydride clusters **174** and **175** [Eq. (149)] (*167,168*). This implies that the reported reaction of **171** with methanol to yield **191** probably proceeds via the initial formation of **175** (*167*). The Os$_3$ cluster **171** also slowly reacts with H$_2$ to form

$$
\begin{array}{c}
\text{H}_2\text{M}_3(\text{CO})_9\{\text{CCO}\} \quad + \quad \text{H}^+ \quad \longrightarrow \quad [\text{H}_3\text{M}_3(\text{CO})_9\{\text{CCO}\}]^+ \quad (149) \\
(\underline{172}, \ \text{M} = \text{Ru}) \qquad\qquad\qquad\qquad (\underline{174}, \ \text{M} = \text{Ru}) \\
(\underline{171}, \ \text{M} = \text{Os}) \qquad\qquad\qquad\qquad (\underline{175}, \ \text{M} = \text{Os})
\end{array}
$$

H$_3$Os$_3$(CO)$_9$(μ_3-CH) (**192**) in 54% yield, but the reaction is inhibited by added CO. The authors suggested the mechanistic path in Eq. (138), a key element of which is the reversibility of the ketenylidene-forming step **185** → **171** [Eq. (150)] (*167*).

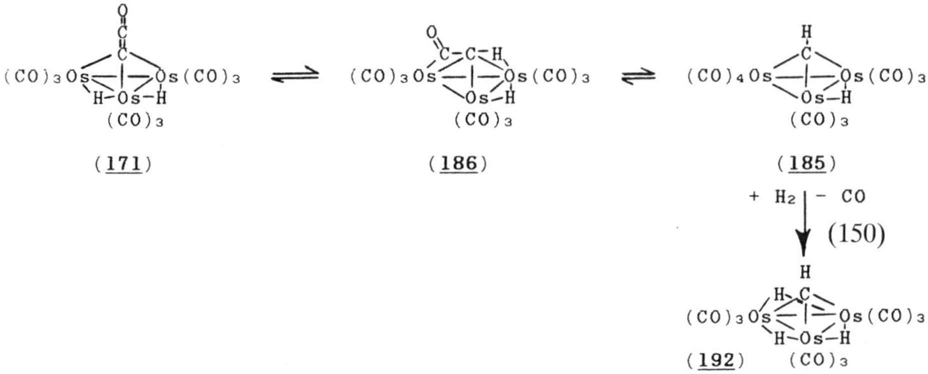

$$ (150) $$

The anionic ketenylidene cluster [Ru$_3$(CO)$_9$(CCO)]$^{2-}$ (**173**) also undergoes protonation at the metal framework to first give [HRu$_3$(CO)$_9$(CCO)]$^-$ and then H$_2$Ru$_3$(CO)$_9$(CCO) (**172**) (*169*). This behavior contrasts with that of [Fe$_3$(CO)$_9$(CCO)]$^{2-}$, which protonates at the ketenylidene α carbon (see below).

2. Nucleophilic Behavior

The anionic clusters **169**, **170**, and **173** show pronounced nucleophilic behavior in their reactivity, with reactions generally occurring at the α carbon of the ketenylidene ligand. For example, **169** and **170** undergo protonation to form an alkylidyne cluster [Eq. (151)] (*162,163*). Even weak acids such as CH$_3$COOH induce this reaction. It has been suggested that the tilt of the ketenylidene ligand exposes the α carbon and

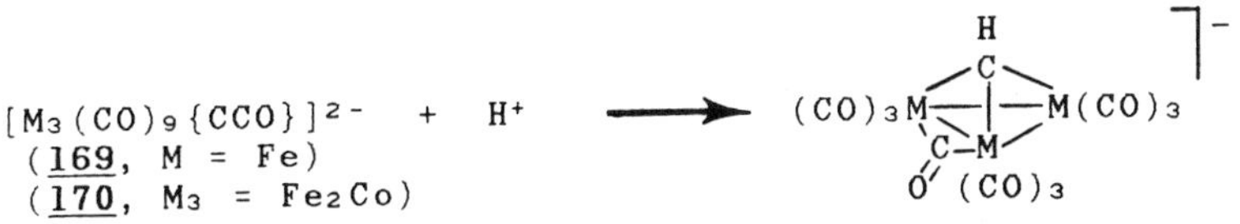

$$[M_3(CO)_9\{CCO\}]^{2-} + H^+ \longrightarrow (CO)_3M\overset{\overset{\displaystyle H}{\underset{\displaystyle C}{\big|}}}{\underset{\overset{\displaystyle C}{\underset{\displaystyle O'(CO)_3}{\big|}}}{}}M(CO)_3 \quad (151)$$

(**169**, M = Fe)
(**170**, M$_3$ = Fe$_2$Co)

facilitates this transformation (*163*). Presumably an intermediate similar to **186** in Eq. (145) is involved in this reaction. Similar reaction of the Fe$_3$ cluster **169** with MeI yields a corresponding ethylidyne cluster (*163*), but both the Ru$_3$ cluster **173** and the Fe$_2$Co cluster **170** are unreactive with this reagent (*164,169*). It may be relevant that the tilt of the ketenylidene in these clusters is less than it is in **169**, thus giving less exposure of the α carbon to the incoming electrophile.

The Ru$_3$ cluster **173** undergoes an interesting transformation with MeOSO$_2$CF$_3$ to form the vinylidene cluster **193** [Eq. (152)] (*169*). This reaction involves addition of methyl cations to both the ketenylidene β carbon and the oxygen atom, but its mechanism is not known.

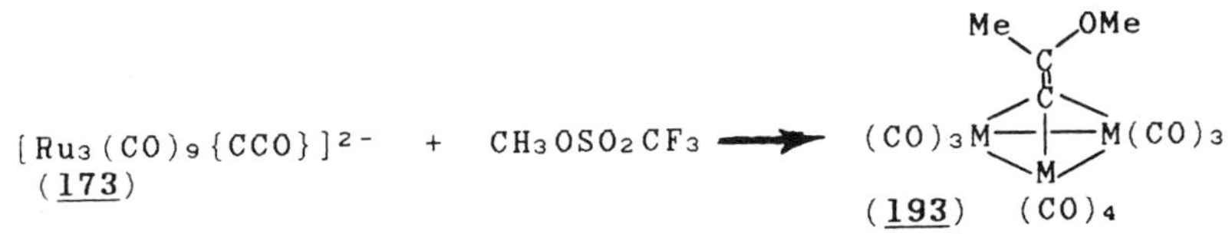

$$[Ru_3(CO)_9\{CCO\}]^{2-} + CH_3OSO_2CF_3 \longrightarrow (CO)_3M\overset{\overset{\displaystyle Me\diagup^{OMe}}{\underset{\displaystyle C}{\big|}}}{\underset{\overset{\displaystyle C}{\underset{\displaystyle M}{\big|}}}{}}M(CO)_3$$

(**173**)   (**193**)   (CO)$_4$

$$(152)$$

D. Cluster Expansion Reactions

Both the anionic Fe$_3$ cluster **170** and the Ru$_3$ cluster **173** have been shown to be useful precursors for the synthesis of mixed-metal carbido clusters formed via the cluster expansion illustrated in Eq. (153) (*164,165,169*). This reaction is believed to involve addition of a coordinatively unsaturated metal to the ketenylidene α carbon followed by

$$M\overset{\overset{\displaystyle O}{\underset{\displaystyle C}{\diagdown}}}{\underset{\underset{\displaystyle M}{\big|}}{}}M + M \longrightarrow M\overset{C}{\underset{M\overset{}{=}M}{\equiv}}M \quad (153)$$

migration of the ketenylidene CO to the cluster framework. The complexes M(CO)$_3$(CH$_3$CN) (M = Cr, W), Fe$_2$(CO)$_9$, [Rh(CO)$_2$Cl]$_2$, [Mn(CO)$_5$(CH$_3$CN)]$^+$, and Ni(CO)$_4$ have been used in these reactions to prepare heteronuclear clusters with Fe$_3$Cr, Fe$_3$W, Fe$_3$Rh, Fe$_3$Mn, Fe$_3$Ni, and Ru$_3$Fe metal frameworks.

VI

SUMMARY

This review has attempted to present a comprehensive summary of the synthesis, characterization, and reactivity of ketene, ketenyl, and ketenylidene complexes as well as the reactions of ketenes with organometallic complexes. Even though all of these areas have been under active investigation for a number of years and significant advances have been made, we feel that there is much yet to be learned. Consider the reactions of ketenes with organometallic complexes, where there are entire classes of organometallic compounds, e.g., carbene, carbyne, and vinylidene complexes, for which no reports of ketene reactions exist and there are only a few scattered examples of ketene insertion reactions. The number of ketenyl and ketenylidene complexes is still rather small and restricted to only a few metals. More general synthetic methods are particularly needed for ketenyl complexes, and all three classes of compounds are in need of more detailed reactivity studies. The application of these compounds in synthetic organic chemistry is virtually unexplored, even though this should be a fertile area since chiral complexes can be prepared and could be useful in stereoselective reactions. We look forward to the development of this chemistry in the years to come.

REFERENCES

1. S. Patai, Ed., "The Chemistry of Ketenes, Allenes, and Related Compounds." Wiley, New York, 1980.
2a. H. Ulrich, "Cycloaddition Reactions of Heterocumulenes." Academic Press, New York, 1967.
2b. H. R. Seikaly and T. T. Tidwell, *Tetrahedron* **42**, 2587 (1986).
3a. H. Staudinger, "Die Ketene." Enke, Stuttgart, 1912.
3b. W. E. Hanford and J. C. Sauer, *Org. React.* **3**, 108 (1946).
3c. R. N. Lacy, Ketenes, *in* "The Chemistry of Alkenes" (S. Patai Ed.), pp. 1161–1227. Wiley (Interscience), New York, 1964.
3d. D. Borrmann, "Methoden der Organische Chemie," Vol 7, Part 4. Thieme, Stuttgart, 1986.
3e. W. T. Brady, *Tetrahedron* **37**, 2949 (1981).
3f. D. P. N. Satchell and R. S. Satchell, *Chem. Soc. Rev.* **4** 231 (1975).
4. I. Fleming, "Frontier Orbitals and Organic Chemical Reactions." Wiley, New York, 1976.
5. E. Lindner and H. Berke, *Z. Naturforsch., Teil B* **29**, 275–276 (1974).
6. F. Ungvary, *J. Chem. Soc. Chem. Commun.* 824 (1984).
7. S. I. Hommeltoft and M. C. Baird, *J. Am. Chem. Soc.* **107**, 2548–2549 (1985).
8. S. I. Hommeltoft and M. C. Baird, *Organometallics* **5**, 190-5 (1986).
9. K. A. Azam, A. J. Deeming, and I. P. Rothwell, *J. Chem. Soc. Dalton Trans.* **91** (1981).

10. S. Gambarotta, S. Strologo, C. Floriani, A. Chiesi-Villa, and C. Guastini, *Inorg. Chem.* **24,** 654–660 (1985).

11. E. A. Jeffery and A. Meisters, *J. Organomet. Chem.* **82,** 315 (1974).

12. C. Blandy and D. Gervais, *Inorg. Chim. Acta* **47,** 197 (1981).

13. C. Blandy and M. Hliwa, *C. R. Seances Acad. Sci., Ser. 2* **296,** 51 (1983).

14. L. Vuitel and A. Jacot-Guillarmod, *Synthesis* 608 (1972).

15. L. Vuitel and A. Jacot-Guillarmod, *Helv. Chim. Acta* **57,** 1703 (1974).

16. R. Hofer, D. Evard, and A. Jacot-Guillarmod, *Helv. Chim. Acta* **68,** 969 (1985).

17a. G. Kuenstle and H. Spes, Ger. Offen. DE 1913097, 24 Sept., 1970.

17b. R. Ruiz de Alda Iturria, Span. ES 420829, 1 Apr., 1976.

17c. H. Hey and H. J. Arpe, Ger. Offen. DE 2241863. 28 Feb., 1974.

17d. H. Hey and H. J. Arpe, Ger. Offen. DE 2165219, 12 July, 1973.

17e. H. Hey and H. J. Arpe, 2,165,219 (CA 79;078115s), 1947.

17f. I. Takasu, M. Higuchi, and Y. Hijioka, Japan. JP 47/30638 [72/30638] 9 Aug., 1972.

17g. I. Takasu, M. Higuchi, and Y. Hijioka, U.S. US 3574728, 13 Apr., 1971.

18. S. Baba and S. Kawaguchi, *Proc. Int. Conf. Coord. Chem., 16th, R37 Univ. Coll., Dublin, Dep. Chem.* (1974).

19. W. A. Herrmann, R. Serrano, U. Kuesthardt, M. L. Ziegler, E. Guggolz, and T. Zahn, *Angew. Chem. Int. Ed. Engl.* **23,** 515 (1984).

19b. W. A. Herrmann, U. Kuesthardt, M. L. Ziegler, and T. Zahn, *Angew. Chem. Int. Ed. Engl.* **24,** 860 (1985).

19c. U. Kuesthardt, W. A. Herrmann, M. L. Ziegler, T. Zahn, and B. Nuber, *J. Organomet. Chem.* **311** 163 (1986).

20. I. S. Kolomnikov, Yu D. Koreshkov, T. S. Lobeeva, and M. E. Vol'pin, *Izv. Akad. Nauk SSSR, Ser. Khim* 1181 (1972).

21. A. Bucheister, P. Klemarczyk, and M. Rosenblum, *Organometallics* **1,** 1679 (1982).

22. M. E. Wright, G. O. Nelson, and R. S. Glass, *Organometallics* **4,** 245 (1985).

23. P. Hong, K. Sonogashira, and N. Hagihara, *J. Organomet. Chem.* **219,** 363 (1981).

24. L. S. Chen, D. W. Lichtenberg, P. W. Robinson, Y. Yamamoto, and A. Wojcicki, *Inorg. Chim. Acta* **25,** 165 (1977).

25. M. Ghavshou and D. A. Widdowson, *J. Chem. Soc. Perkin Trans. 1,* 3065 (1983).

26. Z. Goldschmidt, S. Antebi, D. Cohen, and I. Goldberg, *J. Organomet. Chem.* **273,** 347 (1984).

27. Z. Goldschmidt and S. Antebi, *J. Organomet. Chem.* **259,** 119 (1983).

28. Z. Goldschmidt and S. Antebi, *Tetrahedron Lett.* 271 (1978).

29. Z. Goldschmidt and S. Antebi, *Tetrahedron Lett.* 1225–1228 (1978).

30. M. Sato, K. Ichibari, and F. Sato, *J. Organomet. Chem.* **26,** 267 (1971).

31. M. R. Churchill, B. G. DeBoer, and J. J. Hackbarth, *Inorg. Chem.* **13,** 2098 (1974).

32. D. A. Young *J. Organomet. Chem.* **70,** 95 (1974).

33. W. A. Herrmann, I. Schweizer, P. S. Skell, M. L. Ziegler, K. Weidenhammer, and B. Nuber, *Chem. Ber.* **112,** 2423 (1979).

34. R. J. Angelici and R. G. W. Gingerich, *Organometallics* **2,** 89 (1983).

35. M. McKenna, L. L. Wright, D. J. Miller, L. Tanner, R. C. Haltiwanger, and M. R. DuBois, *J. Am. Chem. Soc.* **105,** 5329 (1983).

36a. M. Ichikawa, K. Sekizawa, K. Shikakura, and M. Kauai, *J. Mol. Catal.* **11,** 167 (1981).

36b. A. Takeuchi and J. R. Katzer, *J. Phys. Chem.* **86,** 2438 (1982).

37. L. H. Polm, G. Van Koten, K. Vrieze, C. H. Stam, and W. C. J. Van Tunen, *J. Chem. Soc. Chem. Commun.* 1177 (1983).

38. O. S. Mills and A. D. Redhouse, *J. Chem. Soc. Chem. Commun.,* 444 (1966).

39. O. S. Mills and A. D. Redhouse, *J. Chem. Soc., A* 1282 (1968).
40. W. A. Herrmann, J. Gimeno, J. Weichmann, M. L. Ziegler, and B. Balbach, *J. Organomet. Chem.* **213**, C26 (1981).
41. I. Bkouche-Waksman, J. S. Ricci, Jr., T. F. Koetzle, J. Weichmann, and W. A. Herrmann, *Inorg. Chem.* **24**, 1492 (1985).
42. D. A. Young, *Inorg. Chem.* **12**, 482 (1973).
43. P. Hong, N. Nishii, K. Sonogashira, and N. Hagihara, *J. Chem. Soc. Chem. Commun.* 993 (1972).
44. T. Yamamato, A. R. Garber, J. R. Wilkinson, C. B. Boss, W. E. Streib, annd L. J. Todd, *J. Chem. Soc. Chem. Commun.* 354 (1974).
45. H. Ueda, Y. Kai, N. Yasuaka, and N. Kasai, *Abstr. Symp. Organomet. Chem, 21st, Sendai, Japan, Oct.* (1973); cited in Ref. 43.
46. A. J. Arce and A. J. Deeming, *J. Chem. Soc. Chem. Commun.* 364 (1982).
47. H. Hoberg and J. Korff, *J. Organomet. Chem.* **152**, C39 (1978).
48. P. Hong, K. Sonogashira, and N. Hagihara, *Nippon Kagaku Zasshi* **89**, 74 (1968).
49. P. Hong, K. Sonogashira, and N. Hagihara, *Tetrahedron Lett.* 1105 (1971).
50. K. Kinugasa and T. Agawa, *Organometl. Chem. Synth.* **1**, 427 (1972).
51. A. Baba, Y. Ohshiro, and T. Agawa, *J. Organomet. Chem.* **110**, 121 (1976).
52. P. Hong, H. Yamazaki, K. Sonogashira, and N. Hagihara, *Chem. Lett.* 535 (1978).
53. H. Yamazaki and P. Hong, *J. Mol. Catal.* **21**, 133 (1983).
54. H. Itatani, M. Kashima, and T. Suehiro (Ube Industries, Ltd.), Eur. Pat. Appl. EP 77542 A1, 27 Apr., 1983 (CHEM ABS: 99(11):87654g.)
55. N. Kenji, F. Shinichi, S. Yashishi, F. Kozo, N. Keigo, and Y. Masayohi (Ube Industries, Ltd.), Eur. Pat. Appl. JEP 6611, 9 Jan, 1980. (CHEM ABS: 93(5):45987p.)
56. (Ube Industries, LTD) Jpn. Kokai Tokkyo Koho JP 59/67243 A2 [84/67243], 16 Apr., 1984. (CHEM ABS: 101(13):110373k.)
57. W. A. Herrmann, *Angew. Chem. Int. Ed. Engl.* **13**, 335 (1974).
58. A. D. Redhouse and W. A. Herrmann, *Angew. Chem. Int. Ed. Engl.* **15**, 615 (1976).
59. W. A. Herrmann and J. Plank, *Angew. Chem. Int. Ed. Engl.* **17**, 525 (1978).
60. W. A. Herrmann, J. Plank, G. W. Kriechbaum, J. L. Ziegler, H. Pfisterer, J. L. Atwood, and R D. Rogers, *J. Organomet. Chem.* **264**, 327 (1984).
61. W. A. Herrmann, J. Plank, M. Ziegler, and K. Weidenhammer, *J. Am. Chem. Soc.* **101**, 3133 (1979).
62. T. W. Bodnar and A. R. Cutler, *J. Am. Chem. Soc.* **105**, 5926 (1983).
63. T. W. Bodner, E. J. Crawford, and A. R. Cutler, *Organometallics* **5**, 947 (1986).
64. A. Cutler, S. Raghu, and M. Rosenblum, *J. Organomet. Chem.* **77**, 381 (1974).
65. H. Hoberg and J. Korff, *J. Organomet. Chem.* **152**, 255 (1978).
66. A. Miyashita, H. Shitara, and H. Nohira, *J. Chem. Soc. Chem. Commun.* 850 (1985).
67. A. Miyashita and R. H. Grubbs, *Tetrahedron Lett.* **22**, 1255 (1981).
68. K. Schorpp and W. Beck, *Z. Naturforsch. Teil B* **28**, 738 (1973).
69. A. Miyashita, H. Shitara, and H. Nohira, *Organometallics* **4**, 1463 (1985).
70. W. R. Roper and A H. Wright, unpublished results; cited in M. A. Gallop and W. R. Roper, *Adv. Organomet. Chem.* **25**, 121 (1986).
71. W. A. Herrmann, J. Weichmann, B. Balbach, and M. L. Ziegler, *J. Organomet. Chem.* **231**, C69 (1982).
72. F. R. Kreissl, M. Wolfgruber, W. Sieber, and K. Ackermann, *J. Organomet. Chem.* **252**, C39 (1983).
73. F. R. Kreissl, M. Wolfgruber, and W. J. Sieber, *J. Organomet. Chem.* **2**, C4 (1984).
74. M. Roper, H. Strutz, and W. Keim, *J. Organomet. Chem.* **219**, C5 (1981).

75a. Y. C. Lin, J. C. Calabrese, and S. S. Wreford *J. Am. Chem. Soc.*, **105**, 1679 (1983).

75b. N. M. Doherty, M. J. Filders, N. J. Forrow, S. A. R. Knox, K. A. Macpherson, and A. G. Orpen, *J. Chem. Soc. Chem. Commun.* 1335 (1986).

76. E. D. Morrison, G. R. Steinmetz, G. L. Geoffroy, W. C. Fultz, and A. L. Rheingold, *J. Am. Chem. Soc.* **105**, 4104 (1983).

77. E. D. Morrison, G R. Steinmetz, G. L. Geoffrey, W. C. Fultz, and A. L. Rheingold, *J. Am. Chem. Soc.* **106**, 4783 (1984).

78a. E. D. Morrison, G. L. Geoffroy, and A. L. Rheingold, *J. Am. Chem. Soc.* **107**, 254 (1985).

78b. E. D. Morrison, G. L. Geoffroy, and A. L. Rheingold, *J. Am. Chem. Soc.* **107**, 3541 (1985).

79a. A. P. Masters, J. S. Sorensen, and T. Ziegler, *J. Org. Chem.* **51**, 3558 (1986).

79b. M. Akita, A. Kondoh, and Y. Moro-oka, *J. Chem. Soc. Chem. Commun.* 1296 (1986).

80. J. S. Holmgren, J. R. Shapley, S. R. Wilson, and W. T. Pennington, *J. Am. Chem. Soc.* **108**, 508 (1986).

81. S. Gambarotta, M. Pasquali, C. Floriani, A. Chiesi-Villa, and C. Guastini, *Inorg. Chem.* **20**, 1173 (1981).

82. P.-B. Hong, K. Sonogashira, and N. Hagihara, *Bull Chem. Soc. Jpn* **39**, 1821 (1966).

83. D. A. Straus and R. H. Grubbs, *J. Am Chem. Soc.* **104**, 5499 (1982).

84. J. O. Meinhart, B. D. Santarsiero, and R. H. Grubbs, *J. Am. Chem. Soc.* **108**, 3318 (1986).

85. S. C. H. Ho, D. A. Straus, J. Armantrout, W. P. Schaefer, and R. H. Grubbs, *J. Am. Chem. Soc.* **106**, 2210 (1984).

86. E. J. Moore, D. A. Straus, J. Armantrout, B. D. Santarsiero, R. H. Grubbs, and J. E. Bercaw, *J. Am. Chem. Soc.* **105**, 2068 (1983).

87a. R. M. Waymouth, B. D. Santarsiero, R. J. Coots, M. J. Bronikowski, and R. H. Grubbs, *J. Am. Chem. Soc.* **108**, 1427 (1986).

87b. R. M. Waymouth, B. D. Sanatarsiero, and R. H. Grubbs, *J. Am. Chem. Soc.* **106**, 4050 (1984).

88. P. T. Barger, B. D. Santarsiero, J. Armantrout, and J. E. Bercaw, *J. Am. Chem. Soc.* **106**, 5178 (1984).

89a. C. P. Casey and J. M. O'Connor, *J. Am. Chem. Soc.* **105**, 2919 (1983).

89b. C. P. Casey, J. M. O'Connor, and K. J. Haller, *J. Am. Chem. Soc.* **107**, 3172 (1985).

90. C. A. Rusik, T. L. Tonker, and J. L. Templeton, *J Am. Chem. Soc.* **108**, 4652 (1986).

91. G. Fachinetti, C. Biran, C. Floriani, A. Chiesi-Villa, and C. Guastini, *J. Am. Chem. Soc.* **100**, 1921 (1978).

92. G. Fachinetti, C. Biran, C. Floriani, A. Chiesi-Villa, and C. Guastini, *Inorg. Chem.* **17**, 2995 (1978).

93. G. S. Bristow, P. B. Hitchcock, and M. F. Lappert, *J. Chem. Soc. Chem Commun* 462 (1982).

94. S. J. Young, H. Hope, and N. E. Schore, *Organometallics* **3**, 1585 (1984).

95. R. Choukroun, F. Dahan, and D. Gervais, *J. Organomet. Chem.* **266**, C33 (1984).

96. L. M. Engelhardt, G. E. Jacobsen, C. L. Raston, and A. H. White, *J. Chem. Soc. Chem. Commun.* 220 (1984).

97. R. M. Waymouth, B. D. Santarsiero, R. J. Coots, M. J. Bronikowski, and R. H. Grubbs, *J. Am. Chem. Soc.* **108**, 1427 (1986).

98. R. B. King, *Inorg. Chem.* **2**, 642 (1963).

99. M. G. Newton, N. S. Pantalea, R. B. King, and C.-F. Chu, *J. Chem. Soc. Chem. Commun.* 10 (1979).

100. A. E. Hill and M. R. Hoffman, *J. Chem. Soc. Chem. Commun.* 574 (1972).
101. J. Klimes and E. Weiss, *Chem. Ber.* 115, 2606 (1982).
102. T.-A. Mitsudo, T. Sasaki, Y. Watanabe, Y. Takegami, S. Nishigaki, and K. Nakatsu, *J. Chem. Soc. Chem. Commun.* 252 (1978).
103. P. Binger, B. Cetinhaya, and C. Kruger, *J. Organomet. Chem.* 159, 63 (1978).
104. J. Klimes and E. Weiss. *Angew. Chem. Int. Ed. Engl.* 21, 205 (1982).
105. K. J. Jens and E. Weiss, *Chem. Ber.* 117, 2469 (1984).
106. J. L. Templeton, R. S. Herrick, C. A. Rusik, C. E. McKenna, J. W. McDonald, and W. E. Newton, *Inorg. Chem.* 24, 1383 (1985).
107. M. Green, J. Z. Nyathi, C. Scott, and F. G. A. Stone, *J. Chem. Soc. Dalton Trans.* 1067 (1978).
108. C. F. Jewell, L. S. Liebeskind, and M. Williamson, *J. Am. Chem. Soc.* 107, 6715 (1985).
109. F. R. Kreissl, M. Wolfgruber, W. Sieber, and H. G. Alt, *Angew. Chem. Int. Ed. Engl.* 22, 149 (1983).
110. F. R. Kreissl, M. Wolfgruber, W. Sieber, and H. G. Alt, *Angew. Chem. Int. Ed. Engl.* 22, 1001 (1983).
111. F. R. Kreissl, M. Wolfgruber, W. Sieber, and K. Ackermann, *J. Organomet. Chem.* 252, C39 (1983).
112. F. R. Kreissl, M. Wolfgruber, and W. J. Sieber, *Organometallics* 2, 1266 (1983).
113. M. Wolfgruber and F. R. Kreissl, *J. Organomet. Chem.* 258, C9 (1983).
114. M. Wolfgruber, W. Sieber, and F. R. Kreissl, *Chem. Ber.* 117, 427 (1984).
115. H. Hornig, E. Walther, and U. Schubert, *Organometallics,* 4, 1905 (1985).
116a. N. Aktogu, H. Felkin., and S. G. Davies, *J. Chem. Soc. Chem. Commun.* 1303 (1982).
116b. N. Aktogu, H. Felkin, G. J. Baird, S. G. Davies, and O. Watts, *J. Organomet. Chem.* 262, 49 (1984).
117a. S. G. Davies and J. C. Walker, *J. Chem. Soc. Chem. Commun.* 209 (1985).
117b. S. G. Davies, R. J. C. Easton, J. C. Walker, and P. Warner, *J. Organomet. Chem.* 296, C40 (1985).
117c. P. W. Ambler and S. G. Davies, *Tetrahedron Lett.* 26, 2129 (1985).
118a. L. S. Liebeskind and M. E. Welker, *Organometallics* 2, 194 (1983).
118b. L. S. Liebeskind and M. E. Welker, *Tetrahedron Lett.* 26, 3079 (1985).
118c. L. S. Liebeskind, M. E. Welker, and V. Goedken, *J. Am. Chem. Soc.* 106, 441 (1984).
118d. L. S. Liebeskind and M. E. Welker, *Tetrahedron Lett.* 25, 4341 (1984).
119. K. G. Moloy, T. J. Marks, and V. W. Day, *J. Am. Chem. Soc.* 105, 5696 (1983).
120. P. J. Fagan, J. M. Manriquez, T. J. Marks, V. W. Day, S. H. Vollmer, and C. S. Day, *J. Am. Chem. Soc.* 102, 5396 (1980).
121. J. Norton, private communication.
122. L. Messerle and M. D. Curtis. *J. Am. Chem. Soc.* 102, 7789 (1980).
123a. W. J. Laws and R. J. Puddephatt, *Inorg. Chim. Acta* 113, L23 (1986).
123b. C. Ruchardt and G. N. Schrauzer, *Chem. Ber.* 93, 1840 (1960).
123c. H. Fischer, *Angew. Chem. Int. Ed. Engl.* 22, 874 (1983).
123d. U. Schubert, H. Hornig, K. U. Erdmann, and K. Weiss, *J. Chem. Soc. Chem. Commun.* 13 (1984).
123e. B. Dorrer and E. O. Fischer, *Chem. Ber.* 107, 2683 (1974).
123f. K. H. Doetz, *Angew. Chem. Int. Ed. Engl.* 18, 954 (1979).
123g. K. H. Doetz and J. Muehlemeier, *Angew. Chem. Int. Ed. Engl.* 21, 929 (1982).
123h. K. H. Doetz and B. Fuegen-Koeste, *Chem. Ber.* 113, 1449 (1980).

123i. W.. D. Wulff and R. W. Kaesler, *Organometallics* **4**, 1461 (1985).

123j. A. Yamashita and T. A. Scahill, *Tetrahedron Lett.* **23**, 3765 (1982).

123k. J. Klimes and E. Weiss, *Chem. Ber.* **115**, 2175 (1982).

123l. M. Franck-Neumann, C. Dietrich-Buchecker, and A. K. Khemiss, *J. Organomet. Chem.* **224**, 133 (1982).

123m. M. Franck-Neumann, C. Dietrich-Buchecker, and A. Khemiss, *Tetrahedron Lett.* **22**, 2307 (1981).

124. W. C. Mercer and G. L. Geoffroy, unpublished observations; W. C. Mercer, Ph.D dissertation, Pennsylvania State University, 1984.

125. B. E. Mann and B. F. Taylor, "^{13}C NMR Data for Organometallic Compounds." Academic Press, New York, 1981.

126. R. M. Silverman, G. C. Gassler, and J. C. Merill, "Spectrometric Identification of Organic Compounds," 4th ed. Wiley, New York, 1981.

127. D. Sutter, L. Charpentier, and H. Dreizler, *Z. Naturforsch. A.* **27A**, 597 (1972).

128. "Tables of Interatomic Distances and Configurations in Molecules and Ions" (L. E. Sutton, Ed.). Special publication No. 11, Chemical Society, London, 1958.

129. S. L. Bassner, E. D. Morrison, G. L. Geoffroy, and A. L. Rheingold, *J. Am. Chem. Soc.* **108**, 5358 (1986).

130. S. L. Bassner, E. D. Morrison, G. L. Geoffroy, and A. L. Rheingold, *Organometallics*, in press (1987).

131a. M. Ichikawa, K. Sekizawa, K. Shikakura, and M. Kauai, *J. Mol. Catal.* **11**, 167 (1981).

131b. A. Takeuchi and J. R. Katzer, *J. Phys. Chem.* **86**, 2438 (1982)

131c. E. L. Muetterties and J. Stein, *Chem. Rev.* **79**, 479 (1979).

131d. A. T. Bell, *Catal. Rev.* **23**, 203 (1981).

131e. W. A. Herrmann, *Angew. Chem. Int. Ed. Engl.* **21**, 117 (1982).

132. F. R. Kreissl, A. Frank, U. Schubert, T. L. Lindner, and G. Huttner, *Angew. Chem. Int. Ed. Engl.* **15**, 632 (1976).

133a. F. R. Kreissl, I. Eberl, and W. Uedelhoven, *Chem. Ber.* **110**, 3782 (1977).

133b. F. R. Kreissl, P. Friedrich, and G. Huttner, *Angew. Chem. Int. Ed. Engl.* **16**, 102 (1977).

134. F. R. Kreissl, I. Eberl, and W. Uedelhoven, *Angew. Chem. Int. Ed. Engl.* **17**, 860 (1978).

135. F. R. Kreissl, W. Uedelhoven, and K. Eberl, *Angew. Chem. Int. Ed. Engl.* **17**, 859 (1978).

136. W. Uedelhoven, K. Eberl, and F. R. Kreissl, *Chem. Ber.* **112**, 3376 (1979).

137. K. Eberl, W. Uedelhoven, H. H. Karsch, and F. R. Kreissl, *Chem. Ber* **113**, 3377 (1980).

138. K. Eberl, W. Uedelhoven, M. Wolfgruber, and F. R. Kreissl, *Chem. Ber.* **115**, 504 (1982).

139. K. Eberl, M. Wolfgruber, W. Sieber, and F. R. Kreissl, *J. Organomet. Chem.* **236**, 171 (1982).

140a. W. J. Sieber, M. Wolfgruber, F. R. Kreissl, and O. Orama, *J. Organomet. Chem.* **270**, C41 (1984).

140b. F. R. Kreissl, G. Reber, and G. Muller, *Angew. Chem. Int. Ed. Engl.* **25**, 643 (1986).

141a. G. L. Hillhouse *J. Am., Chem. Soc.,* **107**, 7772 (1985).

141b. S. Voran and W. Malisch, *Angew. Chem. Int. Ed. Engl.* **22**, 151 (1983).

142. F. R. Kreissl, W. Sieber, and M. Wolfgruber, *Z. Naturforsch. B* **37B**, 1485 (1982).

143. F. R. Kreissl, W. Sieber, and M. Wolfgruber, *Angew. Chem. Int. Ed. Engl.* **22**, 493 (1983).

144. F. R. Kreissl, W. J. Sieber, and H. G. Alt, *Chem. Ber.* **117**, 2527 (1984).
145. E. O. Fischer, A. C. Filippou, H. G. Alt, and K. Ackermann, *J. Organomet. Chem.* **254** C21 (1983).
146. E. O. Fischer, a. C. Filippou, and H. G. Alt, *J. Organomet. Chem.* **276**, 377 (1984).
147a. J. Martin-Gil, J. A. K. Howard, R. Navarro, and F. G. A. Stone, *J. Chem. Soc. Chem. Commun.* 1168 (1979).
147b. M. H. Chisholm, J. C. Huffman, and N. S. Marchant, *J. Chem. Soc. Chem. Commun.* 717 (1986).
148. O. Orama, U. Schubert, F. R. Kreissl, and E. O. Fischer, *Z. Naturforsch. B* **35B**, 82 (1980).
149. C. P. Casey, P. J. Fagan, and V. W. Day *J. Am. Chem. Soc.* **104**, 7360 (1982).
150. C. E. Sumner, J. A. Collier, and R. Pettit, *Organometallics* **1**, 1350 (1982).
151. J. B. Sheridan, G. L. Geoffroy, and A. L. Rheingold, *Organometallics* **5**, 1514 (1986).
152. J. A. K. Howard, J. C. Jeffrey, M. Laguna, R. Navarro, and F. G. A. Stone, *J. Chem. Soc. Chem. Commun.* 1170 (1979).
153. E. O. Fischer and P. Friedrich, *Angew. Chem. Int. Ed. Engl.* **18**, 327 (1979).
154. A. P. Cox, L. F. Thomas, and J. Sheridan, *Spectrochim. Acta* **15**, 542 (1959).
155. F. P. Boer, *J. Am. Chem. Soc.* **90**, 6706 (1968).
156. B Bak, J. Christiansen, K. Kunstmann, L. Nygaard, and J. Rastrup-Andersen, *J. Chem. Phys.* **45**, 883 (1966).
157. J. E. Hallgren, C. S. Eschbach, and D. Seyferth, *J. Am. Chem. Soc.* **94**, 2547 (1972).
158. D. Seyferth, J. E. Hallgred, and C. S. Eschbach, *J. Am. Chem. Soc.* **96**, 1730 (1974).
159. D. Seyferth, *Adv. Organomet. Chem.* **14**, 97 (1976).
160. J. S. Bradley, G. B. Ansell, and E. W. Hill, *J. Am. Chem. Soc.* **101**, 7417 (1979).
161. J. S. Bradley, E. W. Hill, G. B. Ansell, and M. A. Modrick, *Organometallics* **1**, 1634 (1982).
162. J. W. Kolis, E. M. Holt, M. Drezdzon, K. H. Whitmire, and D. F. Shriver, *J. Am. Chem. Soc.* **104**, 6134 (1982).
163. J. W. Kolis, E. M. Holt, and D. F. Shriver, *J. Am. Chem. Soc.* **105**, 7307 (1983).
164. J. W. Kolis, E. M. Holt, J. A. Hriljac, and D. F. Shriver, *Organometallics* **3**, 496 (1984).
165. J. A. Hriljac, P. N. Swepston, and D. F. Shriver, *Organometallics* **4**, 158 (1985).
166. A. C. Sievert, D. S. Strickland, J. R. Shapley, G. R. Steinmetz, and G. L. Geoffroy, *Organometallics* **1**, 214 (1982).
167. J. R. Shapley, D. S. Strickland, G. M. St. George, M. R. Churchill, and C. Bueno, *Organometallics* **2**, 185 (1983).
168a. J. S. Holmgren and J. R. Shapley, *Organometallics* **3**, 1322 (1984).
168b. J. S. Holmgren and J. R. Shapley, *Organometallics* **4**, 793 (1985).
169a. M. J. Sailor and D. F. Shriver, *Organometallics* **4**, 1476 (1985).
169b. A. M. Crespi and D. F. Shriver, *Organometallics* **5**, 1750 (1986).
170. M. Mlekuz, M. F. D'Agostino, J. W. Kolis, and M. J. McGlinchey, *J. Organomet. Chem.* **303**, 361 (1986).
171. E. T. Blues, D. Bryce-Smith, H. Hirsch, and M. J. Simons, *J. Chem. Soc. Chem. Commun.* 699 (1970).
172. E. T. Blues, D. Bryce-Smith, B. Kettlewell, and M. Roy, *J. Chem. Soc. Chem. Commun.* 921 (1973).
173. E. T. Blues, D. Bryce-Smith, I. W. Lawston, and G. D. Wall, *J. Chem. Soc. Chem. Commun.* 513 (1974).
174a. B. E. R. Schilling and R. Hoffman, *J. Am. Chem. Soc.* **101**, 3456 (1979).

174b. M. F. D'Agostina, M. Mlekuz, J. W. Kolis, B. G. Sayer, C. A. Rodger, J. Halet, J. Sailard, and M. J. McGlinchey, *Organometallics* **5**, 2345 (1986).

175a. D. Seyferth, G. H. Williams, and J. E. Hallgren, *J. Am. Chem. Soc.* **95**, 266 (1973).

175b. D. Seyferth, G. H. Williams, and D. D. Traficante, *ibid* **96**, 604 (1974).

176a. L. P. Hammett and A. J. Deyrup, *J. Am. Chem. Soc.* **55**, 1900 (1933).

176b. H. P. Treffers and L. P. Hammett, *ibid* **59**, 1708 (1937).

177. D. Seyferth, G. H. Williams, and C. L. Nivert, *Inorg. Chem.* **16**, 758 (1977).

178. J. B. Keister and T. L. Horling, *Inorg. Chem.* **19**, 2304 (1908).

179. D. A. Roberts and G. L. Geoffroy, *in* "Comprehensive Organometallic Chemistry" (G. Wilkinson, F. G. A. Stone, and E. Abels Eds), Chap. 40. Pergamon, Oxford.

Graphite–Metal Compounds [1,2]

RENÉ CSUK [3], BRIGITTE I. GLÄNZER [3],
and ALOIS FÜRSTNER

Institut für Organische Chemie
Technische Universität Graz
A-8010 Graz, Austria

I

INTRODUCTION

The phenomenon of intercalation, although discovered inadvertently by the Chinese seven centuries B.C. (*1*), has only recently been a focus of thorough studies. The contemporary resurgence of interest in the preparation and properties of graphite intercalation compounds has recently led also to increased knowledge of their fundamental chemistry and physics. There are several reasons for the renewed interest in these materials. On the one hand, it was recognized that graphite intercalates offer the possibility of synthesizing compounds that are lightweight and have high conductivity (*2,3*), thus allowing production of wires and replacement of more expensive metal conductors. On the other hand, graphite and its intercalation compounds provide an interesting area for the examination of current ideas on two-dimensional physics because of their quasi-two-dimensional character (*4*).

One additional reason can easily be identified as a source of interest in graphite compounds. Their increasing use in "normal" chemical reactions as solid-phase supports provided many new and exciting results, although for a long time the study of heterogeneous reactions and of solids was largely neglected (*5*). Several reviews on graphite compounds appeared in the mid-1970s, but since that time our understanding of these materials has grown as several myths concerning the structures and stoichiometries have

---

[1] Dedicated to Professor Dr. H. Weidmann on the occasion of his 60th birthday.
[2] Abbreviations: Bz, benzene; DME, dimethoxyethane; GIC, graphite intercalation compound(s); HOPG, highly ordered pyrolytic graphite; THF, tetrahydrofuran.
[3] Present address: Organisch-Chemisches Institut, Universität Zürich, CH-8057 Zürich, Switzerland.

been disproved (4), thus warranting a review of the most recent and important developments in this field. Emphasis will be put on structural properties and the use of these compounds in organic and organometallic chemistry.

II

STRUCTURE OF GRAPHITE AND ITS LAMELLAR COMPOUNDS

A. *Definitions*

To avoid excessive echoing of other literature in the field, background material will be limited to the minimum commensurate with both the diversity of the readership of this review and the chemical nature of the discussions to follow.

In contrast to alumina, Celite, and silica, graphite has a lamellar structure, with an interlayer distance of 0.335 nm. But the atomic layers in hexagonal graphite, the most stable form, are not directly superimposable. Most often they alternate in the pattern ABAB. Another possibility is ABCABC stacking in rhombohedral graphite, where every third layer is superimposable (6). Electronically, graphite is a semimetal, with a nearly filled π valence band overlapping a nearly empty π conduction band in certain directions in reciprocal space. The graphite sheets may be viewed as huge aromatic macromolecules in which bonding between carbon atoms involves sp^2-hybridized orbitals. The remaining electron per carbon atom enters a delocalized orbital of p-π symmetry, giving rise to weak van der Waals interaction between the planes, the weakness of which is perceived in the excellent lubricity of graphite (7).

As an amphoteric material, graphite can react with either Lewis acids, such as metal halides or oxyhalides, or Lewis bases, such as a alkali metals (8). Intercalation, in a common sense, is the insertion of ions, atoms, or molecules into the voids between the planar sp^2-hybridized carbon sheets of the host's layered bonding network (9).

Bond distances, stacking order, and possibly bond direction may be altered, but the characteristic lamellar identity of the host must be preserved. The interlayer voids are frequently attached, to yield a periodic sequence of filled and empty spaces. It is clear that the ideal formula of graphite intercalation compounds (GIC) cannot be forecast from chemical rules, because their formulas cannot be related to the usual valences of the elements but only to structural data (10). GIC are therefore typical "topochemical" compounds, as opposed to stoichiometric compounds that have experimental compositions in good agreement with their ideal formulas. The *stage* of a compound is defined as the ratio of host layers to guest layers. A first-stage compound is therefore the most concentrated since every interlayer void is filled (see Fig. 1). This stacking phenomenon is a general one in graphite intercalation compounds, even in samples with very dilute intercalate concentrations (*11*).

Depending on the conditions of the preparation, well-defined stages may be obtained from 1 to 8 or 9 or even 12 ± 1 in the case of potassium, for instance. Figure 2 shows for the third-stage compound both the classical Rüdorff and the Herold pleated-layer model. The latter explains almost all of the experimental results concerning the intercalation–deintercalation processes, coherence-domain limitation in the planes, and stage evolution possibilities (*13*).

The stacking of the carbon planes changes through intercalation although they keep their identity. In pure graphite, two neighboring carbon planes stack in the AB configuration since the carbon hexagons do not superimpose in two neighboring planes. In the GIC two carbon planes adjacent to an inserted plane stack in the AA configuration. In higher stages, if two carbon planes are neighboring, they also stack in the AB arrangement (Fig. 3).

Older literature (*14*) claimed well-organized structures for GIC of various stages, including not only MC_8 (M = metal) but also MC_{12n} and

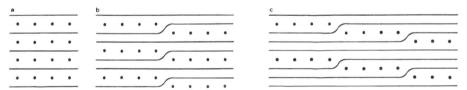

FIG. 1. Sequences of (—) carbon layers and (●) intercalant viewed perpendicular to the *c* axis according to the "nonclassical view" proposed by Herold (*12*). (a) First-stage, (b) second-stage, and (c) third-stage GIC.

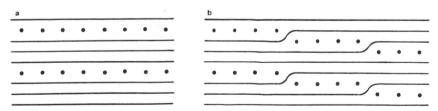

FIG. 2. (a) Classical Rüdorff and (b) nonclassical Herold views of a third-stage GIC. (−) Carbon; (●) intercalant.

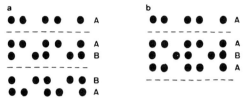

FIG. 3. Stacking in (a) second-stage and (b) third-stage GIC. (●) Layers of carbons; (---) intercalant.

LiC_{6n} for several values of n. However, numerous crystallographic measurements now show that only first-stages MC_8 and LiC_6 have stoichiometrically defined structures. Higher stages are more or less lacunar, allowing atomic diffusion from site to site. The crystallographic structure will therefore depend on temperature, pressure, stoichiometry, and other physical conditions. Transitions may even occur since the inserted lattice may change from ordered and commensurate to the graphite one to ordered and incommensurate, or from ordered to disordered (13). Whereas older reports described graphite acceptor compounds as completely in-plane disordered, recent studies show that transitions similar to those in donor compounds take place (see below). Parry and co-workers (15) pointed out, particularly for cesium–graphite intercalates, that intercalated atoms/ions are not forced to occupy sites at hexagon centers but are merely subject to a rather gentle strain (16) toward such sites. The intercalated heavy metals thereby form liquidlike layers without long-range ordering (10).

B. *Classifications of Graphite Intercalation Compounds with Respect to Synthesis, Structure, and Reactivity*

Besides the large number of reagents that can be intercalated, a number of different types of graphites are used. Naturally occurring *crystal graphite flakes* (e.g., from Ticonderoga, N.Y.) very often have diameters of approximately 1 mm but are only several hundredths of a millimeter in thickness, thus limiting convenient measurement of physical properties. Highly ordered pyrolytic graphite (HOPG), obtained by cracking a hydrocarbon at high temperature followed by heat treatment often combined with the application of presusure, gives samples of larger dimensions. This synthetic graphite is highly oriented along the c axis but in the layer planes consists of a randomly disordered collection of crystallites of about 1 μm average diameter. The electronic structure of HOPG is very nearly the same as for single crystals of graphite (17). *Kish* graphite is obtained by crystallization of carbon from molten steel during the steel manufacturing process. These samples usually are an order of magnitude greater in area and in thickness than natural graphite and exhibit much higher structural ordering than HOPG but are not quite as ordered or as chemically pure as natural single crystal flakes (*11*).

The formation of intercalation compounds is not limited to graphite; it is possible with all carbon having a two-dimensional structure, but the structural faults and the textural disorientation hinder intercalation by a mechanical effect (*14*). Only a small number of reagents can intercalate in the hard or nongraphitable carbons, obtained by pyrolysis of solid-phase organic compounds. The soft or graphitable carbons obtained by liquid-phase pyrolysis or pyrocarbons from pyrolysis of a gaseous phase are better organized and accept the intercalation. The porosity of these materials is low enough to prevent any significant absorption or capillary condensation phenomena, which are found with the hard carbons (*14*). Among these carbons, the fibers best lend themselves to intercalation because of their oriented texture. Their intercalation is under consideration as a method for variation of the adhesive, electrical, and mechanical properties of these commercially important fiber materials.

Historically, three main categories depending on the strength of inter-action between reacting species and graphite have been distinguished (*9*):

1. *Lamellar compounds*, obtained by the attack of moderately strong reductants or oxidants. The aromaticity of graphite is preserved.
2. *Covalent compounds*, arising from the attack of strong oxidizing agents such as fluorine on graphite. A buckled sp^3-hybridized sheet is created and the aromatic planarity of the graphite sheet is destroyed.
3. *Residue compounds*, arising from the decomposition of lamellar compounds by *in vacuo* or thermal treatment. They retain the residue reactant after most of the reactant has been driven off.

This chapter will focus on intercalation compounds and metals *on* graphite. The nomenclature of all these compounds has differed from investigator to investigator (*18*). The term *intercalation compound* in this chapter will include lamellar compounds (*19,20*) as well as residue compounds and compounds with ionic or covalent bonding. Terms such as "crystal compounds" (*21*) and interstitial compounds (*19*) will be avoided whenever possible. "Compounds" in which the foreign species are adsorbed on the surface of the graphite are called *surface compounds* (*22*) and those in which the foreign species occupies substantial sites in the graphite crystal structure are referred to as *substitutional compounds* (*22*).

From these definitions it becomes clear that classical residue compounds also belong to the group of the intercalation compounds. From a chemical point of view another distinction must now be drawn between these compounds (23), depending on three types of reagents to be intercalated:

1. Those agents which, when intercalated, are found to have been involved in electron exchange processes with the graphite alone, so that the reaction X + graphite $\rightleftharpoons$ intercalated X is at least formally reversible. Examples are provided by the alkali metals, pseudohalogens, and Br_2. All or at least most of X can be expected to participate in subsequent reactions, whether as a free atom, an ion, or a mixture of both.

2. Those which need, for intercalation to occur, an irreversible loss of part of their constituent atoms in a certain proportion of their own molecules, as exemplified by the "blue graphite" (24):

$$\text{graphite} + (x + y)H_2SO_4 \rightarrow C^+(HSO_4)^-_x \cdot (H_2SO_4)_y + \frac{x}{2}H_2$$

3. Those which form one or more true covalent bonds between intercalated species and the graphite, such as happens in some fluorine derivatives or "graphite oxide" (23).

Surface compounds can act as reagents or catalytically, but one can expect all regiospecificity to be lost if the surface-bound reagents are removed from the graphite prior to any reaction.

Herold distinguishes only two groups of intercalation compounds, disregarding all other classes of graphite compounds (14):

1. Electron donors as intercalants, e.g., alkali and alkaline earth metals, lanthanides, alloys of these materials, and association of these metals with hydrogen or polar molecules (ammonia, amines, tetrahydrofuran, etc.) or with aromatic molecules. These compounds are easily oxidized when exposed to air and many of them are pyrophoric.

2. Electron acceptors as intercalants, e.g., halogens, mixtures of halogens, halides, oxyhalides of numerous elements in their higher oxidation states, strong Brønsted acids, and acidic oxides. Decomposition and desorption occur easily even at ambient temperature, and some of the graphite acceptor compounds are very hygroscopic.

In general, both chemical affinities and geometric constraints associated with intercalant size and intercalant bonding distances determine whether or not a given chemical species will intercalate (11). Nonintercalation of a given compound may, however, be due to the fact that appropriate experimental conditions for the intercalation have not yet been found (9).

The reactions of the GIC may be classified according to the stages and their changes during the reaction. The "isostage" reactions (reactions in which the stage of the GIC does not change) are bidimensional and often correspond to diffusion processes. Exchange of molecules, atoms, or isotopes between a GIC of constant stage and fluid phase (*25–27*) and some addition reactions, e.g., $C_6Li + nTHF \rightleftharpoons C_6Li(THF)_n$ (*28*) may serve as examples.

The "polystage" reactions (in which the stage changes) are not pure bidimensional processes. In these reactions intercalation and deintercalation progress by the movement of dense layers of the intercalant between the graphite sheets (sliding process) (*29*).

Polystage reactions may proceed with a decrease in stage (*30,31*) or with an increase (*32,33*). Other polystage reactions are exemplified by intercalation–deintercalation processes of alkali metal GIC (*34–36*), displacements of one intercalant by another (*37,38*), deintercalation by chemical reaction (*39,40*), or electrochemical intercalation (*41*). In many cases, catalysis of reactions by GIC leads to partial deintercalation (*29*).

Of special interest are the effects of temperature and pressure on such polystage reactions, since many of them (*42,43*) can be considered (*29*) as consequences of the law of equilibrium displacement of Braun–Le Chatelier.

Whereas older reports made distinctions between ionized and atomic alkali metals on the basis of the quantities of hydrogen and potassium hydroxide released in the reactions of first- and second-stage alkali metal GIC with water (*7,9,44–46*), recent works clearly demonstrated that the quantities of evolved hydrogen and liberated KOH are a function of particle size and no dependence on stage could be detected (*47*). Nearly stoichiometric quantities of H_2 are formed with small particles, but only about 10% of the expected amounts are evolved with large flakes or with HOPG. It was shown that the products of these reactions are trapped and mechanically occluded in shell-like structures, which, on heating, open up and release the occluded gases. These structures also contain small quantities of CH_4, CO, and CO_2, due to some attack of the water on the carbon layers. This was supposed to be caused by activation of the sites of bending of the layers due to mechanical strain. This suggestion is supported by the fact that no CH_4 or CO was observed in the hydrolysis of small particles, whose layers suffer much less mechanical strain. KOH seems to be present in the form of small aggregates encapsulated by carbon sheets (*47*). Since the intercalated potassium is mobile at room temperature, it is not surprising that the quantity of H_2 from sufficiently fine particles of C_8K and $C_{24}K$ is quantitatively the same as from equivalent amounts of potassium metal. However, with larger particles the potassium becomes increasingly inaccessible to the reaction (*47*). Because of these recent results, older reports (*44,45*) on differences in the reactivity of alcohols with C_8K have to be interpreted carefully. These investigations showed the reactivity of C_8K toward alcohols and water to be related to the bulkiness of the latter (*44,45*).

C. Alkali Graphite Intercalation Compounds

The most commonly used potassium–graphite intercalation compound, C_8K, shows an interlayer distance of 5.34 Å (*48*), and all carbon layers are separated by a layer of K (Fig.4). In $C_{24}K$, $C_{36}K$, and $C_{60}K$ the intercalation of the alkali metal occurs in each second, third, fourth, and fifth interlamellar graphite spacing, respectively, but it has been shown (*34,49,50*) that in the higher stages the intercalation structure is less ordered than was assumed at first (*51*). It is now generally accepted that the alkali metal atoms diffuse in the intercalation process from the periphery of the graphite layers into part or all of the interlayer spacings (*52*).

Crystallographic study of these compounds is difficult and remains incomplete because of the nature of the crystals, which contain imperfections, are thermally and chemically fragile, and generally lack proper dimensions for X-ray investigation (*14*). Transmission electron microscopy has contributed to a large extent to our knowledge (*53*), but the low

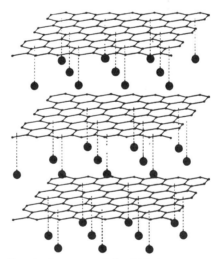

FIG. 4.   Structure of the GIC C_8K; (●) K.

pressures and bombardment by electrons risk perturbing the crystalline organization (*14*). It has been shown by low-energy diffraction and Auger electron spectroscopy that potassium intercalates into the graphite in a layer-by-layer fashion, inward from the surface. The point of intercalation is believed to be at the step edges on the surfaces (*52*).

This process gives rise to a relatively loose exchange of electrons between the graphite and the alkali metal. The electron transfer between guest and host can be taken to be reversible to a large extent (*54*). Because of this reversibility it was believed for quite a long time that two species, alkali metal atoms and their ions, were both potentially available. Different respective modes of action during any subsequent chemical process were proposed (*54*). In a chemical sense the intercalation process is analogous to the formation of ions of aromatic molecules (*55*, so that an "infinite" aromatic anion is formed (*7,56*).

Incorporation of potassium into carbon fibers is also possible, and well-defined intercalation compounds of the first and second stages, C_8K, $C_{16}K$, and $C_{24}K$, are known. The compounds are not formed as easily as in the case of graphite. This seems to be due mainly to the difference in accessibility of the carbon layers (*57*). The color of these compounds is similar to that of the graphite compounds; the incorporation of alkali metals does not affect the pore structure of the fibers, the preferred orientation, or their mechanical properties. The electrical conductivity is greatly enhanced (*57*). These lamellar compounds exhibit metallic behavior and were therefore called "synthetic metals" (*2*). Even superconductivity has been discovered in the first-stage GIC with alkaline metals (*58*). The critical magnetic field seems to depend on the stoichiometry, and some of the results obtained (*58*) are controversial (*59,60*). Graphite–alkali metal intercalates appear to have comparable densities of electrons and holes, rather than being simple *n*-type metals as previously assumed (*61*). Evidence from several sources now confirms that charge transfer is exclusively a surface reaction and that the structure of an intercalation complex may reveal little of the formation mechanism (*15*).

When finely divided crystalline and suitably degassed graphite powder is treated with weighed quantities of K, Rb, or Cs in an evacuated vessel or under inert gas, GIC can be formed with compositions C_xM ($x = 8, 24, 36, 48$, and 60) (*62*). All of the compounds are quite reactive and pyrophoric, igniting in air and reacting explosively with water (*63*). These compounds are stable, however, *in vacuo* at ordinary temperatures, although they lose alkali metal as vapor at higher temperatures (*64*). The direct synthesis of such binary alkali GIC has been known for more than 50 years and was discovered by Weintraub and Fredenhagen (*65–69*).

There are several general ways to synthesize GIC (8,70):

1. Direct heating of known amounts of material under inert gas. The materials obtained by this method are suitable for organic synthesis (71).
2. Heating of graphite and alkali metal to a pressure just above the decomposition pressure of the stage desired.
3. Electrolysis of fused melts (70).
4. Utilization of alkali anion solutions.
5. Exchange from metal radical anions (but solvent molecules will probably cointercalate with the metal) (9).

To obtain exact stoichiometry the isobaric two-bulb synthesis is frequently employed (72–74). An evacuated and sealed system consisting of a bulb containing the graphite is connected to a bulb containing the potassium. The second bulb is then heated, vaporizing the potassium, which then reacts with the graphite. The reaction is terminated by quenching the system at a lower temperature (64).

Maintaining the graphite at a higher temperature than the intercalant avoids condensation of an excess onto the graphite. This system thus easily permits the determination of intercalation or deintercalation isotherms or isobars and also gives information on stage formation and hysteresis effects (9,74–76).

Graphite–alkali metal compounds are relatively simple compared to other intercalation compounds. Their stoichiometry is well defined (9), and the optical properties are those expected from the "alloying" of graphite and alkali metal (9). The C_8M compounds are bronze in reflection, whereas the second-stage compounds $C_{24}M$ are blue. The colors of the intercalation compounds are a useful guideline in monitoring both their preparaion and their subsequent chemistry.

C_8K, $C_{24}K$, C_8Rb, and C_8Cs can also be prepared in an ambient-temperature procedure by a metal transfer reaction in hydrocarbon solution. Cobalt complexes of the type [Co(olefin)(PR$_3$)$_3$], e.g., [Co(C$_2$H$_4$)(PMe$_3$)$_3$] (77), act as metal carriers, dissolving the metal and reversibly forming M[Co(olefin)(PR$_3$)]$_2$. In pentane, intercalates free of solvent are obtained, whereas benzene is intercalated too, yielding compounds M(benzene)$_y$C$_x$ (78). It was found, however, that intercalates prepared by this method are more reactive because of greater disorder in the lattice.

This procedure allows the convenient preparation of $C_n Li$ compounds (77,79,80) but presents unexpected difficulties due to rapid surface passivation of the Li metal. This problem of insufficient access of the carrier molecules can be overcome by using pure Li powder, which yields orange crystals of $Li[Co(C_2H_4)(PMe_3)_3]$. Use of a stoichiometric or a catalytic amount of this complex (in the latter case Li powder must be added) gives the corresponding intercalates. The catalytic mode was shown to be better suited for transfer reactions performed on a preparative scale (77).

The preparation of C_nM by this procedure requires less time for $M = Li$ than for $M = K$, Rb, or Cs, a fact that is certainly not related to the rates of the transfer reactions but rather to the relatively high mobility of Li in C_nLi. This method also seems to be applicable for sodium transfer. Preliminary findings indicate the formation of fairly high stages of C_nNa (77).

Potassium, rubidium, and cesium do not directly form stable M_2C_2 carbides, so the intercalation compounds are the stable species in the carbon–metal system. These metals are volatile, and the vapor pressure easily overcomes the intercalation threshold. In the case of Li, however, the stable species in the carbon–metal system are not the intercalation compounds but the acetylide Li_2C_2 (10,81). In principle a crystalline compound can be formed if the ionization potential is smaller than the electron affinity of graphite. As a consequence, lithium intercalates with a different stoichiometry (59).

Differences between the alkali metals in behavior toward graphite have been explained in terms of electron transfer (82), and this has led to the prediction that concentrated intercalation compounds of Li and Na should be less stable than other alkali metal GIC (62).

The difficulties encountered in the preparation of Li–graphite intercalation compounds are reflected by the fact that, although their synthesis was first described in 1965 (83), almost two decades later their use is still limited (77,84–87). Up to now compounds of the stoichiometry C_6Li (first stage), $C_{12}Li$ (second stage), and $C_{18}Li$ (third stage) (85) have been prepared, but there are also compounds of uncertain composition $C_{15-25}Li$ (86) of stage 2 or $C_{19-33}Li$ of stage 3 (86). The compound $C_{72}Li$ was described as being of eighth stage (85), and for compounds $C_{19-33}Li$ (86) stage 3 was postulated, whereas $C_{17-28}Li$ (87) was found to be of second and third stages, $C_{44}Li$ (86) of third and fourth stages, and $C_{\sim 40}Li$ of fourth stage (87).

These results are in part very conflicting. In addition, a detailed comparison of all these different species is difficult or impossible since they were prepared by different methods, including vapor-phase methods (87,88), powder-pressure methods (86), liquid-phase methods in Li or Li/Na melts at temperatures ranging between 150 and 350°C (87), and metal transfer reactions (77). In many cases lithium carbide, Li_2C_2, was present in the final reaction mixtures. Second-stage Li GIC (79) are not actually stable at room temperature, since they seem to change slowly to a mixture of graphite and the first-stage compound (89). An HOPG-derived second-stage $C_{12}Li$ compound was shown to be stable only in the presence of C_6Li (80). Higher-stage compounds are not stoichiometric (10,80,87).

The sodium GIC compound $C_{64}Na$ has been prepared by heating highly crystalline graphite with 3% sodium under continuous stirring in an atmosphere of pure helium (88,90). The sodium was found to lie in the interstitial planes separated by eight layers of carbons and to increase the interplanar graphite spacing to 4.6 Å. As pointed out by Boersma (62), it must be kept in mind that impurities such as H_2 or O_2 facilate the entry of the metals into the graphite (88,90), probably because their oxides or carbonates act as a spacer or catalyst (91,92). Reaction of graphite and sodium in the presence of nitrogen and hydrogen in a sealed reactor at 400°C allows the preparation of quaternary third-stage compounds $C_{150}Na_{14}N_7H_{14}$ (93).

Although sodium has a smaller tendency to intercalate, it is very easily adsorbed on the surface of amorphous charcoal or alumina to produce highly reactive forms of metals (94,95). Some reactions have been reported and compared to C_8K or potassium on charcoal (95).

Unlike cesium amalgam, which leads to pure Cs GIC, sodium amalgams cannot be used for the synthesis of sodium GIC because they are too stable to react with graphite (81). In addition, potassium and rubidium amalgams react with graphite to form pink first-stage ternaries $KHgC_4$ and $RbHgC_4$, respectively, and blue second-stage compounds K(or Rb)HgC_8 (96). The behavior of these alkali metal amalgams with graphite depends on the relative affinity of the alkali metal for the mercury and for the graphite; the latter increases from Na to Cs, and the former decreases. The high interplanar distance (10.2–10.8 Å) in the ternaries corresponds to a triply intercalated layer in which two sheets of alkali metal atom surround the mercury atoms (81). It has been shown by use of ^{13}C NMR (97) that there is a charge transfer from the alkali metal layers to the graphite and the mercury layers giving rise to a sequence $C^{(-)}M^{(+)}Hg^{(-)}M^{(+)}C^{(-)}M^{(+)}$ along the c axis.

The intercalation occurs in two steps. First, a quasi-selective intercalation with a continuous change of stage occurs, where only the alkali metal penetrates between the graphite planes. Second, a simultaneous cooperative intercalation of alkali metal and mercury takes place into the occupied interlayers, giving the ratio Hg:M = 2:1 (98,99).

D. Graphite Intercalation Compounds of Other Metals

Pure yellow first-stage compounds of Ba and Sr (99,100) with graphite were prepared by direct action of the metal vapor on graphite in metallic tubes sealed under vacuum (*99*). The heating of compressed mixtures of powders of the metal and the graphite appears to be less favorable. For the preparation of the Ca GIC it was important to keep the temperature at about 450°C, since application of higher temperatures led to the formation of CaC_2 (*99*).

By the same methods applied for the alkaline earth metals, the intercalation of lanthanides is possible. For the more volatile metals Sa, Eu, Tm, and Yb the first method is preferred, but for Ce, Pr, Nd, Sm, Eu, and Yb the second method seems more promising (*97,101*). In addition, the reaction of Eu metal and graphite in liquid ammonia (*102*) at −78°C for 4–24 h yields quaternary species having a wide range of stoichiometric phase stability, $Eu_{1-x}(NH_3)_xH_y[C_{4.6-7}]$, depending on the Eu concentration in solution (*103,104*). Other procedures for the synthesis of Eu-containing GIC either use the direct synthesis (*101*) or start from $EuCl_3$ (*105,106*). The direct intercalation of these metals needs a temperature low enough to avoid carburation yielding the interstitial carbides MC_x (for the lanthanide metals) or the acetylides MC_2 (M = Ca, Sr, Ba) and high enough that the vapor pressure overcomes the intercalation threshold (*81*).

E. Interaction of Graphite Intercalation Compounds with Solvents

It has been shown that *n*-donors, THF, DME, and 1,4-dioxane (*30,107,108*), as well as π-donors, benzene, toluene, etc., undiluted or in solutions of cyclohexane or of *n*-donors (*107,109–111*), are able to penetrate into the interlayer spacing (*112*).

In this way C_8K reacts with both THF and furan to yield ternary compounds. These ternarizations of the first-stage C_8K and second-stage $C_{24}K$ appear to occur by a filling up the available space rather than by formation of a stoichiometric compound (*113*). On heating to 60°C decomposition occurred, resulting in the formation of a complex mixture consisting of ethylene, hydrogen, ethane, *n*-butanol, and ethyl acetate (*114*). With THF two $C_{24}K(THF)_{2.2}$ compounds, both of first stage, and two $C_{24}K(THF)_{1.2}$ compounds, one of first and the other of second stage, have been characterized (*113*). In all cases the metal is lodged within cavities between the organic molecules, whose dimensions and orientations are solely responsible for the identity period parallel to the *c* axis (*115*).

Li-THF compounds were obtained from the binary C_nLi (*n* = 6, 12, 18). Whatever the starting GIC, the same first-stage structure is obtained. The identity period $I_c = 12$ Å corresponds to the $[Li(THF)_4]^+$ intercalates with excess Li^+ (*28*).

The capacity of aromatic hydrocarbons to couple in ternary GIC diminishes in the order benzene $>$ toluene $>$ o-xylene (116). In this way toluene and o-xylene are intercalated into K GIC derived from both HOPG and powdered graphite, with the intercalation into $C_{24}K$ (HOPG) being much less pronounced. In contrast to benzene, the intercalation of toluene and o-xylene is reversible (116). The action of solutions of conjugated aromatic hydrocarbons, e.g., phenanthrene or anthracene, in THF on the binary graphite–potassium conpounds C_nK ($n = 8, 24, 36,$ or 48) yields well-defined dark-colored ternary compounds; for instance, C_8K yields the first-stage compounds $C_{24}K(THF)_1$(phenanthrene) and a mixture of the second- and third-stage compounds $C_{48}K(THF)_1$(anthracene) and $C_{72}(THF)_1$(anthracene), respectively (30). It is interesting to note that aromatic compounds penetrate into the interlayer spacing even in cyclohexane solutions, although this solvent cannot penetrate ($107,112$).

The mechanism of the reduction of aromatic compounds by $C_{24}K$ has been investigated and found to proceed in several steps. Sorption of the aromatic compound occurs first, in competition with the sorption of DME or THF. Second, electron transfer takes place from the graphite layer to the aromatic molecule. Finally, there is desorption of the reaction products together with the counterions into the homogeneous phase (112). The high electron-donating ability of $C_{24}K$ is demonstrated by electron transfer to comparatively weak electron aceptors, e.g., biphenyl or α-methylstyrene (112).

With DME and aromatic molecules of different electron affinities $C_{24}K$ yields ternary compounds of the general formula $KC_x(DME)_2$, with x ranging between 27 and 34 (117). These compounds show a crystalline structure very similar to that of the compounds obtained by reduction of graphite by potassium naphthalenide ($30,118,119$). Molecules having high electron acceptor properties yield ternary GIC poor in potassium and with roughly ordered lamellar structure (117).

The interaction between potassium GIC and alcohols or diols in solvating and nonsolvating solvents seems to be different (120). In cyclohexane $C_x(K)_y(RO^-K^+)_z$ compounds are probably formed, whereas in benzene, THF, or DME $C_x(K)_y(RO^-K^+)_z \cdot (solvent)_{y''}$ compounds have been found. These compounds are not able to reduce naphthalene or biphenyl but they do reduce aromatic ketones and nitriles to the corresponding adducts (120). Primary and secondary amines are unreactive under these conditions (45). In addition, the interlamellar complexes of graphitic acid and n-pentanol or n-pentylamine with monoacids of 8 to 12 carbon atoms have been investigated (121). Complexes with pyridine and hexamethylphosphoramide have been described. (122).

III

SYNTHETIC APPLICATIONS

A. Alkali Metal Graphite Intercalation Compounds as Reagents

There seems to be a controversy about the value of C_8K in organic synthesis. Although there are a number of useful applications, Bergbreiter *et al.* (*45*) and McKillop and Young (*123*) regard C_8K as having only limited value as a reducing agent. In contrast to these authors, Umani-Ronchi *et al.* (*124*) reported that it is an useful reagent.

Although C_8K did not seem to offer any particular advantage over the alkali metal dispersion for alkylation of a number of cyclic ketones (*95*), its behavior toward nitriles and esters with an activated methylene group is quite distinctive. As shown in Fig. 5, treatment of aryl or alkyl nitriles **1** and esters **2** with C_8K and alkyl halide (educt : C_8K : alkyl halide = 1:2:2) at $-60°C$ in THF under argon gives monoalkylated products **3** and **4**,

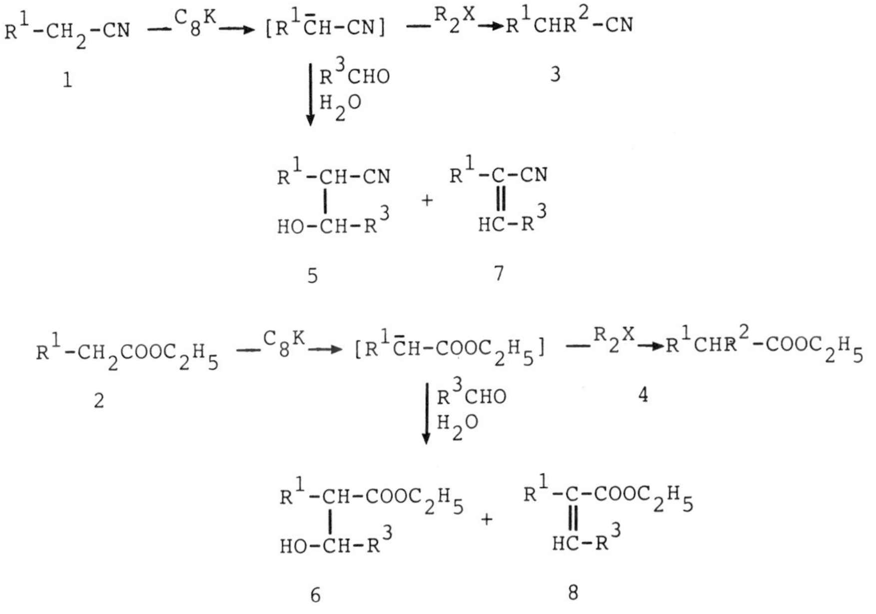

$$R^1-CH_2-CN \xrightarrow{C_8K} [R^1\bar{C}H-CN] \xrightarrow{R^2X} R^1CHR^2-CN$$

$$\mathbf{1} \qquad\qquad\qquad \Big\downarrow \begin{array}{l} R^3CHO \\ H_2O \end{array} \qquad\qquad \mathbf{3}$$

$$\begin{array}{ccc} R^1-CH-CN & & R^1-C-CN \\ | & + & \| \\ HO-CH-R^3 & & HC-R^3 \end{array}$$

$$\mathbf{5} \qquad\qquad\qquad \mathbf{7}$$

$$R^1-CH_2COOC_2H_5 \xrightarrow{C_8K} [R^1\bar{C}H-COOC_2H_5] \xrightarrow{R^2X} R^1CHR^2-COOC_2H_5$$

$$\mathbf{2} \qquad\qquad\qquad \Big\downarrow \begin{array}{l} R^3CHO \\ H_2O \end{array} \qquad\qquad \mathbf{4}$$

$$\begin{array}{ccc} R^1-CH-COOC_2H_5 & & R^1-C-COOC_2H_5 \\ | & + & \| \\ HO-CH-R^3 & & HC-R^3 \end{array}$$

$$\mathbf{6} \qquad\qquad\qquad \mathbf{8}$$

R^1 = H, alkyl, aryl, R^2 = alkyl, X = Cl, Br, I, R^3 = alkyl, aryl
Fig. 5. Alkylation reactions with C_8K (*125*).

respectively, with yields ranging from 40 to 70% (*125*). An advantage of this alkylation reaction is that practically pure monoalkylated products are obtained, whereas only small amounts (0–7%) of the dialkylated products could be isolated. Since Wurtz coupling reactions of the alkyl halides always accompany the alkylations, sufficient amounts of alkyl halide are required. Carbanions are the intermediates of the reaction between the nitriles 1 or esters 2 and C_8K and therefore addition of an aldehyde results in formation of β-hydroxy nitriles 5 or β-hydroxy esters 6 and the corresponding α,β-unsaturated derivatives (**7,8**).

The lower yields observed in the alkylation of nitriles reflect the occurrence of a side reaction, reductive decyanation to the corresponding alkane through an intermediate free radical. This reaction can be enhanced if the metallation of the nitrile is made reversible in the presence of a suitable proton source compatible with C_8K. Using oxalic acid (three equivalents) absorbed on silica (20 wt. %) in pentane, undecane is obtained from 1-undecyl cyanide after 1 at 25°C in 85% yield (*126*).

Interestingly, immediately after mixing of $C_{24}K$ with nitriles or ketones in DME or THF, a strongly exothermic reaction takes place and colored paramagnetic solutions are obtained, showing absorption bands for the corresponding anion radicals (*112*).

Metallation of imines 9 and of 2,4,4,6-tetramethyl-5,6-dihydro-1,3-oxazine 10 in THF at room temperature by C_8K followed by alkylation with a variety of alkyl halides results in good yields of the corresponding carbonyl compounds 11 and 12, respectively (Fig. 6) (*127*). Whereas reaction of the imines 9 can lead to either a ketone or an aldehyde (depending on the nature of R^2), the reaction of 10 leads only to

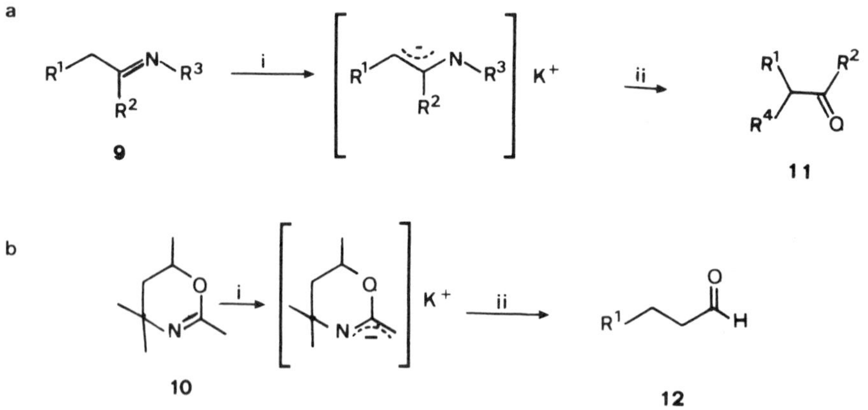

FIG. 6. Imines as starting materials for the synthesis of carbonyl compounds (*127*). (a) i, C_8K/THF; ii, R^4X then H^+/H_2O. (b) i, C_8K/THF; ii, R^1X, $NaBH_4$, then H^+/H_2O.

aldehydes, thus certainly offering interesting possibilities in the synthetic field (*127*).

The alkylation reaction is regioselective. Again, the Wurtz coupling of alkyl halides is a side reaction in THF. This reaction can be suppressed using hexane as a solvent, but in this case the yield of alkylated imine is lower (*127*). Another application in this field is the use of GIC in condensation reactions of carbonyl compounds, as exemplified in Fig. 7 (*128*).

Condensation of aldehydes (e.g., benzaldehyde **13**) with nitromethane **14** yields either the α-hydroxynitroalkane **15** or the corresponding α,β-unsaturated nitroolefins **16**, depending on the reaction conditions (Fig. 8).

In reactions requiring strongly basic conditions alkali metal-rich catalysts prepared from finely divided graphite and suspended in aromatic hydrocarbons gave the best results, but in mildly basic reactions it was usually the high-stage compounds, prepared from coarse graphite and suspended in aliphatic hydrocarbons, that gave the highest yields (*162*).

Usually treatment of alkyl chlorides results in dehalogenations to the corresponding alkanes, whereas alkyl iodides react with C_8K to give Wurtz coupled products. Alkyl bromides were reported to have reactivity patterns intermediate between those of alkyl chlorides and alkyl iodides (*45*). Alkyl sulfonate esters are cleaved to the corresponding alcohols in good yield under the same conditions. On treatment with C_8K, phenyl halides give mixtures of benzene, diphenyl, and phenylcyclohexane; benzyl halides yield dibenzyl; whereas chloroethylbenzene is dehydrogennated and subsequently polymerized to give polystyrene in 80% yield (*129*). Vicinal

$$CH_3-CH_2-O-C(=O)-CH_3 \longrightarrow CH_3-C(=O)-CH_2-C(=O)-O-CH_2-CH_3 \quad 96\ \%$$

$$CH_3-C(=O)-CH_3 \longrightarrow (CH_3)_2-C=CH-C(=O)-CH=C-(CH_3)_2 \quad 86\ \%$$

$$Phe-C(=O)-CH_3 + Phe-C(=O)H \longrightarrow Phe-CH=CH-C(=O)-Phe \quad 100\ \%$$

FIG. 7. Condensation reactions of carbonyl compounds with alkali metal GIC (*128*).

$$Phe-C(=O)H + CH_3NO_2$$
$$13 \qquad\qquad 14$$

93 %   Phe-CH=CH-NO_2
             15

69 %   Phe-CHOH-CH_2-NO_2
             16

FIG. 8. Nitro aldolizations promoted by alkali metal GIC (*128*).

dibromides are readily reduced to the corresponding olefins without further reduction of the carbon–carbon double bond (45,130). In THF as the solvent, partial reduction of the corresponding alkenes was observed (130). The reduction of vicinal dimesylates with C_8K seems to produce rather complicated mixtures of products, including the alkene as the major product. As side reactions, eliminations to form carbonyl compounds and formations of epoxides were reported (45); the latter may undergo partial reduction to alcohols. Intermediate ketones were either reduced to the corresponding alcohols or deprotonated. This behavior of ketones toward C_8K limits their alkylation reactions by C_8K/alkyl halide. The selectivity for monoalkylation is favored by hexane over THF as the solvent (95). For instance, the reaction of α-tetralone with allyl bromide in hexane gives a 98:2% ratio between mono- and dialkylated products, whereas in THF this ratio is 62:38% (95). One possible explanation is that the metal enolate remains adsorbed on the surface in hexane but is somewhat soluble in THF (95,131).

Other reactions mediated by the action of C_8K are reductive couplings. In the reaction of C_8K with benzophenone the distribution of mono-molecular (reduction to alcohol) and bimolecular (dimerization to pinacol) products depends on the C_8K/benzophenone molar ratio as well as on the solvent (132). The solvent dependence was found not to be due to polarity or to radical stabilization. Proton-donating solvents lead to monomolecular reduction, whereas dimerization occurs with non-proton-donating solvents (e.g., benzene) (131,132). Quenching of the reaction with methyl iodide results in the formation of 1-methoxy-1,2-diphenylethane (23). Although preliminary deuterium experiments in the case of cyclohexanone with high-surface sodium-on-carbon indicated that the proton sources for the reduction are the α protons and not the solvent (95), recent investigations (132) show that the proton participating in the reduction process is donated by the solvent, while the hydroxyl group protons of the pinacol are donated by the proton-quenching agent. Dilution experiments indicate that the reduction process does not occur in solution. This is supported by the reaction of ketones (e.g., acetophenone 17) with $C_{24}K$. This intercalation compound immediately changes color when exposed even at room temperature to the vapors of the ketone. $C_{24}K$ is reported (23) to induce self-condensations of carbonyl compounds. Extemely small quantities are needed to induce this coupling and therefore a number of side reactions are suppressed. The reaction of acetophenone 17 is rather complicated (Fig. 9) (23,131), giving rise to at least three products, 18, 19, and 20. At 40°C (17:$C_{24}K$ = 70:1) 40% of 19 is formed, whereas at 95°C (17:$C_{24}K$ = 90:1) 30% of 17 and 22% of 18 and 20 are formed. In comparison, reaction of 17 with sodium (4:1) yields 10–15% of 21 (23).

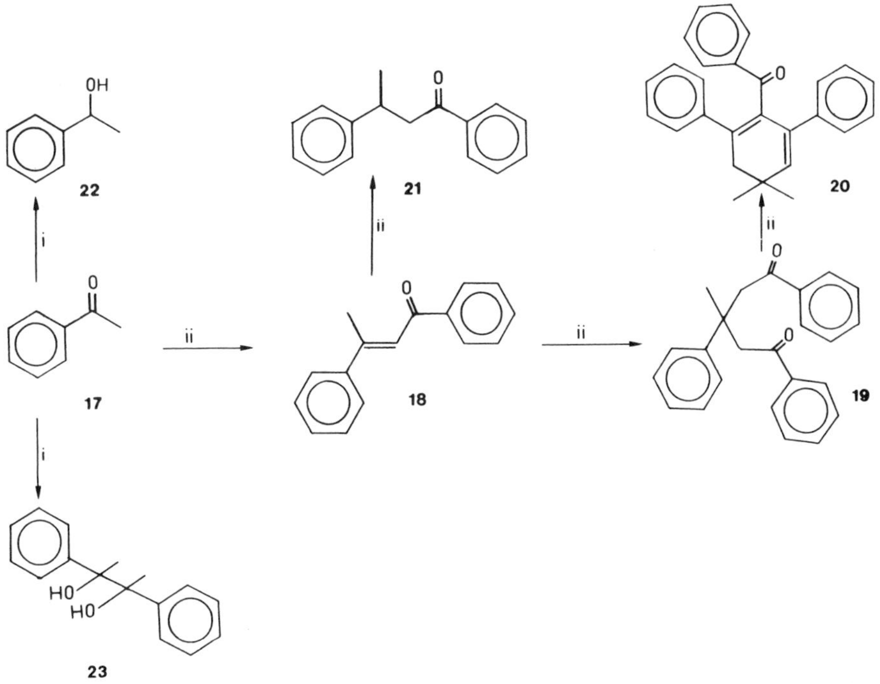

FIG. 9.   Reactions of acetophenone (**17**) with GIC C_8K and $C_{24}K$, respectively (*23,131*). i, C_8K/THF; ii, $C_{24}K/THF$.

The behavior of acetophenone **17** toward $C_{24}K$ was unexpected, since Lalancette *et al.* (*131*) found for the reaction with C_8K only a mixture of alcohol **22** (45%) and pinacol **23** (45%).

The conversion of **18** to **21** corresponds to a kind of heterogeneous Birch-type reaction. Early reports (*131*) of reducing an α,β-unsaturated ketone have now been generalized (*124*). α,β-Unsaturated ketones **34**, carboxylic acids **25**, and Schiff bases **26** can be conveniently reduced to the corresponding saturated compounds **27–29** (Fig. 10). The reduction is performed under mild conditions simply by treating the compounds in THF solution with C_8K at room temperature, or at 55°C in the case of α,β-unsaturated acids **25**. This method fails with α,β-unsaturated esters and yields dimerization products. In the case of α,β-unsaturated ketones the best results were obtained using a 5:1 mixture of THF and hexamethyl-disilazane as the solvent. Although the effect of this additive is not clearly understood, it decreases the amounts of by-products (*124*).

Contrary to the findings of Bergbreiter for C_8K (*45*) and Lalancette (*131*) for a $C_{24}K/C_8K$ mixture, Umani-Ronchi *et al.* (*124*) reported good yields (92%) for the reduction of stilbene to 1,2-diphenylethane by C_8K.

RENÉ CSUK et al.

$R^1-CH=CH-COR^2 \longrightarrow R^1-CH_2-CH_2-C(O)-R^2$ 10-30 min   57-85%   25°C

    24   27

$R^3-CH=CH-COOH \longrightarrow R^3-CH_2-CH_2-COOH$ 45-90 min   85-91%   55°C

    25   28

$R^4-CH=N-R^5 \longrightarrow R^4-CH_2-NH-R^5$ 30 min   83-92%   25°C

    26   29

R^1 = alkyl, aryl, R^2 = alkyl, R^3 = alkyl, aryl, carboxyl,

R^4 = aryl, R^5 = alkyl, aryl (compound 30). [124, 23]

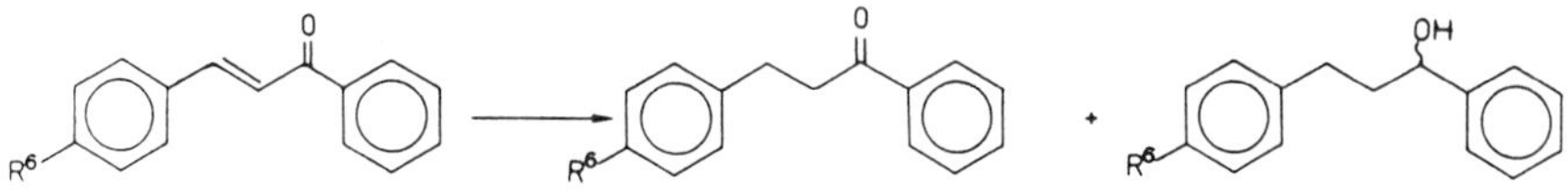

        30   31   32

            R^6 = p-OMe   9 %   91 %

                p-Me   21 %   79 %

                H   34 %   66 %

FIG. 10.   Reactions of α,β-unsaturated carbonyl compounds with C_8K (23,124).

The latter work could not be reproduced in our hands. Even after 3 h only 30% of 1,2-diphenylethane could be isolated (134).

A further application of graphite–alkali metals in organic synthesis is their use in the reductive cleavage of carbon–sulfur bonds of allyl (135) and vinyl (136) sulfones (33, 34) (Fig. 11). This allows a convenient preparation of 2-alkenes 35 either by alkylation of allyl phenyl sulfone (135) or by a Horner–Wittig coupling between phenylsulfonyl methyl phosphonates and carbonyl compounds (136) followed by cleavage of the carbon–sulfur bond with potassium graphite.

All desulfuration methods, however, have the disadvantage of inducing partial $(Z)/(E)$ isomerization. The ratio of (Z) to (E) isomers was found to be temperature-dependent, the (E) isomer being the major product (135,136). In addition, Umani-Ronchi et al. (126) reported that the reaction of 1-arylsulfonyl-2-alkanones, as exemplified by 1-benzene-sulfonyl-5-hydroxypentan-2-one (36), with a huge excess of C_8K in THF at

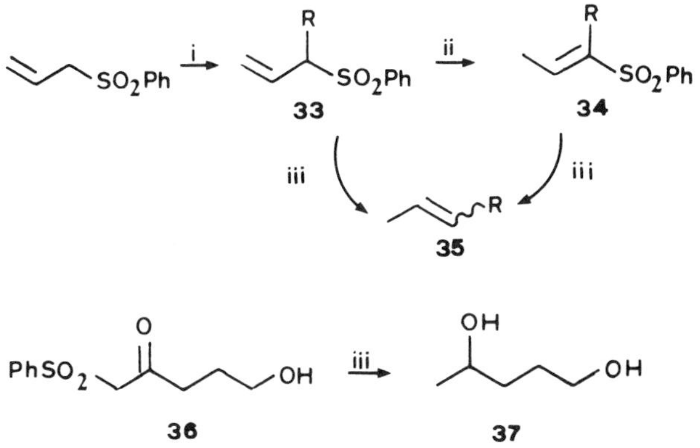

FIG. 11. Use of C_8K for reductive cleavage of carbon–sulfur bonds (135,136). i, BuLi then RX; ii, tert-BuOK; iii, C_8K/THF.

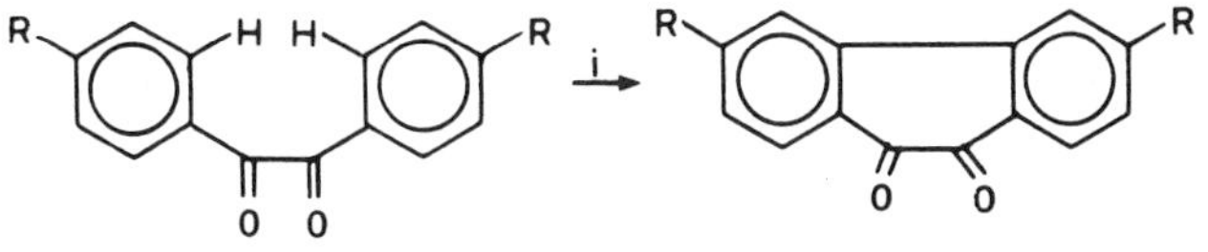

FIG. 12. Synthesis of phenanthrenequinone derivatives (137). Yields were 70% for R = H and 72% for R = Me. i, C_8K/THF.

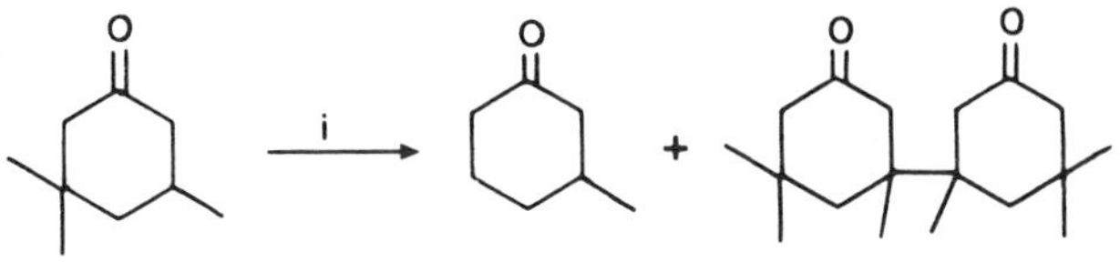

FIG. 13. C–C bond scission (41) and C–C bond formation (42) mediated by C_8K (131). i, C_8K/THF.

25°C reduces the carbonyl group and cleaves the carbon–sulfur bond. Formation of C–C bonds was observed (137) in the reaction of benzil 38, R = H or p,p'-dimethylbenzil 38, R = Me with C_8K in the THF to give excellent yields of the corresponding phenanthrenequinone derivatives (39), whereas the same reaction performed with dispersed potassium gave only traces of products (Fig. 12).

Reaction of 3,3,5-trimethylcyclohexanone (40) with C_8K in THF resulted in both C–C bond formation and C–C bond scission (Fig. 13) (131).

$$2 \; [M(CO)_6] \; + \; 2 \; C_8K \; \longrightarrow \; K_2[M_2(CO)_{10}] \; + \; CO \; + \; C$$

     43   45

with M = Cr, Mo, W

$$2 \; [Fe(CO)_5] \; + \; 2 \; C_8K \; \longrightarrow \; K_2[Fe_2(CO)_8] \; + \; 2 \; CO \; + \; C$$

     44   46

FIG. 14.   Synthesis of binuclear metal carbonyl compounds (*139*).

$$NiX_2[P(OR)_3]_2 \; + \; 2 \; P(OR)_3 \; + \; 2 \; C_8K \qquad Ni[P(OR)_3]_4 \; + \; 2 \; KI \; + \; C$$

     47   48

with X = Cl, Br, I; R = alkyl.

FIG. 15.   Synthesis of Ni(O) complexes by use of C_8K (*140*).

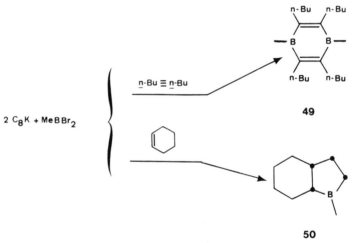

FIG. 16.   Synthesis of boron heterocycles by use of the methylborylene generating system $C_8K/MeBBr_2$ (*141,142*).

In addition, C_8K has been used to prepare the selective reductant for α,β-unsaturated carbonyl compounds, namely monopotassium decacarbonyldichromate, $KHCr_2(CO)_{10}$ from hexacarbonyl chromium (*138*) in THF at 40°C. The reaction between C_8K and Cr, Mo, W, and Fe carbonyls (**43, 44**) provides a facile synthesis of binuclear carbonyls (**45, 46**), according to Fig. 14 (*139*).

     Tetrakis(trialkylphosphite) nickel(0) complexes (**48**) are conveniently prepared from nickel(II) complexes (**47**) and C_8K (Fig. 15) (*140*).

C_8K was used in the synthesis of 1,4-dimethyl-2,3,5,6-tetra-n-alkyl-1,4-diboracyclohexa-2,5-dienes (**49**) from the corresponding di-n-alkylacetylenes with $C_8K/MeBBr_2$, perhaps through the intermediate methylborylene (MeB:) (Fig. 16) (*141*).

The reaction of this system with cyclohexene and other cyclic olefins provides in fairly good yield 2-methyl-2-boratricyclo[7.4.0.0^{3,8}] tridecane (**50**) and analogs thereof, whereas acyclic olefins and conjugated dienes give only the corresponding haloborination products (*142*).

B. Alkali Metal Graphite Intercalation Compounds as Catalysts

1. Hydrogenations

Hydrogenation of ethylene, methylacetylene, 1-butene, and butadiene over potassium–graphite could be readily achieved with GIC (*143*). Selective hydrogenation takes place with dimethylacetylene, rapidly yielding the corresponding alkenes. In the temperature range 70–115°C the hydrogenation of ethylene was found to follow first-order kinetics. $C_{36}K$ proved to be better than $C_{24}K$, which in turn showed more activity than $C_8K_{0.67}$ (*62*). Benzene and toluene could be hydrogenated over C_8K at hydrogen pressures of 100–150 bar, whereas hardly any reaction occurred below a pressure of 10 bar (*62,144*). For the hydrogenation of ethylene it was established that the specific activity of the second-stage compounds is much larger than that of their first-stage counterparts (*145*). Hydrogenation of an activated ethylenic bond seems to be more efficient, as exemplified by the hydrogenation of (E)-stilbene to 1,2-diphenylethane in 90% yield (*127*).

Alkylbenzenes could be hydrogenated to the corresponding cyclohexane derivatives, but dealkylation also took place to yield cyclohexane and methane as by-products (*144*). Butadiene was first hydrogenated rapidly to a mixture of (Z)- and (E)-2-butenes and then to butane (*143*).

2. Isomerizations and Dimerizations

The catalytic behavior of alkali metal GIC in isomerization reactions (and other reactions) can be correlated with structural properties of these compounds. Often the intrinsic activities of second-stage compounds are larger than those of their first-stage counterparts. This is due to the much greater mobility in the former compounds (*145*).

1-Pentene is easily isomerized to (Z)- and (E)-2-pentene with C_8K, $C_{24}K$, $C_{36}K$, and $C_{60}K$ at all temperatures between 125 and 235°C (*64*). For

C_8K an optimum temperature of 180°C was found. The time-dependent deactivation of the catalyst generally increases as the potassium content of the compounds decreases. The deactivation seems to be due to the presence of oxygen and water vapor in the pentene feed, decomposition of the catalyst at higher temperatures, physical adsorption of educt, and the formation of strongly adsorbed polymeric material at low temperatures (*62,64,146*).

The isomerization of 1-butene was found to occur by a base-catalyzed mechanism, with the initial amounts of (Z)-2-butene being significantly larger than those of the (E) isomer (*147*). The initial abstraction of an allylic hydrogen atom in the rate-determining step results in the generation of a π-allylic intermediate that coordinates to an electron-deficient site, the alkali metal species. The hydrogen atom is adsorbed on an electron-rich site, which is provided by the aromatic system (*64,147,148*). The more electron-rich $C_{24}M$ compounds are more active than the corresponding C_8M laminates since they effect the abstraction of the allylic hydrogen atom more readily and thus increase the rate of formation of the π-allylic intermediate (*145,149*). But the apparent catalytic activity is lower than that of either potassium/charcoal or potassium/SiO_2.

The effect of oxygen exposure on catalytic activity is controversial. Alkali metal GIC are partially poisoned by exposure to oxygen but a slight increase in catalytic activity (approximately 10%) occurred with C_8K, $C_{24}K$, and C_8Rb following exposure to small quantities of O_2 (smaller than 0.001 mol/mol); however, virtually complete deactivation was observed when the compounds were exposed to larger amounts of oxygen (*39,145*).

For C_8Cs, $C_{24}Cs$, and $C_{24}Rb$ substantial increases were measured even when relatively large amounts of oxygen (0.25 mol/mol) were employed. In addition, the initial $(Z)/(E)$ product ratios determined over the O_2-exposed alkali metal GIC are much larger than those over the corresponding non-oxygen-treated compounds (*39,145*). The increase of activity seems to be due to trace amounts of alkali oxide species, which are relatively bulky and obviously located along the edges of the graphitic planes, at the same positions as the alkali metal species from which they were derived. This increases the interlayer separation distance and allows facile reaction of additional alkali metal species with the gaseous reactant molecules, which are normally unable to penetrate deeply into the GIC because of the limited interlayer spacings (*145*).

In addition, when the amount of heat released by the exothermic reaction between the alkali metals and the oxygen is large enough to overcome the interaction between the metal layers and the graphitic planes, alkali metal atoms will probably migrate out and become dispersed over the exterior of the graphite particles, giving rise to the formation of a higher-stage intercalation compound (39).

The higher $(Z)/(E)$ ratio is explained in terms of an increase in formation of the anti-π-allylic intermediate at the deeper catalytic sites that are now additionally available. This anti-π-allylic intermediate, which is the precursor of the (Z) product, is therefore formed at the expense of the bulkier syn-π-allylic species, the precursor of the (E) product (145).

The dimerization of propylene with different potassium–graphite compounds as well as with $C_{64}Na$ yielded several different products (150,151). The product distribution was found to depend on the type of catalyst, the ash content of the graphite (152,153), and the temperature applied (150,151). For the dimerization of isobutene with C_xK it was found that for x larger than 24 solvents improve the yield, whereas for x smaller than 24 solvents sometimes inhibit the reaction (154).

With alkali metal catalysts, lower olefins commonly give a variety of products, predominantly dimers and codimers (155).

The olefinic isomers produced are usually not those that are most stable thermodynamically (156). It appears that potassium–graphite-derived catalysts are poor isomerization catalysts unless inorganic impurities are present (155).

2-Alkynes are isomerized to 1-alkynes at 80°C but it is not clear whether the reaction takes place inside or outside the graphite (131). (Z)-Stilbene is isomerized to (E)-stilbene at room temperature (45,136,157).

3. Alkylations

C_8K in toluene, benzene, or isopropylbenzene can act as an alkylation catalyst in the presence of ethylene to yield the corresponding nuclear and side chain-alkylated aromatic hydrocarbons (158,159). No reaction is observed with isobutene/benzene/C_8K even at 200°C and higher pressure (158), whereas toluene gives 50% conversion to 3-phenylpentane. This finding is in contrast to the reactivity of C_8Cs, which supports the formation of tert-butylbenzenes under these conditions (62).

Use of intercalates for alkylation reactions results in a change in the distribution of products from that obtained with metal dispersions. On reaction with ethylene/C_8K/$C_{24}K$, 2-pentenes give 57% of a mixture of linear 3-heptenes and 37% of 3-ethyl-1-pentene, whereas with metal dispersions under similar conditions only 15% of 3-heptenes but 54% of 3-ethyl-1-pentene are formed (155). In this study the rate of the reaction was proportional not to the amount of potassium but to the amount of C_nK. Again, the ash content seems to influence the rate of isomerization of the product olefin, but substantial isomerization of the products occurs when the ash content is approximately 2.4% (*155*).

In most cases it is very difficult to make any predictions about the behavior and reactivity of such systems. Benzene gave no reaction with C_8K at lower temperatures, although with $C_{24}K$ biphenyl was formed (*157,160*). At elevated temperatures, however, formation of several products, including biphenyl (61%), 3-phenylcyclohexene (12%), 1-phenylcyclohexene (11%), 2-phenylcyclohexa-1,3-diene (11%), phenyl-cyclohexane (2%), 3-phenylcyclohexa-1,4-diene (2%), and 4-phenyl-cyclohexane (1%), was reported for the benzene/C_8K system (*110*). Interestingly, no compounds were formed in which both rings had been reduced. On the other hand, benzene vapor reaction on $C_{24}K(THF)_{1.2}$ (*113*) yields intercalated benzene molecules (*161*). For the quaternary compound $KC_{24}(THF)_{1.2}(Bz)_1$ the benzene intercalation was shown to be reversible (*161*). In addition, 1,4- and 1,3-cyclohexadiene are converted partially into benzene, whereas cyclohexene gave no reaction. C_6F_6 gave a highly exothermic and instantaneous reaction leading to the formation of KF (*155*).

4. *Polymerization*

Graphite intercalation compounds can act as mild anionic polymerization initiators (*71,162,163*), offering some benefits in comparison to analogous reactions in homogeneous medium or to metal dispersions. The living polymer end is protected between the graphitic planes (*164*), which allows increased tacticity and slows down some secondary reactions. Normally the living polymer ends can be killed very easily by trace amounts of impurities such as Lewis acids or water. Coordination of the monomers with one or two graphite layers (in addition to diffusion phenomena) can take place, thus permitting a selection of one monomer among several for a specific copolymerization (*162,164,165*). Even stereospecific homopolymerizations of a given monomer can be achieved (*162*). However, the reaction is normally slower, as shown for the polymerization of styrene (*74,163,165–167*), methyl methacrylate (*168*), isoprene (*165,169*), α-methylstyrene, or ϵ-caprolactam (*162*). Some of the effects found for styrene are controversial, since other investigations found no absorptions for carbanionic ends of "living polystyrene" (*112*), whereas α-methylstyrene was found to yield carbanions on contact with $C_{24}K$ in THF or DME (*112*). In addition, polymerizations of different aldehydes (*168*), e.g., methacryl-, croton-, and acetaldehyde (*170*), ethylene (*71*), and cyclic siloxanes (*124*) have been reported.

Because the polymerizations occur in the interlayer spacing the rates of reactions are reduced, corresponding to decreased penetration of the monomers with smaller intercalated atoms. Styrene (*74,166,167*), aldehydes (*166,168,171*) 1,2-epoxy compounds (e.g., ethylene oxide) (*172*), and lactones [e.g., ε-caprolactone (*173*), γ-propiolactone (*174*), pivalolactone (*173*), and γ-valerolactone (*173*)] all show strong interaction with the laminate and high reaction rates, resulting in swelling of the catalyst and finally in exfoliation and consequently deterioration of the catalyst's activity by the polymer growing in the interlayer spacing (*174*). The defective layered structures are filled with large quantities of macromolecular gel, and the amount of solvent in these regions is very small (*175*). The fast deformation in the edge area of the *c* surfaces gives the plates a form known as an ashtray (*176*) or edge effect (*177*), which is due to lower deformation resistance in the regions near the edges of the plate.

Some first-stage graphite–alkali metal laminates are normally not able to initiate such polymerizations because the close packing of the crystalline structure makes monomer penetration very difficult (*162*). Because of their poor reducing power, alkaline earth laminates (e.g., C_6Ba or $C_{12}Ba$) cannot initiate polymerization.

The rate of reaction, yields, and structure of the products formed are influenced mainly (*165,178*) by the graphite, the ratio of initiator to monomer, the solvent, and the temperature. The latter normally increases the percentage of polymer formed since the polymerizations are thermally activated, but the initiator's efficiency depends on the rate at which the graphite separates under the conjugated action of temperature and polymer growth (*165*). In contrast to this temperature dependence, no significant reduction of the polymerization rate with decreasing temperature (50 to -2°C) was found for ε-caprolactam polymerization mediated by $C_{24}K$ (*175*). This reaction proceeds very quickly and quantitatively. Neither monomer concentration, initiator concentration, nor change in the solvent led to significant changes in the reaction rate (*175*).

5. Fischer–Tropsch Synthesis and Formation of Ammonia

Of special interest are applications to the Fischer–Tropsch synthesis (*179,180*) and the synthesis of ammonia. It might be expected that the same principles would be observed during both syntheses, since CO is isoelectronic with N_2 (*179*). Graphite–alkali metal intercalates interact directly with H_2 to varying degrees, ranging from chemical reactions of

C_8K to form $C_8KH_{0.67}$ (*181*) to the physical interaction of MC_{24} (M = K, Rb, Cs) to form compounds such as MH_4C_{24} (*7,182*).

Fischer–Tropsch synthesis, the catalytic hydrogenation of carbon monoxide yielding C_1-C_4 hydrocarbons (*183*), takes place with sodium– or potassium–graphite laminates (*180,184*). The principal products are methane and ethane, but propane and butane are also formed in yields depending on the kind of catalyst used. The yields were low but could be improved by replacing the graphite–alkali metal laminate by a graphite–alkali metal–transition metal chloride complex (*179,185*). In the presence of Co, Fe, Mn, Os, or Ru the principal reaction product is CH_4 (*179,186*), while in the case of Pt, Cu, or Pd ethane is the major product and propane and butane are formed as by-products (*179,187,188*). Nickel and chromium have an intermediate position.

Using CO_2 instead of CO results in lower yields, although the same products are formed (*62*). Use of graphite–Na–$PdCl_2$ resulted in a transformation according to the reaction $CO_2 + H_2 \rightarrow$ Me–O–Me (*189*). Exposure of C_8K, $C_{24}K$, and C_8Rb to CO at temperature higher than 200°C causes extensive decomposition of the intercalates involving the expulsion of interplanar alkali metal species. However, $C_{24}Rb$, $C_{36}Rb$, C_8Cs, and $C_{24}Cs$, because of their stronger metal-to-graphite bonding, are not extensively decomposed and as a result produce primarily methane during the Fischer–Tropsch synthesis (*145*). However, all alkali metal–graphite laminates show only small activities with low initial rates followed by a rapid increase to a maximum and then a continuous decrease. The differences in maximum activity between first- and second-stage compounds seem not to be related to their respective abilities to form ternary compounds with hydrogen (*145*). Experiments with [13]CO confirmed that all hydrocarbons were derived from the conversion of CO and not via direct hydrogenation of carbon of the graphite. Neither H_2O, CO_2, nor unsaturated hydrocarbons were among the gaseous products (*190*). The latter finding is attributed to the catalytic activity of the catalysts for olefin hydrogenation (*145*).

Catalytic formation of ammonia was again, as in the Fischer–Tropsch synthesis, markedly improved by adding a transition metal chloride complex (*62*). Introduction of potassium into this system results in the appearance of activity, which grows as the K:Fe ratio increases and passes through a maximum at a K:Fe value of 6 to 7. The best result was obtained, however, with an $OsCl_3-FeCl_3-K$ (27:2:1) complex (*186,191*). Again, as in the Fischer–Tropsch synthesis and H_2-D_2 equilibrium reactions (*192*), O_2 and CO adsorbed strongly on the catalysts and lowered their activity. For the H_2-D_2 equilibrium reactions no dependence of activity on the stage number could be established (*192*). On heating these graphite–iron–potassium catalysts with nitrogen or a nitrogen/hydrogen mixture, considerable amounts of cyanide are formed (*193*).

In addition, a 3:1 mixture of H_2 and CO is converted with about 20% efficiency and 95% selectivity into acetylene at 100°C over a catalyst generated by partially deintercalating a first-stage GIC C_9FeCl_3 with potassium naphthalenide. When this complex is reduced with other agents, distinctly different catalysts are obtained (*194*). This behavior agrees with results published for other similar catalysts (*195*).

In studies of sorption of gases by alkali metal GIC, some "molecular sieving effects" have been established. With C_8M (M = K, Rb, Cs) there was no sorption of He, H_2, D_2, Ne, N_2, Ar, or CH_4, whereas $C_{24}K$ sorbed H_2, D_2, and small amounts of N_2, but not Ar or CH_4 (*196*). Other studies, however, showed that hydrogen (*197*) and both methane and D_2 are adsorbed (*198*). On saturation at 300°C compounds $C_8KD_{0.7}$ and $C_{24}KD_{0.15}$ have been obtained. Treatment of the GIC with a mixture of methane and D_2 led to the reversible formation of deuterated methanes (*198*). By means of neutron diffraction techniques it was confirmed that these sorptions are due to the occlusion of the gas molecules between the intercalated carbon layers. The sorption characteristics of $C_{24}M$ (M = K, Rb, Cs) are determined mainly by the width of the interlayer vacancy. For C_8M compounds, however, the gas molecules may be sterically hindered from entering the interlayer space (*196*).

C. Graphite Surface Compounds

Seelig pointed out (9) that intercalation compounds of graphite perform interesting and somewhat unexpected chemistry (*199*). There is now good evidence demonstrating the role of these compounds in organic and organometallic chemistry. Although most of these reagents are surface rather than intercalation compounds, true intercalation compounds serve as unique precursors to reagents of outstanding value.

1. Zinc–Graphite

Zinc–graphite (Zn–gr) is easily prepared from C_8K and anhydrous $ZnCl_2$ in refluxing THF under argon according to the equation $2C_8K + ZnCl_2 \rightarrow$ "$C_{16}Zn$" (=Zn–gr) + 2KCl. Although preliminary reports (*200–202*) gave evidence for the existence of lamelar zinc–graphite compounds of stages 1, 3, and 4, later work by the same groups suggested a dispersion of zinc on the graphite surface with only very weak reflections due to intercalated species (*203,204*).

Preliminary investigations by high-resolution electron microscopy give definite evidence that for the Zn–gr reagent most of the zinc is located on the surface of the graphite. This seems to be true for many other metal–graphite reagents (including Mg or Ti) prepared in the same manner by reduction of the corresponding metal salts with C_8K in THF, but different surface structures could be established for each metal (205).

The high activity of this metal-on-graphite dispersion allows quantitative preparation of organozinc reagents by treatment with α-bromoesters at 0°C and with allylic bromides at 20°C. These organozinc reagents have been successfully exploited in Reformatsky reactions (Figs. 17a and 17b) and in the synthesis of homoallylic alcohols (Fig. 17c). Side reactions usually observed in the Reformatsky reaction, e.g., self-condensation of the α-bromoester or carbonyl compound and elimination or retroaldolization of the intermediate α-bromozinc oxyester, are almost completely suppressed. In the reactions of allylic zinc bromides complete allylic rearrangement usually takes place.

Other syntheses using Zn–gr include the reactions of trimethylsilyl α-(bromomethyl) acrylates at 20°C with cyclic ketones to give the corre-

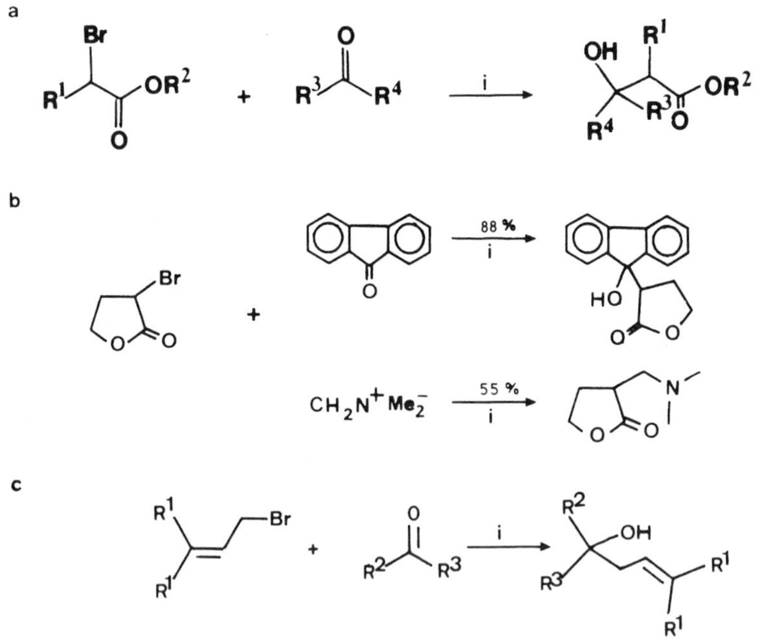

FIG. 17.   (a) Reformatsky and (b and c) Reformatsky-type reactions using Zn–gr (203). (a) i, Zn–gr/THF, 0°C, 86–90% with R^1 = H, Me; R^2 = Et, SiMe$_3$; R^4 = H, alkyl. (b) i, Zn–gr/THF. (c) i, Zn–gr/THF, R^1 = R^2 = R^3 = alkyl, 93–94%.

sponding γ-substituted α-methylene-γ-butyrolactones (*203*); the synthesis of a known cyclopentanone, 2-(6-methoxycarbonylhexyl)-cyclopent-2-en-1-one (**52**), a useful intermediate for prostaglandin synthesis by an acylation reaction (*126,206*); and the monoalkylation of γ-butyrolactone (**51**) by a regioselective palladium-assisted substitution reaction with an allylic ester yielding (**53**) (Fig. 18) (*126*).

Reaction of the α-bromoester-derived Reformatsky reagents (**54**) with (*E*) or (*Z*) configurated allyl acetates (**55**) in palladium-catalyzed substitution reactions yield predominantly (*E*) products (**56, 57**) (Fig. 19). The method of preparing the Reformatsky reagents does not determine the course of the reaction, but the source of the graphite greatly influences the quality of the organozinc species (*204*).

Finally, zinc–graphite has been used for the stereoselective reduction of alkynols in refluxing absolute ethanol. Preliminary results show an excellent (Z):(E) ratio (99:1) and good yields (*126*).

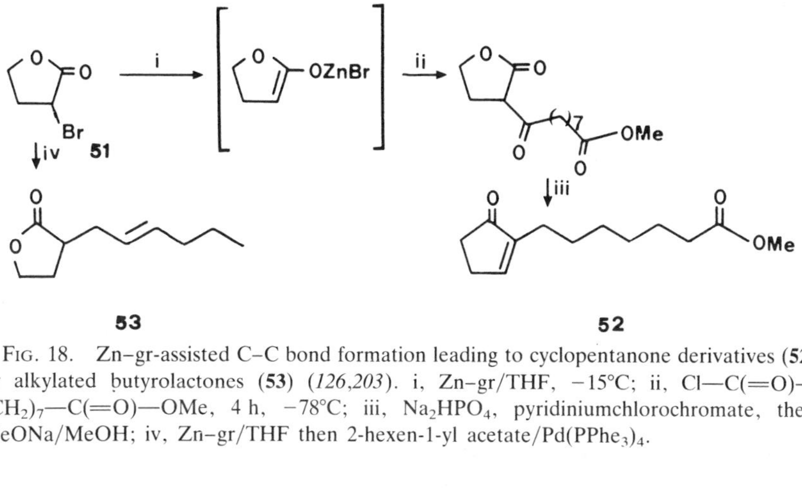

53   **52**

FIG. 18. Zn–gr-assisted C–C bond formation leading to cyclopentanone derivatives (**52**) or alkylated butyrolactones (**53**) (*126,203*). i, Zn–gr/THF, −15°C; ii, Cl—C(=O)—(CH$_2$)$_7$—C(=O)—OMe, 4 h, −78°C; iii, Na$_2$HPO$_4$, pyridiniumchlorochromate, then MeONa/MeOH; iv, Zn–gr/THF then 2-hexen-1-yl acetate/Pd(PPhe$_3$)$_4$.

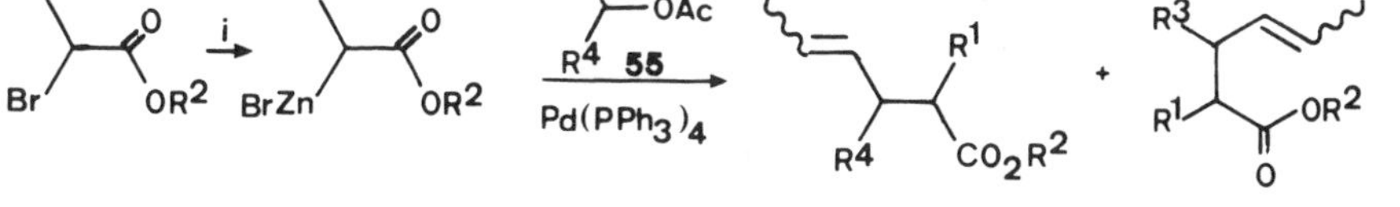

54   **56**   **57**

FIG. 19. Reactions of Zn–gr-derived Reformatsky reagent (**54**) with allyl acetates (**55**) (*204*). i, Zn–gr/THF.

2. Zinc/Silver–Graphite

Although use of Zn–gr now allows Reformatsky reactions to proceed more efficiently, their scope is still limited. To extend this reaction to a wider range of α-halogenoalkanoates and to diastereospecific reactions of cyclic ketones employing, e.g., zinc ethylacetate enolate, zinc of extraordinarily high activity is required. Such activity can be achieved by using a zinc–silver couple, which provides a degree of metal activation far superior to any method previously known (203,207,208).

The zinc/silver–graphite (Zn/Ag–gr) reagent obtained from equimolar amounts of C_8K and $ZnCl_2/AgOAc$ (molar ratio 0.1) (209) allows Reformatsky reactions to proceed even at temperatures as low as −78°C with high (82–92%) yields of products. The essential differences in reactivity between α-bromo- (58, 59) and α-chloroalkanoates (60, 61) previously encountered turned out to be insignificant (Fig. 20).

Chiral cyclic ketones (62, 63) are subject to both diastereospecific Reformatsky reactions (209) (products 64, 65) and Reformatsky-type reactions (products 66, 67) (Fig. 21) (210).

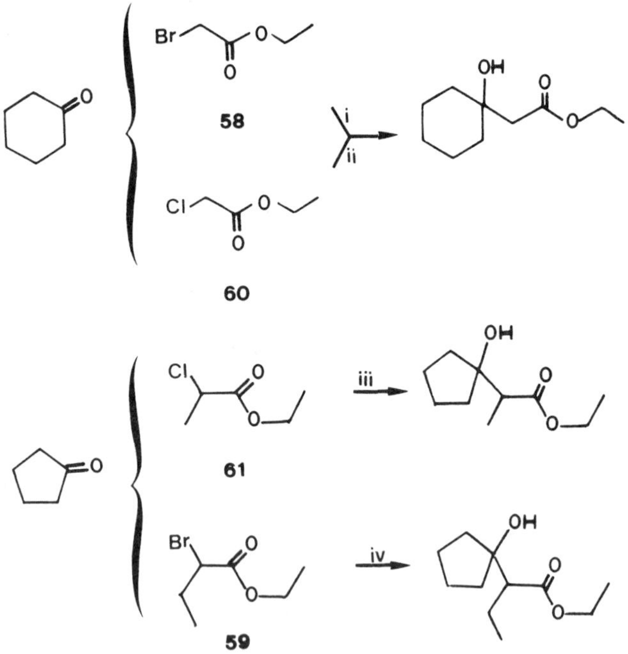

FIG. 20.   Reformatsky reaction Zn/Ag–gr (209). i, Zn/Ag–gr/THF, −78°C, 20 min, 92%; ii, Zn/Ag–gr/THF, −20°C, 10 min, 86%; iii, Zn/Ag–gr/THF, 0°C, 10 min, 82%; iv, Zn/Ag–gr/THF, −10°C, 20 min, 76%.

The high activity of the Zn/Ag–gr reagent therefore permits an extension of the Fischer–Zach glycal synthesis (*211*), the formation of 1,4-(1,5)-anhydro-2-deoxypent(hex)-1-enitols, to aprotic conditions as well as to suitable protected glycosyl halides (**68, 69, 70**) (Fig. 22) (*212,213*). It is of interest to note that Rieke zinc (*207,208*), generally considered to be highly reactive, was found to be unsuitable for glycal formation (products **71–74**) (*212*). Interestingly, **74** was found to be thermodynamically more stable than **73** (*213*).

In addition, "Umpolung" of the anomeric center of carbohydrates (**75**) can be achieved, allowing trapping of the intermediate glycosyl zinc species (**76**) with benzaldehyde (**77**) (Fig. 23) (*214*). This offers a promising route to the synthesis of C-glycosides (**78**) (*215*).

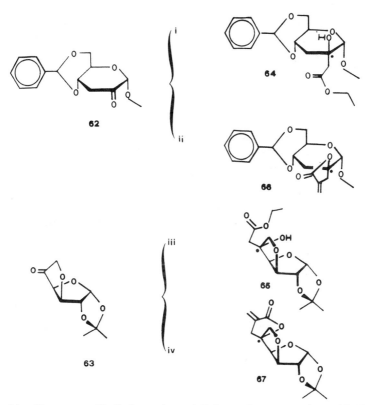

FIG. 21. Diastereospecific Reformatsky and Reformatsky-type reactions with Zn/Ag–gr/THF (*209,210*), i, Zn/Ag–gr/THF then Br—CH₂—C(=O)—OEt, −78°C, 10 min, 90%; ii, Zn/Ag–gr/THF then Br—CH₂—C(=CH₂)—C(=O)—OEt, −78°C, 10 min, 78%; iii, Zn/Ag–gr/THF then Br—CH₂—C(=O)—OEt, −78°C, 10 min, 89%; iv, Zn/Ag–gr/THF then Br—CH₂—C(=CH₂)—C(=O)—OEt, −78°C, 10 min, 92%.

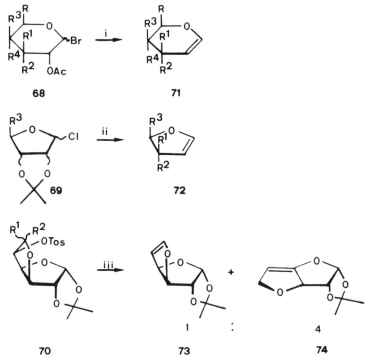

FIG. 22. Synthesis of pyranoid and furanoid glycals from glycosyl halides (*212,213*). i, R^1, R^2 = H, OAc; R^3, R^4 = H, OAc; R^5 = H, CH$_2$OAc, Zn/Ag–gr/THF, −20°C to room temperature, 10–20 min, 83–96%. ii, R^1, R^2 = H, OH; R^3 = CH$_2$—O—CH$_2$—O—CH$_3$, CHO—C(CH$_3$)$_2$—O—CH$_2$, Zn/Ag–gr/THF, −20 to 30°C, 180 min, 81–86%. iii, R^1, R^2 = H, Cl, Zn/Ag–gr/THF, reflux, 4 h, 63%.

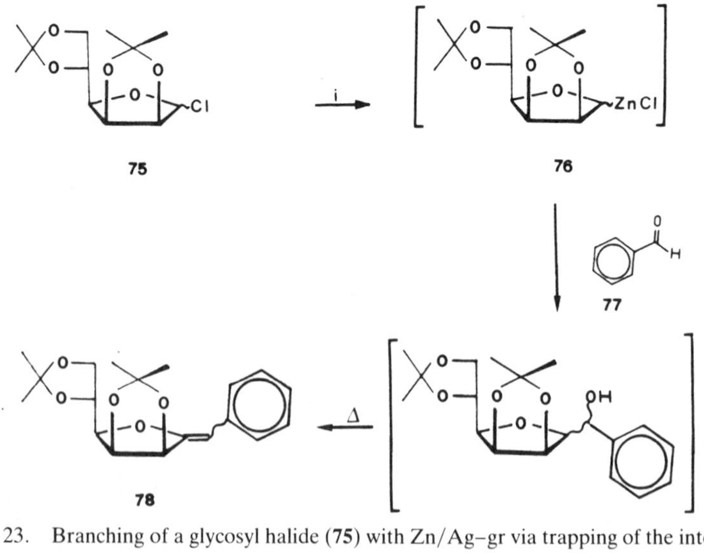

FIG. 23. Branching of a glycosyl halide (**75**) with Zn/Ag–gr via trapping of the intermediate glycosyl zinc species (**76**) with an aldehyde (**77**) (*215*). i, Zn/Ag–gr/THF, −15 to 30°C.

3. Tin–Graphite

In tin–graphite (Sn–gr), easily obtained by reduction of $SnCl_2$ with C_8K (216), all of the metal is claimed to be dispersed on the graphite surface rather than intercalated (126). Sn–gr undergoes oxidative addition with allylic bromides (**79**, **82**) in THF at room temperature to produce diallyltin dibromides (80), which are able to transfer both of the allylic groups in the reaction with aldehydes (Fig. 24) (217). The reaction takes place in good to excellent yields (84–93%) within 3–4 h and proceeds with allylic inversion (products **81** and **83**). Addition of the allyl group to α,β-unsaturated aldehydes **84** gives the 1,2-product **85**.

A very promising result has been reported by Umani-Ronchi and co-workers (126). This group tried to develop an enantioselective synthesis of homoallylic alcohols by forming in situ a chiral tin(IV) complex.

Sequential addition of the monosodium salt of L-(+)-diethyltartrate and **77** to diallyltin dibromide (**86**) in THF at −40°C results in the formation of

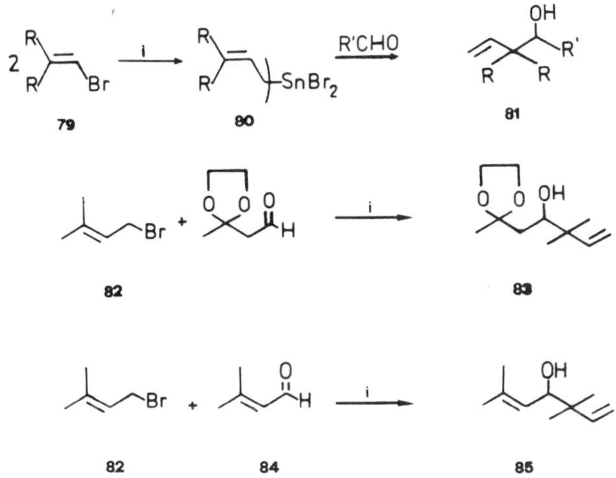

FIG. 24. Reactions of allylic bromides with Sn–gr and aldehydes (217). i, Sn–gr/THF, 80–90%.

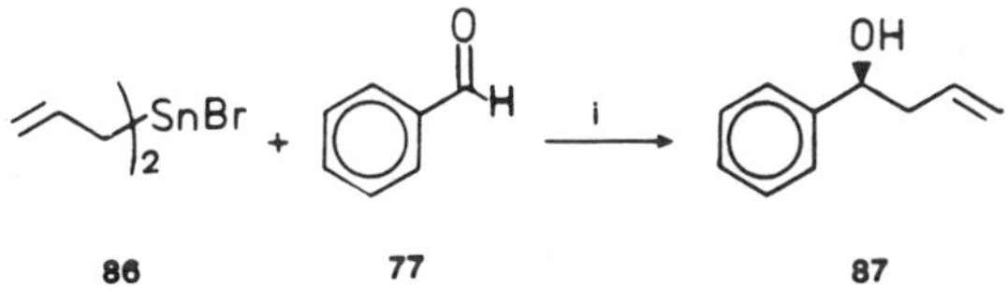

FIG. 25. Enantioselective synthesis of homoallylic alcohols using Sn–gr (126). i, L-(+)-diethyl tartrate (monosodium salt), Sn–gr/THF, −40°C, 65% (enantiomeric excess, 71%).

(*S*)-1-phenyl-3-buten-1-ol (**87**) with an enantiomeric excess of 71% (the yield of the reaction is 65%) (Fig. 25) (*126*).

4. *Palladium–Graphite*

Palladium–graphite (Pd–gr), prepared by the reduction of $PdCl_2$ with C_8K (*218,219*) in DME, has turned out to be a catalyst for the alkenylation

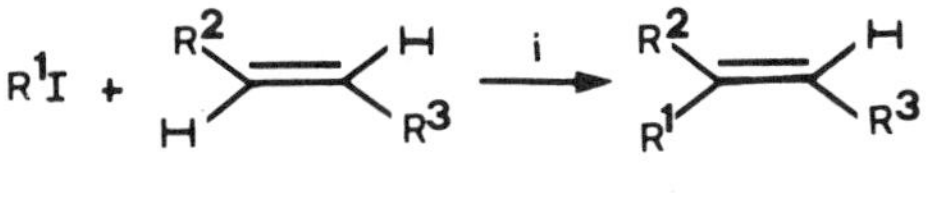

89 **88** **90**

FIG. 26. Heck reaction using Pd–gr (*218*). i, Pd–gr, *n*-Bu_3N, 58–92%; R^1 = Phe, 2-thienyl, hex-1-enyl; R^2 = H, CO_2Et; R^3 = Phe, CO_2Et.

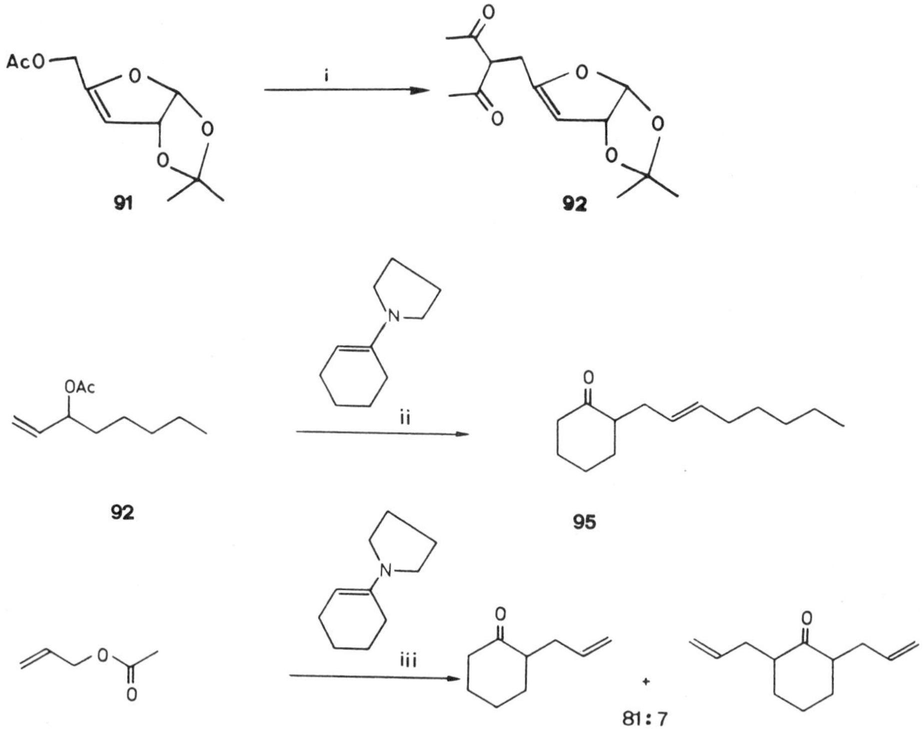

91 **92** **92** **95** **93** **96a** **96b**

FIG. 27. Applications of Pd–gr-mediated nucleophilic substitution of allyl acetates (*218,220*). i, Pd–gr/THF, CH_2—$(CO_2Et)_2$, 35 h, 67%; ii, Pd–gr/THF, 3 h, 72% [91% (*E*)-configurated]; iii, Pd–gr/THF, 8 h, 88%.

or arylation of activated carbon–carbon double bonds (**88**) with vinyl or aryl iodides (**89**) (Heck reaction) (Fig. 26). The Pd in this compound is highly dispersed on the graphite surface and undergoes oxidative addition into the carbon–halogen bond of alkyl and aryl iodides, whereas aryl bromides were found to be unreactive under these conditions. The organopalladium species adds to activated carbon–carbon double bonds, giving the corresponding alkenylated or arylated compounds (**90**) in yields between 58 and 92%.

With monosubstituted olefinic compounds the reaction is usually stereospecific, yielding the substituted carbon–carbon bond with (E) geometry (80–99% E) (*218*). In addition, catalytic amounts (2–10 mol %) of Pd–gr and triphenyl phosphine (2–40 mol %) activate allylic esters (e.g., acetates, phosphates, carbonates) (**91–93**) toward nucleophiles in substitution reactions (products **94–96**) (Fig. 27) (*220*). The nucleophiles employed so far are sodium (or potassium) diethyl malonate, sodium benzenesulfinate, and 1-(1-cyclohexen-1-yl)-pyrrolidine.

Finally, Pd–gr is an effective catalyst (3 mol %) for the hydrogenation of alkynes, alkenes, and anilines (*219*). Suppression of full hydrogenation of alkynes to the corresponding alkanes was almost completely (97–98%) achieved by addition of ethylenediamine to yield predominantly (Z) alkenes ($Z:E$ = 94–98:6–2%) (*219*). The results obtained with this catalyst indicate that it provides an alternative to both palladium on charcoal and the Lindlar catalyst.

Pd–gr prepared in a different manner (*221*) has also been used for the selective reduction of alkynes or 1,3-conjugated dienes.

5. Nickel–Graphite

Nickel–graphite (Ni–gr) of activity 1 (Ni–gr-1) (*222*) obtained by reduction of bis(dimethoxyethane)-dibromonickel with C_8K (*201*) in THF/hexamethylphosphoric triamide (15/1) also belongs to the graphite surface compounds (*126*). Pyrophoric Ni–gr-1 must be prepared and used *in situ* because it partially deteriorates on air exposure. The air exposure affords a modified and less active hydrogenation catalyst, Ni–gr-2 (*223*). This drawback is balanced by large-scale availability and good stability on storage. In addition, Ni–gr-2 displays good selectivity in the reduction of α,β-unsaturated carbonyl compounds (**97**, **98**) and β-diketones (**99**) (products **100–104**) (Fig. 28) (*223*). Aromatic compounds and ethoxycarbonyl moieties are not affected.

In contrast, Ni–gr-1 is an efficient catalyst for the semihydrogenation of alkynes to (Z)-alkenes in the presence of ethylenediamine as catalyst modifier. Stereospecificity of 94–99% is obtained. These hydrogenations

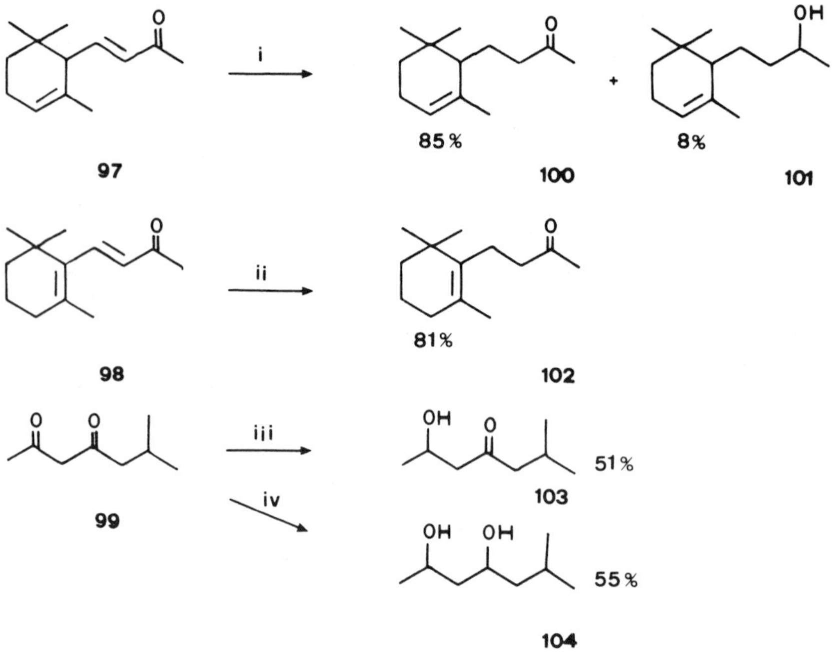

Fig. 28. Selective hydrogenation with Ni–gr-2 (223). i, Ni–gr-2, 20 h, 50°C, 30 atm H$_2$; ii, Ni–gr-2, 15 h, 50°C, 30 atm H$_2$; iii, Ni–gr-2, 22 h, 90°C, 30 atm H$_2$; iv, Ni–gr-2, 18 h, 120°C, 50 atm H$_2$.

can be performed at room temperature and atmospheric pressure. Preferably, the catalyst preparation and the hydrogenation are performed subsequently in the same reaction flask, since the catalyst is very sensitive to air exposure (222).

In addition, Ni–gr-1 has been used successfully for the coupling of allylic bromides at 30°C via a (π-allyl)–nickel halide complex in 60% yield to 1,5-hexadiene. At a higher temperature (70°C) complete conversion of the starting material was achieved, but higher oligomers were obtained as by-products (222).

An Ni–gr compound prepared by reduction of an NiCl$_2$ GIC (195,224) has been shown to be a selective catalyst for alcohol and formic acid dehydrogenation without dehydration. Ethanol yields acetaldehyde, cyclohexanol yields cyclohexanone, and formic acid decomposes to CO$_2$ and H$_2$ (225). Also, methanation (221,226) could be obtained even in the presence of 0.3% S-containing compounds, such as H$_2$S. But in this case, too, high-resolution microscopy has clearly demonstrated the presence of metal outside the graphite (227).

6. Iron–graphite

Iron–graphite (Fe–gr) can be obtained by reduction of $FeCl_3$ (*130,137,228*) or $FeCl_2$ (*228*) with C_8K in THF. There is a controversy (*229,230*) as to whether the iron is intercalated (*200–202,228*) or not (*229*). Umani-Ronchi *et al.* (*228,230*) found weak reflections due to intercalated species, but Schäfer-Stahl (*229*) detected only bulk graphite, iron particles

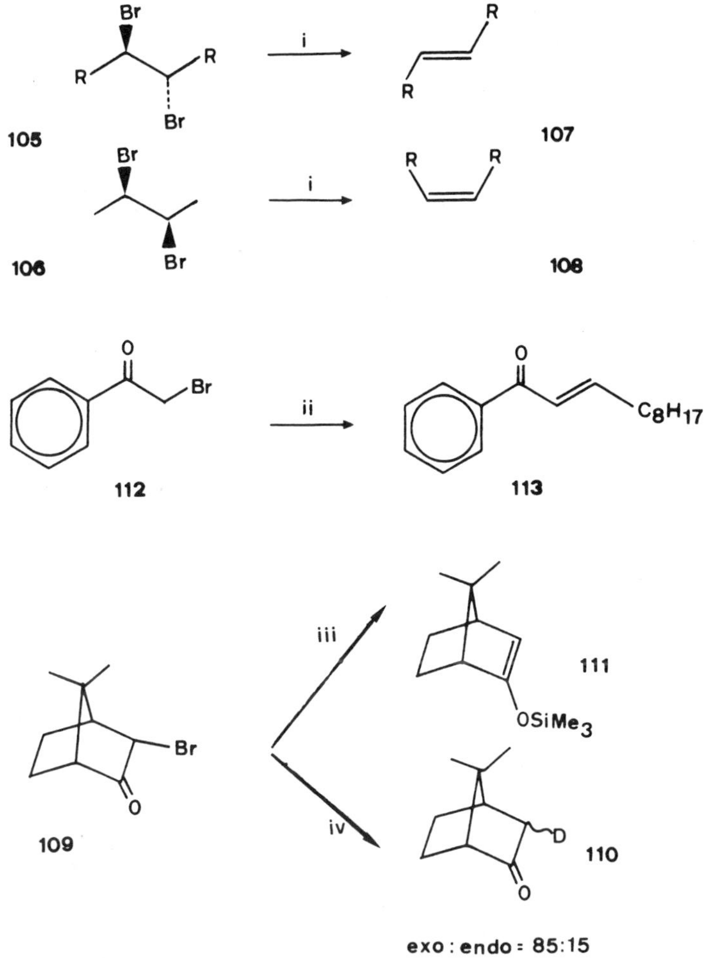

exo : endo = 85:15

FIG. 29.   Reactions with Fe–gr (*126,228*). i, Fe–gr/THF/H_2O, R = alkyl, aryl; ii, Fe–gr/THF/H_2O, then $C_8H_{17}CHO$, 40%; iii, Fe–gr/THF/H_2O, then Me_3SiCl, 36%; iv, Fe–gr/THF/D_2O, 94%.

on the surface of graphite, and amorphous carbon phases instead of the lamellar iron–graphite claimed by the Italian group. The preparation on "true" Fe laminates by another method was previously reported (*195,217,231*).

The product obtained by the reduction of FeCl₃ with C₈K (*130,200,228*), regardless of whether it is an intercalate, can be used for the stereospecific antidebromination of vicinal dibromoalkanes (**105, 106**) (*228*) to alkenes (**107, 108**) and for the reductive dehalogenation of α-bromoketones (Fig. 29). The rates of both reactions were markedly increased by the addition of a small amount of deaerated water. Debromination of α-bromoketones (e.g., **109**) yields the corresponding ketones, probably through intermediate Fe(II) enolates (*126,228*). On addition of deuterium oxide the corresponding α-deuterioketones (**110**) are obtained in good yields (94–96%), whereas addition of chlorotrimethylsilylsilane gives the corresponding O-trimethylsilyl enol ethers (**111**) in only moderate yields (e.g., 36% for camphor). Addition of an aldehyde as the electrophile to the reaction mixture of α-bromoacetophenone (**112**) and Fe–gr gave 1-phenyl-2-nonen-1-one (**113**) in 40% yield.

Iron–graphite can be used for 2-oxoallyl carbonium ion formation from α,α-dibromoketones (**114, 115**) (Fig. 30). These intermediates are trapped with electron-rich compounds, such as enamines (**116**) or dienes (**117**), to yield the corresponding cycloadducts (**118, 119**) (*126,228*).

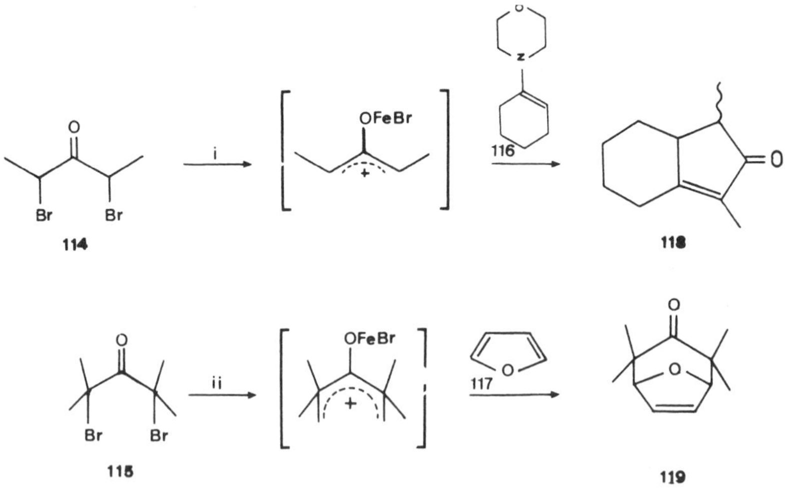

FIG. 30. Cycloadditions with Fe–gr (*126,228*). i, Fe–gr, overall yield 84%, (Z):(E) = 81:5; ii, Fe–gr, overall yield 88%.

7. Magnesium–Graphite

Magnesium–graphite (Mg–gr) has been obtained by the reduction of MgI_2 with C_8K in diethyl ether (201) or THF according to $MgI_2 + 2C_8K \rightarrow C_{16}Mg(KI)_2$. The Mg–gr was reported to be a true first-stage intercalation compound (201), but there is some doubt about this result (134). Reaction of Mg–gr with alkyl or aryl halides allows the preparation of Grignard reagents in excellent yields (95–100%) even at temperatures as low as $-78°C$ with an equilibrium between intercalated (40–60%) and soluble Grignard compounds. Filtration of the Grignard reagent from the reaction mixture left a solid that could be stored without any significant loss of activity for several months. Reactions of the graphite-supported Grignard reagents with ketones, water, or carbon dioxide gave excellent yields (87–96%) (201).

Even fluorobenzene forms the corresponding Grignard compound at a temperature of $-20°C$ (232). The high reactivity of the magnesium–graphite reagent even allows Reformatsky-type reactions with ethyl $\alpha\bar{V}$ halogenoalkanoates (e.g., 58) at $-78°C$ to yield the corresponding Mg ester enolate (120), which is then subject to further controlled reactions (products 121, 122) (Fig. 31). Although the yields are diminished in comparison to those of Reformatsky reactions (using Zn), even branching

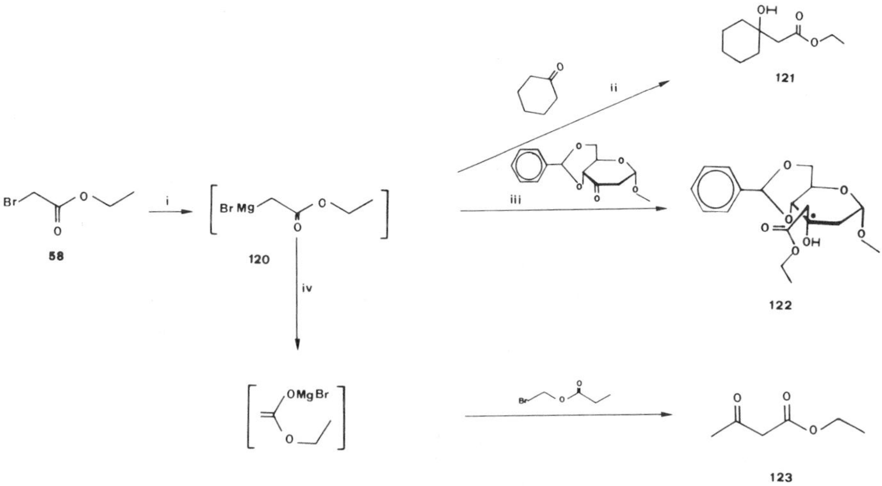

FIG. 31. "Magnesio-Reformatsky" reactions using Mg–gr (232–234). i, Mg–gr/THF, $-78°C$, for paths ii and iii; classical conditions: magnesium turnings in diethyl ether, reflux (for path iv); ii, $-78°C$, 30 min, 71%; iii, $-70°C$, 30 min, 79%; iv, reflux, overall yield of 123 was 75%.

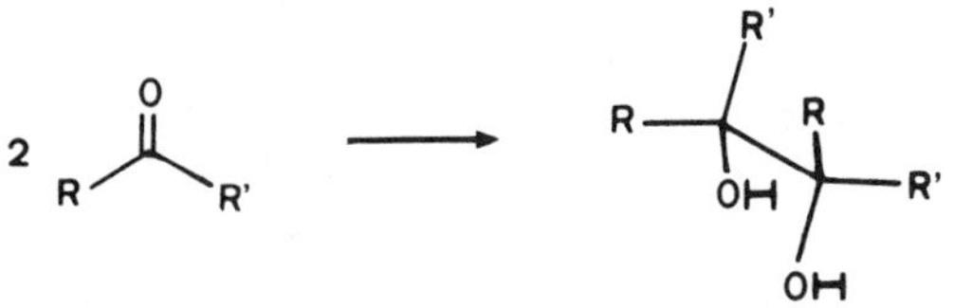

FIG. 32. "Pinacolic" coupling with Mg–gr (*235*). R, R' = alkyl, aryl; 3–4 h, room temperature, 65–91%.

reactions in the carbohydrate field are possible. The general applicability of Mg–gr seems to be restricted (*233*), but (*234*) Rieke Mg (*207,208*) was shown to be insufficiently reactive for this purpose. Classical reactions under these conditions give ethyl acetoacetate (**123**) as the main product.

Among the various reductions of carbonyl compounds, those leading to coupling products are of continuing interest. Mg–gr was found to be a simple, universally applicable, and mercury-free reagent for "pinacolic reductions" (see Fig. 32). It rapidly and invariably reduces all kinds of carbonyl compounds (**124**) with formation of vicinal diols (**125**) (*235*) in excellent yields (65–91%) at room temperature.

8. *Titanium–graphite*

Reduction of Ti(OPr^i)$_4$ with C$_8$K in THF yielded the first- and fourth-stage lamellar compound of titanium (*200,202*), whereas reduction of TiCl$_4$ with C$_8$K in benzene resulted in highly dispersed titanium on the surface of graphite, Ti–gr (*202*), probably with minor amounts of intercalated metal (*126*). Umani-Ronchi *et al.* (*216*) prepared the Ti–gr reagent by reduction of TiCl$_3$ with C$_8$K in THF under reflux for 3 h. This reagent enabled the reductive coupling of acetophenone and cyclohexanone in good yields (79 and 86%, respectively) to 2,3-diphenyl-2-butene and cyclohexylidene–cyclohexane, respectively. It is thought that TiCl$_3$-derived reagents, mainly consisting of Ti$^{(0)}$, generally give alkenes (*236*), while the divalent species obtained from TiCl$_4$ tend to favor pinacol formation (*237*).

In contrast to methods described (*238*), Ti chlorides with either C$_8$K or Zn/Ag–graphite (*209*) rapidly form very reactive low-valent Ti–gr reagents, which, as observed above, gradually deteriorate on prolonged heating. The TiCl$_4$-derived reagent yields alkenes from aromatic and pinacols from aliphatic educts, except for cyclopentanone (which gives rise to a mixture), clearly showing the generally accepted pinacol/alkene dualism in Ti-induced couplings.

A reagent from TiCl$_3$/C$_8$K with a short reduction time (30 min) mediates pinacol formation, whereas on prolonged reduction alkenes are

TABLE I

COUPLING OF CARBONYL COMPOUNDS BY Ti–GR (239)

Educt	Product	Method[a]	Reduction time (min)	Reaction time (min)	Yield (%)
Benzophenone	Tetraphenylethene	A	30	120	80
Benzaldehyde	trans-Stilbene	A	30	50	95
Octanal	Hexadecan-8,9-diol	A	30	15	78
Cyclohexanone	Bicyclohexyl-1,1'-diol	A	30	120	60
Cyclohexanone	Bicyclohexyl-1-1'-diol	A	30	720	55
Cyclohexanone	Bicyclohexyl-1,1'-diol	A	360	120	40
Cyclohexanone	Bicyclohexyl-1,1'-diol	A	900	120	10
Cyclohexanone	Cyclohexylidene-cyclohexane	B	30	15	82
Cyclohexanone	Cyclohexylidene-cyclohexane	B	180	720	79
Acetophenone	trans-2,3-Diphenyl-2-butene	B	180	720	80

[a] A, $TiCl_4/C_8K$; B, $TiCl_3/C_8K$.

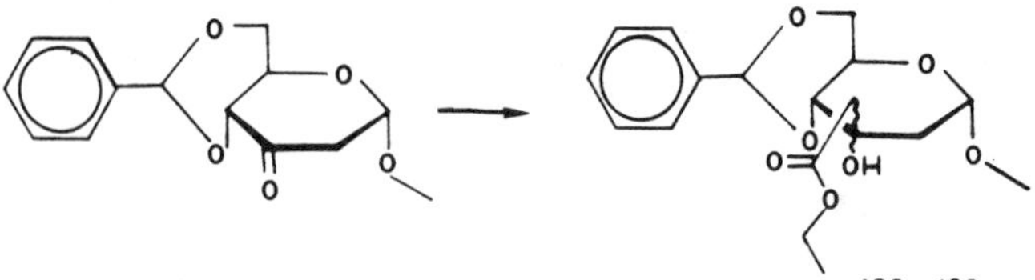

122 + 126

Fig. 33. Aldolizations with Ti–gr ("titano-Reformatsky") (*240*). Ti–gr (from TiCl₄/C₈K), 20°C, 2 h, 56%.

invariably formed from both aliphatic and aromatic carbonyls. This is apparently the first example of a reduction time-dependent carbonyl coupling (Table I) (*239*).

The high reactivity of the Ti–gr reagent from $TiCl_4/C_8K$ allows Reformatsky-type reactions (products **122** and **126**) (Fig. 33) (*240*).

D. *Miscellaneous Nonmetal–Graphite Compounds*

Although most of the recent results concerning the structural properties of GIC have been discussed briefly or at least mentioned above, special aspects have been omitted because there are excellent reviews in the literature.

Electron paramagnetic resonance (EPR) and nuclear magnetic resonance (NMR) spectroscopy of GIC (*13*) give insight and information about atomic motions (*241*), and NMR chemical shifts can be used as microscopic probes for the electronic structure (*242*), universally applicable for all kinds of GIC (*243*).

Of special interest are chemical reactions using GIC. Selective graphite-supported fluorinations using different xenon fluorides have been reported (*244–247*). The GIC obtained by the reaction of the respective xenon (oxy) fluoride have the empirical formulas $C_{19}XeF_2$, $C_{8.7}XeOF_4$, $C_{28.3(\pm2.4)}XeF_4$ (*248*), and $C_{17.8}XeF_4$. These reagents work with or without additional hydrogen fluoride as a catalyst and have the advantage of ease and safety of handling compared with the nonintercalated reagents. Uracil could be fluorinated to the pharmacologically important 5-fluorouracil in outstanding yield as well as β-diketones (*249*). Intercalated TiF_4, TaF_5, and MoF_6 have been shown to be suitable reagents for the difluorination of isolated carbonyl groups to the geminal difluoro compounds (*23*). Organogermanium and organosilicon compounds have been fluorinated by SbF_5 GIC with good yields and selectivity (*250*). The same reagent is more resistant to hydrolysis and easier to handle than neat SbF_5 (*251*). Reaction of ethylene with methane in the presence of this GIC yielded a mixture of ethane and propane (*252*).

Mixtures of cis/trans decalins, cis/trans perhydroindanes, or methyl cyclopentane/cyclohexene, respectively, reach equilibrium in very short reaction times. The isomerizations occur very rapidly and the most stable isomer is formed at low temperatures with only minor side reactions (*253*). Halogen exchange reactions with $C_{24}SbCl_5$, which corresponds to $SbCl_5$ GIC, and halogen alkanes or cycloalkanes (*254*) differ markedly in selectivity from reactions (*255*) with neat $SbCl_5$. Differences between NbF_5 *on* graphite and graphite-intercalated NbF_5 have been encountered (*256*). Of special interest is the use of compounds of this type in Friedel–Crafts reactions (*199,256–258*). AsF_5 GIC leads to graphite compounds of the highest conductivity, whose structure has been investigated by ^{19}F NMR spectroscopy (*259*).

Synthesis (*9,260,261*), properties (*9*), and reactions (*262*) of metal halide GIC have been reviewed (*191*), including transition metal GIC. There is clear evidence that in certain cases metal halide GIC give reagents whose properties are substantially different from those of the nonintercalated materials (*123*). $CuCl_2$ inserted in graphite is used for the high-yield chlorination of ethylene to 1,2-dichloroethane (*263,264*). Graphite hydrogen sulfate, $C_{24}^{+}HSO_4^{-} \cdot 2H_2SO_4$ has been used successfully for esterifications of fatty acids with an esterifying capacity of approximately 0.055 mol of acid and alcohol per gram of graphite hydrogen sulfate, but also for the synthesis of enol esters or acetals. (*E*)/(*Z*)-Geranic acid is cyclized to α-cyclogeranic acid in 90% yield by this reagent (*24,265–267*). Graphite hydrogen sulfate and stoichiometric amounts of nitric acid have been used for nitrations in cyclohexane as the solvent. This reagent gives greater proportions of para substitution than are obtained with homogeneous conditions (*24,265*).

Trifluoromethanesulfonic acid has been intercalated to yield C_9FSO_3H, which is an efficient catalyst for the dimerization of substituted oxiranes leading to 1,4-dioxanes. The same GIC can be used for the synthesis of enol esters (*23*).

GIC have also been developed for chiral synthesis in order to improve the efficiency of homogeneous transition metal complexes. Graphite was partially oxidized to serve as a carrier for carboxyl groups to retain good mechanical stability. By means of such graphite-anchored chiral rhodium reagents, asymmetric hydrosilylation of ketones could be achieved. In almost all cases the absolute configuration of the products obtained was opposite to that observed in homogeneous phase reactions (*268*).

Of special interest are Cr–graphite and other metal–graphite compounds (see above). A Cr GIC was assumed to oxidize primary alcohols to aldehydes (*131*) and found application as a cathode material in high-energy-density batteries (*269–274*). It has been shown, however, that true

GIC $C_{13.6}CrO_3$ will not oxidize alcohols, whereas Seloxcette™, which is not a GIC but a graphite/Cr_2O_5/Cr_3O_8 mixture, will do it with excellent yields (23,267,272,275–277). Other "graphimets™", e.g., Fe, Co, Pt, and Ni, are not GIC, as shown by high-resolution electron microscopy (227). All of the technical applications refer to the product that is not an intercalate, and the true GIC was useless in each case (9).

Unusual stereospecificity in the products of bromination of olefins and aromatics by C_8Br (267,278) or bromine on activated carbon (279) has been established and argues for some influence of the graphite in the chemistry (280).

Finally, it must be pointed out that it is sometimes very difficult to ascertain the identity of the actual graphite–metal compound. Graphitic reagents seem to be more than sponges which gradually release the intercalated species; instead they are new and most promising reagents or catalysts for both inorganic and (metal) organic reactions.

REFERENCES

1. A. Weiss, Angew. Chem. 75, 755 (1963).
2. A. R. Ubbelohde, Carbon 14, 1 (1976).
3. F. L. Vogel, J. Mater. Sci. 12, 982 (1977).
4. S. A. Solin, Adv. Chem. Phys. 49, 455 (1972).
5. J. Bernard, Pure Appl. Chem, 56, 1659 (1984).
6. S. Ergun, Carbon 6, 141 (1968).
7. L. B. Ebert, J. Mol. Catal. 15, 275 (1982).
8. L. B. Ebert, Annu. Rev. Mater. Sci. 6, 181 (1976).
9. H. Selig and L. B. Ebert, Adv. Inorg. Chem. Radiochem. 23, 281 (1980).
10. A. Herold, Springer Ser. Solid State Sci. 38, 7 (1981).
11. M. S. Dresselhaus and G. Dresselhaus, Adv. Phys. 30, 139 (1981).
12. A. Herold and F. L. Vogel, Mater. Sci. Eng. 31 (1977): Proceedings of the Franco-American Conference on Intercalation Compounds of Graphite.
13. H. Estrade-Szwarckopf, Helv. Phys. Acta 58, 139 (1985).
14. A. Herold, in "Intercalated Materials" (F. Levy, Ed.), pp. 323-421. Reidel, Dordrecht, 1979.
15. G. S. Parry, Physica Ser. B 105, 261 (1981).
16. F. Rousseaux, D. Tchoubar, C. Tchoubar, D. Guerard, P. Lagrange, A. Herold, and R. Moret, Synth. Met. 7, 221 (1983).
17. A. W. Moore, in "Chemistry and Physics of Carbon" (P. C. Walker, Jr, and P. A. Thrower, Eds.), Vol. 11, p. 69. Dekker, New York, 1973.
18. D. D. L. Chung, Mater. Sci. Eng. 39, 283 (1979).
19. G. R. Hennig, Prog. Inorg. Chem. 1, 125 (1959).
20. R. C. Croft, Q. Rev. Chem. Soc. 14, 1 (1960).
21. A. R. Ubbelohde and F. A. Lewis, "Graphite and Its Crystal Compounds." Clarendon, Oxford, 1960.
22. G. R. Hennig, Proc. Carbon Conf., 2nd p. 103 (1956).
23. R. Setton, F. Beguin, and S. Piroelle, Synth. Met. 4, 299 (1982).

24. J. P. Alazard, H. B. Kagan, and R. Setton, *Bull. Soc. Chim. Fr.* 499 (1977).
25. A. Aronson, *J. Inorg. Nucl. Chem.* **25**, 907 (1963).
26. P. Lagrange, M.-H. Portmann, and A. Herold, *C. R. Acad. Sci. Ser. C* **283**, 557 (1976).
27. G. Furdin and A. Herold, *Bull. Soc. Chim. Fr.* 1768 (1972).
28. F. Beguin, H. Estrade-Szwarckopf, J. Conard, P. Lauginie, P. Marceau, D. Guerard, and L. Facchini, *Synth. Met.* **7**, 77 (1983).
29. A. Herold, *Mater. Sci. Monogr.* **28A**, 461 (1985).
30. F. Beguin and R. Setton, *Carbon* **13**, 293 (1975).
31. L. Bonnetain, P. Touzain, and A. Hamwi, *Mater. Sci. Eng.* **31**, 45 (1977).
32. M. El Makrini, P. Lagrange, D. Guerard, and A. Herold, *Carbon* **18**, 211 (1980).
33. D. Guerard, P. Lagrange, and A. Herold, *Mater. Sci. Eng.* **31**, 29 (1977).
34. D. E. Nixon and G. S. Parry, *Br. J. Appl. Phys. (J. Phys. D)* **1**, 291 (1968).
35. R. Nishitani and H Suematsu, *Synth. Met.* **7**, 13 (1983).
36. A. Hamwi, P. Touzain, and C. Riekel, *Synth. Met.* **7**, 23 (1983).
37. A. Herold, D. Billaud, D. Guerard, and P. Lagrange, *Mater. Sci. Eng.* **31**, 25 (1977).
38. S. H. Anderson and D. D. L. Chung, *Synth. Met.* **7**, 107 (1983).
39. E. A. Mistryukov and I. K. Korshevets, *Bull. Acad. Sci. USSR Div. Chem.* **2**, 448 (1985).
40. N. Daumas and A. Herold, *C. R. Acad. Sci. Ser. C* **268**, 373 (1969).
41. A. Metrot and J. E. Fischer, *Synth. Met.* **3**, 201 (1981).
42. B. Carton and A. Herold, *Bull. Soc. Chim. Fr.* 1337 (1972).
43. N. Wada, R. Clarke, and S. A. Solin, *Synth. Met.* **2**, 27 (1980).
44. D. E. Bergbreiter and J. M. Killough, *J. Chem. Soc. Chem. Commun.* 913 (1976).
45. D. E. Bergbreiter and J. M. Killough, *J. Am. Chem. Soc.* **100**, 2126 (1978).
46. L. B. Ebert, L. Matty, D. R. Mills, and J. C. Scanlon, *Mater. Res. Bull.* **15**, 251 (1980).
47. R. Schlögl and H. P. Boehm, *Carbon* **22**, 351 (1984).
48. A. Schleede and M. Wellmann, *Z. Phys. Chem. B* **18**, 1 (1932).
49. G. S. Parry, and D. E. Nixon, *Nature (London)* **216**, 909 (1967).
50. G. S. Parry, D. E. Nixon, K. M. Lester, and B. C. Levene, *J. Phys. C* **2**, 2156 (1969).
51. W. Rüdorff and E. Schulze, *Z. Anorg. Allgem. Chem.* **277**, 156 (1954).
52. N. J. Wu and A. Ignatiev, *Phys. Rev. B* **28**, 7288 (1983).
53. E. L. Evans and J. M. Thomas, *J. Solid State Chem.* **14**, 111 (1975)).
54. R. Setton, *J. Mol. Catal.* **27**, 263 (1984).
55. L. B. Ebert and J. C. Scanlon, *Ind. Eng. Chem. Prod. Res. Div.* **19**, 103 (1980).
56. M. E. Vol'pin and Y. N. Novikov, *Top. Nonbenzenoid Aromat. Chem.* **1**, 269 (1973).
57. C. Herinckx, R. Perret, and W. Ruland, *Carbon* **10**, 711 (1972).
58. N. B. Hannay, T. H. Geballe, B. T. Matthias, K. Endres, P. Schmidt, and D. Macnair, *Phys. Rev. Lett.* **14**, 225 (1965)
59. P. Delhaes, *Mater. Sci. Eng.* **31**, 225 (1977).
60. J. Poitrenaud, *Rev. Phys. Appl.* **5**, 275 (1970).
61. D. Guerard, G. M. T. Foley, M. Zanini, and J. E. Fischer, *Nuovo Cimento* **38B**, 410 (1977).
62. M. A. M. Boersma, *Catal. Rev. Sci. Eng.* **10**, 243 (1974).
63. W. Rüdorff, *Adv. Inorg. Chem. Radiochem.* **1**, 223 (1959).
64. D. M. Ottmers and H. F. Rase, *Ind. Eng. Chem. Fundam.* **5**, 302 (1966).
65. K. Fredenhagen and G. Cadenbach, *Z. Anorg. Allg. Chem.* **158**, 249 (1926).
66. W. Rüdorff, *Angew. Chem.* **71**, 487 (1959).
67. H. Schäfer-Stahl and K. Knoll, *Carbon* **22**, 183 (1985).
68. E. Weintraub, U.S. Patent 922.645; *Chem. Abstr.* **3**, 2040 (1909).
69. E. Weintraub, Fr. Demande 585.185 (1923).

132   RENÉ CSUK et al.

70. Y. N. Novikov and M. Vol'pin, *Russ. Chem. Rev.* **40**, 733 (1971).
71. H. Podall, W. E. Foster, and A. P. Giraitis, *J. Org. Chem.* **23**, 82 (1958).
72. F. Hulliger, *Phys. Chem. Mater. Layered Struct.* **5**, 52 (1976).
73. A. Herold, C. R. *Acad. Sci. Sec. C* **232**, 1489 (1951).
74. A. Herold, *Bull. Soc. Chim. Fr.* 999 (1955).
75. J. G. Hooley, *Carbon* **11**, 225 (1973).
76. J. G. Hooley and M. W. Bartlett, *Carbon* **5**, 417 (1967).
77. J. O. Besenhard, H. Witty, and H.-F. Klein, *Carbon* **22**, 97 (1984).
78. H.-F. Klein, J. Groß, and J. O. Besenhard, *Angew. Chem.* **92**, 476 (1980); *Angew. Chem. Int. Ed. Engl.* **19**, 441 (1980).
79. D. Guerard and A. Herold, *Carbon* **13**, 337 (1975).
80. D. Billaud, E. McRae, J. F. Mareche, and A. Herold, *Synth. Met.* **3**, 21 (1981).
81. A. Herold, D. Billaud, D. Guerard, P. Lagrange, and M. El Makrini, *Physica Ser. B* **105**, 253 (1981).
82. K. Tamaru, *Adv. Catal.* **20**, 327 (1969).
83. R. Juza and V. Wehle, *Naturwissenschaften* **32**, 560 (1965).
84. H.-F. Klein, *Kontakte (Merck)* **3**, 3 (1982).
85. M. Bagouin, D. Guerard, and A. Herold, *C. R. Acad. Sci. Ser. C* **262**, 557 (1966).
86. D. Guerard and A. Herold, *C. R. Acad. Sci. Ser. C* **275**, 571 (1972).
87. P. Pfluger, V. Geiser, S. Stoltz, and H. J. Güntherodt, *Synth. Met.* **3**, 27 (1981).
88. R. C. Asher and S. A. Wilson, *Nature (London)* **181**, 409 (1958).
89. J. Conard, H. Estrade-Szwarckopf, P. Lauginie, and G. Hermann, *Springer Ser. Solid State Sci.* **38**, 264 (1981).
90. R. C. Asher, *J. Inorg. Nucl. Chem.* **10**, 238 (1959).
91. G. R. Hennig, *Prog. Inorg. Chem.* **1**, 184 (1959).
92. G. R. Hennig and L. Meyer, *Phys. Rev.* **87**, 459 (1952).
93. D. Billaud and A. Herold, *Carbon* **16**, 301 (1978).
94. National Distillers Chemical Co. booklet on the preparation of "High Surface Sodium" (1953), no longer available; cf. Ref. 95.
95. H. Hart, B.-L. Chen, and C.-T. Peng, *Tetrahedron Lett.* **36**, 3121 (1977).
96. M. El Makrini, G. Furdin, P. Lagrange, J. F. Mareche, E. McRae, and A. Herold, *Synth. Met.* **2**, 197 (1980).
97. M. El Makrini, D. Guerard, P. Lagrange, and A. Herold, *Carbon* **18**, 203 (1980).
98. P. Lagrange, M. El Makrini, D. Guerard, and A. Herold, *Synth. Met.* **2**, 191 (1980).
99. D. Guerard, M. Chaabouni, P. Lagrange, M. El Makrini, and A. Herold, *Carbon* **18**, 257 (1980).
100. D. Guerard and A. Herold, *C. R. Acad. Sci. Ser. C* **279**, 455 (1974).
101. D. Guerard and A. Herold, *C. R. Acad. Sci. Ser. C* **281**, 929 (1975).
102. W. Rüdorff, *Chimia* **19**, 489 (1965).
103. H. Schäfer-Stahl and G. von Eynatten, *Synth. Met.* **7**, 73 (1983).
104. W. E. Craven, "Intercalation of the Rare Earth Elements into Graphite and Dichalcogenides." M. S. Thesis, U.S. Air Force Institute of Technology, Air University, Wright-Patterson Airforce Base, Ohio, 1965.
105. E. Stumpp and G. Nietfeld, *Z. Anorg. Allg. Chem.* **456**, 261 (1979).
106. D. Guerard, C. Zeller, and A. Herold, *C. R. Acad. Sci. Ser. C* **283**, 437 (1976).
107. I. Rashkov, G. Merle, C. Mai, J. Gole, and I. M. Panayotov, *C. R. Acad. Sci. Ser. C* **283**, 339 (1976).
108. G. Merle, J. M. Letoffe, I. B. Rashkov, and P. Claudy, *J. Therm. Anal.* **13**, 293 (1978).
109. G. Merle, C. Mai, J. Gole, and I. B. Rashkov, *Carbon* **15**, 243 (1976).
110. F. Beguin and R. Setton, *J. Chem. Soc. Chem. Commun.* 611 (1976).

111. G. Merle, I. B. Rashkov, C. Mai, and J. Gole, *Mater. Sci. Eng.* **31**, 39 (1977).
112. I. B. Rashkov, I. M. Panayotov, and V. C. Shishkova, *Carbon* **17**, 103 (1979).
113. R. Setton, F. Beguin, J. Jegoudez, and C. Mazieres, *Rev. Chim. Min.* **19**, 360 (1982).
114. F. Beguin, J. Jegoudez, C. Mazieres, and R. Setton, *C. R. Acad. Sci. Paris Ser.* 2 **293**, 969 (1981).
115. F. Beguin, R. Setton, A. Hamwi, and P. Touzain, *Mater. Sci. Eng.* **40**, 167 (1979).
116. Y. V. Isaev, Y. N. Novikov, M. E. Vol'pin, I. Rashkov, and I. Panayotov, *Synth. Met.* **6**, 9 (1983).
117. I. Rashkov, V. Shishkova, I. Panayotov, G. Merle, and J. M. Letoffe, *Mater. Sci. Eng.* **57**, 155 (1983).
118. J. Amiell, P. Delhaes, F. Beguin, and R. Setton, *Mater. Sci. Eng.* **31**, 243 (1977).
119. C. Minh Duc, C. Mai, R. Riviere, and J. Gole, *J. Chim. Phys.* 991 (1972).
120. I. B. Rashkov, I. M. Panayotov, and V. C. Shishkova, *Carbon* **17**, 479 (1979).
121. A. de Andres Gomez de Barreda, M. Fernandez de Lezeta, and F. Aragon de la Cruz, *An. Quim.* **77**, 401 (1981).
122. D. Ginderow and R. Setton, *Carbon* **6**, 81 (1968).
123. A. McKillop and D. W. Young, *Synthesis* 481 (1979).
124. M. Contento, D. Savoia, C. Trombini, and A. Umani-Ronchi, *Synthesis* 30 (1979).
125. D. Savoia, C. Trombini, and A. Umani-Ronchi, *Tetrahedron Lett.* 653 (1977).
126. D. Savoia, C. Trombini, and A. Umani-Ronchi, *Pure Appl. Chem.* **57**, 1887 (1985).
127. D. Savoia, C. Trombini, and A. Umani-Ronchi, *J. Org. Chem.* **43**, 2907 (1978).
128. U. Rochus and R. Kickuth, Ger. Patent 1,095,832 (1957); *Chem Abstr.* **56**, 910076 (1962).
129. F. Glockling and D. Kingston, *Chem. Ind. (London)* 1037 (1961).
130. M. Rabinovitz and D. Tamarkin, *Synth. Commun.* **14**, 377 (1984).
131. J.-M. Lalancette, G. Rollin, and P. Dumas, *Can. J. Chem.* **50**, 3058 (1972).
132. D. Tamarkin and M. Rabinovitz, *Synth. Met.* **9**, 125 (1984).
133. G. Furdin, P. Lagrange, A. Herold, and C. Zeller, *C. R. Acad. Sci. Ser.* C **282**, 563 (1976).
134. A. Fürstner, Projected Ph.D. thesis, Techn. Univ., Graz (1987).
135. D. Savoia, C. Trombini, and A. Umani-Ronchi, *J. Chem. Soc. Perkin Trans.* 1, 123 (1977).
136. P. O. Ellingsen and K. Undheim, *Acta Chem. Scand. Sect. B* **33**, 528 (1979).
137. D. Tamarkin, D. Benny, and M. Rabinovitz, *Angew. Chem.* **96**, 594 (1984); *Angew. Chem. Int. Ed. Engl.* **23**, 642 (1984).
138. G. P. Boldrini and A. Umani-Ronchi, *Synthesis* 596 (1976).
139. C. Ungurenasu and M Palie, *J. Chem. Soc. Chem. Commun.* 388 (1975).
140. K. A. Jensen, B. Nygaard, G. Elisson, and P. H. Nielsen, *Acta Chem. Scand.* **19**, 768 (1965).
141. S. M.van der Kerk, P. H. M. Budzelaar, A. L. M. van Eekeren, and G. J. M. van der Kerk, *Polyhedron* **3**, 271 (1984).
142. S. M. van der Kerk, J. C. Roos-Venekamp, A. J. M. van Beijnen, and G. J. M. van der Kerk, *Polyhedron* **2**, 1337 (1983).
143. M. Ichikawa, M. Soma, T. Onishi, and K. Tamaru, *J. Catal.* **9**, 418 (1968).
144. M. Ichikawa, Y. Inoue, and K. Tamaru, *J. Chem. Soc. Chem. Commun.* 928 (1972).
145. M. P. Rosynek and Y.-P. Wang, *J. Mol. Catal.* **27**, 277 (1984).
146. D. M. Ottmers, "Catalytic Properties of Potassium Graphite Compounds with Reference to Their Structural and Electrical Characteristics." *Diss. Abstr.* **B27**, 155 (1966).
147. M. P. Rosynek, J. S. Fox, and J. L. Jensen, *J. Catal.* **71**, 64 (1981).

148. S. Tsuchiya, T. Misumi, N. Ohuye, and H. Imamura, *Bull. Chem. Soc. Jpn.* **55**, 3089 (1982).
149. D. G. Onn, G. M. T. Foley, and J. E. Fischer, *Mater. Sci. Eng.* **31**, 271 (1977).
150. A. A. Yeo and J. K. Hambling, Brit. Patent 912,822 (1962); *Chem. Abstr.* **58**, 11214 (1963).
151. A. A. Yeo and J. K. Hambling, Brit. Patent 912,825 (1962); *Chem. Abstr.* **58**, 10074H (1963).
152. A. A. Yeo, J. K. Hambling, and G. W. Alderson, Brit. Patent 912,823 (1962); *Chem. Abstr.* **58**, 10074 E (1963).
153. A. A. Yeo, J. K. Hambling, and G. W. Alderson, Brit. Patent 912,824 (1962); *Chem. Abstr.* **58**, 10074 C (1963).
154. A. A. Yeo, J. K. Hambling, and G. W. Alderson, Brit. Patent 912,821 (1962); *Chem. Abstr.* **59**, 5020 (1963).
155. J. B. Wilkes, *Ind. Eng. Prod. Res. Dev.* **21**, 585 (1982).
156. H. Pines and W. M. Stalich, *in* "Base Catalyzed Reactions of Hydrocarbons and Related Compounds" (H. Pines and W. M. Stalick, Eds.), p. 19. Academic Press, New York, 1977.
157. J.-M. Lalancette and R. Roussel, *Can. J. Chem.* **54**, 2110 (1976).
158. H. Podall and W. E. Foster, *J. Org. Chem.* **23**, 401 (1958).
159. W. E. Foster, U.S. Patent 3,160,670 (1964).
160. J. Jegoudez, C. Mazieres, and R. Setton, *Synth. Met.* **7**, 85 (1983).
161. A. Hamwi, P. Touzain, and L. Bonnetain, *C. R. Acad. Sci. Ser.* 2 **299**, 1385 (1984).
162. J. Gole, G. Merle, and J. P. Pascault, *Synth. Met.* **4**, 269 (1982).
163. J. Parrod and G. Beinert, *J. Polym. Sci.* **53**, 99 (1961).
164. J. Gole, *Mater. Sci. Eng.* **31**, 309 (1977).
165. E. Loria, G. Merle, J. P. Pascault, and I. B. Rashkov, *Polymer* **22**, 95 (1981).
166. I. M. Panayotov and I. B. Rashkov, *J. Polym. Sci. Polym. Chem. Ed.* **11**, 2615 (1973).
167. I. M. Panayotov and I. B. Rashkov, *Makromol. Chem.* **175**, 3305 (1974).
168. I. B. Rashkov, S. L. Spassov, and I. M. Panayotov, *Makromol. Chem.* **170**, 39 (1973).
169. G. Merle, J. P. Pascault, Q. T. Pham, C. Pillot, R. Salle, J. Gole, I. B. Rashkov, I. M. Panayotov, D. Guerard, and A. Herold, *J. Polym. Sci. Polym. Chem. Ed.* **15**, 2067 (1977).
170. Charbonnages de France, Fr. Patent 1,566796 (1969); *Chem. Abstr.* **71**, 125189 (1969).
171. C. Stein, Fr. Patent 2,067,543 (1969).
172. I. M. Panayotov, I. V. Berlinova, and I. B. Rashkov, *J. Polym. Sci. Polym. Chem. Ed.* **13**, 2043 (1975).
173. I. B. Rashkov and I. Gitsov, *J. Polym. Sci.* **22**, 905 (1984).
174. I. Rashkov, I. Panayotov, and I. Gitsov, *Polym. Bull.* **4**, 97 (1981).
175. I. B. Rashkov, I. Gitsov, I. M. Panayotov, and J. P. Pascault, *J. Polym. Sci.* **21**, 923 (1983).
176. G. A. Saunders, Ph. D. thesis, University of London, 1962.
177. I. B. Rashkov, I. Gitsov, and I. M. Panayotov, *J. Polym. Sci.* **21**, 937 (1983).
178. E. Loria, J. P. Pascault, G. Merle, J. Gole, Q. T. Pham, I. B. Rashkov, and I. M. Panayotov, *Meet. Ionic Polym., Straβbourg, Feb.* Preprint 146 (1978).
179. V. I. Mashinskii, V. A. Postnikov, Y. N. Novikov, A. L. Lapidus, M. E. Vol'pin, and Y. T. Eidus, *Isz. Akad. Nauk. SSSR, Ser. Khim.* **9**, 2018 (1976).
180. M. P. Rosynek, *ERDA Rep.* FE-2467-1, FE-2467-2, (Nov. 1976) (Avail. NTIS).
181. P. Lagrange, A. Metrot, and A. Herold, *C. R. Acad. Sci. Ser. C* **278**, 701 (1974).
182. K. Tamaru, *Am. Sci.* **60**, 474 (1972).

183. C. Masters, *Adv. Organomet. Chem.* **17**, 61 (1979).
184. M. Ichikawa, M. Sudo, M. Soma, T. Onishi, and K. Tamaru, *J. Am. Chem. Soc.* **91**, 1538 (1969).
185. M. Ichikawa, S. Naito, K. Kawase, T. Kondo, and K. Tamaru, Offenlegungsschrift 2,149,161 (1972); *Chem. Abstr.* **76**, 158880 (1972).
186. M. Ichikawa, T. Kondo, K. Kawase, M. Sudo, T. Onishi, and K. Tamaru, *J. Chem. Soc. Chem. Commun.* 176 (1972).
187. K. I. Aika, H. Kori, and A. Ozaki, *J. Catal.* **22**, 424 (1972).
188. K. Urabe, K. I. Aika, and A. Ozaki, *J. Catal.* **32**, 108 (1974).
189. S. Natio, O. Ogawa, M. Ichikawa, and K. Tamaru, *Chem. Commun.* 1266 (1973).
190. M. P. Rosynek and J. B. Winder, *J. Catal.* **56**, 258 (1979).
191. Y. N. Novikov and M. E. Vol'pin, *Physica Ser. B* **105**, 471 (1981).
192. N. Tomotsu, I. Kojima, and I. Yasumori, *J. Catal.* **86**, 280 (1984).
193. V. A. Postnikov, L. M. Dmitrienko, R. F. Ivanova, N. L. Dobrolyubova, M. A. Golubeva, T. I. Gapeeva, Y. N. Novikov, V. B. Shur, and M. E. Vol'pin, *Izv. Akad. Nauk SSSR, Ser. Khim.* **12**, 2642 (1975).
194. W. Jones, R. Schlögl, and J. M. Thomas, *J. Chem. Soc. Chem. Commun.* 464 (1984).
195. M. E. Vol'pin, Y. N. Novikov, N. D. Lapkina, V. I. Kasatochkin, Y. T. Struchkov, M. E. Kazakov, R. A. Stukan, V. A. Povitskij, Y. S. Karimov, and A. V. Zvarikina, *J. Am. Chem. Soc.* **97**, 3366 (1975).
196. K. Watanabe, T. Kondow, M. Soma, T. Onishi, and K. Tamaru, *Proc. R. Soc. London Ser. A.* **333**, 51 (1973).
197. D. Guerard, C. Takoudjou, and F. Rousseaux, *Synth. Met.* **7**, 43 (1983).
198. M. Ichikawa, K. Kawase, and K. Tamaru, *J. Chem. Soc. Chem. Commun.* 177 (1972).
199. M. A. M. Boersma, *in* "Advanced Materials in Catalysis" (J. J. Burton and R. L. Garten, Eds.), Chap. 3, p. 67. Academic Press, New York, 1977.
200. D. Braga, A. Ripamonti, D. Savoia, C. Trombini, and A. Umani-Ronchi, *J. Chem. Soc. Chem. Commun.* 927 (1978).
201. C. Ungurenasu and M. Palie, *Synth. React. Inorg. Met.-Org. Chem.* **7**, 581 (1977).
202. D. Braga, A. Ripamonti, D. Savoia, C. Trombini, and A. Umani-Ronchi, *J. Chem. Soc. Dalton Trans.* 2026 (1979).
203. G. P. Boldrini, D. Savoia, E. Tagliavini, C. Trombini, and A. Umani-Ronchi, *J. Org. Chem.* **48**, 4108 (1983).
204. G. P. Boldrini, M. Mengoli, E. Tagliavini, C. Trombini, and A. Umani-Ronchi, *Tetrahedron Lett.* **27**, 4223 (1986).
205. R. Csuk, A. Fürstner, F. Hofer, and H. Weidmann, unpublished results.
206. C. Boga, D. Savoia, C. Trombini, and A. Umani-Ronchi, *J. Chem. Res. (M)* 2461 (1985).
207. R. D. Rieke, *Top. Curr. Chem.* **59**, 1 (1975).
208. R. D. Rieke and S. J. Uhm, *Synthesis* 452 (1975).
209. R. Csuk, A. Fürstner, and H. Weidmann, *J. Chem. Soc. Chem. Commun.* 775 (1986).
210. R. Csuk, A. Fürstner, H. Sterk, and H. Weidmann, *J. Carbohydr. Chem.* **5**, 459 (1986).
211. E. Fischer and K. Zach, *Sitz. Ber. Kgl. Preuss. Akad. Wiss.* **16**, 311 (1913).
212. R. Csuk, A. Fürstner, B. I. Glänzer, and H. Weidmann, *J. Chem. Soc. Chem. Commun.* 1149 (1986).
213. R. Csuk, B. I. Glänzer, A. Fürstner, H. Weidmann, and V. Formacek, *Carbohydr. Res.* **157**, 235 (1986).
214. R. Csuk, A. Fürstner, and H. Weidmann, *Int. Carbohydr. Symp., 13th Ithaca, N. Y.* 124, A115 (1986).

215. U. Hacksell and G. D. Daves, *Prog. Med. Chem.* **22**, 1 (1985).

216. G. P. Boldrini, D. Savoia, E. Tagliavini, C. Trombini, and A. Umani-Ronchi, *J. Organomet. Chem.* **280**, 307 (1985).

217. A. Knappwost and W. Metz, *Naturwissenschaften* **56**, 85 (1969).

218. D. Savoia, C. Trombini, A. Umani-Ronchi, and G. Verardo, *J. Chem. Soc. Chem. Commun.* 541 (1981).

219. D. Savoia, C. Trombini, A. Umani-Ronchi, and G. Verardo, *J. Chem. Soc. Chem. Commun.* 540 (1981).

220. G. P. Boldrini, D. Savoia, E. Tagliavini, C. Trombini, and A. Umani-Ronchi, *J. Organomet. Chem.* **268**, 97 (1984).

221. J.-M. Lalancette, U.S. Patent 3,804,916; *Chem. Abstr.* **80**, 145363 (1974).

222. D. Savoia, E. Tagliavini, C. Trombini, and A. Umani-Ronchi, *J. Org. Chem.* **46**, 5340 (1981).

223. D. Savoia, E. Tagliavini, C. Trombini, and A. Umani-Ronchi, *J. Org. Chem.* **46**, 5344 (1981).

224. J.-M. Lalancette, Can. Patent 979.914; *Chem. Abstr.* **84**, 179627 (1976).

225. A. A. Slinkin, Y. N. Novikov, N. A. Pribitkova, L. J. Leznover, A. M. Rubinstein, and M. E. Vol'pin, *Kinet. Katal.* **14**, 633 (1973).

226. J.-M. Lalancette, U.S. Patent 3,847,963; *Chem. Abstr.* **83**, 181889 (1975).

227. D. J. Smith, R. M. Fisher, and L. A. Freeman, *J. Catal.* **72**, 51 (1981).

228. D. Savoia, E. Tagliavini, C. Trombini, and A. Umani-Ronchi, *J. Org. Chem.* **47**, 876 (1982).

229. H. Schäfer-Stahl, *J. Chem. Soc. Dalton Trans.* 328 (1981).

230. D. Braga, A. Ripamonti, D. Savoia, C. Trombini, and A. Umani-Ronchi, *J. Chem. Soc. Dalton Trans.* 329 (1981).

231. H. Klotz and A. Schneider, *Naturwissenschaften* **49**, 448 (1962).

232. R. Csuk, A. Fürstner, M. Hasslacher, and H. Weidmann, unpublished results.

233. R. Csuk and B. I. Glänzer, unpublished results.

234. R. Csuk, A. Fürstner, and H. Weidmann, unpublished results.

235. R. Csuk, A. Fürstner, and H. Weidmann, *J. Chem. Soc. Chem. Commun.* 1802 (1986).

236. E. J. McMurry, M. P. Fleming, K. L. Kees, and L. R. Krepski, *J. Org. Chem.* **43**, 3255 (1978).

237. E. J. Corey, R. L. Danheiser, and S. Chandrasekaran, *J. Org. Chem.* **41**, 260 (1976).

238. R. Dams, M. Malinowski, I. Westdorp, and H. Y. Geise, *J. Org. Chem.* **47**, 248 (1982).

239. R. Csuk, A. Fürstner, C. Rohrer, and H. Weidmann, submitted (1987).

240. R. Csuk, M. Hasslacher, and H. Weidmann, unpublished results (1986).

241. D. D. Dominguez, H. A. Resing, C. F. Poranski, Jr., and J. S. Murday, *Mater. Res. Soc. Symp. Proc.* **20**, 363 (1983).

242. J. E. Fischer, *Mater. Sci. Eng.* **31**, 211 (1977).

243. T. Tsang and H. A. Resing, *Solid State Commun.* **53**, 39 (1985).

244. M. Rabinovitz, I. Agranat, H. Selig, C.-H. Lin, and L. Ebert, *J. Chem. Res. (S)* 216 (1977).

245. I. Agranat, M. Rabinovitz, H. Selig, and C.-H. Lin, *Synthesis* 267 (1977).

246. H. Selig and O. Gani, *Inorg. Nucl. Chem. Lett.* **11**, 75 (1975).

247. M. Rabinovtiz, I. Agranat, H. Selig, C.-H. Lin, and L. Ebert, *J. Chem. Res. (M)* 2350 (1977).

248. H. Selig, M. Rabinovitz, I. Agranat, C.-H. Lin, and L. Ebert, *J. Am. Chem. Soc.* **99**, 953 (1977).

249. S. S. Yemul, H. B. Kagan, and R. Setton, *Tetrahedron Lett.* **21**, 277 (1980).

250. R. J. P. Corriu, J. M. Fernandez, and C. Guerin, *J. Organomet. Chem.* **192**, 347 (1980).

251. J.-M. Lalancette and J. Lafontaine, *J. Chem. Soc. Chem. Commun.* 815 (1973).
252. G. A. Olah, J. D. Felberg, and K. Lammertsma, *J. Am. Chem. Soc.* **105**, 6529 (1983).
253. K. Laali, M. Muller, and J. Sommer, *J. Chem. Soc. Chem. Commun.* 1088 (1980).
254. J. Bertin, J. L. Luche, H. B. Kagan, and R. Setton, *Tetrahedron Lett.* **9**, 763 (1974).
255. J. L. Luche, J. Bertin, and H. B. Kagan, *Tetrahedron Lett.* **9**, 759 (1974).
256. G. A. Olah and J. Kaspi, *J. Org. Chem.* **42**, 3046 (1977).
257. G. A. Olah, J. Kaspi, and J. Bukala, *J. Org. Chem.* **42**, 4187 (1977).
258. J.-M. Lalancette, M.-J. Fournier-Breault, and R. Thiffault, *Can. J. Chem.* **52**, 589 (1974).
259. H. A. Resing, M. J. Moran, G. R. Miller, L. G. Banks, C. F. Poranski, Jr., and D. C. Weber, *Mater. Res. Soc. Symp. Proc.* **20**, 355 (1983).
260. G. Bewer, N. Wichmann, and H. P. Boehm, *Mater. Sci. Eng.* **31**, 73 (1977).
261. J. G. Hooley, *Mater. Sci. Eng.* **31**, 17 (1977).
262. Y. N. Novikov, V. A. Postnikov, A. V. Nefed'ev, and M. E. Vol'pin, *Isz. Akad. Nauk SSSR, Ser. Khim.* **10**, 2381 (1975).
263. Wacker Chemie Ges., French Patent 1,533,567 (1968); *Chem. Abstr.* **71**, 38323 (1969).
264. H. Brinkel, H. Derleth, and H. Fischer, French Patent 2,104487 (1972); *Chem. Abstr.* **79**, 23968 (1973).
265. R. Setton, *Mater. Sci. Eng.* **31**, 303 (1977).
266. J. Bertin, H. B. Kagan, J. L. Luche, and R. Setton, *J. Am. Chem. Soc.* **96**, 8113 (1974).
267. H. B. Kagan, *Chem. Tech.* 510 (1976).
268. H. B. Kagan, T. Yamagishi, J. C. Motte, and R. Setton, *Isr. J. Chem.* **17**, 274 (1978).
269. J. M. Adams, J. M. Thomas, and M. J. Walter, *J. Chem. Soc. Dalton Trans.* 1459 (1975).
270. M. Armand, *in* "Fast Ion Transport in Solids" (W. van Gool, Ed.), p. 665. North Holland Publ., Amsterdam, 1973.
271. R. G. Gunther, U.S. Appl. 453252 (1974); *Chem. Abstr.* **84**, 47134 (1976).
272. L. B. Ebert, R. A. Huggins, and J. I. Brauman, *Carbon* **12**, 199 (1974).
273. J. G. Hooley and M. Reimer, *Carbon* **13**, 401 (1975).
274. N. Platter and B. de la Martiniere, *Bull. Soc. Chim. Fr.* 177 (1961).
275. L. B. Ebert and H. Seelig, *Mater. Sci. Eng.* **31**, 177 (1977).
276. G. Eichinger and J. O. Besenhard, *J. Electroanal. Chem.* **72**, 1 (1976).
277. L. B. Ebert, *Prepr. Am. Chem. Soc. Petr. Div.* **22**, 69 (1977).
278. A. Page-Lecuyer, J. L. Luche, H. B. Kagan, G. Colin, and C. Mazieres, *Bull. Soc. Chim. Fr.* 1690 (1973).
279. S. H. Stoldt and A. Turk, *J. Org. Chem.* **34**, 2370 (1969).
280. L. B. Ebert and L. Matty, Jr., *Synth. Met.* **4**, 345 (1982).

Nucleophilic Activation of Carbon Monoxide: Applications to Homogeneous Catalysis by Metal Carbonyls of the Water Gas Shift and Related Reactions

PETER C. FORD AND ANDRZEJ ROKICKI

Department of Chemistry
University of California
Santa Barbara, California 93106

I

INTRODUCTION

The goal of this review will be to summarize and discuss the roles played by certain nucleophile adducts of metal-coordinated carbon monoxide:

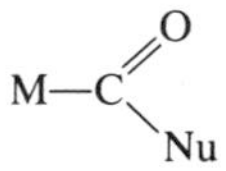

$$M-C\overset{\displaystyle O}{\underset{\displaystyle Nu}{\diagup}}$$

(Nu = a nucleophile) in the stoichiometric and catalytic chemistry of metal carbonyl complexes. The scope of this topic is sufficiently large to preclude a truly comprehensive review of all such species, so this article will focus on relatively recent developments for systems where oxygen or nitrogen bases have been used as the nucleophiles. The syntheses, spectral characterizations, and chemical reaction studies of various nucleophile–carbonyl complexes will be described with the goal of drawing attention to how such adduct formation not only activates the carbonyl directly involved for further reaction but also influences the reactivities of the balance of the complex. This topic has not been the subject of comprehensive review for some time, although reactions of metal carbonyls with oxygen and nitrogen nucleophiles have been addressed in several short reviews (*1–5*) and a rather perceptive discussion of the reactions of nucleophile–carbonyl adducts appears in a 1983 Ph.D. dissertation by Gross (*6*).

139

The potential importance of nucleophilic activation in homogeneous catalysis can be illustrated with the water gas shift reaction (WGSR) [Eq. (1)], which can be catalyzed by alkaline solutions of a number of metal carbonyls, e.g., $Fe(CO)_5$ and $Ru_3(CO)_{12}$ (1). A logical first step for the activation of CO via this system is the reaction between a coordinated carbonyl and hydroxide in solution to give a hydroxycarbonyl adduct [Eq. (2)], which undergoes subsequent decarboxylation to a metal hydride. The prototype for this conversion is Eq. (3), which was reported by Hieber and Leutert over five decades ago (7). Related mechanistic studies on the homogeneous oxidations of CO by aqueous metal ions were first carried out by Halpern and co-workers in the early 1960s (2).

$$H_2O + CO \rightleftharpoons H_2 + CO_2 \tag{1}$$

$$M-CO + OH^- \rightleftharpoons M-CO_2H^- \rightarrow M-H^- + CO_2 \tag{2}$$

$$Fe(CO)_5 + 2OH^- \rightarrow HFe(CO)_4^- + HCO_3^- \tag{3}$$

Nucleophilic activation of coordinated CO has further scope in homogeneous catalysis given that nucleophiles are employed in a variety of other metal carbonyl-catalyzed reactions such as the reductive carbonylation of nitroaromatics, the oxidations and reductions of CO, and the hydrogenations and hydroformylations of alkenes by CO/H_2O mixtures (see below). In addition, hydroxycarbonyls $M-CO_2H$ and related complexes have been often invoked as likely intermediates in proposed schemes for electrochemical and photochemical reductions of carbon dioxide mediated by transition metal complexes (8). Thus, understanding the properties of such nucleophile–carbonyl adducts and elucidating the quantitative mechanisms by which these species react are essential to characterizing the mechanisms for a number of important catalytic transformations.

II

PREPARATION AND PROPERTIES OF NUCLEOPHILE–CARBONYL ADDUCTS

Nucleophile–carbonyl adducts of the types illustrated in Fig. 1 have been characterized as stable complexes, and a number have been detected as spectrally observable intermediates, while the existence of others can only be inferred from kinetic data or products formed. The formation of such adducts is a truly general phenomenon given the number of binary and substituted, mononuclear and cluster, metal carbonyls that have been shown to participate in such reactions. The list of nucleophiles for which adducts have been characterized is similarly broad, including the hydride

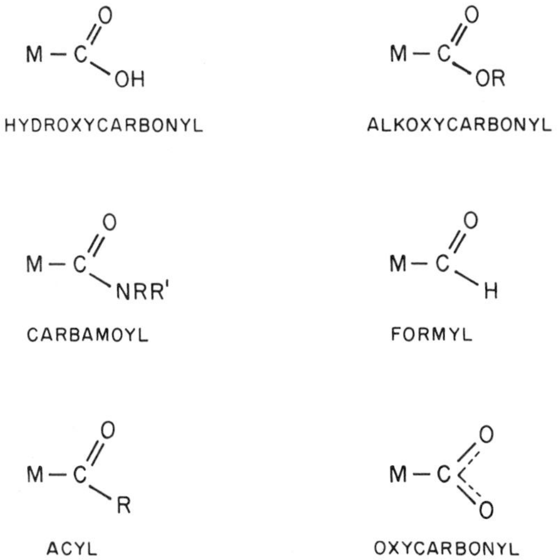

$$M - C \overset{\displaystyle O}{\underset{\displaystyle OH}{\big\langle}}$$

HYDROXYCARBONYL

$$M - C \overset{\displaystyle O}{\underset{\displaystyle OR}{\big\langle}}$$

ALKOXYCARBONYL

$$M - C \overset{\displaystyle O}{\underset{\displaystyle NRR'}{\big\langle}}$$

CARBAMOYL

$$M - C \overset{\displaystyle O}{\underset{\displaystyle H}{\big\langle}}$$

FORMYL

$$M - C \overset{\displaystyle O}{\underset{\displaystyle R}{\big\langle}}$$

ACYL

$$M - C \overset{\displaystyle O}{\underset{\displaystyle O}{\big\langle}}$$

OXYCARBONYL

FIG. 1.   Names and formulas for different types of nucleophile–carbonyl adducts.

anion, various carboanions, a variety of anionic and neutral oxygen and nitrogen nucleophiles, and halides.

A. Synthesis Procedures

Preparation by direct addition of a nucleophile anion Nu⁻ to a coordinated CO has been a widely applicable approach. Selected typical examples of stable, at least partially characterized nucleophile–carbonyl complexes prepared according to Eq. (4) are collected in section A of Table I (9–18). The limitation of this method is the requirement that the combination of the electrophilicity of the coordinated CO and the nucleophilicity of Nu⁻ must exceed some threshold value that depends on a wide variety of parameters discussed below.

$$L_n M(CO)_m + Nu^- \rightarrow L_n M(CO)_{m-1} \left(C \overset{\displaystyle O}{\underset{\displaystyle Nu}{\big\langle}} \right)^- \tag{4}$$

Equations (5) and (6) are variations of this approach where the nucleophile is bound as either the conjugate Brønsted acid or a main group organometallic (sections B and C, Table I) (19–39). Reaction of NuH with metal carbonyls [Eq. (5)] is restricted to strongly activated, usually cationic, metal carbonyl complexes since NuH is always a much weaker nucleophile than the conjugate base. The reaction can be shifted toward products by addition of a base such as NEt₃; however, under such circumstances the actual reaction pathway may indeed be Eq. (4).

TABLE I

EXAMPLES OF CHARACTERIZED NUCLEOPHILE–CARBONYL METAL COMPLEXES LISTED ACCORDING TO
METHOD OF PREPARATION[a]

Complex	Characterization	Ref.

A. $L_nM(CO)_m + Nu^- \rightleftharpoons L_nM(CO)_{m-1}\left(C{\overset{\displaystyle O}{\underset{\displaystyle Nu}{\diagup\atop\diagdown}}}\right)$ [Eq. (4)]

Complex	Characterization	Ref.
$M(CO)_4C(O)OMe]^-$	IR, ^1H NMR, kinetics	9
M = Fe, Ru, Os		
$Co(CO)_4C(O)OR$	IR, ^1H NMR	10
R = Me, Et, iPr, cHx, CH_2Ph		
$ReCp^*(CO)(p-N_2C_6H_4OMe)C(O)OMe$	IR, ^1H NMR, MS	11
$Ru_3(CO)_{11}C(O)OMe]^-$	IR, ^1H and ^{13}C NMR	12
$FeCp(PPh_3)C(O)OH$	IR, ^1H and ^{13}C NMR	13,14
$MoCp(CO)_2(PPh_3)C(O)OH$	IR, ^1H and ^{13}C NMR	14
$ReCp(CO)(p-N_2C_6H_4X)C(O)OH$	IR, ^1H NMR, MS	11
X = Me, MeO, NEt_2		
$Os_3(CO)_{11}C(O)NMe_2]^-$	IR	15
$Fe(1,5-\eta-C_6H_7)(CO)_2C(O)I$	IR	16

$(CO)_3Fe{-}C{\overset{\displaystyle O}{\underset{\displaystyle X}{\diagup\atop\diagdown}}}$ IR, ^1H and ^{13}C NMR, X   17

$R\underset{\uparrow}{\overset{}{\quad\;}}$

X = CR_2, O, NR

$CH_2{\overset{\displaystyle CH_2}{\diagup}}{\underset{\displaystyle O{-}C{\diagdown O}}{\diagdown}}M(CO)_2PR_3$ IR, ^1H and ^{31}P NMR, X   18

M = Mo, W; R + Ph, CH_2Ph

B. $L_nM(CO)_m^{p+} + NuH \rightleftharpoons L_nM(CO)_{m-1}\left(C{\overset{\displaystyle O}{\underset{\displaystyle Nu}{\diagup\atop\diagdown}}}\right)^{(p-1)+} + H^+$ [Eq. (5)]

p = 0, 1, 2

Complex	Characterization	Ref.
$trans$-$PtXL_2C(O)OR$	IR, ^1H NMR, VIS	19–21
X = Cl, Br, I		
R = Me, Et, Pr, i-Pr, etc.		
L = PPh_2Me, PPh_3		
$trans$-$NiC_6Cl_5L_2C(O)OR$	IR, ^1H NMR	22
R = Me, Et		
L = PPh_2Me, $PPhMe_2$		
cis-$IrCl(dppe)_2C(O)OMe]^+$	IR, ^1H, ^{31}P NMR	23
$trans$-$PtCl(PEt_3)_2C(O)OH$	IR, ^1H, ^{13}C, and ^{31}P NMR	24
$FeCp(CO)_2C(O)NHR$	IR, ^1H and ^{13}C NMR	25
R = $(CH_2)_2NMe_2$, $(CH_2)_2OH$		
$[FeCp(CO)_2]_2[\mu$-$C(O)NR(CH_2CH_2)_2NRC(O)]$	IR, ^1H and ^{13}C NMR, X	25
R = H, Me		

TABLE I (*continued*)

Complex	Characterization	Ref.
$MCp(CO)LC(O)NH_2$	IR, ^{1}H and ^{13}C NMR, X	*26,27*
M = Ru, Os		
L = CO, MeCN, MeNC, PPh_3, PEt_3		
cis-$[PtCl_2(CO)C(O)NiPr_2]^-$	IR, X	*28*
$[MnCp(CO)(NO)C(O)NH_2]^+$	IR, X	*29,30*

$$\text{C.}\quad L_nM(CO)_m{}^{p+} + NuY \rightleftharpoons L_nM(CO)_{m-1}\left(C\begin{matrix}{\nearrow}O\\ \searrow Nu\end{matrix}\right)^{(p-1)+} + Y^+ \quad [\text{Eq. (6)}]$$

Y = Li, MgX, HgR, $C(NMe_3)_3$, $HB(OR)_2$, $B(OR)_3$
p = 0, 1, 2
r = 0, 1

Complex	Characterization	Ref.
$Mn(CO)_4(L)C(O)CH_2Ph$	IR, ^{1}H NMR	*31*
L = PPh_3, $PPhMe_2$, $P(OPh)_3$		
$[ReBr(CO)_3(PPh_3)C(O)Me]^-$	IR, ^{1}H and ^{13}C NMR	*32*
$Rh(triphos)(CO)C(O)R$	IR, ^{31}P NMR	*33*
R = H, Me; triphos = $CH(CH_2PPh_2)_3$		
$L_nM[C(O)R]_2{}^{2-}$	IR, ^{1}H NMR, X	*34*
M = Mn, Re, Fe		
R = Me, Et, $PhCH_2$, *i*-Pr, EtO		
L = CO, RNC, Cp		
$ReCp^x(NO)(CO)C(O)H$	IR, ^{1}H NMR, MS	*35*
$[MM'(CO)_9C(O)H]^-$	IR, ^{1}H and ^{13}C NMR	*36*
M, M' = Mn, Re		
$[Ru(diphos)(CO)C(O)H]^-$	IR, ^{1}H and ^{13}C NMR, X	*37*
diphos = dppe, dppm		
$[MoCp(Me)(CO)_2C(O)H]^-$	IR, ^{1}H and ^{13}C NMR	*38*
$M(CO)_nC(O)R^-$	IR, ^{1}H and ^{13}C NMR	*39*
M = Fe, n = 4; M = Ni, n = 3		
R = NMe_2, OMe, OEt		

$$\text{D.}\quad MNu(CO)_m + L \rightleftharpoons LM(CO)_{m-1}C\begin{matrix}{\nearrow}O\\ \searrow Nu\end{matrix} \quad [\text{Eq.(7)}]$$

Complex	Characterization	Ref.
$Mn(CO)_4(^{13}CO)C(O)Me$	IR, kinetics	*40*
$Pt_2(\mu\text{-}X)_2L_2(PhC(O))_2$	IR, ^{31}P NMR	*41*
L = PEt_3, PMe_2Ph, $PMePh_2$, PPh_3, $P(c\text{-}Hx)_3$		
X = Cl, Br, I		
$FeCp(P)(L)C(O)Et$	IR, ^{1}H NMR	*42*
P = PPh_3, $P(OCH_2)_3CMe$		
L = CO, $C_6H_{11}NC$		

TABLE I (*continued*)

Complex	Characterization	Ref.

$$\text{E.} \quad L_nMNu + CO \;\rightleftharpoons\; L_nM-C\overset{\displaystyle O}{\underset{\displaystyle Nu}{\Big\langle}} \qquad [\text{Eq. (8)}]$$

Pt(dppe)(Me)C(O)OMe	IR, ^{1}H and ^{31}P NMR	43
Pt(diphos)(L)C(O)R	IR, ^{1}H, ^{13}C, and ^{31}P NMR	44
diphos = vdpp, dppe, dppp, dppb		
L = Cl, AcO		
R = Me, Et, Ph, c-C$_6$H$_9$		
Ptdiphos(c-C$_6$H$_9$)C(O)OR	IR, ^{1}H, ^{13}C, and ^{31}P NMR	45
diphos = vdpp, dppe, dppp, dppb		
R = H, Me		
trans-PtPh(PEt$_3$)$_2$C(O)OH	IR, ^{1}H, ^{13}C, and ^{31}P NMR, X	46
Ir(CO)$_2$(PPh$_3$)$_2$C(O)OR	IR, ^{1}H NMR	47
R = Me, n-Pr, t-Bu, Ph		

$$PhO(O)C\diagdown\;\begin{array}{c} \diagup CH_2 \diagdown \\ PMe_2 \qquad PMe_2 \\ | \qquad\qquad | \\ Pd \qquad\quad Pd \\ \diagup | \diagdown CO \diagup | \diagdown C(O)OPh \\ PMe_2 \qquad PMe_2 \\ \diagdown CH_2 \diagup \end{array}$$

(structure above)	IR, ^{1}H and ^{13}C NMR	48
Rh(OEP)C(O)H	IR, ^{1}H and ^{13}C NMR, MS, X	49
OEP = octaethylporphyrin		
ThCp$_2'$(OR')(η^2-C(O)R)	IR, ^{1}H and ^{13}C NMR,	50
R = Me, n-Bu, CH$_2t$-Bu	kinetics	
R' = t-Bu, CHt-Bu$_2$		
M(Cp*)$_2$Y(η^2-C(O)NR$_2$),	IR, ^{1}H and ^{13}C NMR, X	51
M = Th, U; R = Me, Et		
Y = Cl, NR$_2$, η^2-C(O)NR$_2$		
ThCp$_3$(η^2-C(O)R)	IR, ^{1}H and ^{13}C NMR	52
R = Me, i-Pr, n-Bu, *neo*-Pe, *sec*-Bu, NEt$_2$		

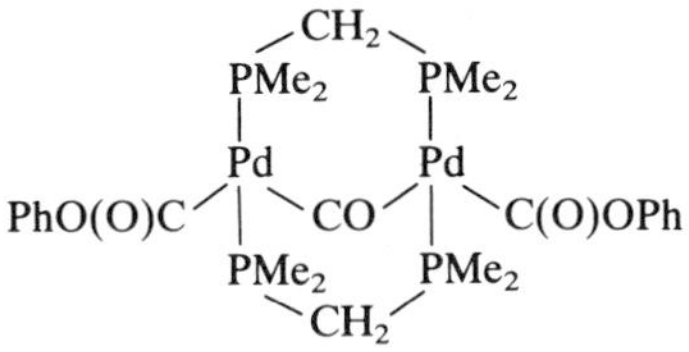

$$\text{F.} \quad L_nM^{m-} + NuC\overset{\displaystyle O}{\underset{\displaystyle X}{\Big\langle}} \;\rightarrow\; LnM\left(C\overset{\displaystyle O}{\underset{\displaystyle Nu}{\Big\langle}}\right)^{(m-1)-} + X^- \qquad [\text{Eq. (10)}]$$

$$m = 0, 1, 2$$
$$X = Cl, Br, MeC(O)O$$

FeCp(CO)$_2$C(O)NR$_2$	IR, ^{1}H NMR	55
R = Me, Et		
Fe(CO)$_4$C(O)H]$^-$	IR, ^{1}H and ^{13}C NMR	56,5?
Co(CO)$_4$C(O)Me	IR	58
Co(CO)$_3$(PPh$_3$)C(O)Et	IR	59
Co(CO)$_4$C(O)OMe	IR, NMR^b	60

TABLE I (*continued*)

G. $L_nM + NuC\overset{O}{\underset{X}{\diagup}} \rightleftharpoons L_n(X)MC\overset{O}{\underset{Nu}{\diagup}}$ [Eq. (11)]

$X = Cl, Br, H$

Complex	Characterization	Ref.
$IrCl_2(CO)L_2C(O)OR$ $L = PMe_2Ph, AsMe_2Ph, PMe_3$ $R = Me, Et, Ph$	IR, ^1H NMR	61
$IrClL_2(CO)(H)C(O)OH$ $L = PMe_2Ph, PEt_2Ph$	IR	62
$MCl(PPh_3)C(O)OR$ $M = Ni, Pd; R = OMe, OEt$	IR, ^1H NMR	63
$PdCl(t\text{-}BuNC)_2C(O)R$ $R = OMe, OEt, OBz$	IR, ^1H NMR	63
$PdClL_3C(O)OR$ $L = PPh_3, PPh_2Me, PEt_3$ $R = Me, Et$	IR, ^1H NMR	64
$OsClL_2(CO)(H)C(O)H$ $L = PPh_3$	IR, ^1H NMR	65
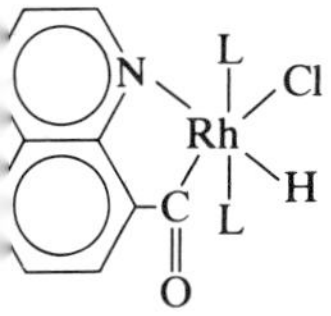 $L = PPh_3$	IR, ^1H NMR	66
$MCl_2(PMe_2Ph)_3C(O)R$ $M = Rh, Ir; R = Me, n\text{-}Pr, i\text{-}Pr$	IR, ^1H and ^{31}P NMR, X	67
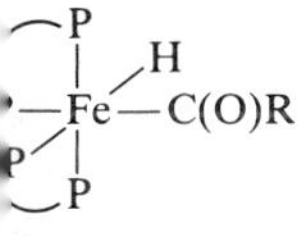 $PP = dmpe; R = Ph, Et$	IR, ^1H and ^{31}P NMR	68
$IrL_3(X)(H)C(O)H$ $X = H, Me, BH_4$ $L = PMe_3$	IR, ^1H, ^{13}C, and ^{31}P NMR	69
$IrL_3Cl(H)C(O)OMe$ $L = PMe_3$	IR, ^1H, ^{13}C, and ^{31}P NMR	69

H. $L_nM{-}Y + CO_2 \rightarrow L_nM{-}C\overset{O}{\underset{OY}{\diagup}}$ [Eq. (13)]

$Co(N_2)(PPh_3)_3C(O)OH$	IR	70

TABLE I (*continued*)

Complex	Characterization	Ref

I. $L_nM(CO_2)^{p-} + R^+ \rightarrow LnMC\overset{\displaystyle O}{\underset{\displaystyle OR}{\diagdown}}{}^{](p-1)-}$ [Eq. (14)]

$p = 1, 0$
$R = Me, H$

$IrCl(dmpe)_2C(O)OMe^+$	IR, X	71

$RNC\diagdown\overset{\displaystyle L}{\underset{\displaystyle L}{\vert}}\diagup\overset{O}{\underset{\,}{}}\diagdown Os\diagup\overset{\displaystyle L'}{\underset{\displaystyle L'}{}}$ ... $\overset{]2+}{}$

with $\underset{\underset{OMe}{\overset{\Vert}{C}-O}}{}$

| 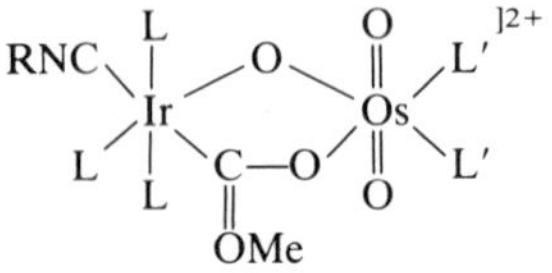 | IR, ^{13}C and ^{31}P NMR, X | 72a |

$L = PPh_3$
$L' = 4\text{-}t\text{-}Bu\text{-pyridine}$
$R = t\text{-}Bu$

| $Fe(Cp)(CO)_2C(O)OMe$ | IR | 72b |

J. $L_nM{-}R + R'NCO \rightarrow L_nM{-}C\overset{\displaystyle O}{\underset{\displaystyle NRR'}{\diagdown}}$ [Eq. (15)]

| $WCp(CO)_3C(O)NHCH_3$ | IR | 73 |
| $Os(\mu\text{-}H)(CO)_{10}(\mu\text{-}p\text{-}MeC_6H_4HC(O)H)$ | IR, X | 74 |

K. $L_nM(CN{-}R)^{p+} + OH^- \rightarrow L_nM{-}C\overset{\displaystyle O}{\underset{\displaystyle NHR'}{\diagdown}}{}^{](p-1)+}$ [Eq. (16)]

$R = H$

| $trans\text{-}Pt(PPh_3)_2(CNMe)C(O)NHMe]^+$ | IR | 75 |

L. $L_nMC\overset{\displaystyle O}{\underset{\displaystyle Nu}{\diagdown}} + Nu'Y^{p-} \rightarrow L_nMC\overset{\displaystyle O}{\underset{\displaystyle Nu'}{\diagdown}} + NuY^{p-}$ [Eq. (17)]

$Y = H, MgBr, BR_3$
$p = 0, 1$

$Pt(dppp)(c\text{-}C_6H_9)C(O)NRR'$	IR, ^{13}C and ^{31}P NMR	45a
$R = R' = Me; R = H, R' = t\text{-}Bu$		
$ReCp(No)[C(O)R](PPh_3)$	IR, ^{1}H and ^{13}C NMR	76
$R = Me, Et, Ph, etc.$		
$IrCl_2(CO(PMe_2Ph)_2C(O)OH$	IR, ^{1}H NMR	77
$FeCp(CO)_2C(O)OMe$	IR	78

^{a} Ligand abbreviations: $Cp^x = Cp, Cp^*$; $Cp = \eta^5\text{-}C_5H_5$; $Cp^* = \eta^5\text{-}C_5Me_5$; cHx = 1-cyclohexenyl diphos $= Ph_2P(CH_2)_nPPh_2$, $n = 1$ dppm, $n = 2$ dppe, $n = 3$ dppp, $n = 4$ dppb.
^{b} No actual data reported.

Synthesis via Eq. (6) has been most frequently used in the preparation of metal acyls and formyls, but some metal alkoxycarbonyls and carbamoyls have also been prepared in this manner.

$$L_nM(CO)_m^{P+} + NuH \rightleftharpoons L_nM(CO)_{m-1}\left(C\overset{\displaystyle O}{\underset{Nu}{\Big\langle}}\right)^{(p-1)+} + H^+ \tag{5}$$

$$L_nM(CO)_m^{P+} + NuY \rightleftharpoons L_n\ddot{M}(CO)_{m-1}\left(C\overset{\displaystyle O}{\underset{Nu}{\Big\langle}}\right)^{(p-1)+} + Y^+ \tag{6}$$

$$Y = Li, MgX, C(NMe_3)_3$$

An alternative approach to preparation of the nucleophile–carbonyl ligand would be insertion of a CO into a previously formed M–Nu bond via migratory insertion [Eq. (7); section D, Table I] (40–42) or bimolecular insertion [Eq. (8); section E, Table I] (43–52). Equation (7) is a common

$$NuM(CO)_m + L \rightleftharpoons LM(CO)_{m-1}\left(C\overset{\displaystyle O}{\underset{Nu}{\Big\langle}}\right) \tag{7}$$

$$L_nMNu + CO \rightleftharpoons L_nM\left(C\overset{\displaystyle O}{\underset{Nu}{\Big\langle}}\right) \tag{8}$$

pathway to acyl complexes, but for the analogous formyls the migratory insertion route is exceedingly rare (53) despite considerable speculation regarding such a pathway in various reaction mechanisms (54). The difference between these two methods may be more apparent than real since the intimate mechanism for Eq. (8) may be stepwise formation of an $L_nM(Nu)CO$ species followed by migratory insertion, e.g., Scheme 1.

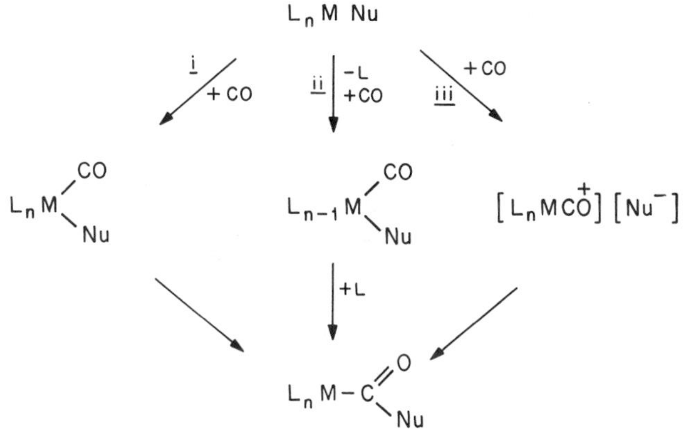

SCHEME 1

The carbonylations of complexes of the type Pt(diphos)(L)R [Table I, section E; Eq. (9)] are examples of such ambiguity. There is direct spectral evidence (low-temperature IR and NMR) (44) for the formation of a carbonyl intermediate consistent with the operation of route *ii* of Scheme 1; however, it was not possible to exclude unequivocally reaction via route 1-*i* as well. For example, carbonylation of the square planar complex Pt(dppe)(OCH$_3$)CH$_3$ to give Pt(dppe)(CO$_2$CH$_3$)CH$_3$ occurs via a rate law first order in [CO], and low-temperature ^{13}C NMR spectra suggest the formation of a five-coordinate intermediate (43). On the other hand, carbonylation of complexes of the type *trans*-Ir(OR)(CO)(PPH$_3$)$_2$ has recently been shown (47) to proceed by the displacement or prior dissociation of the nucleophile RO$^-$, i.e., route 1-*iii*.

$$cis\text{-PtCl}(c\text{-C}_6\text{H}_9)(\text{dppp}) + \text{CO} \rightleftharpoons [cis\text{-Pt}(c\text{-C}_6\text{H}_9)(\text{CO})(\text{dppp})]\text{Cl}$$

$$\rightarrow cis\text{-PtCl}(\text{C(O)}(c\text{-C}_6\text{H}_9)(\text{dppp}) \qquad (9)$$

$$c\text{-C}_6\text{H}_9 = 1\text{-cyclohexenyl; dppp} = 1,3\text{-bis(diphenylphosphino)propane}$$

Another approach to nucleophile–carbonyl adducts is the use of an electron-rich transition metal complex to displace a halide or other leaving group from an organic carbonyl compound [Eq. (10); section 1, Table I] (55–60). This method, which is formally an oxidative addition, allows the preparation of such -C(O)Nu adducts beginning with metal complexes at lower oxidation states than would be expected to be active toward nucleophilic attack via reactions such as Eqs. (4)–(6). Metal acyls, formyls, alkoxycarbonyls, and carbamoyls have all been prepared in this manner.

$$L_n M^{p-} + \text{NuC}\overset{\displaystyle O}{\underset{\displaystyle X}{\diagdown}} \longrightarrow L_n M\left(C\overset{\displaystyle O}{\underset{\displaystyle \text{Nu}}{\diagdown}}\right)^{(p-1)-} + X^- \qquad (10)$$

A related approach, but one requiring a metal complex of lower coordination number, is the oxidative addition represented by Eq. (11), which has been used for the preparation of metal acyls, formyls, and alkoxycarbonyls (Table I, section G) (61–69). For example, the reaction of the formaldehyde with Os(CO)$_3$(PPh$_3$)$_2$ gives the dihapto complex Os(CO)$_3$(PPh$_3$)$_2$(η^2-CH$_2$O), which, when heated, gives a hydride formyl complex [Eq. (12)]. The examples also include the only report of the preparation of a metal hydroxycarbonyl complex via oxidative addition of the C–H bond of formic acid to a metal center (62). Methyl oxalyl chloride also has been used successfully for the preparation of alkoxycarbonyls by both types of oxidative addition (211). These reactions involve the intermediacy of carboalkoxycarbonyl complexes MC(O)C(O)OR, which must undergo decarbonylation to give the alkoxycarbonyl.

$$L_nM + NuC{\overset{O}{\underset{X}{\diagup}}} \longrightarrow L_n(X)M{\overset{O}{\underset{Nu}{\diagup}}} \tag{11}$$

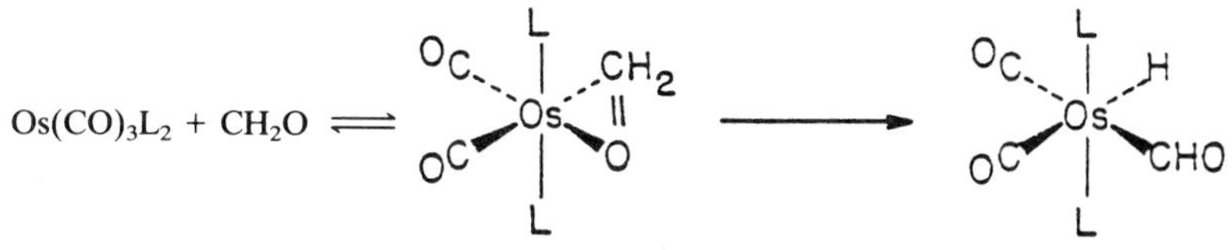

$$Os(CO)_3L_2 + CH_2O \rightleftharpoons \quad\longrightarrow \tag{12}$$

L = PPh$_3$

Apart from the general methods discussed above, a number of other potentially useful approaches to the synthesis of nucleophile–carbonyl complexes have been reported. One interesting approach is the insertion of carbon dioxide into a metal heteroatom bond as in Eq. (13). These reactions are of particular interest given that they represent a means to CO_2 activation. However, only a few alkoxycarbonyl complexes have been synthesized in this way (70), and only indirect evidence has been presented for formation of a hydroxycarbonyl complex via this pathway. The limitation of this route is the requirement for the unfavorable "abnormal" CO_2 insertion to give the carbon-coordinated oxycarbonyl isomer as opposed to the more common insertion yielding oxygen-coordinated carboxylates. Alkoxycarbonyl complexes can also be achieved by O-alkylation of carbon-coordinated CO_2 [Eq. (14)] (71,72) and it should be noted that a hydroxycarbonyl is the conjugate acid of such coordinated CO_2. However, preparations according to Eq. (14) are restricted by the limited number of stable η^1-CO_2 complexes not stabilized by bridging to another metal ion.

$$L_nM{-}Y + CO_2 \longrightarrow L_nMC{\overset{O}{\underset{OY}{\diagup}}} \tag{13}$$

Y = H, R, MgX, Li, K

$$L_nM(CO_2)^{p-} + R^+ \longrightarrow L_nMC{\overset{O^{(p-1)-}}{\underset{OR}{\diagup}}} \tag{14}$$

R = H, CH$_3$; p = 1, 0

Conceptually close to the above are the two methods of preparation of carbamoyls depicted by Eqs. (15) and (16) (73,74). The potential generality of these reactions as synthetic routes remains largely unexplored, as only a few carbamoyls have been prepared via these routes.

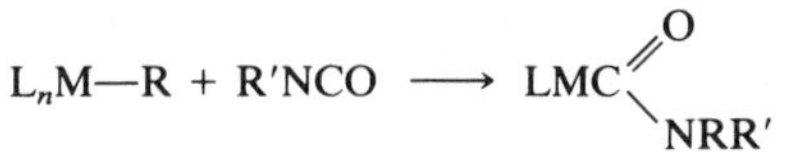

$$L_nM—R + R'NCO \longrightarrow LMC{\overset{\displaystyle O}{\underset{NRR'}{}}} \qquad (15)$$

$$L_nM(CN—R)^{p+} + OH^- \longrightarrow \longrightarrow L_nMC{\overset{\displaystyle O^{(p-1)+}}{\underset{NRH'}{}}} \qquad (16)$$

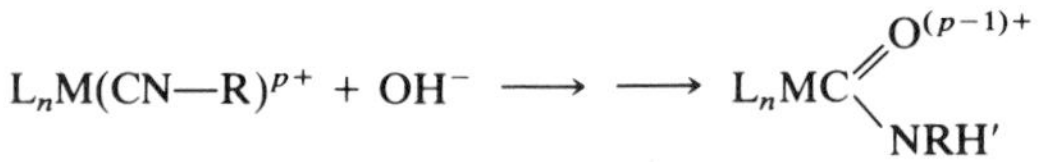

Lastly, the nucleophile–carbonyl complexes themselves can be used as starting points for preparation of other nucleophile carbonyl complexes [Eq. (17); section G, Table I]. Metal acyls, alkoxy- and hydroxycarbonyls, and carbamoyls have been prepared in this manner (50,75–78). In many cases, the chemistry may be dominated by dissociation to afford the metal carbonyl [i.e., the reverse of Eq. (4)] followed by attack of the new nucleophile.

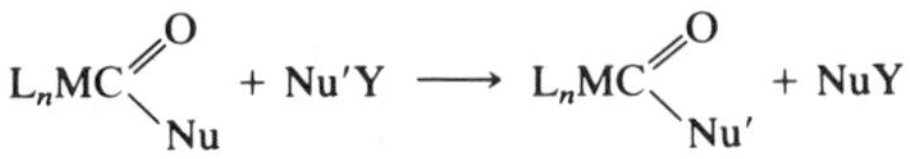

$$L_nMC{\overset{\displaystyle O}{\underset{Nu}{}}} + Nu'Y \longrightarrow L_nMC{\overset{\displaystyle O}{\underset{Nu'}{}}} + NuY \qquad (17)$$

B. Structural Properties

The structure of an η^1-coordinated nucleophile–carbonyl ligand adduct can largely be described in terms of two principal canonical forms A and B, although if Nu has an unshared electron pair, for example, an amide, a third cannonical form C may be a significant contributor.

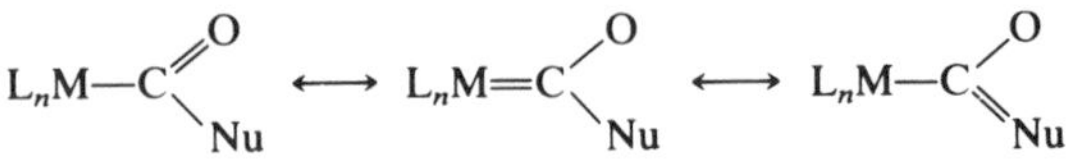

$$L_nM—C{\overset{\displaystyle O}{\underset{Nu}{}}} \longleftrightarrow L_nM{=}C{\overset{\displaystyle O}{\underset{Nu}{}}} \longleftrightarrow L_nM—C{\overset{\displaystyle O}{\underset{Nu}{}}}$$

$$A \qquad\qquad B \qquad\qquad C$$

A would be analogous to the structures expected for organic homologs where M is replaced by a carbon atom. (This analogy has led some authors to refer to hydroxycarbonyl complexes as "metallocarboxylic acids," a term the present authors prefer to avoid. However, to be consistent, formyls probably should be "hydridocarbonyls," acyls "alkyl- or arylcarbonyls," and carbamoyls "amidocarbonyls.") Similarly, the delocalization of electron pair(s) on the Nu group via C would be delocalization seen in organic analogs such as amides. The canonical form B, which is much like that represented for the d_π–p_π backbonding of metallocarbene complexes, would thus be a key feature distinguishing these complexes from the analogous organic systems.

These bonding representations imply an sp^2-hybridized central carbon atom for the M–C(O)Nu group. Consistent with this are the M–C–O bond angles, which generally fall in the 125–130° range, although angles as large as 140° and as small as 115° have been observed (Table II). Moreover, the M–C(O)Nu group is essentially planar (excluding atoms in Nu other than the attachment point) with all relevant atoms generally lying within several picometers of the least-squares plane (67–79,80,82,83,93a,104).

Bond lengths in these L_nM–C(O)Nu groups depend on the nature of the L_nM metal center and of Nu. As expected, the C–O bond lengths of the adducts are greater than those of terminal carbonyls, which lie in the 114–116-pm range. The normal range for the various adducts is 118–122 pm, but values outside this have been observed (Table II). The M–C distances in such species are in general longer than found for similar terminal metal carbonyls, which implies that the extent of π-backbonding from the metal is less for the M–C(O)Nu groups.

Another interesting structural aspect would be effects on other metal–ligand bonds. Substantial elongation of bonds trans to the η^1-C(O)Nu group has been observed for several d^8 square planar complexes, indicating a strong structural trans effect of this function. For example, the Pd–Cl bond of trans-PdCl(η^1-C(O)Nu)(PPh$_3$)$_2$ is 244.6 pm for Nu = n-C$_3$H$_7$ (96a), 243.0 pm for n-C$_6$H$_{13}$ (96b), 243.0 pm for -C(O)CO$_2$CH$_3$ (64b), and 240.7 pm for OCH$_3$ (84), all values much longer than that predicted from the sum of the covalent radii and well within the range of the Pd–Cl distances opposite to other strong trans-effect substituents (117).

Structures have also been determined for the -C(O)Nu functionality in η^2 complexes where both the oxygen and the carbon are coordinated either at a single metal (D) or as a bridge between two metals (E). For such coordination, there appears to be a small systematic lengthening of the C=O bond relative to η^1 complexes, especially for systems where both atoms are coordinated to the same oxophilic metal center. Some general trends are summarized in Table II.

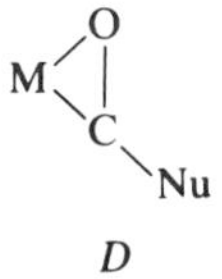

$$D \qquad E$$

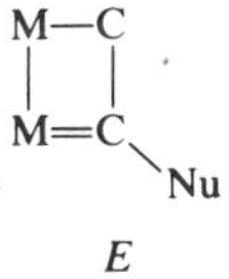

C. Spectroscopic Properties

Nuclear magnetic resonance (especially ^{13}C NMR) and infrared spectroscopy have proved the most used and successful diagnostic tools in characterizing the formation and transformations of various L_nM–(CO)Nu complexes and related species. Table III summarizes some general ranges

TABLE II

Selected Structural Data for Nucleophile Adducts of
Metal Carbonyls and Organic Analogs

Parameter[a]	Nucleophile–carbonyl			Organic analog
	Alkoxycarbonyls[b]			Esters[c]
	η^1	η^2	μ, η^2	
BL(C=O)	113–130	—	126–134	122
BL(C—O)	127–139	—	131–142	133
A(MC=O)	124–133	—	120–130	—
	Hydroxycarbonyls[d]			Carboxylic acids[e]
	η^1	η^2	μ, η^3	
BL(C=O)	123.8	—	128.5	123
BL(C—O)	133.4	—	129.4	136
A(MC=O)	126.7	—	124.3	—
	Formyls[f]			Aldehydes[g]
	η^1	μ, η^2	μ, η^2, η^2	
BL(C=O)	110–122	135	149.6	121
A(MCO)	128–140	133.7	68.1–69.3	124
	Acyls[h]			Ketones[i]
	η^1	η^2	μ, η^2	
BL(C=O)	118–129	118–125	126	124
A(MCO)	120–134	73–87.5	≈120	—
	Carbamoyls[j]			Carboxyamides[k]
	η^1	η^2	μ, η^2	
BL(C=O)	118–136	127–137	125–131	124
BL(C—N)	130–144	124–135	133–136	134
A(MCO)	115–140	70–75	110–124	115

[a] BL, Bond length (picometers); A, angle (degrees).
[b] Refs. (17,18,43,60,72a,79–87).
[c] Ref. (88, vol. 2, p. 909).
[d] Refs. (46,87).
[e] Ref. (88, Vol. 2, pp. 628–629).
[f] Refs. (49b,90–95).
[g] Ref. (88, Vol.2, pp. 944, 945).
[h] Refs. (40,67,81,96–103).
[i] Ref. (88, Vol. 2, p. 1019).
[j] Refs. (25,27,28,30,51,74,104–116).
[k] Ref. (88, Vol. 2, pp. 986–990).

for the infrared stretching frequencies (ν) and ^{13}C chemical shifts (δ_C) for various nucleophile adducts. Comparisons are also made for organic compounds R–C(O)Nu having homologous formulas.

As would be expected, given the bond order reduction for the metal carbonyl group on formation of the nucleophile adduct, the ν_{CO} stretching vibrations of the η^1-C(O)Nu group are shifted by some 300 to 500 cm^{-1} to

TABLE III

SELECTED IR DATA (ν IN cm^{-1}) AND NMR DATA (δ_c AND δ_H IN ppm DOWNFIELD FROM TMS) FOR VARIOUS NUCLEOPHILE–CARBONYL COMPLEXES AND ORGANIC ANALOGS

Parameter	Nucleophile–carbonyl			Organic analog[a]
	Alkoxycarbonyls[b]			Esters
	η^1	μ, η^2		
ν(C=O)	1580–1703	1255		1720–1765
ν(C—O)	1000–1080	—[c]		1180–1290
δ_c	158–215	221		160–180
	Hydroxycarbonyls[d]			Carboxylic acids
	η^1	μ, η^3		
ν(O—H)	3435–2650	3700		3350[e]
				2400–3000[f]
ν(C=O)	1565–1658	1595		1680–1760
δ_c	173–218	—[c]		160–185
δ_H	7.8–11	—[c]		10–13
	Formyls[g]			Aldehydes
	μ^1	η^2	μ, η^{2h}	
ν(C=O)	1550–1700	1450–1480	1160–1190	1695–1725
ν(C—H)	2500–2720[i]	—[c]	—[c]	2720–2830
δ_c	225–310[j]	355–372[k]	265.5	190–205
δ_H	11–17[j]	—[c]	11–13.3	9.3–10
	Acyls[l]			Ketones
	η^1	η^2	μ, η^2	
ν(C=O)	1500–1685	1425–1620	1430–1530	1650–1750
δ_c	216–307[m]	254–362[k]	295–360	185–220
	Carbamoyls[n]			Carboxyamides
	η^1	η^2	μ, η^2	
ν(N—H)	3120–3495	—[c]	3340–3440	3380–3540
ν(C=O)	1480–1630	1490–1560	1235–1514	1640–1680
δ_C	183–235	248.5	189–250	160–180
	Halocarbonyls[o]			Acid halides
ν(C=O)	1655–1735			1765–1810
δ_C	—			159–175

[a] IR data, Ref. (118a); ^{13}C NMR data, Ref. (119); ^{1}H NMR data, Ref. (88).

[b] Refs. (9–12,17,18,22,23,39,43,45,47,48,58,61,62,64,69,71,72,80–82,84–86,120–133).

[c] Not reported.

[d] Refs. (9,11a,13,14,23,24,35,45a,46,48,61,62,77,116,134–140).

[e] Free.

[f] H-bound.

[g] Refs. (42,49,56,57,65,90–95,116,141–145).

[h] For μ, η^2, η^2 (CHO), δ_C = 150–190 ppm, η_H = 5–7.5 ppm, Ref. (94b).

[i] Two or even three bands in this region have been reported as a result of Fermi resonance of ν(C–H) and first overtone of δ(C–H).

[j] Usual range, much lower values have been observed in Rh(OCP)C(O)H, δ_C = 194.4 ppm and δ_H = 2.9 ppm, due to field effects of porphyrin ligand, Ref. (49a, c).

[k] Contribution of oxycarbene form (M–O–CR) predominates, Ref. (50–52).

[l] Refs: (32,40,50–52,57,67,68,76,97,100–103,116,143,146).

[m] Values of ν_{cO} = 1490 cm^{-1} and δ_C = 340.1 ppm recorded for fac-Re(CO)$_3$diphos-[C(O)SiPh$_3$] (147).

[n] Refs. (25,26,28,30,51,74,104–116,148–150).

[o] Refs. (16,151,152).

lower frequency relative to the ν_{CO} bands in the parent carbonyls. The ν_{CO} of η^1-C(O)Nu appears in the 1735–1480-cm^{-1} range with the exact position depending on the nature of Nu and L_nM, the formal charge on the complex, and the physical state of the sample. The ν_{CO} values parallel those for the analogous organic compounds (118b) but as a rule appear at about 70–100 cm^{-1} lower frequency, consistent with contributions from canonical structure B to the bonding. The intensity of this transition is generally less than normally found for terminal metal carbonyls but is still strong compared to those of other bands in the spectra of such complexes. Coordination in the η^2 mode lowers ν_{CO} further, with the largest shifts being observed for mononuclear η^2 complexes of actinide (51,99) and for the dinuclear complex $W(Cp)_2(\mu\text{-}\eta^2\text{-}CHO)ZrH(Cp^*)_2$. Very low ν_{CO} values have therefore been taken as evidence for η^2 nucleophile–carbonyl complexes. However, the recent report (103) that the structurally characterized complex $FeCp(P(CH_3)_3)_2(\eta^1\text{-}C(O))C_2H_5$ displays a very low ν_{CO} value of 1520 cm^{-1} suggests that the correlation is equivocal.

The ^{13}C chemical shifts of the central carbon of the η^1-C(O)Nu group have been found to fall in the 216–310-ppm frequency range for metal acyls and formyls and the lower-frequency 158–218-ppm range for alkoxycarbonyls, hydroxycarbonyls, and carbamoyls (Table III). The ^{13}C chemical shift range extends as high as 362 ppm for acyls η^2 coordinated to strongly oxophilic metal centers and 372 ppm for η^2 formyls (50). Similarly, η^2 coordination of a carbamoyl gives a higher δ_C; for example, a value of 248.5 ppm was recorded for $Cp_2^*ThCl(\eta^2\text{-}C(O)NEt_2)$ (51). Invariably, δ_C for a nucleophile-carbonyl adduct is larger than that of the parent carbonyl. Qualitatively there appears to be an inverse correlation between the δ_C and ν_{CO} values for various -C(O)Nu functions.

As noted above, there is structural evidence for a strong trans effect of the η^1-nucleophile–carbonyl ligand in certain square planar d^8 complexes. This has also been probed by spectroscopic techniques. In the family of complexes Pt(dppp)(R)Y (where R = 1-cyclohexenyl), the coupling constant (J_{PtP}) between the platinum and the phosphorus trans to Y falls in the order for various Y: Cl (4327 Hz) > CH$_3$O (3466) > CO (3323) > CN (2916) > PPh$_3$ (2910) > -CH$_2$CN (2477) > -C(O)OCH$_3$ (1882) > -C(O)OH (1874) > CH$_3$ (1841) > Ph (1758) > -C(O)N(CH$_3$)$_2$ (1698) (45c), the lower values of J_{PtP} indicating increased trans influence (153). A similar series has been noted for the Ir–Cl stretching frequency ν_{Ir-Cl} for various Y in the complexes trans-IrCl(Y)(dppe)$_2^{n+}$, which follow the order for various Y: CO (310 cm^{-1}) > -C(O)OCH$_3$ (270) > H (255) > (CO)H (220) (23). An analogous order has been observed for the ν_{Ir-H} stretching frequency for the series trans-IrH(Y)(dppe)$_2^{n+}$: Cl (2220 cm^{-1}) > CO (2160) > -C(O)OH (2080) > -C(O)H (1940) (23). These sets of observations are self-consistent and indicate the very strong trans effect of the nucleophile–carbonyl ligands in d^8 square planar complexes, with the formyl, acyl, and carbamoyls being somewhat stronger in this regard than the hydroxy- and alkoxycarbonyls.

D. Alkoxycarbonyl, Hydroxycarbonyl, and Carbamoyl Complexes

Figure 2a indicates the transition metals for which alkoxycarbonyl complexes have been structurally and/or spectrally characterized. The methods of preparation are diverse, including those illustrated by Eqs. (4), (5), (8), (10), (11), (17), (6), (13), (14), and (16) in order of relative

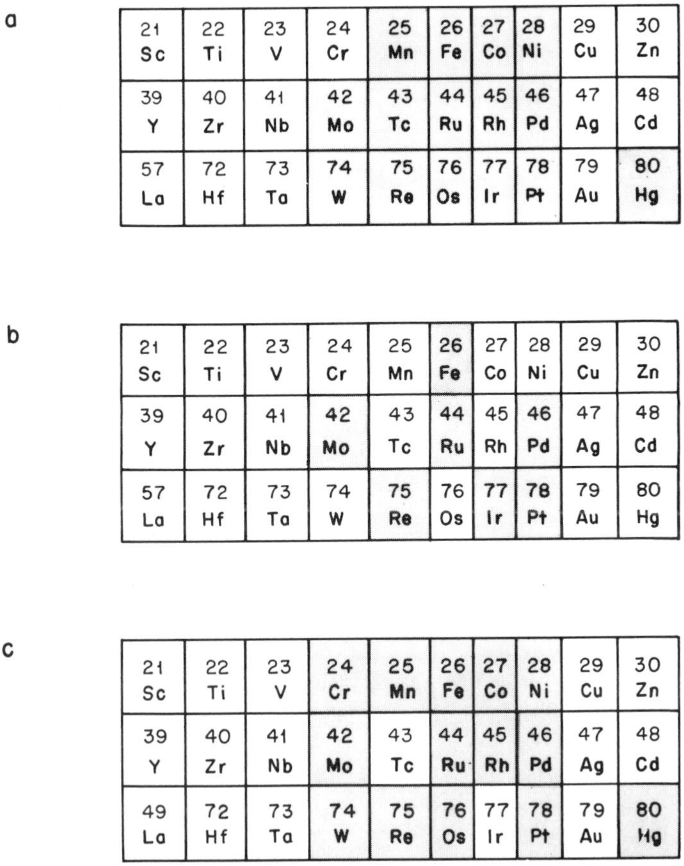

a

21 Sc	22 Ti	23 V	24 Cr	25 Mn	26 Fe	27 Co	28 Ni	29 Cu	30 Zn
39 Y	40 Zr	41 Nb	42 Mo	43 Tc	44 Ru	45 Rh	46 Pd	47 Ag	48 Cd
57 La	72 Hf	73 Ta	74 W	75 Re	76 Os	77 Ir	78 Pt	79 Au	80 Hg

b

21 Sc	22 Ti	23 V	24 Cr	25 Mn	26 Fe	27 Co	28 Ni	29 Cu	30 Zn
39 Y	40 Zr	41 Nb	42 Mo	43 Tc	44 Ru	45 Rh	46 Pd	47 Ag	48 Cd
57 La	72 Hf	73 Ta	74 W	75 Re	76 Os	77 Ir	78 Pt	79 Au	80 Hg

c

21 Sc	22 Ti	23 V	24 Cr	25 Mn	26 Fe	27 Co	28 Ni	29 Cu	30 Zn
39 Y	40 Zr	41 Nb	42 Mo	43 Tc	44 Ru	45 Rh	46 Pd	47 Ag	48 Cd
49 La	72 Hf	73 Ta	74 W	75 Re	76 Os	77 Ir	78 Pt	79 Au	80 Hg

FIG. 2. Transition metals (shaded) for which (a) alkoxycarbonyl, (b) hydroxycarbonyl, and (c) carbamoyl complexes have been characterized.

importance. In general, stable η^1-CO_2R complexes are most numerous for the third-row late transition elements, especially platinum, presumably owing to the greater thermodynamic or kinetic stabilities of these species. Methoxycarbonyls are (unsurprisingly) the most numerous. There are a few examples of structurally characterized μ-η^2 alkoxycarbonyl complexes but none with η^2 coordination of the -CO_2R to a single metal.

Spectrally, the ν_{CO} and δ_C values for alkoxycarbonyl groups vary over much wider ranges than do the analogous parameters for organic esters, reflecting largely the diversity of the metal centers and of their coordination spheres. For example, substitution of a more electron-donating ligand shifts the alkoxycarbonyl ν_{CO} to lower frequency, cf. $Co(CO)_4CO_2CH_3$ (1691 cm^{-1}) versus $Co(CO)_3(PPh_3)CO_2CH_3$ (1669 cm^{-1}) ($10b$). Similarly, for isoelectronic complexes, those which have the lower oxidation state display the lower ν_{CO} frequencies, e.g., $Co(CO)_4CO_2CH_3$ (1691 cm^{-1}) versus $Fe(CO)_4CO_2CH_3^-$ (1621 cm^{-1}) ($9,10b$). In contrast, the differences in this parameter are small for homologous complexes of a specific triad such as $M(CO)_4CO_2CH_3^-$ (M = Fe, Ru, or Os) (9) or $MCl(PPh_3)_2CO_2C_2H_5$ (M = Ni, Pd, or Pt) ($142,150,152$).

Figure 2b indicates the metals for which the less abundant hydroxycarbonyl complexes have been characterized. These tend to be grouped even more distinctly at the late transition metals. The differences may be in large part due to the lower stabilities of the hydroxycarbonyls, which have several pathways for decomposition, principally decarboxylation, not available to the alkoxycarbonyls (see below). Methods of preparation include those illustrated in Eqs. (4), (5), (8), and (17) with less conclusive evidence for Eqs. (11) and (12). Table IV lists the hydroxycarbonyl complexes for which reasonably convincing characterizations have been reported. The most easily accessible spectral parameters ν_{CO} and δ_C are essentially indistinguishable from those of analogous alkoxycarbonyls. The δ_H values for the hydroxycarbonyl hydrogen in the proton NMR has proved difficult to identify owing to exchange with traces of H_2O or other proton sources in solution, although these problems have been met by using a solvent such as DMSO-d_6 ($46b$) or recording spectra at low temperature (75). The ν_{OH} vibration of the hydroxycarbonyl group, although weak, proved informative in certain cases. For example, Bennett and Rokicki ($45a,46,134$) showed that the solid-state IR spectra of Pt(II) hydroxycarbonyl complexes indicated that the -CO_2H groups are strongly hydrogen-bonded (ν_{OH} = 2643–2690 cm^{-1}), probably due to the formation of dimers. In dichloromethane solution, dissociation to monomers was apparent (ν_{OH} 3428–3440 cm^{-1}) although the extent of such dissociation was a function of the coordination sphere.

The only X-ray crystal structure of a hydroxycarbonyl complex reported to date is of *trans*-PtPh(PtEt$_3$)$_2$CO$_2$H (*46*). This clearly shows that the complex in the solid state is dimerized via medium-strength (O$\cdots$H distance of 270 pm) hydrogen bonds, verifying the conclusions drawn from the IR data. The bond lengths and angles of the -CO$_2$H group are consistent with those seen for organic carboxylic acids (Table II). However, the Pt–C bond length for Pt–CO$_2$H at 205.0 pm is shorter than that for Pt–Ph at 207.1 pm, a small reduction that is evidence supporting the contribution of the carbene resonance structure *B* to the overall bonding. Similarly shortened metal–carbon bonds have been noted for the alkoxycarbonyl complexes (*17b*), consistent with the analogies drawn between these systems.

Figure 2c illustrates the broad collection of transition metals for which carbamoyl complexes M–C(O)NRR' have been characterized. In addition such complexes of thorium and uranium have been reported (*51*); thus carbamoyl complexes are approximately as widespread and abundant as the alkoxycarbonyls. There is no obvious correlation between the stabilities of the carbamoyl complexes and the natures of R and R' except when one or both are protons. In such cases, reaction with base leading to formation of isocyanates [Eq. 18] is a potential decomposition pathway analogous to the base-catalyzed decarboxylation of hydroxycarbonyls (see below). Methods by which carbamoyl complexes have been prepared include Eqs. (5), (4), (17), (6), (10), (15), (16), and (8) in order of relative importance.

$$L_nMC{\overset{\displaystyle O}{\underset{\displaystyle NHR}{\Big\backslash}}} + B: \rightarrow L_nM^- + RNCO + BH^+ \qquad (18)$$

Spectral properties of carbamoyl complexes parallel those of analogous alkoxycarbonyl complexes, although the carbonyl stretching vibrations ν_{CO} are usually found at lower frequencies (1480–1630 cm^{-1}) for the η^1 carbamoyls. A much lower value (1235 cm^{-1}) was found for the μ-η^2 carbamoyl in the binuclear complex [Fe(CO)$_4$C(O)NR$_2$]Ni (*111*). The ^{13}C NMR spectra exhibit δ_C values in the 183–250-ppm range, shifted to considerably higher frequency from those for the organic amides in analogy with other nucleophile–carbonyl adducts (*145*). Proton NMR spectra show the two substituents to be nonequivalent at lower temperatures, an indiction of the slow rotation around the carbon–nitrogen bond and consistent with the contribution of the canonical structure *C* to the overall bonding. The coalescence temperature for these nonequivalent signals varies significantly with the nature of the complex (*3,26,130*).

TABLE IV

Isolated Hydroxycarbonyl Metal Complexes

		Spectral data				
		IR ν (cm^{-1})[a]		NMR δ (ppm)[b]		
No.	Compound	OH	C=O	H	C	
1	FeCp(PPh$_3$)(CO)CO$_2$H	2700	1565	9.36	214.3	*13, 138*
2	Fe(1,5-η-C$_6$H$_7$)(CO)$_2$CO$_2$H	—	1658[c]	—	—	*137*
3	MoCp(CO)$_2$(PPh$_3$)CO$_2$H	—	1616	—	209.5	*14*
4	RuCp(CO)(PPh$_3$)CO$_2$H	—	1593	—	199.0	*138*
5	Ru(bpy)$_2$(CO)CO$_2$H	3070[d]	1587[d]	—	—	*139*
6	ReCp(NO)(PPh$_3$)CO$_2$H	—	1591[e]	8.8[f]	197.1[f]	*140*
7	ReCp(NO)(CO)CO$_2$H	2960, 2860, 2705, 2690	1631	9.47	—	*35*
			1585	8.6[h]	—	*136*
8	ReCp(CO)(p-N$_2$C$_6$H$_4$R)CO$_2$H	2240[i,j]	1662	7.8	—	*11*
	R = Me, OMe, NEt$_2$	$-$1550[e,k]		$-$8.9[l]		
9	ReCp*(CO)(p-N$_2$C$_6$H$_4$R)CO$_2$H	—	1623	9.41[m]	—	*11a*
	R = OMe		$-$1587[e]·k			
10	Re$_3$(CO)$_{14}$(μ-η^3-CO$_2$H)	3700[g]	1595	—	—	*89*
11	IrCl$_2$(CO)(PMe$_2$Ph)$_2$CO$_2$H	3313[g]	1663[g]	—	—	*61*
		3295[n]	1677[n]	—	—	*61*
		3295	1670	—	—	*61*
			1670[e]	9.4	—	*77*
						77
12	IrCl$_2$(CO)(AsMe$_2$Ph)$_2$CO$_2$H	3306[n]	1662[g]	—	—	*61*
13	IrCl(PR$_2$Ph)$_2$(CO)(H)CO$_2$H					
	R = Me	3380[d]	1635[d]	—	—	*62*
	R = Et	3380[d]	1660[d]	—	—	
14	*trans*-IrH(dppe)$_2$CO$_2$H$^+$	3435[e]	1610	—	—	*23*

158

No.	Compound					Ref.
15	PtR(diphos)CO$_2$H					
	R = Me, diphos = dppe	—	1635m^o, 1601s^o	—	—	134
	R = Me, diphos = dppp	2680w^o	1635s^e, 1600m^e	—	200.5	46
	R = Ph, diphos = dppp	3440w^e, 2680vw^e, 3438w^e, 2688w^o	1611s^o, 1601s^e	—	197.7	46
16	PtR(diphos)CO$_2$H R = C$_6$H$_9$, C(O)C$_6$H$_9$; diphos = vdpp, dppe, dppp, dppb	2670–2690w^o, 3428–3441w^e	1630–1651s^o, 1615–1656s^e	9.26–9.82^p	195.5–197.9^r	45a
17	PtR(vdpp)CO$_2$H					128
	R = CF$_3$	2685	1610	—	—	
	R = CH$_3$	2685	1604	—	—	
18	trans-PtCl(PEt$_3$)$_2$CO$_2$H	3140, 2240^j, 2040^j	1595	8.5^h, 8.8^t, 11.0^u	173.4^s	24
19	trans-PtPh(PEt$_3$)$_2$CO$_2$H	2643w^n, 3433w^e, 2653w^e	1627w^n, 1592s^n, 1583s^e, 1583s^e	8.1, 9.8^u, 10.8^t	205.1, 208.6^s	46

[a] In Nujol except where indicated.
[b] In CD$_2$Cl$_2$ except where indicated.
[c] Hexane.
[d] Not reported.
[e] CH$_2$Cl$_2$.
[f] Personal communication from J. Gladzs.
[g] CHCl$_3$.
[h] CD$_3$CN.
[i] Fluorolube.
[j] ν_{OD}.
[k] Two or three bands due to strongly coupled ν(NN) + ν(CO) + δ(COH) vibrations.
[l] CDCl$_3$.
[m] (CD$_3$)$_2$CO.
[n] Hexachlorobutadiene.
[o] KBr.
[p] DMSO-d_6.
[r] Various solvents.
[s] 5:1 (vol/vol) (CD$_3$)$_2$CO/H$_2$O.
[t] C$_6$D$_6$.
[u] C$_6$D$_5$CH$_3$.

III

EQUILIBRIA AND KINETICS OF ADDUCT FORMATION

A. Thermodynamic and Theoretical Studies

Carbon monoxide in its free state is rather unreactive. For example, the uncatalyzed reaction of CO with aqueous hydroxide to give formate ion [Eq. (19)] is rather slow, with a rate constant of only $2.7 \times 10^{-5}\,\mathrm{atm^{-1}\,s^{-1}}$ at 62°C (153) despite having a very favorable free energy [$\Delta G^0(25°C) = -9.6\,\mathrm{kcal/mol}$] and exothermic enthalpy change [$\Delta H^0(25°C) = -16.6\,\mathrm{kcal/mol}$] (154). Even allowing for the low solubility of CO in water [$\sim 6 \times 10^{-4}$ mol/liter-atm (155), which gives a calculated second-order rate constant of about 0.05 $M^{-1}\,\mathrm{s^{-1}}$ under these conditions], the rate is relatively sluggish. The reactivity of CO toward such nucleophiles is clearly accelerated by coordination. For example, the rate constant for formation of the hydroxycarbonyl adduct $Ru_3(CO)_{11}(CO_2H)^-$ depicted in Eq. (20) has been determined to be $1.2 \times 10^4\ M^{-1}\ \mathrm{s^{-1}}$ in 10/90 (vol/vol) H_2O/THF at 25°C (156), many orders of magnitude higher than the rate of reaction of free CO with hydroxide. However, it is important to note that the two reactions and the conditions under which they have been studied are not the same.

$$OH^-(aq) + CO(g) \ \rightarrow \ HCO_2{}^-(aq) \qquad (19)$$

$$Ru_3(CO)_{12} + OH^- \ \rightarrow \ Ru_3(CO)_{11}(CO_2H)^- \qquad (20)$$

Such enhanced reactivity can be attributed to coordination-induced changes in the electronic distribution within the CO as well as to the ability of the metal complex to act as an electron sink throughout the reaction pathway. For example, theoretical analysis (157) of the photoelectron spectra for metal carbonyls such as $Fe(CO)_5$ or $Fe(CO)_4(\eta^2\text{-}C_2H_4)$ led to the conclusions that the atomic charges on both the metal (about +1.1) and the CO carbons (about +0.16) are positive, whereas the carbonyl oxygens are negatively charged (about −0.38). In comparison, the dipole moment of free CO is small, 0.12 D (158), and the calculated charge separation is only 0.02e, the carbon possibly being negative. [Similarly, the enhanced reactivity of the carbonyl oxygens toward electrophiles (159) may be rationalized in terms of this electronic redistribution and a low-valent metal's ability to act as an electron source.] For the uncatalyzed hydrolysis of CO to give formate ion, the logical intermediate would be the unsupported hydroxycarbonyl anion CO_2H^-. The calculated energy of this species is about 35 kcal/mol greater than that of the formate ion product

HCO_2^- (*160*), so one can estimate an energy change of about $+20$ kcal/mol to form CO_2H^- in the first step of CO hydrolysis, an estimate nearly equal to the apparent E_a (*22* kcal/mol) reported for the uncatalyzed reaction (*153*). Thus, the metal atom activates the CO for reaction with hydroxide by stabilizing the high-energy CO_2H^- intermediate. However, it is also apparent that this stabilization changes the course of the reaction, which under mild conditions stops at the metal–hydroxycarbonyl complex $M-CO_2H^-$ or proceeds via subsequent decarboxylation as in Eq. (2).

The susceptibility of a coordinated carbonyl to nucleophilic attack in a reaction such as Eq. (4) can be qualitatively correlated with the electrophilicity of the carbonyl carbon. Thus, a more electron-withdrawing L_nM fragment would be expected to make the carbonyl more positive and more susceptible reaction with nucleophiles. For example, the cationic Re(I) complex $Re(CO)_6^+$ undergoes facile ^{18}O enrichment of the carbonyls by exchange with the oxygens of ^{18}O-labeled water, presumably via intermediacy of a hydroxycarbonyl species formed by the attack of water on the coordinated CO [Eq. (21)] (*4,161–163*). In contrast, the isoelectronic but uncharged tungsten complex $W(CO)_6$ does not undergo such exchange under analogous condition (*162*). Similarly, modification of the ligand field by, for example, replacing one or more carbonyls with more electron-donating phosphines will be reflected in the decreased reactivity of the remaining carbonyls. While one should be cautious to consider also the consequences of steric effects, it is clear that such modifications have important implications with regard to the effective electrophilicity of the carbonyl reaction site. For example the rates of H_2O/CO oxygen exchange for the Re(I) and Mn(I) cations follow the order $M(CO)_6^+ > M(CO)_5L^+ \gg M(CO)_4L_2^+$. Another example of this effect is the demonstration by Pettit and co-workers that the dicarbonyl complex $(\eta^5-C_5H_5)Fe(CO)_2(PPh_3)^+$ reacts readily with hydroxide to form a hydroxycarbonyl [Eq. (22)], but the bis-phosphine derivative $(\eta^5-C_5H_5)Fe(CO)(dppe)^+$ [dppe = 1,2-diphenylphosphino)ethane] does not (*13*).

$$Re(CO)_6^+ + H_2O^* \rightleftharpoons Re(CO^*)_6^+ + H_2O \qquad (21)$$

$$(\eta^5-C_5H_5)Fe(PPh_3)(CO)_2^+ + OH^- \rightarrow (\eta^5-C_5H_5)Fe(PPh_3)(CO)(CO_2H) \qquad (22)$$

Similar qualitative and quantitative observations were the basis of empirical correlations proposed by Darensbourg and Darensbourg (164) and by Angelici and Beacic (*165*) and aimed at predicting the relative reactivities of metal carbonyls toward nucleophilic addition. These correlations are based on the positions of the terminal CO stretching frequencies ν_{CO} in the vibrational spectra or, more precisely, on the force constants

k_{CO} of these transitions. The premise of this treatment is that a higher k_{CO} indicates a more positive carbonyl carbon. The correlation proved particularly successful in rationalizing the stereochemical site of the CO subject to such adduct formation. For example, analysis of the IR spectra of the Group VI carbonyls $M(CO)_5L$ (M = Cr, Mo, or W; L = a phosphine or phosphite ligand) shows that the carbonyls cis to L have the higher k_{CO} values, and alkylation with the Grignard reagent $C_6H_5CH_2MgCl$ in THF gives the cis acyl product except when L is the very bulky tri-o-toluylphosphine. Qualitatively, it was also observed that the reaction rate of Grignard addition is lower for complexes having lower k_{CO} values (Table V); for example, the hexacarbonyls are considerably more reactive than the various substituted $M(CO)_5L$ species. In a related development, Bush and Angelici (166) have recently shown that the relative electrophilicity of the species L_nM estimated from the CO force constants of $L_nM(CO)$ can be used to predict relative reactivities of the L' of L_nML' toward nucleophiles, where L' is ethylene or another unsaturated hydrocarbon.

TABLE V

RATE CONSTANTS[a] AND CARBONYL FORCE CONSTANTS k_{CO} FOR
REACTION A IN 27°C THF[b]

(A) $LM(CO)_5 + PhCH_2MgCl \rightarrow cis\text{-}LM(CO)_4 \left(C \overset{\text{O}}{\underset{CH_2Ph}{\diagup\diagdown}} \right) + MgCl^+$

M	L	$k \times 10^3$ (m/s)	k_{CO} (trans)	k_{CO} (cis)[c]
W	CO	108	16.41	16.41
W	$P(OEt)_3$	25.7	15.70	15.89
W	$P(CH_3)_3$	19.1	15.53	15.77
W	PPh_3	5.0	15.57	15.89
W	$P(c\text{-}C_6H_{11})_3$	1.4	15.33	15.68
Mo	CO	91.4	16.52	16.52
Mo	$P(CH_3)_3$	14.8	15.59	15.86
Mo	$P(n\text{-}Bu)_3$	11.0	15.56	15.88
Mo	PPh_3	4.8	15.49	15.96
Cr	CO	53.0	16.49	16.49
Cr	$P(OEt_3)_3$	3.2	15.74	15.88
Cr	$P(CH_3)_3$	3.0	15.54	15.75
Cr	PPh_3	3.25	15.51	15.85

[a] $[PhCH_2MgCl] = 0.12\ M$.
[b] Data from Ref. (164).
[c] In dynes per angstrom.

Ab initio LCAO–MO–SCF calculations have been used by Dedieu and Nakamura (*167*) to address the question of nucleophilic addition to carbonyl ligands. These authors concluded that the prototype reaction of H^- (free hydride) and the trigonal bipyramidal $Fe(CO)_5$ to give $Fe(CO)_4CHO^-$ would occur at the axial carbonyl, would have no activation barrier, and would be highly exothermic [ΔH(calc) = -69.2 kcal/mol]. Such calculations refer to the reaction thermodynamics in the absence of solvent effects etc., which will be discussed below. They also concluded that the high exothermicity was the result of involvement of the σ orbital of the CO in a bonding interaction with the metal and the presence of a low-lying orbital (in this case the d_{z^2} orbital) to stabilize the bonding combination of the π^*_{CO} and S_H orbitals. The minimum approach angle was concluded to be about 120° from the Fe–C bond, and this would occur simultaneously with bending of the attacked carbonyl to relieve the repulsive interaction between the negative oxygen atom and the hydride as well as allowing optimal mixing of the π^*_{CO}, S_H, and d_{z^2} orbitals. The importance of the interaction with the metal center is illustrated by the conclusion that the analogous hypothetical reaction of H^- plus CO to form the HCO^- ion is exothermic by only 12.5 kcal/mol (*167*). Similar calculations carried out to address the nucleophilic reactions of OH^- and CH_3^- with $Fe(CO)_5$ led to the conclusion that the exothermicity of the reactions follows the order $H^- < OH^- < CH_3^-$, which is the same as but somewhat attenuated from that calculated for the reactions of these nucleophiles with free CO (*167*).

Squires and co-workers (*168*) used a flowing afterglow apparatus to determine relative binding energies of a series of anions Nu^- to $Fe(CO)_5$ in the gas phase at 300 ± 2 K. The measured binding energies vary over a range of 40 kcal/mol, decreasing in the order $H^- > OH^- > CH_3O^- > C_2H_5O^- > t\text{-}BuO^- > F^- > CH_3S^- > HS^- > CH_3CO_2^-$. The experimental values for H^- (56.1 ± 4 kcal/mol) and for OH^- (53.1–60.3 kcal/mol) are indistinguishable; however, the collection of anions displayed linear free-energy relationships of their $Fe(CO)_5$ affinities with their proton affinities [$PA(Nu^-)$] and their acetyl ion affinities (Table VI). From these correlations, it was concluded that the thermodynamic stabilities of the $Fe(CO)_4(CONu)^-$ adducts to heterolytic C–Nu bond cleavage should follow the order $OH^- < H^- < CH_3^-$, but the differences between the first two were not large. Despite the quantitative differences, the experimental and thereoretical studies are qualitatively in agreement regarding the substantial intrinsic stabilities of these adducts.

It is clear that the thermodynamic stability of an adduct such as that in Eq. (4) depends on a variety of factors determining the electrophilicity of

TABLE VI

COMPARISONS OF GAS-PHASE ANION AFFINITIES FOR $(Fe(CO)_5$, H^+, AND
CH_3CO^+ [SELECTED VALUES FROM REF. (168)] (kcal/mol)

X^-	$D[(CO)_4FeCO-X^-]$	$PA(X^-)$	$D(CH_3CO^+-X^-)$
H^-	56.1 ± 4	400.4	231.3
OH^-	$53.1 \le x \le 60.3$	390.7	227.5
CH_3O^-	$39.1 \le x \le 51.0$	381.4	222.5
EtO^-	≤ 45.7	376.1	217.5
$t\text{-}BuO^-$	≤ 42.1	373.3	215.4
F^-	40.9 ± 3	371.3	203.4
HS^-	≥ 23.1	353.5	181.7
$CH_3CO_2^-$	≤ 29.5	348.5	172.1

the coordinated carbonyls as well as some fundamental properties describing the nucleophilicity of Nu^-. In solution, both features are unquestionably modified by the dielectric medium, and Eq. (4) should be rewritten as

$$L_nM(CO)_m(\text{sol}) + Nu^-(\text{sol}) \rightarrow L_nM(CO)_{m-1}\left(C\overset{O}{\underset{Nu}{\diagup}}\right)^-(\text{sol}) \qquad (23)$$

Given that the neutral carbonyl and each of the anions are unlikely to be solvated equivalently, moving from the gas phase to solution or from one solvent to another should lead to changes in the equilibrium constant for Eq. (23) and changes in the reaction kinetics. A solvent that is a good acceptor (solvates anions) (169) and has a relatively high dielectric constant, such as methanol, would solvate small anions such as Nu^- more strongly than the larger, more delocalized adduct $L_nM(CO)_{m-1}(CO_2Nu)^-$. Thus, the equilibrium constant for Eq (23) would be expected to be much smaller in methanol than in a poorer acceptor solvent such as THF. An example of this effect can be drawn from stopped-flow kinetics studies of the reaction between $Ru_3(CO)_{12}$ and methoxide [Eq. (24)] in methanol and in various methanol/THF mixtures (156). Both the equilibrium constant K_{24} and the forward rate constant k_{24} were found to be markedly dependent on the composition of the solvent. For example, K_{24} increases by a factor of 30 from $\sim 1 \times 10^3 \ M^{-1}$ to $3 \times 10^4 \ M^{-1}$ in going from methanol to 10/90 (vol/vol) CH_3OH/THF with an even greater increase in k_{24} from 2×10^3 to $2 \times 10^5 \ M^{-1} \ s^{-1}$.

$$Ru_3(CO)_{12} + CH_3O^- \overset{K_{24}}{\rightleftharpoons} Ru_3(CO)_{11}(CO_2CH_3)^- \qquad (24)$$

Some information regarding the relative reactivities of various nucleophiles can be extrapolated from the investigations of related systems. For example, the thermodynamic linear free-energy correlation between the gas-phase $Fe(CO)_5$ and acetyl ion affinities for different anionic nucleophiles (Table VI) suggests that a similar relationship might exist for the reaction dynamics as well. Kinetics studies have been carried out for acyl systems using a pulsed ion cyclotron resonance spectrometer to determine a kinetic nucleophilicity order for the gas-phase displacement of halide from acyl halides $[X^- + RC(O)Y \rightarrow Y^- + RC(O)X]$ for a collection of different X, Y, and R (170). The relative nucleophilicities $CH_3O^- > F^- > CN^- \sim SH^- > Cl^-$ parallel the gas-phase proton affinities. A similar correspondence between nucleophilic activity and Brønsted basicity was observed for the same reactions in protic solvents, although the gas-phase rates were much greater than the corresponding solution-phase rates. These observations were argued to be due to the strongly polar nature of the transition state (and intermediate), which would be a "tetrahedral" alkoxylike species formed by addition of X^- to the carbonyl of $RC(O)Y$. Both X^- and this intermediate should be strongly solvated, a similarity which suggests that the energy differences between reactants and transition states would follow a like relative order for the list of nucleophiles in different solvents. The analogy between the nucleophile reactions with acyl halides and those with coordinated CO is imperfect, but it is notable that both involve structural rearrangement owing to rehybridization of the carbon and transfer of charge from Nu^- to the oxygen, so some correlation between kinetics behaviors of the two systems would be a reasonable expectation.

Reactions of uncharged nucleophiles with coordinated CO provide a different situation. The initially formed adduct would be strongly dipolar owing to the charge transfer from Nu to the carbonyl oxygen, which should be facilitated by a polar solvent medium. Such reactions have not been observed in the gas phase, presumably because they are energetically unfavorable in the absence of such stabilization. However, there is ample evidence for the operation of this pathway in polar media (*see below*).

The equilibrium and rate constants for carbonyl adduct formation with nucleophile anions may also be affected by the counter cations which can ion-pair with either Nu^- or the resulting adduct (171). Generally, ion pairing reduces the reactivity of the nucleophile, a problem especially evident in solvents of low polarity. The use of cryptands or crown ethers to complex alkali metal cations (thus reducing ion pairing and increasing solubility in such solvents) can significantly enhance reactivity, as does the use of large organic cations like tetra(*n*-butyl)ammonium (TBA^+) or bis(triphenylphosphoranylidene) ammonium (PPN^+). A related procedure is phase transfer catalysis, which employs immiscible mixtures of the metal carbonyl in an organic solvent and an aqueous solution of an alkali metal salt M'^+Nu^- plus an organic soluble onium salt or a cryptand or crown ether to carry Nu^- into the organic solvent. The latter subject has been reviewed recently (172).

B. Reaction Dynamics Investigations

Kinetics studies of systems for which characterizable nucleophile–carbonyl adducts have been isolated are relatively few. Rates of acyl $[M(CO)R]$ formation by the transfer of benzyl anions from the Grignard reagent $PhCH_2MgCl$ to several W, Mo, and Cr $M(CO)_5L$ species as well as to several iron carbonyls $Fe(CO)_4L$ in THF have been studied by Darensbourg, Darensbourg, and co-workers (164a). Similar studies have been carried out by Dobson et al. for the transfer of CH_3^- from methyl lithium to several Group VI carbonyls in either (173). In addition, the kinetics of the formation of alkoxycarbonyl adducts has been studied for the reactions of the platinum(II) cation $PtCl(PPh_3)_2(CO)^+$ with various alcohols (20) and for the reactions of the mononuclear and trinuclear iron triad carbonyls $M(CO)_5$ and $M_3(CO)_{12}$ (M = Fe, Ru, or Os) with $NaOCH_3$ in methanol (9,156,174). While these studies have explored the effects of changing the metal center and the ligand field somewhat, investigations of the reactivities of different nucleophiles under comparable conditions or of the influence of the solvent medium have been even more limited.

The kinetics studies noted above (164) for the alkyl anion transfers from $PhCh_2MgCl$ in THF to several Group VI complexes [Eq. (25)] demonstrated a rate law first order in each reactant, although the Grignard reagent concentration range was limited, 0.06–0.12 M. Activation parameters gave small positive $\Delta H^\dagger$ and large negative $\Delta S^\dagger$ values, consistent with a bimolecular transition state. The measured rate constant k_{obs} at a fixed $[PhCH_2MgCl]$ (0.12 M) proved to be modestly dependent on the nature of both M and L (Table V). Rates of homologous complexes follow the order W > Mo ≫ Cr, but the total range is no more than a factor of 10. With regard to the effect of varying L, the order L = CO > $P(OC_2H_5)_3$ > $P(CH_3)_3$ > PPh_3 > $P(c\text{-}C_6H_{11})_3$ was found to be consistent with the deactivation of the remaining CO's by replacing one CO with a more electron-donating L. However, although the site of nucleophilic attack was subject to electronic control in the absence of excessively bulky ligands, the k_{obs} values proved to be far more dependent on the steric bulk of L, as reflected by the Tolman cone angles (175), than on the electronic factors reflected by the CO stretching force constants. The order with respect to varying M was also argued to be steric in nature, with the smaller (more crowded?) Cr complexes displaying the least reactivity.

$$M(CO)_5L + PhCH_2MgCl \xrightarrow{k_{obs}} \left[cis\text{-}M(CO)_4L \left(C \diagup \begin{matrix} O \\ CH_2Ph \end{matrix} \right) \right] MgCl \qquad (25)$$

The kinetics of the much faster reaction of methyl lithium in ether with some of the same group VI carbonyls displayed a similar reactivity order for the different $M(CO)_6$ homologs but showed a much greater difference in rate from the less reactive monosubstituted $M(CO)_5L$ derivatives. In these cases the rate data proved to be consistent with a mechanism where the reactive methyl transfer agent is $LiCH_3$ in equilibrium with the more prevalent tetrameric form. Initial coordination of the $LiCH_3$ as a Lewis acid at the more electron-rich *trans* carbonyl oxygen of $W(CO)_5L$ was proposed to precede CH_3^- transfer to the *cis* carbonyl (173).

The reaction of $PhCH_2MgCl$ with $Fe(CO)_4L$ in THF gives the *trans* adduct [Eq. (26)], which is also consistent with the trans CO having the highest-frequency CO stretch (164). The second-order reaction rates display activation parameters with modest positive $\Delta H^{\dagger}$ and large negative $\Delta S^{\dagger}$ values similar to those measured for the analogous reactions with the Group VI carbonyls. In this case a wider range of rates for various L was observed but there was no correlation with the ligand cone angles. The reaction rates were sharply dependent on the electronic character of the unique ligand and followed the order $CO \gg P(OPh)_3 > P(OCH_3)_3 > PPh_3 > P(n\text{-}C_4H_9)_3$. The ν_{co} values for the trans CO followed the same order.

$$Fe(CO)_4L + PhCH_2MgCl \xrightarrow{k_{26}} trans\text{-}[Fe(CO)_3L(C \diagup \begin{matrix} O \\ CH_2Ph \end{matrix})]MgCl \qquad (26)$$

The kinetics investigation (20) of the reversible reaction of the Pt(II) cation $PtCl(PPh_3)_2(CO)^+$ with various neutral alcohols in CH_2Cl_2/DMF (92/8, vol/vol) [Eq. (27)] provides another example where steric effects apparently dominate the relative reactivities in an analogous series. The second-order rate constants for the forward reaction range from 11.1 M^{-1} s^{-1} for methanol to 0.68 M^{-1} s^{-1} for isopropanol while the corresponding equilbrium constants range from 1.2×10^{-3} to 6.8×10^{-5}, respectively, with the reverse rates relatively insensitive to the nature of ROH. The more sterically demanding alcohols display the smaller forward rate and equilibrium constants.

$$PtCl(PPh_3)_2(CO)^+ + ROH \rightleftharpoons PtCl(PPh_3)_2(CO_2R) + H^+ \qquad (27)$$

Reaction rates of methoxycarbonyl formation for the iron triad complexes $M(CO)_5$ and $M3(CO)_{12}$ (M = Fe, Ru, or Os) in methanol [Eq. (28)] were studied quantitatively by Gross, Trautman, and Ford (9,156,174) as models for the activation step for the water gas shift catalysis by these metal carbonyls in alkaline solution. Each of the reactions displayed second-order kinetics behavior for the forward step with rates sufficiently large to make stopped-flow spectrophotometer kinetics techniques necessary. For the mononuclear $M(CO)_5$ complexes, adduct formation is at the axial carbonyl, as expected from the higher ν_{CO} frequencies at that site. The second-order rate constants for methoxycarbonyl formation in methanol follow the order M = Os > Ru > Fe with a range of roughly a factor of 10 (Table VII). Strikingly, the rates of adduct formation for the trinuclear clusters $M_3(CO)_{12}$ under the same conditions follow the opposite order, as do the equilibrium constants $K_{28} = k_{28}/k_{-28}$ for these clusters (Table VII). Notably, for ruthenium and osmium, the rates, but not the equilibrium constants, of methoxycarbonyl formation are larger for the mononuclear complexes but the opposite is true for the iron analogs. The reactivity differences for the $M(CO)_5$ species have been rationalized in part by Dedieu and Nakamura on the basis of orbital energy differences of

TABLE VII

RATE CONSTANTS AND EQUILIBRIUM CONSTANTS FOR REACTIONS OF THE IRON TRIAD CARBONYLS $M(CO)_5$ AND $M_3(CO)_{12}$ WITH METHOXIDE ACCORDING TO EQ. (28)[a] (25°C)

Metal carbonyl	Medium	$k_f{}^b \times 10^{-3}$	$K^c \times 10^{-3}$
$Fe_3(CO)_{12}$	MeOH	11.3	1.4
$Ru_3(CO)_{12}$	MeOH	2.05	1.0
	50:50 MeOH/THF	18.0	~5
	30:70 MeOH/THF	47.0	
	10:90 MeOH/THF	204.	30
$Os_3(CO)_{12}$	MeOH	0.61	0.8
$Fe(CO)_5$	MeOH	1.1	0.007
	50:50 MeOH/THF	7.8	1.9
	10:90 MeOH/THF	98	~300
$Ru(CO)_5$	MeOH	7	0.1
	50:50 MeOH	~30	~8
$Os(CO)_5$	MeOH	14	—

[a] Data from Refs. (9,156,174).
[b] Second-order rate constant (in $M^{-1}\,s^{-1}$) for formation of the methoxycarbonyl complex.
[c] Equilibrium constant for formation of the methoxycarbonyl complex.

the metals (*167*). According to the theoretical arguments discussed above, attack of a nucleophile at the axial carbonyl of the $M(CO)_5$ would be very significantly affected by the energy of the "d_{z^2}" orbital pointing directly at that carbonyl. The lower energy of this orbital in the heavier metal complexes was argued to favor such adduct formation relative to the homologous iron compound. Related arguments concerned with the symmetries of the cluster orbitals were also offered to rationalize the reactivity differences $Fe_3(CO)_{12} > Ru_3(CO)_{12}$ and $Ru(CO)_5 > Ru_3(CO)_{12}$, but these authors agreed that these arguments were insufficient to reconcile all the kinetics observables (*167*).

$$M_m(CO)_n + CH_3O^- \xrightarrow[k_{-28}]{k_{28}} M_m(CO)_{n-1}(CO_2CH_3)^- \qquad (28)$$

When the reaction kinetics for Eq. (28) were examined for $Ru_3(CO)_{12}$ (*156*) and $Fe(CO)_5$ (*9*) in methanol/THF solutions, it was shown in each case that both the rate and equilibrium constants for methoxycarbonyl formation increased dramatically as the volume percentage of THF in the solvent correspondingly increased. As noted above, these differences can be attributed to the increasing nucleophilicity of CH_3O^- in the less polar and less protic media. In the same context, the rate for Eq. (24) was an order of magnitude higher in 10/90 (vol/vol) CH_3OH/2-propanol than in neat methanol. In 2-propanol solutions, it was also possible to probe the relative nucleophilicities of different alkoxide anions and the reactivity order $CH_3O^- > OH^- > C_2H_5O^- > 2\text{-PrO}^-$ was observed. Similar relative reactivities for methoxide and hydroxide were seen in THF solutions and are consistent with the generally observed greater nucleophilicity of the methoxide ion (*176*). However, the lesser reactivities of ethoxide and isopropoxide, both of which are stronger Brønsted bases (*177*), may reflect the influence of steric factors for these larger anions as found for Eq. (27).

Substitution of $P(OCH_3)_3$ for one CO on either $Ru_3(CO)_{12}$ or $Ru(CO)_5$ strongly deactivates the remaining carbonyls to nucleophilic attack (*156,174*). The substituted cluster $Ru_3(CO)_{11}P(OCH_3)_3$ displayed both a rate constant and an equilibrium constant two orders of magnitude smaller than for the unsubstituted analog. Given that for the monsubstituted cluster the nucleophile has the option of reaction with a CO on an unsubstituted ruthenium, this extent of deactivation demonstrates the efficiency of the metal–metal bonds in transmitting electronic effects.

Kinetics studies of adduct formation have also been reported for several nitrogen nucleophiles. The reaction of $M(CO)_6$ (M = Cr, Mo, W) with tetraethylammonium azide in acetone (*178*) leads to isocyanate complexes [Eq. (29)]. It was proposed that reactions occur via rate-limiting formation of azidocarbonyls $(CO)_5M–C(O)N_3^-$, which undergo rapid rearrangement to the isocyanate. The E_a values decrease in the order Cr > Mo > W, and this was attributed to smaller steric effects owing to the increasing metal atom radii. However, $\Delta S^{\dagger}$ becomes less favorable over the same sequence, an effect opposite to that expected if steric constraints are decreasing.

$$M(CO)_6 + N_3^- \rightarrow M(CO)_5(NCO)^- + N_2 \qquad (29)$$

The rate law for carbamoyl formation via the reversible reaction of cationic Re(I) and Mn(I) carbonyls plus various primary amines [Eq. (30)] (179) is first order in carbonyl and second order in amine. The third-order behavior was explained in terms of the amine attack requiring general base catalysis by the second amine. [In this context, it is interesting to note that the rate of $Re(CO)_6^+/H_2O$ oxygen exchange in aqueous acetonitrile was reported to be second order in H_2O (163).] In general, the relative rates and K values for Eq. (30) follow patterns similar to those described for other nucleophiles. The dominating theme appears to be steric effects with the reactivity orders Re(I) > Mn(I), for various R's n-butyl > cyclohexyl > isopropyl > sec-butyl and for various L's $PPh(CH_3)_2$ > $PPh_2(CH_3)$ > PPh_3 (for all but R = n-butyl), consistent with steric control but opposite to the expected electronic effects of the different L's.

$$trans\text{-}ML_2(CO)_4^+ + 2RNH_2 \rightleftharpoons trans,mer\text{-}ML_2(CO)_3(C(O)NHR) + RNH_3 \qquad (30)$$

M = Re or Mn

In brief summary, some general remarks might be made regarding the studies discussed in this section. It is clear that electronic, steric, and media effects have a major influence on the nature and stability of the nucleophile carbonyl adducts that are formed. Such adduct formation is favored by electronic properties of the complex that decrease the electron density at the carbon site of nucleophilic attack, obvious features being the oxidation state of the metal center and the other ligands attached to it. Steric bulk of the attacking nucleophile and of ligands adjacent to the CO in question has a significant effect on the dynamics and thermodynamics of the reaction. While such effects might be expected for any bimolecular reaction, they may also reflect the most favored angle of nucleophile attack being about 60° from the axis described by the M–CO bond and the necessity for concomitant bending of the M–C–O angle from 180° to roughly 120°. Lastly, the reaction dynamics are markedly dependent on the nucleophilicity of Nu, a parameter strongly dependent not only on the identity of Nu but also on the composition of the medium, ion-pairing effects, and the like. It is clear that further systematic quantitative investigations of these reactions are in order.

IV

REACTIVITIES OF ADDUCTS

Scheme 2 illustrates a variety of stoichiometric reactions that have been observed for nucleophile–carbonyl complexes. Reactions 2-*i* and 2-*ii* in this scheme are simply the reverse of the direct nucleophile attack and of CO insertion into an M–Nu bond, respectively, which were discussed above as synthesis procedures for the adduct. The decarboxylation illustrated in reaction 2-*iii* is, among other things, for Y = H a key step in proposed mechanisms for homogeneous catalysis of the water gas shift reaction (see below). Reaction 2-*iv*, the hydrogenation of these adducts, has been observed for several examples; however, it is likely that the intimate mechanism requires prior dissociation of a ligand, i.e., reaction 2-*v*. Reaction 2-*v* is simply the labilization of one ligand leading to an unsaturated $L_{n-1}M_m(\eta^1\text{-C(O)Nu})$ intermediate, which can decay by several optional pathways. These pathways include coordination by another ligand L' to give the substituted product $L'L_{n-1}M_m(\eta^1\text{-C(O)Nu})$, mono-hapto-to-dihapto linkage isomerization of η^1-C(O)Nu to give an η^2-C(O)Nu complex, or migration of Nu to a metal coordination site (the reverse of migratory insertion). Reaction 2-*vi* is the alkylation of η^1-C(O)Nu to give the carbenoid function η^1-C(OR)Nu. Several other

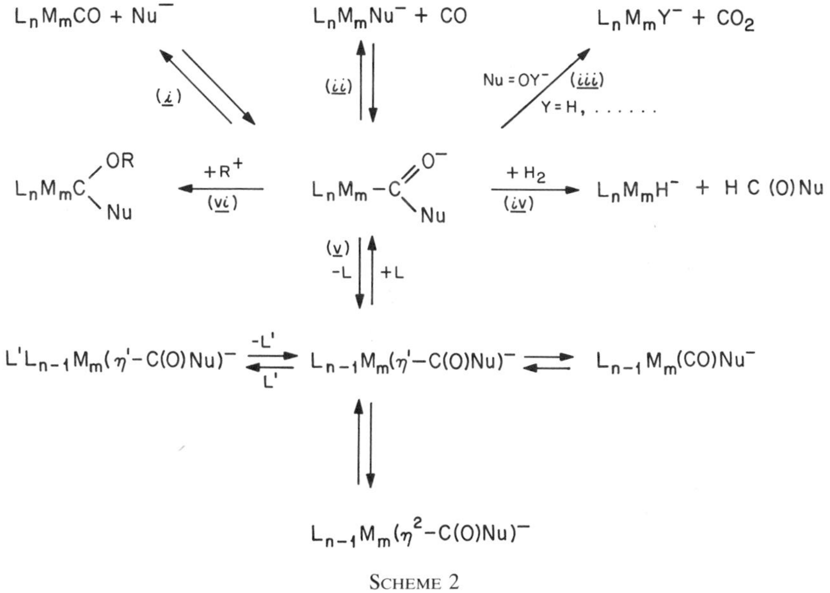

$$L_nM_mCO + Nu^- \qquad L_nM_mNu^- + CO \qquad L_nM_mY^- + CO_2$$

$$Nu = OY^-\ (\underline{iii})$$
$$Y = H, \ldots \ldots$$

$$(\underline{i}) \qquad (\underline{ii})$$

$$L_nM_mC\begin{smallmatrix}OR\\ \\Nu\end{smallmatrix} \xleftarrow[(\underline{vi})]{+R^+} L_nM_m-C\begin{smallmatrix}O^-\\ \\Nu\end{smallmatrix} \xrightarrow[(\underline{iv})]{+H_2} L_nM_mH^- + HC(O)Nu$$

$$(\underline{v}) \quad \begin{smallmatrix}-L\\+L\end{smallmatrix}$$

$$L'L_{n-1}M_m(\eta^1\text{-C(O)Nu})^- \underset{L'}{\overset{-L'}{\rightleftharpoons}} L_{n-1}M_m(\eta^1\text{-C(O)Nu})^- \rightleftharpoons L_{n-1}M_m(CO)Nu^-$$

$$L_{n-1}M_m(\eta^2\text{-C(O)Nu})^-$$

SCHEME 2

reactions not pictured in Scheme 2 have been observed or proposed for the η^1-C(O)Nu ligand. These include deprotonation of the hydroxycarbonyl group [Eq. (31)], insertion of an alkene or alkyne [Eq. (32)], decomposition of the proposed azido carbonyl η^1-C(O)N$_3^-$ to give dinitrogen plus coordinated NCO$^-$, and associative addition of another nucleophile Nu' to give a tetrahedral carbon-coordinated η^1-C(O)(Nu)Nu' species. The subsequent discussion in this section will focus largely on reactions for which some quantitative details are available.

$$M—CO_2H \rightleftharpoons M\text{-}CO_2^- + H^+ \tag{31}$$

$$M—C\overset{O}{\underset{Nu}{\diagdown}} + \overset{|\ \ |}{\underset{|\ \ |}{C=C}} \longrightarrow M—\overset{|\ \ |}{\underset{|\ \ |}{C-C}}-C\overset{O}{\underset{Nu}{\diagdown}} \tag{32}$$

A. Substitution Reactions

It has been known for several decades that the presence of Lewis bases markedly enhances the substitution lability of metal carbonyl complexes in solution. Such observations also led to the speculation that the formation of nucleophile–carbonyl adducts was responsible for this enhanced lability. For example, Basolo and Morris (180) reported that the reaction of Fe(NO)$_2$(CO)$_2$ with various ligands [Eq. (33)] occurs by simple bimolecular kinetics in toluene or dichloromethane but that a first-order pathway is additionally operable in THF or methanol. Furthermore, in the latter solvents, the substitution is catalyzed by different anionic bases with relative effectiveness N$_3^-$ > Cl$^-$ > Br$^-$ > I$^-$. They proposed that the catalysis pathway was the result of anion addition to a carbonyl producing a labile intermediate.

$$Fe(NO)_2(CO)_2 + L \longrightarrow Fe(NO)_2(CO)L + CO \tag{33}$$

There have been an number of indications from the literature that bases are involved in the catalytic labilization of ligands in carbonyl complexes. Examples include the reports that Mn(CO)$_5$(CH$_3$CN) reacts with pyridine to give fac-Mn(CO)$_3$py$_3$ at rates orders of magnitude higher than that of the triphenylphosphine reaction with the same substrate (181), the use of NaBH$_4$ to catalyze the substitutions of the Group VI hexacarbonyls (182), the use of phase transfer catalysis with aqueous NaOH to catalyze the same substitutions (183), the use of NaOCH$_3$ to catalyze the exchange of ^{13}CO with Ru$_3$(CO)$_{12}$ in methanol (132), the effects of trace water or alcohols on such substitution kinetics in nonaqueous solvents (184), and the observation of catalytic activity of highly dissociated salts (e.g., [PPN]Cl or

[PPN][CH$_3$CO$_2$]) on ligand substitution reactions of both mononuclear (*185*) and polynuclear (*186*) metal carbonyls. Several of these observations prompted Brown and Bellus (*187*) to comment that nucleophilic attack at a carbonyl followed by labilization at a position cis to the η^1-C(O)Nu ligand may be a fairly general phenomenon in such systems.

Quantitatively, there have been few investigations of the ligand substitution rates of a well-characterized nucleophile carbonyl adduct. The acyl complex Mn(CO)$_5$C(O)CH$_3$ has been reported to be much more substitution labile than the hexacarbonyl analog Mn(CO)$_6^+$ (*188*), but the reactivity ratio is not known. Perhaps the most extensively investigated example is the methoxycarbonyl complex Ru$_3$(CO)$_{11}$(CO$_2$CH$_3$)$^-$, which readily undergoes replacement of one CO by a P(OCH$_3$)$_3$ in THF solution [Eq. (34)] (*12*). The reaction kinetics have been interpreted mechanistically in terms of the limiting dissociation of one CO [Eq. (35)] followed by the competitive trapping of the resulting intermediate by either CO or P(OCH$_3$)$_3$ [Eqs. (35) and (36)]. Under conditions of high [P(OCH$_3$)$_3$] and low [CO], a limiting unimolecular rate constant of 5.8 s^{-1} (25°C) was measured (*12*). The measured volume of activation $\Delta V^{\ddagger}$ for this reaction rate proved to be +16 ml/mol (*189*), a value that confirms the proposed limiting dissociative mechanism. For comparison, the rate constant for the analogous CO dissociation from Ru$_3$(CO)$_{12}$ in THF has been measured to be about 3×10^{-6} s^{-1} (25°C) (*190*), roughly the same as the value estimated from earlier substitution studies in hydrocarbon solvents (*191*). Thus, one can see that simply forming the methoxide adduct serves to labilize the system by about six orders of magnitude.

$$\text{Ru}_3(\text{CO})_{11}(\text{CO}_2\text{CH}_3)^- + \text{P(OCH}_3)_3 \; \rightleftharpoons$$

$$\text{Ru}_3(\text{CO})_{10}(\text{P(OCH}_3)_3)(\text{CO}_2\text{CH}_3)^- + \text{CO} \quad (34)$$

$$\text{Ru}_3(\text{CO})_{11}(\text{CO}_2\text{CH}_3)^- \underset{k_{-35}}{\overset{k_{35}}{\rightleftharpoons}} \text{Ru}_3(\text{CO})_{10}(\text{CO}_2\text{CH}_3)^- + \text{CO} \quad (35)$$

$$\text{Ru}_3(\text{CO})_{10}(\text{CO}_2\text{CH}_3)^- + \text{P(OCH}_3)_3 \overset{k_{36}}{\longrightarrow} \text{Ru}_3(\text{CO})_{10}(\text{P(OCH}_3)_3)(\text{CO}_2\text{CH}_3)^- \quad (36)$$

At first thought, it is not obvious why modifying a carbonyl ligand by forming the nucleophile adduct would serve to labilize other ligands, especially other CO's as seen in Eqs. (33) and (34). Since π-backbonding is such an important component of the metal–CO interaction, one might presume that increasing the electron density on the metal center would decrease, not increase, CO lability. Given the observations in Section III that for such M(CO)$_5$L complexes the CO stretching frequencies are higher for the cis CO's than for the trans CO, such π-backbonding increases should be greater at the trans than at the cis position. Thus, one might conclude from a superficial ground-state argument that there would be a site preference for cis substitution for complexes where L = η^1-C(O)Nu but no acceleration over the unsubstituted $M(CO)_6$ complexes. However, since there is considerable labilization by the η^1-C(O)Nu ligand, other factors must play the dominant role in determining the reaction dynamics. Molecular orbital calculations directed at investigating cis labilization for the $Mn(CO)_5Br$ complex have indicated that the rationale for the rate acceleration must lie in the relative energies of the pentacoordinate transition state (intermediate) formed as the result of cis-CO labilization (192). A particularly important factor is the π-donor ability of the unique ligand that stabilizes these high-energy species.

Qualitatively, one can easily see that, on a relative basis, the η^1-C(O)Nu ligand should be a weaker π acceptor than is CO in the same site, if for no other reason than that one of the π^* orbitals of the CO was sacrificed in forming the M–Nu bond. However, in the case of η^1-C(O)Nu another feature may be of greater importance, i.e., the ability of this ligand to coordinate as either a two-electron donor monohapto *F* or a four-electron donor dihapto ligand *G*. There are well-characterized examples of η^2-C(O)Nu coordination in both mononuclear and polynuclear complexes. Thus, it is possible, but by no means established, that ligand dissociation from the position cis to the η^1-C(O)Nu ligand is assisted by simultaneous rearrangement from the monohapto to the dihapto configuration. Clearly the subject warrants further insightful theoretical and experimental study.

$$M-C\overset{\displaystyle O}{\underset{\displaystyle Nu}{\diagdown}} \qquad M\overset{\displaystyle O}{\underset{\displaystyle C-Nu}{\diagup}}$$

F   *G*

The role of nucleophile–carbonyl adducts in labilizing ligands may have widespread connotations with regard to the other reactions of metal carbonyls. Hydrogenation and other oxidative additions require an open coordination site on the reactive metal center; thus the dissociative lability of $Ru_3(CO)_{11}(CO_2CH_3)^-$ [Eq. (35)] is the probable origin of this species reactivity toward H_2 (see below) (33). The labilizing effect may also carry over to surface chemistry. It has been noted that the room temperature absorption of the triangular clusters $Fe_3(CO)_{12}$ and $Ru_3(CO)_{12}$ on the oxide supports Al_2O_3 and Cab-O-Sil gives discrete species that are dramatically more labile to CO exchange than are the parent clusters (193a). Although the mechanisms are not understood, it is quite likely that the origin of this activation would be the formation of oxy- and hydroxycarbonyl groups from reaction with the oxides or hydroxides on the support surface (*193b,c*). In this context it is notable that Weinberg and co-workers (*193d*) demonstrated by inelastic electron tunneling spectroscopy that $Mo(CO)_6$ absorbed (reversibly) on a hydroxylated aluminum oxide surface displays vibrational bands in the region $1200-1650$ cm^{-1}, which they interpret as evidence for formation of the hydroxycarbonyl complex $Mo(CO)_5(CO_2H)^-$.

Lastly, it is notable that Halpern and co-workers (*194*) recently concluded that the apparently unimolecular migratory insertion depicted in Eq. (37) is catalyzed by nucleophilic solvents and other nucleophiles. Previously it was thought that the kinetic effect of nucleophiles on related migratory insertion chemistry was due to trapping of the coordinatively unsaturated product of this step; however, these workers concluded that the effects are due to catalysis of Eq. (37) itself. Again, the intermediacy of some form of nucleophile–carbonyl adduct may be the origin of this catalysis.

$$Mn(CO)_5(CH_2Ph) \overset{Nu}{\rightleftharpoons} Mn(CO)_4(C(O)CH_2Ph) \qquad (37)$$

B. *Hydrogenations*

The reactions of nucleophile–carbonyl complexes with dihydrogen represent key steps in a number of mechanistic proposals for homogeneous catalysis by metal carbonyls. The majority of these involve metal acyl or formyl intermediates. For example, H_2 reduction of η^1-coordinated $C(O)Nu$, e.g., Eq. (38), has been proposed in mechanisms for the Fischer–Tropsch catalytic hydrogenation of CO (*195–197*). However, although analogous reductions have been carried out stoichiometrically in homogeneous solution using more active reactants such as aluminum or boron hydrides (*35,69,198–199*), the present reviewers are not aware of demonstrated examples of direct hydrogenation of an η^1 nucleophile–carbonyl ligand in this manner. For the reduction by the Group III hydrides, it is likely that Lewis acid coordination at the oxygen further activates the carbonyl (*200*). Thus, η^2 coordination to an oxophilic metal should serve to activate a formyl or acyl ligand toward H_2 reduction and is the likely explanation for the ready conversion of $Cp_2^*Zr(CO)H_2$ into $Cp_2^*Zr(OCH_3)H$ (*202*).

$$L_nM(\eta^1\text{-}C(O)Nu) + H_2 \rightarrow L_nM(\eta^1\text{-}CH(OH)Nu) \qquad (38)$$

An alternative mode of hydrogenation is described by Eq. (39). Such reactions for metal acyls have been proposed to be crucial in mechanisms for the catalytic homologation of alcohols and for the hydroformylation and hydroxymethylation of alkenes (*201,202*). A similar reaction of metal formyls to give formaldehyde as an intermediate has been proposed as an alternative mechanism for Fischer–Tropsch catalysis (*197,204,205*), and reductions of alkoxycarbonyls and carbamoyls are likely to be mechanistically significant in the syntheses of formates and formamides via catalytic CO hydrogenation (*206*). With regard to the latter cases, it should be noted that, while Eq. (39) may be one step in a possible scheme for facilitating the reduction of CO to hydrocarbon or derivatized hydrocarbon products, the combination of adduct formation [Eq. (4)] and reaction with H_2 [Eq. (39)] constitutes a reduction of the metal center, not of the carbonyl. For example, if Nu is OH^- the formate produced is a hydrolysis, not a reduction, product of CO.

$$L_nM(\eta^1\text{-}C(O)Nu) + H_2 \rightarrow L_nM\text{—}H + HC(O)Nu \qquad (39)$$

Despite the significance of these reactions, there have been but a few quantitative investigations of the H_2 reductions of well-characterized nucleophile–carbonyl complexes. Doxsee and Grubbs (*206a*) studied as a model for this type of reaction the H_2 reduction of the lithium dimethylamide adduct of $Cr(CO)_6$ in THF [Eq. (40), conditions: 2.5 atm H_2, 130°C]. The products were dimethyl formamide plus traces of methanol shown by labeling experiments to originate from the CO. In other solvents trimethylamine, formed by reduction of the initially formed DMF, was also a product. A similar study was carried out by Martin and Baird (*207*) with the acyl complex $Co(CO)_3L(\eta^1\text{-}C(O)CH_3)$ [Eq. (41), L = $PhMe_2P$]. Hydrogenation (35 atm) in 20°C toluene gave, over a period of hours, the products methanol (trace), acetaldehyde ($\sim18\%$), ethyl formate ($\sim30\%$), and ethanol ($\sim12\%$), but it was concluded that the primary product of this reduction is acetaldehyde, the other products being formed from this.

$$(CO)_5Cr\left(C\diagdown^{O}_{N(CH_3)_2}\right)^{-} + H_2 \rightarrow (CO)_5CrH^- + HC\diagup^{O}_{\diagdown N(CH_3)_2} \qquad (40)$$

$$Co(CO)_3L(\eta^1\text{-}C(O)CH_3) + H_2 \rightarrow HCo(CO)_3L + CH_3CHO \qquad (41)$$

Ungváry and Markó (*208*) recently studied the reaction kinetics and mechanism of hydrogenation of the related ethoxycarbonyl cobalt complex $Co(CO)_4(CO_2C_2H_5)$ to ethyl formate in *n*-octane solution [Eq. (42)]. The rate law had the form $-d[M(CO_2R]/dt = k[M(CO_2R)][H_2][CO]^{-1}$ ($k = 1.2 \times 10^{-5}$ s^{-1} at 25°C), consistent with a mechanism where the initial complex must first dissociate CO [Eq. (43)] in order to open a rate constant for CO dissociation, k_{43}, was determined to be $1.7 \times 10^{-3}\,s^{-1}$.

The ethoxycarbonyl complex also reacts with the cobalt hydride $HCo(CO)_4$ to give ethyl formate [Eq. (45)] according to an analogous mechanism. Although these studies did not allow measurement of the absolute rate constants for the reactions of various species with the "unsaturated" complex $Co(CO)_3(CO_2C_2H_5)$, the following order of relative reactivities was obtained: $CO(1.0) > HCo(CO)_4$ (0.084) $> H_2$ (0.0067) (208). The greater reactivity of the cobalt hydride has been used to argue for the greater importance of this reactant (rather than H_2) as the key reductant in mechanisms for reactions such as the cobalt carbonyl-catalyzed hydroformylation of alkenes (209). A mechanism analogous to Eqs. (43) and (45) has also been shown (210) to be the principal pathway for the $HMn(CO)_5$ reduction of the ethoxycarbonyl complex $Mn(CO)_5(CO_2C_2H_5)$.

$$2Co(CO)_4(CO_2C_2H_5) + H_2 \xrightarrow{k_{42}} Co_2(CO)_8 + 2HCO_2C_2H_5 \qquad (42)$$

$$Co(CO)_4(CO_2C_2H_5) \rightleftharpoons Co(CO)_3(CO_2C_2H_5) + CO \qquad (43)$$

$$Co(CO)_3(CO_2C_2H_5) + H_2 \rightleftharpoons Co(H_2)(CO)_3(CO_2C_2H_5) \rightarrow \cdots \qquad (44)$$

$$Co(CO)_4(CO_2C_2H_5) + HCo(CO)_4 \rightarrow Co_2(CO)_8 + HCO_2C_2H_5 \qquad (45)$$

Mechanistic studies have also been carried out for the dihydrogen reduction of the methoxide adduct of $Ru_3(CO)_{12}$ [Eq. (46)] (12) in THF. The rate behavior was analogous to that determined for the cobalt ethoxycarbonyl complex discussed above, and it was concluded that again the mechanism must involve reversible dissociation of CO [Eq. (35)] followed by reaction of the unsaturated cluster with H_2. A kinetic deuterium isotope effect of 1.4 was observed, consistent with this mechanism; however, the proposed intermediate $H_2Ru_3(CO)_{10}(CO_2CH_3)^-$ was not detected. Analysis of the kinetics led to the conclusion that the relative reactivities of CO and H_2 toward the "unsaturated" cluster $Ru_3(CO)_{10}(CO_2CH_3)^-$ are 1.0 and 10^{-5}, respectively, which indicates the much greater selectivity of this anion. The ruthenium hydride $HRu(CO)_4^-$ did not reduce $Ru_3(CO)_{11}(CO_2CH_3)^-$ under comparable conditions. The unsubstituted parent $Ru_3(CO)_{12}$ is only slowly reduced by H_2 at elevated temperatures, and the likely reason for the nucleophilic activation of H_2 reduction is the markedly greater lability of the anionic cluster toward ligand substitution reactions.

$$Ru_3(CO)_{11}(CO_2CH_3)^- + H_2 \rightarrow HRu_3(CO)_{11}^- + HCO_2CH_3 \qquad (46)$$

In this context it is notable that an active alkene hydrogenation catalyst has been reported (186b) to result from the addition of the PPN^+ salts of NCO^-, Br^-, or Cl^- to $Ru_3(CO)_{12}$ in THF. The catalytic activity apparently is the consequence of the remarkably greater lability of the CO's owing to interactions of the cluster with the anions.

C. Reactions with Alkenes

Insertions of an alkene into the metal–carbon bond of a nucleophile–carbonyl adduct [Eq. (32)] has possible connotations with regard to catalysis mechanisms of such species (60). Insertion of simple alkenes to give isolable metalloorganic products does not appear to be particularly facile. However, reactions of trans-PdCl(PPh$_3$)$_2$(CO$_2$CH$_3$) with styrene or methyl acrylate at 100–180°C afforded, respectively, methyl cinnamate and dimethyl fumarate in low yield (148) with accompanying reduction of Pd(II) to Pd(0). A logical pathway for these reactions would be insertion of the alkene into the Pd–CO$_2$CH$_3$ bond followed by β-elimination of the metal hydride. A much more facile reaction is the addition of butadiene to Co(CO)$_4$CO$_2$CH$_3$ at 25°C to give a π-allylic complex [Eq. (47)] (17). the analogous reaction of Co(CO)$_3$(PPh$_3$)CO$_2$CH$_3$ is slower, presumably because of the lower lability of this species to CO dissociation. The diethylcarbamoyl complex NiI(Et$_2$NH)$_2$(C(O)NEt$_2$) has been shown to undergo insertions of allene and diphenylacetylene to give stable complexes [Eq. (48) and (49)] (150). Acidification of the products gives N,N-diethylmethylacrylamide and cis-N,N-diethyl-2-phenylcinnamide, respectively.

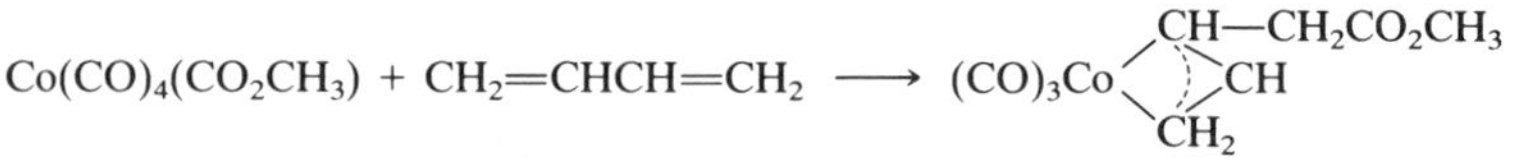

$$Co(CO)_4(CO_2CH_3) + CH_2{=}CHCH{=}CH_2 \longrightarrow (CO)_3Co\underset{CH_2}{\overset{CH-CH_2CO_2CH_3}{\diagup}}CH \quad (47)$$

$$(Et_2NH)_2Ni\underset{I}{\overset{\displaystyle C-NEt_2}{\diagdown}} \quad \xrightarrow{+H_2C=C=CH_2} \quad Ni\underset{Et_2NH}{\overset{I}{\diagdown}}\ \ C{-}C{-}NEt_2 \quad (48)$$

$$\xrightarrow{+Ph-C{\equiv}C-Ph}\quad \underset{Et_2N-C}{\overset{Ph}{\diagup}}C{=}C\underset{Ni(Et_2NH)}{\overset{Ph}{\diagdown}} \quad (49)$$

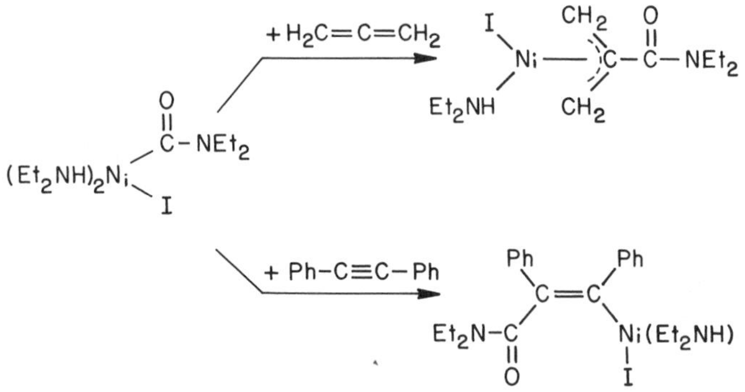

D. Reductive Eliminations

Another reaction of potential importance in catalysis mechanisms is the reductive elimination of the nucleophile–carbonyl ligand with another group, e.g., a hydrogen as suggested above as part of the hydrogenation mechanism or an alkyl group [Eq. (50)]. An example of this is derived from the work of Petz (*39c*) who demonstrated that the methylation of the methoxycarbonyl adduct $Fe(CO)_4(CO_2CH_3)^-$ led to formation of an iron methylated complex, $Fe(CO)_4(CH_3)(CO_2CH_3)$, rather than the expected Fischer-type carbene [Eq. (51)]. This was stable only at low temperature ($< -30°C$) and decomposed at 25°C to give methyl acetate plus iron carbonyls including $Fe_3(CO)_{12}$. Somewhat analogous reductive elimination from $Pd(PPh_3)_2(CO_2CH_3)_2$ occurs in 50°C methanol under CO to give dimethyl oxalate plus unidentified Pd(0) carbonyls (*121*). The bis(ethoxycarbonyl) complex $Rh(Cp^*)(CO)(CO_2Et)_2$ proved more robust but did react with I_2 in CH_2Cl_2 to give a mixture of products including $(CO_2Et)_2$ (30–40%), Cp^*CO_2Et (40%), and traces of diethyl carbonate. In Et_2O, Cp^*CO_2Et was the sole organic carbonyl product of the decomposition (*82*). Similar reductive eliminations from bis(nucleophile–carbonyl) adducts are apparently responsible for the formation of oxalate and oxamide "double carbonylation" products via either stoichiometric or catalytic reactions of metal complexes from CO and alcohols or amines, respectively (*150,211–214*).

$$L_nM \overset{C(O)Nu}{\underset{R}{\big\langle}} \longrightarrow L_nM + RC(O)Nu \qquad (50)$$

$$(CO)_4FeCO_2CH_3 \xrightarrow{CH_3SO_3F} (CO)_4Fe \overset{CH_3}{\underset{CO_2CH_3}{\big\langle}} \qquad (51)$$

$$(CO)_4Fe \overset{CH_3}{\underset{CO_2CH_3}{\big\langle}} \longrightarrow CH_3CO_2CH_3 + Fe_x(CO)_y \qquad (52)$$

E. *Decarboxylations and Related Processes*

Decarboxylation is a facile decomposition pathway for hydroxycarbonyl complexes and is a key step in cyclic mechanisms for catalysis of the water gas shift reaction and other oxidations of CO by various aqueous metal ions, e.g., Eq. (53) (*1,2*). Two mechanisms have been the subject of most discussion: Eq. (54), concerted hydrogen transfer to the metal by a "β-elimination" type of mechanism, and Eqs. (55)–(57), a stepwise mechanism via deprotonation to give the η^1-oxycarbonyl complex $M–CO_2^-$ (i.e., a C-coordinated carbon dioxide), which decays by CO_2 loss followed by protonation to give the hydride complex M–H. A third mechanism involving hydrolysis of the hydroxycarbonyl ligand to a C-coordinated carbonic acid or bicarbonate [Eq. (58)] has also been suggested (215,216). Although evidence in support is not compelling, this pathway has analogy in the hydrolysis of acyls and related processes (below).

$$Hg^{2+} + CO + H_2O \rightarrow Hg(0) + CO_2 + 2H^+ \tag{53}$$

$$L_nM—CO_2H \rightarrow L_nM—H + CO_2 \tag{54}$$

$$L_nM—CO_2H \rightleftharpoons L_nM—CO_2^- + H^+ \tag{55}$$

$$L_nM—CO_2^- \rightarrow L_nM^- + CO_2 \tag{56}$$

$$L_nM^- + H^+ \rightleftharpoons L_nM—H \tag{57}$$

$$L_nM—CO_2H + OH^- \rightleftharpoons L_nM—CO_3H_2^- \rightarrow L_nM^- + H_2CO_3 \tag{58}$$

Experimental evidence provides support for both the concerted elimination of CO_2 [Eq. (54)] and the stepwise deprotonation/decarboxylation [Eqs. (55)–(57)] mechanisms in different systems, and it appears that the dominance of one or the other will depend on the natures of the metal, other ligands, and reaction conditions. One might speculate on what parameters of these system would favor one or the other pathway. In analogy to the β-elimination of alkenes from metal alkyl complexes (217), an open coordination site on M may be a necessary (but not sufficient) condition for effective concerted O → M transfer of a hydrogen via Eq. (54). The open site may be a feature of the original complex or result from ligand dissociation or the change in hapticity of a ligand such as an arene or nitrosyl. Concerted β-elimination should be less sensitive to the electronic nature of the ancillary ligands than the stepwise deprotonation/decarboxylation because the former replaces a largely σ-donor :CO_2H^- ligand with a sigma donor H^-. One would expect the concerted pathway to be favored by factors leading to a stronger M–H bond, i.e., those making the L_nM^- species a stronger base, while the stepwise mechanism would be favored by parameters making the L_nM fragment more acidic. The overall rate constant for the latter would be $k_{56}K_{55}[H^+]^{-1}$, where K_{55} is the equilibrium constant for the Brønsted acid dissociation and k_{56} the rate constant for loss of the Lewis acid CO_2 from the conjugate base $L_nM—CO_2^-$. Furthermore, since Eq. (56) is formally a reductive elimination, i.e., $[(L_nM^+)(CO_2^{2-})]^- \rightarrow L_nM^- + CO_2$, those L's favoring the lower metal oxidation state should enhance the stepwise mechanism. However, the two effects are interrelated, given that π-acceptor ligands favoring a lower oxidation state would also make the L_nM fragment more acidic. With regard to ligand steric effects, concerted β-elimination of CO_2 should be much more sterically demanding than Eq. (56) unless prior dissociation of L from $L_nM–CO_2H$ is a prerequisite for Eq. (54). The intermediate proposed for decarboxylation via Eq. (58) would also be disfavored by bulky ligands.

Some qualitative observations are particularly relevant with regard to the above mechanisms. The hydroxycarbonyl complex first isolated (*61*) is the iridium(III) species $IrCl_2L_2(CO)(CO_2H)$ [L = $P(CH_3)_2Ph$]. This is moderately stable in chloroform but slowly decomposes in solution or in the solid to give the hydride $HIrCl_2(CO)L_2$. In alkaline methanol solution, decarboxylation to $IrCl(CO)L_2$ is rapid. A closely related complex, *trans,mer*-$IrCl_2L_3(CO_2H)$ (*77*), is considerably more robust but does decarboxylate in refluxing THF. Reaction with $Li[Si(CH_3)_3]$ in dry THF gives the spectrally characterized oxycarbonyl anion *trans,mer*-$IrCl_2L_3(CO_2)^-$, which decarboxylates within several hours at room temperature to a mixture of products. Thus, for both these hexacoordinate Ir(III) complexes, the stepwise mechanism appears to be dominant.

Similar conclusions have been drawn from studies of the exchange of oxygen between CO and solution H_2O of the Group V and Group VI hexacarbonyls $M(CO)_6^+$ and $M(CO)_6$ (*218*) and some substituted derivatives. Such exchange, which presumably proceeds via a hydroxycarbonyl intermediate, is catalyzed by base, but the hydride formation via the competing decarboxylation pathway is enhanced to an even greater extent. Base catalysis has also been observed for decarboxylations of several other characterized hydroxycarbonyl complexes including $CpRe(CO)$-$(ArN_2)(CO_2H)$ (*11a*), $CpRe(NO)(CO)(CO_2H)$ (*35,136*), and *trans*-IrCl-(dppe)$(CO_2H)^+$ (*23*).

The behavior of the iron hydroxycarbonyl complex $CpFe(PPh_3)$$(CO)(CO_2H)$ proved to be entirely different. This species was prepared and isolated by addition of KOH to an aqueous benzene solution of $CpFe(PPh_3)(CO)_2^+$ and the K^+ salt of the conjugate base $CpFe(PPh_3)$$(CO)(CO_2)^-$ was prepared by using excess KOH (*13*). The latter reaction and the reversibility of the initial adduct formation [Eq. (59)] demonstrate nicely the amphoteric nature of the hydroxycarbonyls. The oxycarbonyl species proved to be unreactive toward decarboxylation even when heated in formamide to 100°C. In contrast, the hydroxycarbonyl decomposed readily [Eq. (60)] on warming in benzene. Thus, in this case, concerted β-elimination from the hydroxycarbonyl [Eq. (54)] is the much more facile mechanism.

$$CpFeL(CO)_2^+ + OH^- \rightleftharpoons CpFeL(CO)(CO_2H) \rightleftharpoons CpFeL(CO)(CO_2)^- + H^+ \quad (59)$$

$$CpFeL(CO)(CO_2H) \rightarrow CpFeL(CO)H + CO_2 \quad (60)$$

$$(L = PPh_3)$$

A similar conclusion appears to be valid for two other hydroxycarbonyl complexes, the molybdenum species $CpMo(PPh_3)(CO)_2(CO_2H)$ (*14*) and the ruthenium(II) cation *cis*-$Ru(bpy)_2(CO)(CO_2H)^+$ (bpy = 2,2'-bipyridine) (*139*). The former decarboxylates to $CpMo(PPh_3)(CO)_2H$

on prolonged stirring in methanol but does not display a base-catalyzed pathway for this reaction. The latter, which has a pK_a of 9.6, reacts in pH 8.1 aqueous solution under N_2 at 100°C to give CO_2, H_2, and Ru(II) decarboxylation products, but appreciable thermolysis is not observed for similar treatment at pH 11.0, where the cis-Ru(bpy)$_2$(CO)(CO$_2$) species is predominant. In the latter case the controlling factor may be the unwillingness of the cis-Ru(bpy)$_2$CO coordination sphere to assume the d^8 electronic configuration.

With regard to kinetics investigations, the rate laws for the oxidation of CO by aqueous metal ions are largely dominated by the carbonyl complex and hydroxycarbonyl adduct formation steps (2). An important early study of a well-defined carbonyl complex is the oxidation of the CO of Co(CN)$_2$(PEt$_3$)$_2$(CO)$^-$ by aqueous Fe(CN)$_6^{3-}$ (219). The initial steps produced a species believed to be the cobalt(III) hydroxycarbonyl YCo– CO$_2$H [YCo = (Fe(CN)$_5$(μ-CN)Co(CN)$_2$(PEt$_3$)$_2^{4-}$], which underwent decomposition to Co(I) derivatives plus CO_2. The reaction rates were consistent with the model described by Eq. (61) with the calculated K_{61}, k_h, and k_o having the values 3.4×10^{-2} M ($pK_a = 11.5$), 1×10^{-3} s^{-1}, and 3.4 s^{-1}, respectively. Thus, according to this model, the kinetics data indicate that the oxycarbonyl is about three orders of magnitude more reactive than the hydroxycarbonyl conjugate acid. Of course, the relative importance of the two pathways would be pH-dependent, the pK_a value indicating that the k_h pathway should be dominant at $[H^+] > \sim 10^{-8}$ M.

$$YCo^{III}(CO_2H) \;\overset{K_{61}}{\rightleftharpoons}\; YCo^{III}(CO_2)^- + H^{3+} \qquad (61)$$
$$\downarrow k_h \qquad\qquad\qquad\qquad \downarrow k_o$$
$$[YCo^I]^- + H^+ + CO_2 \qquad [YCo^I]^- + CO_2 + H^+$$

The continuing interest in homogeneous WGSR catalysis has been the stimulus of several extended investigations of the reductions of binary metal carbonyls by aqueous bases to give anionic metal hydrides. One such species is the Ru$_3$(CO)$_{12}$ cluster, the precursor of a ruthenium-based WGSR catalyst in alkaline solution (220) (see below). Gross and Ford, (156) have carried out kinetics investigations of the sequential reactions Eqs. (62) and (63), the first step being quite rapid [$k_{51} = 1.3 \times 10^4$ M^{-1} in 25°C 10/90 (vol/vol) H_2O/THF solution], the second much slower. Decarboxylation kinetics were investigated in aqueous methanol, conditions closer to those used for WGSR catalysis and where conversion to HRu$_3$(CO)$_{11}^-$ proved to be quantitative. Reaction of Ru$_3$(CO)$_{12}$ with base in aqueous methanol gives initially a mixture of the hydroxycarbonyl and methoxycarbonyl adducts. (Note that the methoxycarbonyl cluster does not undergo decarboxylation itself but is in labile equilbrium with the hydroxycarbonyl analog.) Gross and Ford's careful analysis of the response of the kinetics to the variables [base], $[H_2O]$, and P_{CO} led to the conclusion that $Ru_3(CO)_{11}(CO_2H)^-$ decarboxylation is independent of [base], hence does not occur via formation of the oxycarbonyl conjugate base. Furthermore, the reaction is much faster under argon than under CO, strongly suggesting decarboxylation from an unsaturated cluster [Eqs. (64) and (65)]. Such CO dissociation is consistent with the demonstrated dissociative lability of the methoxycarbonyl analog [Eq. (35)].

$$Ru_3(CO)_{12} + OH^- \overset{k_{62}}{\rightleftharpoons} Ru_3(CO)_{11}(CO_2H)^- \tag{62}$$

$$Ru_3(CO)_{11}(CO_2H)^- \longrightarrow HRu_3(CO)_{11}^- + CO_2 \tag{63}$$

$$Ru_3(CO)_{11}(CO_2H)^- \rightleftharpoons Ru_3(CO)_{10}(CO_2H)^- + CO \tag{64}$$

$$Ru_3(CO)_{10}(CO_2H)^- \longrightarrow HRu_3(CO)_{10}^- + CO_2 \overset{+CO}{\longrightarrow} HRu_3(CO)_{11}^- \tag{65}$$

The hydroxide reduction of $Fe(CO)_5$ [Eq. (3)], also a WGSR catalyst (221–225), has been investigated in several laboratories (226–228). Early studies of the WGSR kinetics in aqueous methanol displayed a dependence of the catalysis rates on base concentration (225), and this dependence has been interpreted (226) in terms of rate-limiting decarboxylation of $Fe(CO)_4(CO_2H)^-$ via deprotonation, i.e., Eqs. (55) and (56). Further kinetics investigation of this reaction [and that of the $Ru(CO)_5$ homolog] led to the reinterpretation (9) that the base dependence for the WGSR catalysis is instead the result of the small equilibrium constant for Eq. (66) under the catalysis conditions. The decarboxylation rates proved to be independent of P_{CO}, which argues against prior CO dissociation; however, the pentacoordinate iron complexes are less likely to require such a pathway for the β-elimination. A potentially crucial observation was that decarboxylation rates are higher for solutions containing higher percentages of protonic solvents (water or methanol). One interpretation (9) would be that the protonic solvent may be required to mediate the transfer of hydrogen from the hydroxycarbonyl oxygen to the metal as illustrated in H.

$$Fe(CO)_5 + OH^- \rightleftharpoons Fe(CO)_4(CO_2H)^- \tag{66}$$

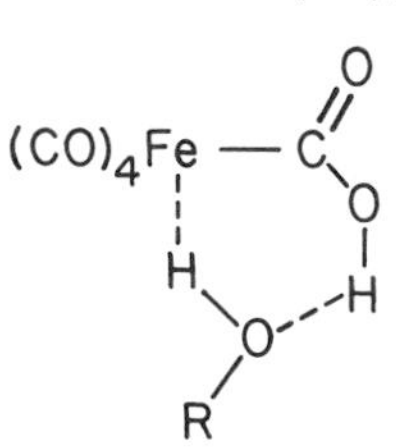

H

Squires and co-workers (*227,228*) examined the reaction of $Fe(CO)_5$ with OH^- in the gas phase using the flowing afterglow method. The reaction with bare OH^- is so energetic that decarbonylation is the result [Eq. (67)] but hydrated OD^- gives the ion product believed to be the hydroxycarbonyl [Eq. (68)]. Despite the mildly favorable energetics [ΔH(est) < -17 kcal/mol], unimolecular decarboxylation of $Fe(CO)_4CO_2H^-$ to $HFe(CO)_4^-$ plus CO_2 does not occur in the gas phase over the milliseconds time scale, suggesting a significant intrinsic barrier for this process. These researchers noted that reaction of the gas-phase cluster ion $(OH)NH_3^-$ with $Fe(CO)_5$ does lead to $HFe(CO)_4^-$ formation, and they suggested that this may be the result of an NH_3-mediated decarboxylation within an ion–molecule pathway analogous to that illustrated by *H*, provided excess internal energy from the initial OH^- attachment is available. Why a similar reaction is not seen with $OD(D_2O)^-$ is puzzling but emphasizes the importance of considering some rather subtle aspects of these mechanisms.

$$Fe(CO)_5(g) + OH^-(g) \rightarrow Fe(CO)_3OH^- + 2CO \tag{67}$$

$$Fe(CO)_5(g) + OD(D_2O)_n^- \rightarrow Fe(CO)_5OD^- + nD_2O \quad (n = 1, 2, 3, \text{ or } 4) \tag{68}$$

Related decarboxylations must also be key steps in carbonyl oxidations such as decarbonylations by trimethylamine oxide [Eq. (69)] (*229*) as well as CO oxidations by other nitrogen oxides or dioxygen (*230*).

$$Fe(CO)_5 + ONMe_3 \rightleftharpoons Fe(CO)_4(C(O)ONMe_3) \rightarrow Fe(CO)_4NMe_3 + CO_2 \tag{69}$$

In principle, decarboxylation of an alkoxycarbonyl to give a metal alkyl is also a possibility. No such reactions have been reported for cases where R is a saturated alkyl. However, the allyloxycarbonyl $Co(CO)_4$-$(CO_2CH_2CH:CH_2)$ readily decarboxylates at 20°C in THF according to Eq. (70) (*10b*). Given the thermal lability of various analogous alkyl- and aryloxycarbonyls toward substitutions of one CO (*10b,208*), it is probable that the decarboxylation of the allyl derivative occurs via CO dissociation with subsequent coordination of the allyl group double bond assisting the CO_2 elimination step. However, no mechanistic investigation of this reaction has been published. Attempts to decarboxylate thermally the benzyl derivative $Co(CO)_4(C(O)CH_2C_6H_5)$ were unsuccessful. Two similar examples for Pt(II) and Fe(II) allyloxycarbonyls have been reported (*21,17c*).

$$Co(CO)_4(CO_2CH_2CH:CH_2) \rightarrow Co(CO)_3(\eta^3\text{-}C_3H_5) + CO + CO_2 \tag{70}$$

Although CO_2 elimination from other alkoxycarbonyls is unknown, there is some precedent for the alcoholysis of the carbon–metal bond to give an organic carbonate. Dimethyl carbonate is one product of the carbonylation of methanol by palladium(II) acetate in the presence of added phosphines and amine bases *(231)*. This has been proposed to occur via the reaction of a Pd(II) methoxycarbonyl with methoxide to give a tetrahedral intermediate *I*, which then reductively eliminates the dimethyl carbonate, e.g., Eq. (71), but the mechanistic details of this system were not demonstrated. Species similar to *I* are also possible intermediates in the exchange of one OR group for another OR′. For example, many alkoxycarbonyl complexes readily undergo such exchange in the presence of the alcohol R′OH or hydrolysis in the presence of water (see above). Such a tetrahedral intermediate finds analogy in the mechanisms for the hydrolysis and alcoholysis of organic esters. However, the alkoxycarbonyl exchanges may be the result of a process that does not find precedent in the organic systems, namely the reversible dissociation of OR′ to reform the carbonyl $(M\!-\!CO_2R^- \rightleftharpoons M\!-\!CO + OR^-)$ *(13)*.

$$\left(Pd^{II}\!-\!C\!\!\underset{OCH_3}{\overset{O}{\diagdown}} \right)^+ + CH_3O^- \;\rightleftharpoons\; Pd^{II}\!-\!\underset{OCH_3}{\overset{\overset{\textstyle O}{\|}}{C}}OCH_3 \;\longrightarrow\; Pd^0 + (CH_3O)_2CO \quad (71)$$

$$I$$

Two other analogies to the decarboxylation reactions should indeed be noted here. The first would be the base-initiated hydrolysis and alcoholysis of the metal–carbon bond of acyl complexes *(232)* leading to apparent reductive elimination of an ester or acid [Eq. (72)] from a tetrahedral intermediate similar to *I*. The reactions of CH_3O^- with various metal acetyls were second order, a kinetics behavior consistent with the reversible formation of this intermediate. For different L_nM fragments, the rates followed the order $Co(CO)_4 \gg Mn(CO)_5 > Co(CO)_3PPh_3 \gg CpFe(CO)PPh_3$, the inverse of the basicities of the L_nM^- anions. In other words, the more acidic metal centers promote the reactions by favoring formation of the tetrahedral intermediate and/or by serving as a more facile leaving group in the second step. It was also noted that the trifluoroacetyl derivative $Mn(CO)_5(C(O)CF_3)$ reacted with methoxide much faster than the acetyl analog, presumably because the electron-withdrawing CF_3 group made the carbonyl carbon more electrophilic.

$$L_nM\!-\!C\!\!\overset{O}{\underset{R}{\diagdown}} + R'O^- \;\rightleftharpoons\; L_nM\!-\!\underset{R}{\overset{\overset{\textstyle O^-}{|}}{C}}\!-\!OR' \;\longrightarrow\; L_nM^- + RCO_2R' \quad (72)$$

A problem with the above mechanism is that the already negative nucleophile–carbonyl ligand ($-C(O)Nu'$)$^-$ appears to be an unfavorable site for anion attack compared to other carbonyls on the complex. Although reaction via an energetically less favorable intermediate is not excluded by such a consideration, an alternative mechanism ought to receive consideration. This would be for the second nucleophile Nu^- to coordinate the metal center, as in Eq. (73), prior to reductive elimination of the $NuC(O)Nu'$ moiety. This sequence appears to be more attractive for reaction of the 16-electron complex in Eq. (71) than for the 18-electron species $Co(CO)_4C(O)R$ or $Mn(CO)_5C(O)R$, but it cannot be excluded even for these given the now recognized enhanced labilities of nucleophile–carbonyl complexes toward ligand lability.

$$L_nM-C(O)Nu' + Nu \rightleftharpoons L_nM\begin{matrix} \diagup C(O)Nu' \\ \diagdown Nu \end{matrix} \longrightarrow L_nM + NuC(O)Nu' \quad (73)$$

Another analogy to decarboxylation is Eq. (74), the formation of ureas from the reaction of metal carbonyl cations and organic amines (73). With limited primary amine, the carbamoyl complex $CpM(CO)_3(C(O)NHR)$ is formed. This species reacts with excess primary amine to give the urea product but with added triethylamine to give the isocyanate RNCO. These observations led to the conclusion that the decomposition of the carbamoyl complex occurs via initial deprotonation followed by elimination of the isocyanate from the metal center [Eq. (75)]. The isocyanate subsequently reacts with amine to form urea. Notably, Eq. (75) is closely analogous to the deprotonation mechanism for CO_2 elimination from hydroxycarbonyls [Eqs. (55) and (56)].

$$CpM(CO)_4^+ + 4RNH_2 \rightarrow CpM(CO)_3^- + (RNH)_2CO + 2RNH_3^+ \quad (74)$$

$$CpM(CO)_3(C(O)NHR) + Et_3N \rightleftharpoons Et_3NH^+ + CpM(CO)_3(C(O)NR)^- \quad (75)$$

$$\rightarrow CpM(CO)_3^- + RNCO$$

It is clear that there is a spectrum of mechanisms operable for decarboxylations and related processes. For hydroxycarbonyls both the β-elimination and the stepwise deprotonation/CO_2 elimination pathways have been demonstrated, the former requiring prior ligand dissociation in some, but not necessarily all, cases. The third pathway, involving nucleophile attack at the hydroxycarbonyl carbon, is more problematic but has a strong analogy to the reductive alcoholysis or hydrolysis of metal acyl complexes. The kinetics behavior of the two principal mechanisms differ in terms of their response to solution pH, the presence of excess ligand, etc.; hence when the two are competitive for the same hydroxycarbonyl complex, the conditions will determine the competition between the two. Other compelling factors are the basicity of the metal center and associated ligands, the readiness of these to undergo two-electron reduction, and possible stereochemical demands on the transition states or key intermediates along the reaction coordinate. Unfortunately, the studies cited above are not sufficiently comprehensive or complete to provide a convincingly predictive model for the mechanistic predilection of a specific system.

V

HOMOGENEOUS CATALYSIS OF THE WATER GAS SHIFT AND RELATED REACTIONS

There has been considerable interest in homogeneous catalysis of the water gas shift reaction over the past decade, beginning with independent reports from three different groups in 1977 describing WGSR homogeneous catalysts based on $Ru_3(CO)_{12}$ in KOH/2-ethoxyethanol (221), on a series of Group VIII carbonyls in aqueous amine solutions (222), and on rhodium halide in aqueous acetic acid (233). These reports were preceded several decades earlier by suggestions from Reppe and others (221,234) regarding possible WGSR catalysis by alkaline solutions of iron carbonyl plus several reports of WGSR catalysts in the patent literatures (235). The subject has been reviewed several times (1,2,236). Therefore, this section will not provide a comprehensive review but will focus on a more general discussion of WGSR catalysis and related processes where metal carbonyls and nucleophiles serve as cocatalysts.

A. The Water Gas Shift Reaction

1. WGSR Catalysis in Alkaline Solutions

The geneology of the Reppe chemistry as well as the more recent discoveries of WGSR catalysis by metal carbonyls in alkaline solutions can be traced to the initial investigations by Hieber and co-workers (7), who demonstrated that carbonyls are reduced by aqueous base to the corresponding anions or hydrides [e.g., Eq. (3)]. Acidification of these solutions released both CO_2 and H_2, the former from the neutralization of carbonate, the latter from protonation of the reduced metal species and subsequent reductive elimination of H_2. However, the WGSR in this manner requires stoichiometric consumption of base and then of acid. The cycle

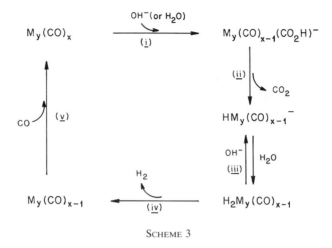

SCHEME 3

illustrated in Scheme 3 was proposed as a logical WGSR mechanism as catalyzed by metal carbonyls in alkaline solution. This cycle consists of four key steps: (3-*i*) nucleophilic activation of CO by water or OH$^-$ to give the hydroxycarbonyl intermediate; (3-*ii*) decarboxylation to the metal anion or metal hydrides, which would be in labile equilibrium with each other (3-*iii*) with concentrations dependent on the metal hydride pK_a value and the effective pH of the solution; and (3-*iv*) reductive elimination of H$_2$ to give a coordinatively unsaturated metal; and (3-*v*) reaction with CO to reform the metal carbonyl. Ample precedent exists for each of these steps; however, such precedent does not ensure the validity of this mechanism.

The breadth of compounds that have been demonstrated to be precursors of homogeneous WGSR catalysts in alkaline solutions or aqueous solutions containing organic amines is remarkable. Catalysis of the shift reaction appears to be a truly general property of binary carbonyls $M_x(CO)_y$ and a number of other carbonyls in such solutions. Table VIII lists the conditions under which various metal complexes have been shown to be precursors for WGSR catalysts (*237–271*). Comparative reactivities are not listed, given that there has been relatively little uniformity in the conditions used by different researchers in this area. The first such catalysts reported displayed modest turnover frequencies (a few moles of H$_2$ per mole of catalyst per day) under mild conditions ($T = 100°C$, $P_{CO} = 1$ atm) (*222,233*); however, more active catalysts have been the subject of recent reports (see below).

Quantitative studies of the WGSR catalysis by Fe(CO)$_5$ plus KOH in aqueous methanol have concluded that this system follows a mechanism analogous to that shown in Scheme 3. This catalyst is only modestly active, a feature that Pearson and Mauermann (226) have attributed to the contradictory pH demands of different steps in the catalysis cycle. Under the buffered conditions of the mature catalyst (CO_2 neutralization and formate formation lower the pH), the only iron complexes observable by *in situ* infrared spectroscopic experiments were $HFe(CO)_4^-$ and $Fe(CO)_5$ and the catalysis rates are proportional to the base concentration (225) but independent of P_{CO}. This kinetics behavior was attributed to rate-limiting formation of $HFe(CO)_4^-$ by the reaction of $Fe(CO)_5$ plus OH^- to give $HFe(CO)_4^-$, presumably via equilibrium formation of the intermediate $Fe(CO)_4CO_2H^-$ followed by decarboxylation. As noted above, analysis of kinetics and equilibrium studies led Trautman, Gross, and Ford to conclude (9) that the base dependence of the catalysis rate was a consequence of the small formation constant for the iron hydroxycarbonyl [Eq. (76)]. The relatively low pH in the bicarbonate/carbonate-buffered mature catalysis solution therefore makes formation of $HFe(CO)_4^-$ rate-limiting. However, raising the pH to accelerate this step would prove unproductive given that $HFe(CO)_4^-$ is a relatively weak base [$pK_a = 5.9$ in 25°C 70% aqueous methanol (226)], and formation of H_2 from $HFe(CO)_4^-$ [Eq. (77)] would then become the slow step under such conditions. Thus, the contradictory demands of the two halves of the cycle with respect to the solution pH in combination with the values of the key equilibrium constants K_{76} and K_{77} restrict the potential activity of this system.

$$Fe(CO)_5 + OH^- \underset{K_{76}}{\rightleftharpoons} Fe(CO)_4CO_2H^- \rightarrow HFe(CO)_4^- + CO_2 \qquad (76)$$

$$HFe(CO)_4^- + H_2O \underset{K_{77}}{\rightleftharpoons} H_2Fe(CO)_4 + OH^- \xrightarrow{+CO} H_2 + Fe(CO)_5 \qquad (77)$$

The $Ru_3(CO)_{12}/KOH$/ethoxyethanol WGSR catalyst, which was demonstrated to be catalytic in both metal and base (221), is another system with modest activity at $T = 100°C$, $P_{CO} = 1.0$ atm. Initial spectral and kinetics studies (237) under batch reactor conditions (as opposed to flow reactor conditions) demonstrated the presence of both triruthenium and tetraruthenim cluster anions, the principal ones being the hydrides $HRu_3(CO)_{11}^-$ and $H_3Ru_4(CO)_{12}^-$. Catalysis rates proved to be first order in P_{CO} and in [Ru], and WGSR rates proved to be independent of whether the initial ruthenium source was $Ru_3(CO)_{12}$ or $H_4Ru_4(CO)_{12}$ or whether the initial base was KOH, K_2CO_3, or $KHCO_3$; under the catalysis conditions, the pH is largely determined by the HCO_3^-/CO_3^{2-} buffer and the various ruthenium clusters are in labile equilibrium.

Subsequent studies (156,272) have led to the conclusions that, while various catalysis cycles involving ruthenium carbonyls of different nuclearities are likely to be operating, the principal cycle under the relatively mild

TABLE VIII

Examples of Homogeneous Water Gas Shift Catalysts[a]

Metal precursor	Medium	P_{CO} (atm)	T (°C)	Refs.
Alkaline solutions				
$Ru_3(CO)_{12}$	$KHCO_3$ or KOH/aq. 2-$EtOC_2H_4OH$	0.4–7.5	80–135	221,237
$Ir_4(CO)_{12}$	K_2CO_3/aq. 2-$EtOC_2H_4OH$	0.9–2.0	90–130	223,239
$Fe(CO)_5$	KOH/aq. 2-$EtOC_2H_4OH$ or aq. MeOH	1–28	100–160	240,242
$Rh_6(CO)_{16}$	KOH/aq. 2-$EtOC_2H_4OH$	0.9–75	100–135	223,243
$Cr(CO)_6$	KOH/aq. MeOH	7.8	140	240,244
$W(CO)_6$	KOH/aq. MEOH	7.7	95–170	240,244
$RuCl_3$	KOH/aq. 2-$EtOC_2H_4OH$	0.3	90–100	245
$Ru_3(CO)_{12}$	Na_2S/aq. MeOH	28	160	246
$[Ru(bpy)_2COCl]PF_6$	KOH/H_2O	3–20	70–150	247
Amine solutions				
$Ru_3(CO)_{12}$	aq. Me_3N/THF	25–80	100–150	248
$Rh_6(CO)_{16}$	aq. Me_3N/THF	25	125	248
$Rh_6(CO)_{16}$	aq. en/2-$EtOC_2H_4OH$	0.8	100	249
$Pt(P(i-Pr)_3)_3$	Pyridine/aq. acetone	20	100–153	250
$H_4Ru_4(CO)_4/Fe(CO)_5$	Pyridine/aq. 2-$EtOC_2H_4OH$ (or piperidine)	0.9	100	223
$Ru_3(CO)_{12}/Fe(CO)_5$	aq. pyridine	0.4–0.6	100	251
$Rh(CO)(OH)(P(i-Pr)_3)_2$	aq. pyridine	20	100	252
$Ru_3(CO)_{12}$	Bipyridine–H_2O	0.5–0.8	100–150	253
$RhCl_3$	2,9-dmphen-S/H_2O	1	100	254
$RhCl_3$	aq. 4-picoline	0.5–2.0	70–130	255
$IrCl_3$	2,9-dmphen-S/H_2O	1	100	254
$RhCl(CO)Z$	Z/H_2O	3.5	85	256

Neutral or acidic solutions

[Rh(CO)$_2$Cl]$_2$	aq. HCl/NaI/acetic acid	0.3–0.8	60–100	257,258
Ru$_3$(CO)$_{12}$	aq. H$_2$SO$_4$/diglyme	0.3–6.6	90–140	243,259
HRh$_2$(CO)$_3$(dppn)$_2^+$	aq. n-PrOH	1	90	271
K$_2$PtCl$_4$/SnCl$_4$	aq. HCl/acetic acid	0.4	88	262
Ni(pnp)Cl$^+$	aq. n-BuOH	0.9	75	263
Ir(COD)(2,9-dmphen-S)	H$_2$O, pH 2–12	1.0	100	254
Ir(COD)(dppe)	aq. dioxane	6–10	140	264
Ir(COD)phen	aq. dioxane	6–10	140	264
Rh(COD)(PPh$_3$)$_2$	aq. dioxane	30	155	265
Pt$_2$H$_3$(dppm)$_2^+$	aq. MeOH	0.1–8	100	261
Ru(TPPS)CO^{4-}	H$_2$O	1	100	266
NiCl$_2$(PMe$_3$)$_2$	aq. EtOH	1	130	267
Rh(bpy)$_2^+$	aq. EtOH	1	90	268
Rh$_6$(CO)$_{16}$	DMBA-P/aq. THF	25–50	100	269
Os$_3$(CO)$_{12}$	Zeolite (HX)	1	140	270
HRu$_3$(CO)$_{11}^-$	Silica/H$_2$O(g)	0.25	100–150	271

[a] Ligand abbreviations: en, ethylenediamine; 2,4-dmphen-S, 2,9-dimethyl-4,7-diphenyl-1,10-phenanthroline disodium sulfate; Z, water-soluble derivatives of bis(diphosphenylethyl)amine; dppm, bis(diphenylphosphino)methane; COD, cyclooctadiene; pnp, 2,6-bis(diphenylphosphinomethyl)pyridine; TPPS; $meso$-tetra(4-sulfonatophenyl)porphyrine; DMBA-P, polymer-bound dimethylbenzylamine.

conditions described above involves triruthenium clusters. Shore *et al.* demonstrated that when H_2 was removed continuously by using a palladium thimble, the WGSR activity increased significantly and the overwhelming ruthenium species present proved to be $HRu_3(CO)_{11}^-$ (*272*). The first-order dependence on P_{CO} is in contrast to the WGSR catalysis by $Fe(CO)_5$ and would not be consistent with a cycle such as Scheme 3, where the only step involving free CO would be reaction with a coordinated unsaturated complex, which should be rapid, not rate-limiting. In this context, Scheme 4 was proposed for the triruthenium species, the rate-limiting step being reaction between CO and $HRu_3(CO)_{11}^-$ to give $Ru_3(CO)_{12}$.

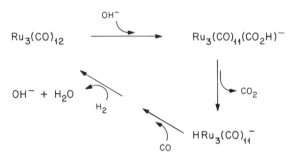

SCHEME 4

A closer examination of the properties of the triruthenium carbonyls reveals why a different mechanism is required in this case. Although reaction of $HRu_3(CO)_{11}^-$ with a strong acid in the presence of CO does lead to the formation of $Ru_3(CO)_{12}$ plus H_2, presumably via the formation of a dihydride (*237,273*), this step is precluded under the conditions of the alkaline catalysis solutions (pH ~ 9) since $HRu_3(CO)_{11}^-$ is an exceedingly weak base (*274*). Thus, an alternative pathway, perhaps direct attack of CO on the anionic cluster, becomes necessary to close the cycle. The mechanism of this step has been the subject of several investigations (*275*), but the specific nature of the key intermediates along the reaction coordinate to H_2 formation remains the subject of speculative interpretation (*237,275,276*). In this context, it is notable that mixed metal catalysts with 3:1 or 2:1 Ru/Fe ratios proved to be significantly more active than analogous systems based on either iron or ruthenium alone (*238,253*). It was suggested that the mixed metal clusters in this case may release hydrogen more readily (*238*).

An alternative mechanism for WGSR catalysis in alkaline solution which must be considered is the production of ionic formate by OH^- reaction with CO followed by the catalyzed decomposition of that substrate [Eqs. (78) and (79)] (*1,243*). In the strongly alkaline starting conditions of several WGSR catalysts, the uncatalyzed reaction to give HCO_2^- is facile (*153*).

$$CO + OH^- \rightleftharpoons HCO_2^- \tag{78}$$

$$HCO_2^- + H_2O \xrightarrow{\text{cat.}} H_2 + CO_2 + OH^- \tag{79}$$

Although various studies have concluded that this catalysis pathway is insignificant for the iron triad carbonyls (*1*), the formate mechanism has received support from several authors for WGSR catalysis by the chromium triad carbonyls $M(CO)_6$ (*240*). In addition, analogous pathways have been proposed for heterogeneous WGSR catalysis (*277*). For the Group VI carbonyls, it was shown that rates of WGSR catalysis by $M(CO)_6$ (M = Cr, Mo, or W) in alkaline aqueous methanol are inversely dependent on P_{CO} and display the respective E_a values 35, 35, and 32 kcal/mol over the temperature range 130–180°C (*240*). Thus, it was argued that a necessary step for the WGSR catalysis is CO dissociation to give $M(CO)_5$, followed by Eq. (80) to give the formate complex. In addition, the tungsten species $W(CO)_5(HCO_2)^-$ has been shown to undergo decarboxylation at rates sufficient to be non-limiting for the WGSR catalysis by the alkaline tungsten carbonyl system (*278*). That such a high activation energy for CO substitution would be required under these conditions is surprising given that formation of nucleophile–carbonyl adducts (even with formate) might be expected to activate the $M(CO)_6$ species toward CO labilization.

$$M(CO)_5 + HCO_2^- \rightarrow M(CO)_5(O_2CH)^- \rightarrow HM(CO)_5^- + CO_2 \tag{80}$$

The Group VIII carbonyls were also studied by other workers (*216*), who concluded that a mechanism similar to Scheme 3 was operable. However, the conditions were different, the solvent being 2-ethoxyethanol and the temperatures being much lower. The latter workers found that the hydride intermediates $HM(CO)_5^-$, key to both mechanisms, were formed at significantly lower temperatures from the reaction of $M(CO)_6$ plus OH^- than from the reaction of $M(CO)_6$ plus HCO_2^- and they reasoned that effective CO activation most likely proceeded via formation of a hydroxycarbonyl complex. Nonetheless, given the HCO_2^- ubiquitous to these reaction mixtures, it seems likely that both pathways are operable and that one or the other dominates in different temperature regimes.

Alkaline aqueous 2-ethoxyethanol solutions prepared using the cluster $Ir_4(CO)_{12}$ also proved to be active WGSR catalysts (*239*). In the context of the above discussion, it is of interest that this iridium carbonyl cluster system decomposes formate catalytically under a dinitrogen atmosphere (the ruthenium cluster catalyst does not). However, this decomposition is inhibited by CO, and the uptake of CO by the reaction with base in the absence of iridium is significantly slower than the WGSR in the presence of iridium carbonyl. Thus, the predominant contribution to WGSR catalysis by alkaline solutions of iridium cannot involve the formate mechanism, but instead must proceed through a pathway similar to that illustrated in Scheme 3.

A final observation related to the WGSR catalysis in alkaline solution is concerned with the relative sulfur tolerance of such systems. Several studies (1,246) have shown that an iron-based homogeneous catalyst is deactivated by sulfur impurities while catalysis by ruthenium is not. The tolerance of the ruthenium system to sulfur was spectacularly demonstrated by King and co-workers, who substituted Na_2S for KOH as the base in studies carried out at 160°C aqueous methanol and found the Ru-based catalyst to retain at least 60% of its activity (246). These workers also found that the Group VI metal carbonyls and osmium carbonyls remain catalytically active under such conditions but that the iron-based system is completely deactivated.

2. WGSR Catalysis in Aqueous Organic Amine Solutions

A number of metal complexes have been demonstrated to be effective WGSR catalysts in solutions of aqueous aromatic or aliphatic amines, a few examples of which are summarized in Table VIII. Such systems include those described by Pettit and co-workers (222), who demonstrated reactivities of a number of Group VIII metal carbonyls in aqueous Me_3N/THF for the WGSR as well as for hydroformylations of alkenes and reductions of nitroaromatics (see below). Prior reports (235a) in the patent literature also claim WGSR activity for several platinum metal salts with a broad range of different aqueous amines used as the solvent.

Catalysis by organic amines is not surprising, given the equilibrium concentration of OH^- from hydrolysis. The large concentrations of amine, especially when used as the cosolvent with water, provide a large buffer capacity and allow the amine and its conjugate acid to act as a general base or acid as necessary. Another feature is that the amines, particularly the aromatic or chelating amines, may coordinate the metal to give new species of greater (or lesser) catalytic activity. Generally, tertiary or aromatic amines would be preferable for WGSR purposes owing to the tendency of primary or secondary amines to form formamides or ureas. Other reactions such as the reduction of pyridine (279) and hydrogen exchanges of amine alkyl protons (280) have been noted as competing processes under conditions of WGSR catalysis. Nonetheless, despite such side reactions and despite the absence of careful comparisons of amine and alkaline base cocatalysts with the metal complexes under strictly analogous conditions, amine solutions appear to be generally the more active for WGSR catalysis.

Several systems listed in Table VIII deserve special comment. One is that prepared from $Ru_3(CO)_{12}$ in aqueous Me_3N/THF solution. Slegeir et al. (248) examined this under wider ranges of $[Ru]_{total}$, P_{CO}, and T and found very high turnover frequency values for conditions of high P_{CO} and T and, especially, for low $[Ru]$. They concluded that under these conditions the predominant active metal catalysts are Ru_1 species, i.e., $Ru(CO)_5$ and its derivatives. The fragmentation of the cluster to $Ru(CO)_5$ under high P_{CO} and T is well known, so the question remains whether catalysis by these mononuclear species should indeed be much higher than by the ruthenium clusters, especially in the context of mononuclear iron being but a modest WGSR catalyst in both alkaline and amine solutions.

Examination of the reaction dynamics of the individual steps of the cycle illustrated in Scheme 3 for $M_y(CO)_x = Ru(CO)_5$ appears to be consistent with the higher activity of the mononuclear ruthenium carbonyl system. The reaction of $Ru(CO)_5$ with OH^- to give the hydroxycarbonyl has a much higher equilibrium constant than for the iron analog (9); thus, formation of $HM(CO)_4^-$, which is apparently rate-limiting for M = Fe (225), is much faster for M = Ru. Furthermore, $HRu(CO)_4^-$ has been shown to be a much stronger base than $HFe(CO)_4^-$ (281); therefore, both halves of the proposed catalysis cycle, CO activation and H_2 elimination, are apparently much more facile for the mononuclear ruthenium species than for the corresponding iron system. For .the triruthenium clusters, hydrogen formation is rate-limiting and slow because the exceedingly weak basicity of $HRu_3(CO)_{11}^-$ precludes a simple protonation/H_2 elimination mechanism as shown in Scheme 2, and an alternative CO-dependent mechanism for H_2 formation becomes necessary (156,237).

Also of particular note is the report by Sauvage annd co-workers that solutions prepared from equimolar amounts of $RhCl_3$ or $IrCl_3$ and various substituted phenanthrolines proved to have very high activities for the WGSR (254). The particular activities of these systems are apparently due to the formation of 1:1 (Rh or Ir)/phenantholine complexes, which indeed have been demonstrated to be catalytically active in neutral solution (see below). The role of the aromatic amine appears in this case to be largely as a ligand rather than as a Brønsted base. In contrast, both roles are undoubtedly played by the amine in WGSR catalysis by solutions such as those prepared from $RhCl_3$ (255), $Pt(i-Pr_3P)_3$ (250), or $Rh(CO)(OH)$ $(i-Pr_3P)_2$ (252) in aqueous pyridine.

3. WGSR Catalysis in Acidic or Neutral Solutions

Although the bulk of the early WGSR homogeneous catalysts had Brønsted bases as cocatalysts, several of these were used in acidic solution. In acid the CO must be siginificantly activated by coordination to be susceptible to nucleophilic attack by water [Eq. (81)] rather than by hydroxide. Such might be the case when the metal is in a high oxidation state. Alternatively, the reaction might be subject to acid catalysis by protonation of the metal center [Eq. (82)], a step that serves to raise by two the formal oxidation state of the metal. If indeed the CO is sufficiently activated, then the low pH of such systems would be expected to enhance the production of H_2 from the reduced metal centers or metal hydrides formed as the result of hydroxycarbonyl decarboxylation.

$$M—CO^{n+} + H_2O \rightleftharpoons M—CO_2H^{(n-1)+} + H^+ \tag{81}$$

$$M—CO \overset{H^+}{\rightleftharpoons} HM—CO^+ \overset{H_2O}{\rightleftharpoons} HM—CO_2H + H^+ \tag{82}$$

Although there are certainly examples of WGSR catalysis by metal carbonyls in acids, this is much less general than in basic solutions. For example, ruthenium carbonyls form a fairly active WGSR catalyst in aqueous diglyme solutions of H_2SO_4 (*238,243,259*), and protonated complexes of some nuclearity [e.g., Eq. (82)] appear to be responsible. However, similar solutions of $Fe(CO)_5$ and $Ir_4(CO)_{12}$ are completely inactive under the same conditions. All three of these carbonyls are precursors of active WGSR catalysts in aqueous alkaline alcohols or aqueous organic amine solutions (Table VIII).

An extensively investigated catalyst (*233,257,258*) is that prepared by Eisenberg from $[Rh(CO)_2Cl]_2$, glacial acetic acid, concentrated HCl, plus NaI in water. Under CO, the active solutions contained only mononuclear rhodium species, principally $RhI_2(CO)_2^-$, $RhI_5(CO)^{2-}$, and $RhI_4(CO)_2^-$. The two key processes are proposed to be CO reduction of Rh(III) to Rh(I) and oxidation of Rh(I) back to Rh(III) by the acidic medium [Eqs. (83) and (84)]. The WGSR kinetics displayed bimodal behavior with respect to variables such as P_{CO}, $[I^-]$, and T, and this was explained in terms of Eq. (83) being the rate-limiting process at $T > 80°C$ [E_a (apparent) = 9.3 kcal/mol] but H_2 formation [Eq. (84)] being rate-limiting at lower temperature [E_a (apparent) = 25.8 kcal/mol].

$$Rh(III) + CO + H_2O \rightarrow Rh(I) + CO_2 + 2H^+ \tag{83}$$

$$Rh(I) + 2H^+ \rightarrow Rh(III) + H_2 \tag{84}$$

Homogeneous WGSR catalysis by rhodium(I) bipyridine complexes in aqueous ethanol has been studied by Creutz and co-workers (*268*), who found the activity to be optimal at pH~3 and to fall off slowly at higher pH and more sharply at lower pH. This behavior was explained in terms of rate-limiting oxidation of the CO on the Rh(III) hydride $HRh(bpy)_2(CO)^{2+}$, which has a pK_a of ~3 [Eq. (85)]. CO activation was concluded to be inversely dependent on $[H^+]$; hence, reaction at significantly lower pH is inhibited. Equation (86) represents a logical mechanism for this step, given that CO coordinated to Rh(III) should be strongly activated toward reaction with nucleophiles. The decreased WGSR rates at higher pH values (the rates drop by roughly an order of magnitude between pH 3 and pH 7) cannot be explained simply on the basis of the rates and equilibria of Eqs. (85) and (86) but are likely the result of changes in the complexes owing to CO displacement of bipyridines under these conditions.

$$HRh(bpy)_2(CO)^{2+} \rightleftharpoons Rh(bpy)_2(CO)^+ + H^+ \qquad (85)$$

$$HRh(bpy)_2(CO)^{2+} + H_2O \rightleftharpoons HRh(bpy)_2(CO_2H)^+ + H^+ \qquad (86)$$

$$\rightarrow CO_2 + H_2Rh(bpy)_2^+$$

Although several WGSR catalysts have been reported to be effective in roughly neutral aqueous solutions (Table VIII), perhaps the most interesting are those described by Sauvage and co-workers (*254*). The catalyst precursors are substituted phenanthroline complexes of iridium $IrL(COD)^+$ (L = substituted phenanthroline, COD = cyclooctadiene) and are among the most active and stable WGSR catalysts reported to operate under mild conditions, e.g., P_{CO} = 1 atm, T = 100°C, pH 2–12 (turnover frequencies of nearly 10^3 mol H_2/mol Ir per day). The key may be the substituents on the phenanthroline (see structure *J* for an example), sulfonate groups to enhance aqueous solubility, and sterically bulky groups in the 2 and 9 positions. The role of the bulky groups is not clear, but one possibility is that they prevent ligand redistribution among the iridium centers to give less active bis(phenanthroline) complexes. A mechanism analogous to that described by Scheme 3, with dihydrogen elimination rate-limiting, was proposed by the authors (*254*), although the experimental rate data did not provide a convincing case for this. The most confusing feature of these data is the observation that the WGSR rates are first order in CO at $P_{CO} < 1.0$ atm but are inhibited by CO at pressures above 7.5 atm.

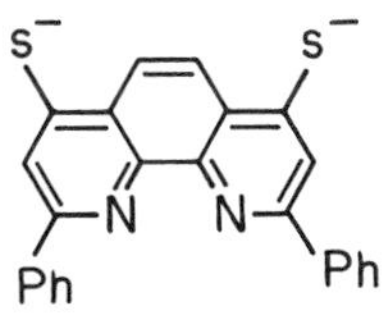

4. *Other WGSR Catalysts*

It is notable that several metal complexes have been attached to solid supports and demonstrated to perform as WGSR catalysts under mild conditions. One such system has been described by Doi and co-workers (*271*), who anchored the triruthenium anion $HRu_3(CO)_{11}^-$ to silica via ammonium or pyridinium function groups and found this to be an effective WGSR catalyst under mild conditions. The advantage of such supported species would be in the design of flow reactors and possibly in stabilizing catalysts that may degrade by bimolecular pathways, for example, the ruthenium carbonyls, which form higher-order clusters in alkaline solution. For small gaseous substrates and products like CO and H_2, the normal advantage of ease of separation of the catalyts from reactants and products is of less importance, although some of the spin-off chemistry may be significant.

B. *Related Reactions*

1. *Hydroformylation*

An important aspect of research into possible homogeneous WGSR catalysts has always been potential applications of such catalysts in related transformations of various substrates, such as reductive carbonylations and hydrogenations. Such applications trace back to the early work of Reppe and others who used basic solutions of iron carbonyl as catalysts for the hydroformylation/hydroxymethylation of alkenes [Eqs. (87) and (88)] (*221,282*). The generality was further demonstrated by Pettit and co-workers (*222*), who carried out similar hydroformylations of propene using amine solutions of iron, ruthenium, osmium, iridium, and platinum carbonyls, Ru, Rh, and Ir being the most active, under relatively mild conditions (aqueous Me_3N/THF, $P_{CO} = 25$ atm, 110–180°C). Laine (*283*) subsequently showed that similar hydroformylation of *n*-pentene could be effected by rhodium or ruthenium carbonyls in alkaline solutions (KOH/aqueous methanol, 135°C, $P_{CO} = 55$ atm). The ruthenium catalyst proved to be very selective for straight-chain products under both sets of conditions; in alkaline solution the products were 97% hexanal and only 3% 2-methylpentanal. In contrast, the rhodium catalyst gave roughly a 1:1 ratio of linear to branched-chain products and proved much more active for the subsequent reduction to the corresponding alcohols.

$$RHC{=}CH_2 + 2CO + H_2O \rightleftharpoons RCH_2CH_2CHO + CO_2 \qquad (87)$$

$$RCH_2CH_2CHO + CO + H_2O \rightleftharpoons RCH_2CH_2CH_2OH + CO_2 \qquad (88)$$

A logical mechanism for hydroformylation under these conditions is illustrated in Scheme 5, in which WGSR, ligand reduction, and reductive

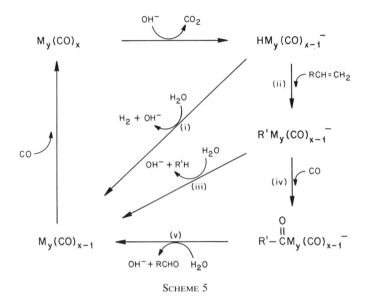

SCHEME 5

carbonylation are competitive processes. The respective catalytic efficiencies would depend on the competition between reaction 5-*i*, which leads to dihydrogen formation, and 5-*ii*, whereby the hydride captures the alkene RCH:CH$_2$, and the competition between 5-*iii*, protonation of the alkyl complex to give the hydrogenated product RCH$_2$CH$_3$, and 5-*iv*, migratory insertion of CO to give eventually the aldehyde RCH$_2$CH$_2$CHO. Notably, the protonation of the alkyl or acyl ligands indicated in steps 5-*iii* and 5-*v* could proceed via initial protonation of the metal center followed by reductive elimination of R′H or R′C(O)H, respectively.

The alkene hydroformylations described above occur under conditions where these metal carbonyl systems are active for the WGSR in the absence of alkene. However, Laine observed that the ruthenium catalysts were more active for hydroformylation of 1-pentene than for the WGSR (*283*). Since the rate-limiting step of the WGSR in this is the formation of H$_2$ from HRu$_3$(CO)$_{11}^-$ (*272*), the key step of the hydroformylation cycle must be interception of the trinuclear hydride by the alkene (5-*ii*). [Consistent with this proposal, other workers (*285*) have indeed noted that HRu$_3$(CO)$_{11}^-$ is an active catalyst for the hydroformylation of alkenes by CO plus H$_2$]. According to kinetics studies of the HRu$_3$(CO)$_{11}^-$ ligand substitution, reaction with alkene should be quite facile but should be inhibited by increasing P_{CO}; however, this point has not been exhaustively tested.

A related development is the discovery by Murata and co-workers (286) that cobalt carbonyl in the presence of 1,2-bis(diphenylphosphino)ethane (dppe) is an active catalyst for homogeneous hydroformylation of propene under conditions where these reactants are active WGSR catalysts in the absence of alkene. In aqueous triethylamine ($T = 135°C$, $P_{CO} = \sim12$ atm, $P_{propene} = \sim9$ atm) both C_4 aldehydes and hydrogen were produced, indicating that WGSR and hydroformylation are competitive. However, when the reaction was investigated in aqueous THF, dioxane, or diglyme in the absence of added base, C_4 aldehydes proved to be the dominant products and only traces of H_2 were detected. No hydroformylation was observed in the absence of water, while the use of D_2O gave deuterated aldehydes. In higher-temperature aqueous dioxane (165°C), C_4 alcohols, C_4 acids, and dipropyl ketones were also formed (287). The alcohols can be attributed to reduction of the aldehyde products, while the butyric acid can be attributed to base-catalyzed hydrolysis of an acyl intermediate $C_3H_7C(O)Co(CO)_m(dppe)_x$, proposed to be formed via steps analogous to 5-*ii* and 5-*iv*. (Indeed, when pyridine was added, hydrocarboxylation of propene to butyric acid became the predominant reaction.) At higher propene concentrations, dipropyl ketone became the major product [Eq. (89)], and when ethylene was used as the substrate, optimization of the conditions led to selectivity for diethyl ketone formation as high as 99% (1% propanal) (287). The ketone/aldehyde product ratio proved to be a complicated function of the reaction parameters, including P_{CO} and the nature of the diphosphine ligand added as a cocatalyst. Qualitatively, these results were explained in terms of competitive trapping of the cobalt acyl species by reaction with a cobalt hydride to give the aldehyde or by insertion of a second alkene into the acyl–metal bond followed by reaction with a hydrogen donor to give the ketone. For this system, protonation of the cobalt acyl by water in a reaction analogous to 5-*v* appears to be less likely owing to the strongly acidic character of the cobalt carbonyls.

$$2CH_3CH{=}CH_2 + CO + H_2O \rightarrow C_3H_7C(O)C_3H_7 \qquad (89)$$

Several other catalysts for alkene hydroformylation under WGSR conditions have been described. Okano and co-workers (288) utilized a catalyst based on $RhH_2(CO_3H)(P(i\text{-}pr)_3)_2$ in aqueous THF to effect hydroformylation of alkenes such as styrene. The active intermediate was concluded to be *trans*-$RhH(CO)(P(i\text{-}Pr)_3)_2$ formed by CO reaction with *trans*-$Rh(OH)(CO)(P(i\text{-}Pr)_3)_2$. The same catalyst favors hydrogenation rather than hydroformylation of alkenes that have electron-withdrawing substituents. Another rhodium-based catalyst is the "A-frame" $Rh_2(\mu\text{-}$

CO)(CO)$_2$(dppm)$_2$ [dppm = bis(diphenylphosphino)methane], which is an active WGSR catalyst in acidic aqueous 1-propanol or bis(2-ethoxyethyl) ether (*260*). Under the same conditions (90°C, 0.65 atm CO) the system also proved active for the hydroformylation of ethene to propanal but was not very selective, with WGSR and hydroformylation operating at comparable rates and hydrogenations of ethene to ethane and propanal to propanol somewhat slower. Overall hydrohydroxymethylation of C$_n$ alkenes to C$_{n+1}$ alcohols was also observed by Kaneda and co-workers (*289*) for a variety of rhodium complexes, using both alkyl and aromatic amines as cocatalysts in aqueous 2-ethoxyethanol.

2. Hydrogenations and Related Reductions

In Reppe-type hydroformylation and hydrohydroxymethylation of alkenes, the carbon monoxide serves both as a reductant (as evidenced by formation of CO$_2$) and as a C$_1$ source to extend the carbon chain, while the water serves as the proton source. Furthermore, in accord with Scheme 5, water, hydroxide, or another base serves to activate the CO by forming a nucleophile–carbonyl adduct, generally the hydroxycarbonyl, which decarboxylates to give a reduced metal center or metal hydride. This is the principal role of nucleophilic activation in the WGSR and related processes such as the reduction of alkenes [Eq. (90)] depicted by Scheme 5, although it is likely that the substitution lability of the nucleophile–carbonyl complexes may play an important role in certain reactions. The intermediacy of a metal hydride provides an optional pathway for the WGSR catalyst to serve as a simple reductant or hydrogenation agent for the appropriate substrate. For example, insertion of an alkene into the M–H bond (step 5-*i*) gives the metal alkyl, which competitively may insert CO to give the metal acyl or react with H$_2$O to produce alkane. Examples of substrates reduced catalytically by CO plus H$_2$O under conditions appropriate for WGSR catalysis include alkenes (*260,262,265,288,290*), nitroarenes (*269,291–293*), nitrogen oxides (*294*), ketones and aldehydes (*295,296*), quinones (*297*), and even aromatic heterocycles (*298*). The activities and selectivities of such hydrogenations are complex functions of the metal catalyst, the substrates present and the reaction conditions. A recent review (*236c*) summarizes many of these observations.

$$RCH{=}CH_2 + CO + H_2O \rightarrow RCH_2CH_3 + CO_2 \qquad (90)$$

Among the more broadly investigated substrates for reduction under WGSR conditions are the nitroarenes, especially nitrobenzene [Eq. (91)]. An early investigation by Pettit and co-workers (*291*) demonstrated that carbonyls of Fe, Ru, Os, Rh, Ir, and Pt were all catalysts for the CO/H$_2$O reduction of nitrobenzene in aqueous trimethylamine/THF solutions. With $Fe(CO)_5$, this could be carried out at room temperature ($P_{CO} = {\sim}120$ atm), and it was concluded that the key reductant is the $HFe(CO)_4^-$ anion produced by reaction of $Fe(CO)_5$ with base. However, the iron system was unstable due to oxidative degradation of the catalyst, Rhodium, iridium, and osmium clusters proved to be much more robust, operated effectively at higher temperatures, and in each case were specific for nitrobenzene reduction as opposed to H_2 production via the WGSR, for which each is an active catalyst in the absence of substrate. Indeed, for the $Os_3(CO)_{12}$ catalyst, the reduction of $PhNO_2$ by CO plus H_2O is two orders of magnitude faster than the WGSR under analogous conditions. This rate enhancement can undoubtedly be attributed to the very slow production of H_2 from the $HOs_3(CO)_{11}^-$ intermediate (*272*), which is apparently intercepted by the nitroarene. In contrast, the catalyst based on $Ru_3(CO)_{12}$ reduced $PhNO_2$ and produced H_2 via the WGSR at comparable rates.

$$PhNO_2 + H_2O + 3CO \rightarrow PhNH_2 + 3CO_2 \qquad (91)$$

A particularly interesting extension of this study was the demonstration by Pettit and co-workers (*292*) that an aqueous trimethylamine/THF solution of $Rh_6(CO)_{16}$ or $Ir_4(CO)_{12}$ catalyzed the reduction of nitroarenes by either CO or H_2, and thus could utilize synthesis gas. The model proposed is shown in Scheme 6, where the metal species L_xM can either

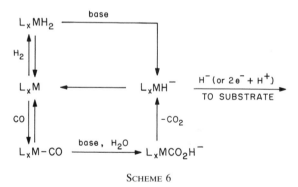

SCHEME 6

activate H_2 by oxidative addition followed by deprotonation to give the L_xMH^- reductant or coordinate CO to give an M–CO species that reacts with base to give the same L_xMH^-. Whereas some catalysts may be effective with either reductant separately, the second requirement for effective use of synthesis gas is that both cycles operate at competitive rates in the presence of both substrates. Thus $Os_3(CO)_{12}$ meets the first requirement but not the second, since CO inhibits its reaction with H_2.

However, these conditions are apparently met by the rhodium and iridium carbonyl catalysts in the approximate temperature range 100–190°C, with the rhodium catalyst giving a reactivity ratio of unity at 137°C for a variety of nitroarene substrates. Under these conditions, the rhodium catalyst is able to effect consumption of both CO and H_2 at rates proportional to their partial pressures, and thus can be used effectively for any CO/H_2 mixture.

The use of phase transfer techniques to effect the CO/H_2O reduction of nitroarenes to amines with alkaline solutions of iron and ruthenium carbonyls as catalysts has been investigated by Alper and co-workers (299). Much milder conditions than used for the homogeneous solutions were effective. In a related study (300), the nitroarenes were heated (60°C) under CO in an anhydrous 2-methoxyethanol/THF solution in the presence of catalytic amounts of $Ru_3(CO)_{12}$ and NaOMe. The arene formamide ArNHCHO was the major product. A likely pathway is the reduction of $ArNO_2$ to $ArNH_2$ by CO with the alcohol serving as the proton source, followed by carbonylation to the formamide (below). When H_2 was used instead as the hydrogen source the products were largely the formamide plus the amine with a somewhat higher overall efficiency. For the latter reaction, using $Fe_3(CO)_{12}$ instead of $Ru_3(CO)_{12}$ as the catalyst led to the carbamate esters $ArNHCO_2CH_3$ and ureas $(ArNH)_2CO$ as the major products. Although the specific reasons for these differences are as yet unknown, it should be pointed out that, while the cluster anion $Ru_3(CO)_{11}(CO_2CH_3)^-$ remains intact in the presence of excess base and under CO, the iron cluster fragments to the mononuclear adduct $Fe(CO)_4(CO_2CH_3)^-$ (156,179); thus the catalysts are not structurally analogous.

3. Carbonylations

Hydrocarboxylation [Eq. (92)] was noted above as one pathway competing with the hydroformylation of alkenes by cobalt carbonyls under WGSR conditions and is a specific example of a series of Reppe-type carbonylations illustrated by Eq. (93), where RXH is an amine, thiol, alcohol, mercaptan, etc. Carbalkoxylation or hydroamidation result when RXH is an alcohol or amine, respectively. Other catalytic carbonylations occurring via nucleophile-carbonyl adducts include the oxidative carbonylations of alcohols to give organic carbonates or oxalates and the analogous reactions of amines leading to ureas. Much of this chemistry has been reviewed elsewhere (301).

$$CH_3CH{=}CH_2 + CO + H_2O \rightarrow C_3H_7CO_2H \qquad (92)$$

$$R'CH{=}CH_2 + CO + RXH \rightarrow R'C_2H_4C\overset{\displaystyle O}{\underset{\displaystyle XR}{{\Large\diagdown}}} \qquad (93)$$

Hydrocarboxylation may be catalyzed under conditions where WGSR catalysis might also be expected; however, it should be noted that the two processes are not closely related since neither oxidation of CO to CO_2 nor reduction of H_2O to H_2 accompanies the former reaction. A likely key step in hydrocarboxylation is hydrolysis of a coordinated acyl formed from the metal hydride as in Scheme 5 to give carboxylic acid and reform the metal hydride [Eq. (94)]. This reaction, which was discussed in Section IV, E, is analogous to the hydrolyses of organic carbonyls such as esters. However, it should be emphasized that the polarity of the addition of the H^+ and OH^- fragments across the metal–carbon bond is opposite to that suggested in steps *ii* and *v* of Scheme 5 for the hydrolysis of metal alkyls to alkanes and metal acyls to aldehydes. The hydrolysis of acyl to carboxylate may proceed through the intermediacy of a tetrahedral species like that illustrated in Eq. (73); however, the resulting elimination of an L_xM^- should then be more favorable for more acidic metal centers (*232*). An alternative mechanism for hydrolysis of the metal acyl would be for the addition of OH^- to occur first at the metal center, followed by reductive elimination of the acyl hydroxide, i.e., the carboxylic acid [Eq. (95)]. Unlike direct attack on the acyl carbonyl, such a pathway may require dissociation of a ligand from the metal to provide a site for OH^- coordination, a step that can be subjected to mechanistic evaluation.

$$L_xM—C{\overset{O}{\underset{R}{}}} + H_2O \rightarrow L_xMH + HO—C{\overset{O}{\underset{R}{}}} \qquad (94)$$

$$L_xM—C{\overset{O}{\underset{R}{}}} + OH^- \rightleftharpoons \left[L_xM{\overset{OH}{\underset{C(O)R}{}}} \right]^- \rightarrow L_xM^- + HO—C{\overset{O}{\underset{R}{}}} \qquad (95)$$

A completely different mechanism for hydrocarboxylation would have the sequence of molecular events reversed: formation first of the hydrocarbonyl complex by reaction of OH^- with metal carbonyl, followed by insertion of alkene into the metal–carbon bond [Eq. (32)] and then hydrolysis to give the carboxylic acid [Eq. (96)]. While it was noted in Section IV,C that such alkene insertions do not appear to be particularly facile, there certainly are precedents for this pathway, including the probable mechanism for the cobalt carbonyl-catalyzed formation of diethyl ketone from ethene, CO, and H_2O under WGSR conditions (*282*)

$$L_xM—\overset{|}{\underset{|}{C}}—\overset{|}{\underset{|}{C}}—CO_2H + H_2O \rightarrow L_xM^+ + H—\overset{|}{\underset{|}{C}}—\overset{|}{\underset{|}{C}}—CO_2H + OH^- \qquad (96)$$

$$CH_2{=}CH_2 + CO + \tfrac{1}{2}O_2 \rightarrow CH_2{=}CHCO_2H \qquad (97)$$

(above). A similar mechanism was proposed by Fention *et al.* (*302*) for the Wacker-type $PdCl_2/CuCl_2$-catalyzed synthesis of acrylic acid from ethene [Eq. (97)]. Analogous mechanisms can be proposed for the carbalkoxidations and hydroamidations of alkenes.

The oxidative carbonylations of alcohols to dialkyl oxalates and to dialkyl carbonates were discussed in Sections IV,D and IV,E, respectively. Similarly, the oxidative carbonylation of amines to ureas were discussed in Section IV,E. Alper and Hartstock recently reported (*304*) a homogeneous catalyst for the oxidative carbonylations of amines to carbamate esters [Eq. (98)] using a Wacker-type $PdCl_2/CuCl_2$ catalyst at room temperature and atmospheric pressure. Although mechanistic studies were not reported, reaction via either an alkoxycarbonyl or a carbamoyl complex appears likely.

$$ArNH_2 + CO + ROH \rightarrow ArNHCO_2R \tag{98}$$

Simple carbonylations of amines are also catalyzed by metal carbonyl complexes, presumably via the intermediacy of nucleophile–carbonyl complexes. An example is the carbonylation of organic amines to formamides as illustrated in Eq. (99). This reaction as well as the hydroamidation of alkenes [Eq. (100)] is catalyzed by $Ru_3(CO)_{12}$ in benzene (120–180°C, 40 atm CO) (*305*). Some incomplete evidence was presented supporting the formation of the hydridocarbamoyl cluster HRu_x-$(CO)_yC(O)NHR$; however, mechanistic investigations of this and related amine carbonylations have been limited.

$$PhNH_2 + CO \rightarrow H-C\overset{\displaystyle O}{\underset{\displaystyle NHPh}{}} \tag{99}$$

$$RCH{=}CH_2 + PhNH_2 + CO \rightarrow RCH_2CH_2C\overset{\displaystyle O}{\underset{\displaystyle NHPh}{}} \tag{100}$$

VI

CONCLUDING REMARKS

In this article the syntheses, characterizations, and reactions of metal nucleophile–carbonyl adducts for various oxygen and nitrogen nucleophiles were reviewed with the goal of illustrating how such adducts may influence the reactions of metal carbonyl complexes. Both stoichiometric and catalytic processes were described; however, the magnitude of the subject has limited the presentation largely to relatively well-defined systems involving hydroxide, alkoxide, and amine nucleophiles. For example, the role of hydroxycarbonyl adducts in the oxidation of CO by reaction with OH^- (or H_2O) was discussed extensively in the context of the homogeneous catalysis of the water gas shift reaction and related spin-off technology. However, it is clear that the scope of both stoichiometric and catalytic reactions involving nucleophilic activation of metal carbonyls is much larger. Such reactions include the oxidation of coordinated CO by O_2 (306) or its complexes (307) and by other oxygen-centered nucleophiles such as amine oxides (308), other nitrogen oxides (5)', nitrite (309), DMSO (310), and iodosobenzene (140) as well as the conversion of CO to COS by reaction with HS^- (311), to phosgene by reaction with Cl_2 (312), and to isocyanate by reaction with azides or hydrazine (116). Although the evidence is often speculative or circumstantial, the intermediacy of nucleophile carbonyl adducts in these and other aspects, such as the solid-support-induced changes (193d) in the catalytic activities of metal carbonyls and the promoting effects of halide ions on certain catalytic systems, is probable. Further mechanistic investigations of the reactions of well-defined nucleophile–carbonyl complexes will provide a more quantitative basis for the roles of such species in key catalytic systems (313).

ACKNOWLEDGMENTS

We thank the U.S. Department of Energy, Office of Basic Energy Science, for support of research in these laboratories on catalysis of the water gas shift reaction and other chemistry related to the nucleophilic activation of carbon monoxide and for support during the preparation of this article. A number of former associates of these laboratories contributed to our understanding of the nucleophilic activation of carbon monoxide. Of special note are Drs. David C. Gross, Martin Anstock, and Douglas Taube.

REFERENCES

1. P. C. Ford, *Acc. Chem Res.* **14**, 31 (1981).
2. J. Halpern, *Comments Inorg. Chem,* **1**, 3 (1981).
3. R. J. Angelici, *Acc. Chem. Res.* **5**, 335 (1972).
4. D. J. Darensbourg, *Isr, J. Chem.* **15**, 247 (1977).
5. R. Eisenberg and D. E. Hendricksen, *Adv. Catal.* **28**, 79 (1979).
6. D. C. Gross, Ph.D. Dissertation, Univ. of California, Santa Barbara, 1983.
7. W. Hieber and F. Leutert, *Z. Anorg. Allg. Chem.* **204**, 145 (1932).
8a. J. Wu and C. P. Kubiak, *J. Am. Chem. Soc.* **105**, 7456 (1983).
8b. J. M. Lehn and R. Zeissel, *Proc. Natl. Acad. Sci. U.S.* **99**, 701 (1982).
8c. C. T. Tso and A. R. Cutler, *J. Am. Chem. Soc.* **108**, 6069 (1986).
9. R. J. Trautman, D. C. Gross, and P. C. Ford, *J. Am. Chem. Soc.* **107**, 2355 (1985).
10a. M. Tasi, A. Sisak, F. Ungváry, and G. Pályi, *Monatsh. Chem.* **116**, 1103 (1985).
10b. M. Tasi and G. Pályi, *Organometallics* **4**, 1523 (1985).

11a. C. F. Barrientos-Penna, A. B. Gilchrist, A. H. Klahn-Oliva, A. J. L. Hanlan, and D. Sutton, *Organometallics* **4**, 478 (1985).
11b. C. F. Barrientos-Penna, A. H. Klahn-Oliva, and D. Sutton, *Organometallics* **4**, 367 (1985).
11c. C. F. Barrientos-Penna, A. B. Gilchrist, and D. Sutton, *Organometallics* **2**, 1265 (1983).
12a. M. Anstock, D. Taube, D. C. Gross, and P. C. Ford, *J. Am. Chem. Soc.* **106**, 3696 (1984).
12b. D. J. Taube, A. Rokicki, M. Anstock, and P. C. Ford, *Inorg. Chem.*, **26**, 526 (1987).
13. N. Grice, S. C. Kao, and R. Pettit, *J. Am. Chem. Soc.* **101**, 1627 (1979).
14. D. H. Gibson, K. Owens, and T.-S. Ong, *J. Am. Chem. Soc.* **106**, 1125 (1984).
15. A. Mayr, Y. C. Lin, N. M. Boag, and H. D. Kaesz, *Inorg. Chem.* **21**, 1704 (1982).
16. G. R. John, L. A. P. Kane-Maguire, and R. Kanitz, *J. Organomet. Chem.* **312**, C21 (1986).
17a. R. Aumann, H. Ring, C. Krüger, and R. Goddard, *Chem. Bar.* **112**, 3644 (1979).
17b. M. R. Churchill and K.-N. Chen, *Inorg. Chem.* **15**, 788 (1976).
17c. G. Annis, S. V. Ley, C. R. Self, and R. SivaramKrishnan, *J. Chem. Soc. Chem. Commun.* 299 (1980).
18. T. S. Coolbaugh, B. D. Santasiero, and R. H. Grubbs, *J. Am. Chem. Soc.* **106**, 6310 (1984).
19. H. C. Clark, K. R. Dixon, and W. J. Jacobs, *J. Am. Chem. Soc.* **91**, 1346 (1969).
20. J. E. Byrd and J. Halpern, *J. Am. Chem. Soc.* **93**, 1634 (1971).
21. H. Kurosawa, *Inorg. Chem.* **14**, 2148 (1975).
22. M. Wada and K. Oguro, *Inorg. Chem.* **15**, 2346 (1976).
23. M. A. Lilga and J. A. Ibers, *Organometallics* **4**, 590 (1985).
24. M. Catellani and J. Halpern, *Inorg. Chem.* **19**, 566 (1980).
25. A. J. Lindsay, S. Kim, R. A. Jacobson, and R. J. Angelici, *Organometallics* **3**, 1523 (1984).
26. H. Behrens and A. Jungbauer, *Z. Naturforsch.* **34b**, 1477 (1979).
27. H. Wagner, A. Jungbauer, G. Thiele, and H. Behrens, *Z. Naturforsch.* **34b**, 1487 (1979).
28. D. B. Dell'Amico, F. Calderazzo, and G. Pelizzi, *Inorg. Chem.* **18**, 1165 (1979).
29. N. G. Connelly and L. F. Dahl, *Chem. Commun.* 880 (1970).
30a. D. Messer, G. Landgraf, and H. Behrens, *J. Organomet. Chem.* **172**, 349 (1979).
30b. H. Behrens, *Adv. Organomet. Chem.* **18**, 1 (1980).
31. D. Drew, M. Y. Darensbourg, and D. J. Darensbourg, *J. Organomet. Chem.* **85**, 73 (1975).
32. D. W. Parker, M. Marsi, and J. A. Gladysz, *J. Organomet. Chem.* **194**, C1 (1980).
33. G. G. Johnston and M. C. Baird, *J. Organomet. Chem.* **314**, C51 (1986).
34. C. M. Lukehart, *Acc. Chem. Res.* **14**, 109 (1981) and references therein.
35. J. R. Sweet and W. A. G. Graham, *J. Am. Chem. Soc.* **104**, 2811 (1982).
36. W. Tam, M. Marsi, and J. A. Gladysz, *Inorg. Chem.* **22**, 1413 (1983).
37. G. Smith and D. J. Cole-Hamilton, *J. Chem. Soc. Dalton Trans.* 2501 (1983).
38. J. T. Gauntlett, B. F. Taylor, and M. J. Winter, *J. Chem. Soc. Chem. Commun.* 420 (1984).
39a. W. Petz, *J. Organomet. Chem.* **105**, C19 (1976).
39b. Ibid **90**, 223 (1975).
39c. W. Petz, *Organometallics* **2**, 1044 (1983).
40. F. Calderazzo, *Angew. Chem., Int. Ed. Engl.* **16**, 299 (1977).
41. G. K. Anderson and R. J. Cross, *J. Chem. Soc. Dalton Trans.* 712 (1980).

42. T. C. Flood, K. D. Campbell, H. H. Downs, and S. Nakanishi, *Organometallics* **2**, 1590 (1983).
43a. H. E. Bryndza, S. A. Kretchmar, and T. H. Tulip, *Chem. Commun.* 977 (1985).
43b. H. E. Bryndza, *Organometallics* **4**, 1686 (1985).
44. M. A. Bennett and A. Rokicki, in preparation (1987).
45a. M. A. Bennett and A. Rokicki, *Organometallics* **4**, 180 (1985).
45b. M. A. Bennett and A. Rokicki, *J. Organomet. Chem.* **244**, C31 (1983).
45c. M.. A. Bennett and A. Rokicki, *Aust. J. Chem.* **38**, 1307 (1985).
46a. M. A. Bennett and A. Rokicki, *Proc. 191st ACS Nat. Meet. 191st, New York* (1986).
46b. M. A. Bennett, G. Robertson, A. Rokicki, and W. Wicramasinghe, in preparation (1987).
47a. W. M. Rees and J. D. Atwood, *Organometallics* **4**, 402 (1985).
47b. W. M. Rees, M. R. Churchill, J. C. Fettinger, and J. D. Atwood, *Organometallics* **4**, 2179 (1985).
47c. D. J. Darensbourg, K. M. Sanchez, and A. L. Rheingold, *J. Am. Chem. Soc.* **109**, 290 (1987).
48. M. L. Kullberg and C. P. Kubiak, *Organometallics* **3**, 632 (1984).
49a. B. B. Wayland and B. A. Woods, *J. Chem. Soc. Chem. Commun.* 700 (1981).
49b. B. B. Wayland, B. A. Woods, and R. Pierce, *J. Am. Chem. Soc.* **104**, 302 (1982).
49c. B. B. Wayland, A. Duttaahmed, and B. A. Woods, *J. Chem. Soc. Chem. Commun.* 142 (1983).
49d. H. W. Bosch and B. B. Wayland, *J. Organomet. Chem.* **317**, C5 (1986).
50. K. G. Moloy and T. J. Marks, *J. Am. Chem. Soc.* **106**, 7051 (1984).
51a. P. J. Fagan, J. M. Manriquez, S. H. Vollmer, C. S. Day, V. W. Day, and T. J. Marks, *J. Am. Chem. Soc.* **103**, 2206 (1981).
51b. P. A. Kongshaug and R. G. Miller, *Organometallics* **6**, 372 (1987).
52a. D. C. Sonnenberger, E. A. Mintz, and T. J. Marks, *J. Am. Chem. Soc.* **106**, 3484 (1984).
52b. G. Paolucci, G. Rossetto, P. Zanella,, K. Yünlü, and R. D. Fischer, *J. Organomet. Chem.* **272**, 363 (1984).
53. P. J. Fagan, K. G. Moloy, and T. J. Marks, *J. Am. Chem. Soc.* **103**, 6959 (1981).
54a. H. Pichler and H. Schulz, *Chem. Eng. Tech.* **42**, 1162 (1970).
54b. G. Henrici-Olivé and S. Olive, *Angew. Chem.* **88**, 144 (1978).
54c. R. G. Pearson, H. W. Walker, H. Mauermann, and P. C. Ford, *Inorg. Chem.* **20**, 2741 (1981).
55. R. B. King, *J. Am. Chem. Soc.* **85**, 1918 (1963).
56. J. P. Collman and S. R. Winter, *J. Am. Chem. Soc.* **95**, 4089 (1973).
57a. P. A. Kongshaug, K. R. Haugen, and R. G. Miller, *J. Am. Chem. Soc.* **104**, 627 (1982).
57b. P. A. Kangshaug and R. G. Miller, *Organometallics* **6**, 372 (1987).
58. R. F. Heck, *J. Organomet. Chem.* **2**, 195 (1964).
59. W. Hieber and H. Duchatsch, *Chem. Ber.* **98**, 1744 (1965).
60. D. Milstein and J. L. Huckaby, *J. Am. Chem. Soc.* **104**, 6150 (1982).
61. A. J. Deeming and B. L. Shaw, *J. Chem. Soc. (A)* 443 (1969).
62. I. S. Kolomnikov, V. P. Kukolev, Yu. D. Koreshkov, V. A. Mosin, and M. E. Volpin, *Izv. Akad. Nauk SSSR, Ser. Khim.* **10**, 2371 (1972).
63. S. Otsuka, A. Nakamura, T. Yoshida, M. Naruto, and K. Ataka, *J. Am. Chem. Soc.* **95**, 3180 (1973).
64a. E. D. Dobrzynski and R. J. Angelici, *Inorg. Chem.* **14**, 59 (1975).

64b. J. Fayos, E. Dobrzynski, R. J. Angelici, and J. Clardy, *J. Organomet. Chem.* **59**, C33 (1973).
65. K. L. Brown, G. R. Clark, Ch. E. L. Headford, K. Marsden, and W. R. Ropper, *J. Am. Chem. Soc.* **101**, 503 (1979).
66. J. W. Suggs, *J. Am. Chem. Soc.* **100**, 640 (1978).
67. M. A. Bennett, J. C. Jeffery, and G. B. Robertson, *Inorg. Chem.* **20**, 323 (1981).
68. C. A. Tolman, S. D. Ittel, A. D. English, and J. P. Jesson, *J. Am. Chem. Soc.* **101**, 1742 (1979).
69. D. L. Thorn, *Organometallics* **1**, 197 (1982).
70. I. S. Kolomnikov, G. Stepovska, S. Tyrlik, and M. E. Volpin, *Zh. Obsh. Khim.* **42**, 1652 (1972).
71. R. L. Harlow, J. B. Kinney, and T. Herskovitz, *J. Chem. Soc. Chem. Commun.* 813 (1980).
72a. J. D. Audett, T. J. Collins, B. D. Santarsiero, and G. H. Spies, *J. Am. Chem. Soc.* **104**, 7352 (1982).
72b. T. Forschner, K. Menard, and A. Cutler, *J. Chem. Soc. Chem. Commun.* 121 (1984).
72c. C. T. Tso and A. R. Cutler, *J. Am. Chem. Soc.* **108**, 6069 (1986).
73. W. Jetz and R. J. Angelici, *J. Am. Chem. Soc.* **94**, 3799 (1972).
74. R. D. Adams, N. M. Golembeski, and J. P. Selegue, *Inorg. Chem.* **20**, 1242 (1981).
75. P. M. Treichel. W. J. Knebel, and R. W. Hess, J. Am. Chem. Soc. **93**, 5425 (1971).
76. W. E. Buhro, A. Wong, J. H. Merrifield, G.-Y. Lin, A. C. Constable, and J. A. Gladysz, *Organometallics* **2**, 1852 (1983).
77. K. Bowman, A. J. Deeming, and G. P. Proud, *J. Chem. Soc. Dalton Trans.* 857 (1985).
78. R. B. King, M. B. Bisnette, and A. Fronzaglia, *J. Organomet. Chem.* **5**, 341 (1966).
79a. V. G. Albano, P. I. Bellon, and M. Sansoni, *Inorg. Chem.* **8**, 298 (1969).
79b. P. I. Bellon, M. Manassero, F. Porta, and M. Sansoni, *J. Organomet. Chem.* **80**, 139 (1974).
80a. L. Garlaschelli, S. Martinengo, P. Chini, F. Canziani, and R. Bau, *J. Organomet. Chem.* **213**, 379 (1981).
80b. S. Martinengo, H. Fumagelli, P. Chini, and V. G. Albano, *J. Organomet. Chem.* **116**, **116**, 333 (1976).
81a. G. Ciani, A. Sironi, P. Chini, and S. Martinengo, *J. Organomet. Chem.* **213**, C37 (1981).
81b. P. Chini, S. Martinengo, and G. Giordano, *Gazz. Chim. Ital.* **102**, 330 (1972).
82. P. L. Burk, D. Van Engen, and K. S. Campo, *Organometallics* **3**, 493 (1984).
83a. G. M. Reisner, I Bernal, H. Brunner, and M. Muschiol, *Inorg. Chem.* **17**, 783 (1978).
83b. G. Cardaci, G. Bellachioma, and P. Zanazzi, *J. C. S. Dalton*, 473 (1987).
84a. L. N. Zhir-Lebed', L. G. Kuz'mina, Yu. T. Struchkov, O. N. Temkin, and V. A. Golodov, *Koord. Khim.* (English translation) **4**, 795 (1978).
84b. G. del Piero and M. Cesari, *Acta Crystallogr.* **B35**, 2411 (1979).
85. A. J. Carty, *Adv. Chem. Ser.* **196**, 163 (1982).
86a. C. Bauer, E. Guggolz, W. A. Herrmann, G. Kriechbaum, and M. Ziegler, *Angew. Chem. Int. Ed. Engl.* **21**, 212 (1982).
86b. C. Bauer, E. Guggolz, W. A. Herrmann, G. Kriechbaum, and M. Ziegler, *Angew. Chem. Suppl.* 434 (1982).
87. B. F. G. Johnson, J. Lewis, W. J. H. Nelson, J. N. Nicholls, M. D. Vargas, D. Braga, K. Henrick, and M. McPartlin, *J. Chem. Soc. Dalton Trans.* 1809 (1984).
88. "Comprehensive Organic Chemistry" (I. O. Sutherland, Ed.). Pergamon, Oxford, 1979.

89a. B. K. Balbach, F. Helus, F. Oberdorfer, and M. L. Ziegler, *Angew. Chem. Int. Ed. Engl.* **20,** 470 (1981).

89b. F. Oberdorfer, B. Balbach, and M. L. Ziegler, *Z. Naturforsch.* **37b,** 157 (1982).

90. W.-K. Wong, W. Tam, C. E. Strouse, and J. A. Gladysz, *J. Chem. Soc. Chem. Commun.* 530 (1979).

91. C. P. Casey, S. M. Neuman, M. A. Andrews, and D. R. McAlister, *Pure Appl. Chem.* **52,** 625 (1980).

92. G. Smith, D. J. Cole-Hamilton, M. Thornton-Pett, and M. B. Hursthourse, *J. Chem. Soc. Dalton Trans.* 2501 (1983).

93. G. O. Nelson and Ch. E. Sumner, *Organometallics* **5,** 1983 (1986).

94a. M. R. Churchill and H. J. Wasserman, *Inorg. Chem.* **21,** 226 (1982).

94b. P. A. Belmonte, F. Geoffrey, N. Cloke, and R. R. Schrock, *J. Am. Chem. Soc.* **105,** 2643 (1983).

95. P. T. Wolczanski, R. S. Threlkel, and J. E. Bercaw, *J. Am. Chem. Soc.* **101,** 218 (1979).

96a. R. Bardi, A. Del Pra, A. M. Piazzesi, and L. Toniolo, *Inorg. Chim. Acta* **35,** L345 (1979).

96b. R. Bardi, A. M. Piazzesi, A. Del Pra, G. Cavinato, and L. Toniolo, *Inorg. Chim. Acta* **102,** 99 (1985).

97. E. Carmona, L. Sánchez, J. M. Marín, M. L. Poveda, J. L. Atwood, R. D. Priester, and R. D. Rogers, *J. Am. Chem. Soc.* **106,** 3214 (1984).

98. W. R. Roper, G. E. Taylor, J. M. Waters, and L. J. Wright, *J. Organomet. Chem.* **182,** C46 (1979).

99. P. J. Fagan, J. M. Manriquez, T. J. Marks, V. W. Day, S. H. Vollmer, and C. S. Day, *J. Am. Chem. Soc.* **102,** 5393 (1980).

100a. M. D. Curtis, K. B. Shiu, and W. M. Butler, *Organometallics* **2,** 1475 (1983).

100b. F. R. Kreissl, W. Sieber, H. Keller, J. Riede, and Wolfgruber, *J. Organomet. Chem.* **320,** 83 (1987).

101a. J. R. Blickensderfer, C. B. Knobler, and H. Kaesz, *J. Am. Chem. Soc.* **97,** 2686 (1975).

101b. J. R. Blickensderfer and H. Kaesz, *J. Am. Chem. Soc.* **97,** 2681 (1975).

102. J. Powell, A. Kukis, Ch. J. May, S. C. Nyburg, and S. J. Smith, *J. Am. Chem. Soc.* **103,** 5941 (1981).

103. H. G. Alt, M. E. Eichner, B. M. Jansen, and U. Thewalt, *Z. Naturforsch.* **37B,** 1109 (1982).

104a. G. L. Breneman, D. M. Chipman, C. J. Galles, and R. A. Jacobson, *Inorg. Chim. Acta* **3,** 447 (1969).

104b. D. M. Chipman and R. A. Jacobson, *Inorg. Chim. Acta* **1,** 393 (1967).

105. Y. Becker, A. Eisenstandt, and Y. Shvo, *Tetrahedron* **34,** 799 (1978).

106. R. D. Adams, D. F. Chodosh, and N. M. Glembeski, *Inorg. Chem.* **17,** 266 (1978).

107. L. Maresca, G. Natile, A.-M. Mantti-Lanfredi, and A. Tripicchio, *J. Am. Chem. Soc.* **104,** 7661 (1982).

108a. A. S. Batsanov and Yu. T. Struchkov, *J. Organomet. Chem.* **248,** 101 (1983).

108b. P. Jermakoff and N. J. Cooper, *J. Am. Chem. Soc.* **109,** 2173 (1987).

109. A. Muller, U. Seyer, and W. Eltzner, *Inorg. Chim. Acta* **32,** L65 (1979).

110. E. Keller, A. Trenkle, and H. Vahrenkamp, *Chem. Ber.* **110,** 441 (1977).

111. W. Petz, C. Kruger, and R. Goddard, *Chem. Ber.* **112,** 3413 (1979).

112. R. Szostak, C. E. Strouse, and H. D. Kaesz, *J. Organomet. Chem.* **191,** 243 (1980).

113. R. Adams, D. A. Katahira, and J. P. Selegue, *J. Organomet. Chem.* **213,** 259 (1981).

114a. N. M. Boag, C. B. Knobler, and H. D. Kaesz, *Agnew. Chem., Int. Ed. Engl.* **22,** 249 (1983).

114b. N. M. Boag, C. B. Knobler, and H. Kaesz, *Angew. Chem. Suppl.* 198 (1983).
115a. K. Porschke, G. Wilke, and C. Kruger, *Angew. Chem. Int. Ed. Engl.* **22**, 547 (1983).
115b. K. Porschke, G. Wilke, and C. Kruger, *Angew. Chem. Suppl.* 786 (1983).
116. Q.-B. Bao, A. L. Reingold, and T. B. Brill, *Organometallics* **5**, 2259 (1986).
117a. D. L. Weaver, *Inorg. Chem.* **9**, 2250 (1970).
117b. M. A. Bennett, R. N. Johnson, G. B. Robertson, I. B. Tomkins, and P. O. Whimp, *J. Am. Chem. Soc.* **98**, 3514 (1976).
117c. A. G. Constable, W. S. McDonald, and B. L. Shaw, *J. Chem. Soc. Dalton Trans.* 2282 (1980).
118a. J. B. Lambert, H. F. Shurvell, L. Verbit, R. G. Cooks, and G. H. Stout, "Organic Structural Analysis," Pt. 2, Ch. 4, pp. 225-252. Macmillan New York, 1976.
118b. L. J. Bellamy, "The IR Spectra of Complex Molecules," 2nd Ed., Vol. 2, Ch. 5, pp. 128–194. Chapman & Hall, New York, 1980.
119. G. C. Levy, R. L. Lichter, and G. L. Nelson, "Carbon-13 Nuclear Magnetic Resonance" 2nd Ed., Ch. 5, pp. 136–170. Wiley (Interscience), New York, 1980.
120. T. Kruck and M. Noack, *Chem. Ber.* **97**, 1693 (1964).
121. P. Fitton, M. P. Johnson, and J. E. McKeon, *J. Chem. Soc. Chem. Commun.* 6 (1968).
122. W. Beck and K. v. Werner, *Chem. Ber.* **104**, 2901 (1971).
123. V. I. Sokolov, G. Z. Suleimanov, A. A. Musaev, and O. A. Reutov, *Izv. Akad. Nauk SSSR, Ser. Khim.* 2093 (1980).
124a. H. C. Clark and K. v. Werner, *Synth. React. inorg. Met.-Org. Chem.* **4**, 355 (1974).
124b. H. C. Clark and K. v. Werner, *Chem. Ber.* **110**, 667 (1977).
125. W. A. Herrmann, G. W. Kriechbaum, Ch. Bauer, B. Koumbouris, H. Pfisterer, E. Guggolz, and M. L. Ziegler, *J. Organomet. Chem.* **262**, 89 (1984).
126. W. J. Cherwinski, B. F. G. Johnson, J. Lewis, and J. R. Norton, *J. Chem. Soc. Dalton Trans.* 1156 (1975).
127. W. Hieber, F. Lux, and C. Herget, *Z. Naturforsch.* **20B**, 1159 (1965).
128. F. Rivetti and U. Romano, *J. Organomet. Chem.* **154**, 323 (1978).
129. T. Yoshida, Y. ueda, and T. Otsuka, *J. Am. Chem. Soc.* **100**, 3941 (1978).
130. J. Daub, A. Hasenhundl, K. P. Krenkeler, and J. Schmetzer, *Liebigs ann. Chem.* 997 (1980).
131. J. H. Merrifield, Ch. E. Strouse, and A. Gladysz, *Organometallics* **1**, 1204 (1982).
132. D. J. Darensbourg, R. L. Gray, and M. Pala, *Organometallics* **3**, 1928 (1984).
133. A. Sacco, P. Giannoccaro, and G. Vasapollo, *Inorg. Chim. Acta* **83**, 125 (1984).
134. T. G. Appleton and M. A. Bennett, *J. Organomet. Chem.* **55**, C88 (1983).
135. R. A. Michelin, M. Napoli, and R. Ros, *J. Organomet. Chem.* **175**, 239 (1979).
136. C. P. Casey, M. A. Andrews, and J. E. Rinz, *J. Am. Chem. Soc.* **101**, 741 (1979).
137. J. G. Atton and L. A. P. Kane-Maguire, *J. Organomet. Chem.* **246**, C23 (1983).
138. D. H. Gibson and T.-S. Ong, *Organometallics* **3**, 1911 (1984).
139. H. Ishida, K. Tanaka, M. Morimoto, and T. Tanaka, *Organometallics* **5**, 724 (1986).
140. W. Tam, G.-Y. Lin, W.-K. Wong, W. A. Kiel, V. K. Wong, and J. A. Gladysz, *J. Am. Chem. Soc.* **104**, 141 (1982).
141. J. A. Gladysz, *Adv. Organomet. Chem.* **20**, 1 (1982).
142. G. Erker and F. Rosenfeldt, *Angew. Chem. Int. Ed. Engl.* **17**, 605 (1978).
143. R. C. Schoening, J. L. Vidal, and A. Fiato, *J. Organomet. Chem.* **206**, C43 (1981).
144. S. G. Davies, J. Hibberd, S. J. Simpson, S. E. Thomas, and O. Watts, *J. Chem. Soc. Dalton Trans,* 701 (1984).
145. D. S. Barratt and D. J. Cole-Hamilton, *J. Chem. Soc. Chem. Commun.* 458 (1985).
146. R. B. Hitam, R. Narayanaswamy, and A. J. Rest, *J. Chem. Soc. Dalton Trans.* 615 (1983).

212   PETER C. FORD and ANDRZEJ ROKICKI

147. J. R. Anglin and W. A. G. Graham. *J. Am. Chem. Soc.* **98**, 4678 (1976).
148. W. H. de Roode, D. G. Prins, A. Oskam, and K. Vrieze, *J. Organomet. Chem.* **154**, 273 (1978).
149. R. J. Angelici and T. Formanek, *Inorg. Chim. Acta* **76**, L9 (1983).
150. H. Hoberg and F. J. Fañanás, *Angew. Chem., Int. Ed. Engl.* **24**, 325 (1985).
151. K. Noack, *J. Organomet. Chem.* **13**, 411 (1968).
152. M. F. Farona and G. R. Camp, *Inorg. Chim. Acta* **3**, 395 (1969).
153. M. Iwata, *Chem. Abstr.* **70**, 76989 (1969).
154. Calculated from data in F. T. Wall, "Chemical Thermodynamics." Freeman, San Francisco, 1958.
155. W. Braker and A. L. Mossman, "The Matheson Unabridged Gas Databook: A Compilation of Physical and Thermodynamic Properties of Gases," [Vol. 1], Matheson Gas Products, E. Rutherford. N.J.
156. D. C. Gross and P. C. Ford, *J. Am. Chem. Soc.* **107**, 585 (1985). **108**, 6100 (1986).
157a. D. B. Beach, S. P. Smit, and W. P. Jolly, *Organometallics*, **3**, 556 (1984).
157b. D. B. Beach and W. L. Jolly, *Inorg. Chem.* **22**, 2137 (1983), and references therein.
158. T. E. Huheey, "Inorganic Chemistry," 2nd ed., p. 154. Harper, New York, 1978.
159. C. P. Horwitz and D. F. Shriver, *Adv. Organomet. Chem.* **23**, 219 (1984).
160. J. Chandrasekha, J. G. Andrade, and P. von R. Schleyer, *J. Am. Chem. Soc.* **103**, 5612 (1981).
161. E. L. Muetterties, *Inorg. Chem.* **4**, 1841 (1965).
162. D. J. Darensbourg, B. J. Baldwin, and J. A. Froelich, *J. Am. Chem. Soc.* **102**, 4688 (1980).
163. R. L. Kump and L. J. Todd, *Inorg. Chem.* **20**, 3715 (1981).
164a. M. Y. Darensbourg, H. L. Conder, D. J. Darensbourg, and C. Hasday, *J. Am. Chem. Soc.* **95** 5919 (1973).
164b. D. J. Darensbourg and M. Y. Darensbourg, *Inorg. Chim. Acta* **5**, 247 (1971).
165. R. J. Angelici and L. J. Blacik, *Inorg. Chem.* **11**, 1754 (1972).
166. R. C. Bush and R. J. Angelici, *J. Am. Chem. Soc.* **108**, 2735 (1986).
167. A. Dedieu and S. Nakamura, *Nouv. J. Chim.* **8**, 317 (1984).
168. K. R. Lane, L. Sallans, and R. R. Squires, *J. Am. Chem. Soc.* **108** 4368 (1986); **107**, 5369 (1985).
169. V. Gutmann, "The Donor–Acceptor Approach to Molecular Interactions," Ch. 2. Plenum, 6 New York, 1978.
170. O. I. Asubiojo and J. I. Brauman, *J. Am. Chem. Soc.* **101**, 3715 (1979).
171. M. Y. Darensbourg, *Prog. Inorg. Chem.* **33**, 221 (1985).
172. H. des Abbayes, *Isr. J. Chem.* **26**, 249 (1985).
173. G. R. Dobson and J. R. Paxson, *J. Am. Chem. Soc.* **95**, 5925 (1973).
174. D. C. Gross and P. C. Ford, *Inorg. Chem.* **21**, 1702 (1982).
175. C. A. Tolman, *J. Am. Chem. Soc.* **92**, 2956 (1970).
176. M. L. Bender and W. A. Glasson, *J. Am. Chem. Soc.* **81**, 1590 (1959).
177. J. Hine and M. Hine, *J. Am. Chem. Soc.* **74**, 5266 (1952).
178. W. Werner. W. Beck, and H. Engelmann, *Inorg. Chim. Acta* **3**, 331 (1969).
179. R. J. Angelici and R. W. Brink, *Inorg. Chem.* **12**, 1067 (1973).
180a. D. E. Morris and F. Basolo, *J. Am. Chem. Soc.* **90**, 2536 (1968).
180b. F. Basolo, *Inorg. Chim. Acta* **50**, 65 (1981).
181. D. Drew, D. J. Darensbourg, and M Y. Darensbourg, *Inorg. Chem.* **14**, 1579 (1975).
182. J. Chatt, G. J. Leigh, and N. Thankarajan, *J. Organomet. Chem.* **29**, 105 (1971).
183. K. Y. Hui and B. L. Shaw, *J. Organomet. Chem.* **124**, 262 (1977).
184. H. Elias, H.-T. Macholdt, K. J. Wannowius, M. J. Blandamer, J. Burgess, and B. Clark, *Inorg. Chem.* **25**, 3048 (1986).

185a. G. R. Dobson, *Acc. Chem. Res.* **9**, 300 (1976).
185b. J. E. Pardue, M. N. Memering, and G. R. Dobson, *J. Organomet. Chem.* **71**, 407 (1974).
186a. G. Lavigne and H. D. Kaesz, *J. Am. Chem. Soc.* **106**, 4647 (1984).
186b. J. L. Zuffa, M. L. Blohm, and W. A. Gladfelter, *J. Am. Chem. Soc.* **108**, 552 (1986).
187. T. L. Brown and P. A. Bellus, *Inorg. Chem.* **17**, 3726 (1978).
188. J. D. Atwood and T. L. Brown, *J. Am. Chem. Soc.* **98**, 3160 (1976).
189. D. J. Taube, R. van Eldik, and P. C. Ford, *Organometallics* **6**, 125 (1987).
190. T. Chin and P. C. Ford, in preparation (1987).
191. A. Poé and M. V. Twigg, *J. Chem. Soc. Dalton Trans.* 1860 (1974).
192. D. L. Lichtenberger and T. L. Brown, *J. Am. Chem. Soc.* **100**, 366 (1978).
193a. I. Boszormenyi and L. Guczi, *Inorg. Chim. Acta* **112**, 5 (1986).
193b. M. Laniechi and R. L. Burwell, *J. Colloid Interface Sci.* **75**, 95 (1980).
193c. T. L. Brown, *J. Mol. Catal.* **12**, 41 (1981).
193d. G. J. Gajda, R. H. Grubbs, and W. H. Weinberg, *J. Am. Chem. Soc.* **109**, 66 (1987).
194. S. L. Webb, C. M. Giandomenico, and J. Halpern, *J. Am. Chem. Soc.* **108**, 345 (1986).
195. W. A. Herrmann, *Angew. Chem. Int. Ed. Engl.* **21**, 117 (1982).
196. W. Keim, *in* "Catalysis in C_1 Chemistry" (W. Keim, Ed.), Ch. 1 D. Reidel, Dordrecht, 1983.
197. G. Henrici-Olivé and S. Olivé, "Catalyzed Hydrogenation of Carbon Monoxide." Springer-Verlag, Berlin, 1984.
198. J. A. Gladysz, *Adv. Organomet. Chem.* **20**, 1 (1982).
199. C. P. Casey, M. A. Andrews, D. R. McAlister, and J. E. Rinz, *J. Am. Chem. Soc.* **102**, 1927 (1980).
200. C. Masters, "Homogeneous Transition Metal Catalysis," Ch. 3.2. Chapman & Hall, London, 1981; *Adv. Organomet. Chem.* **17**, 61 (1979).
201. J. M. Manriquez, D. R. McAlister, R. D. Sanner, and J. E. Bercaw, *J. Am. Chem. Soc.* **100**, 2716 (1978).
202. D. W. Slocum, *in* "Catalysis in Organic Synthesis" (W. H. Jones, Ed.). Academic Press, New York, 1980.
203. R. F. Heck and D. S. Breslow, *J. Am. Chem. Soc.* **83**, 4023 (1961).
204. D. R. Fahey, *J. Am. Chem. Soc.* **103**, 136 (1981).
205. H. M. Feder and J. W. Rathke, *Ann. N. Y. Acad. Sci.* **333**, 45 (1980).
206a. K. M. Doxsee and R. H. Grubbs, *J. Am. Chem. Soc.* **103**, 7696 (1981).
206b. J. S. Bradley, *J. Am. Chem. Soc.* **101**, 7419 (1979).
207. J. T. Martin and M. C. Baird, *Organometallics* **2**, 1073 (1983).
208. F. Ungváry and L. Markó, *Organometallics* **2**, 1608 (1983).
209. Ref. 197, p. 222.
210. I. Kovacs, C D. Hoff, F. Ungváry, and L. Markó, *Organometallics* **4**, 1347 (1985).
211a. J. T. Chen and A. Sen, *J. Am. Chem. Soc.* **106**, 1506 (1984).
211b. A. Sen, J. T. Chen, W. M. Vetter, and R. R. Whittle, *J. Am. Chem. Soc.* **109**, 148 (1987).
212a. F. Ozawa, T. Sugimoto, Y. Yuasa, M. Santra, T. Yamamoto, and A. Yamamoto, *Organometallics* **3**, 683 (1984).
212b. F. Ozawa, T. Sugimoto, T. Yamamoto, and A. Yamamoto, *Organometallics* **3**, 692 (1984).
213. D. M. Fenton and P. J. Steinwand, *J. Org. Chem.* **39**, 701 (1974).
214. F. Rivetti and U. Romano, *L Chim Ind.* 627 (1980).
215. H. C. Clark and W. J. Jacobs, *Inorg. Chem.* **9**, 1229 (1970).
216. D. J. Darensbourg and A. Rokicki, *Organometallics* **1**, 1685 (1982).

217. J. P. Collman and L. S. Hegedus, "Principles and Applications of Organotransition Metal Chemistry," p. 73. Univ. Science Books, Mill Valley, CA, 1980.
218. D. J. Darensbourg and J. A. Froelich, *Inorg. Chem.* **17**, 3300 (1978).
219. J. E. Bercaw, L. Y. Goh, and J. Halpern, *J. Am. Chem. Soc.* **94**, 6534 (1972).
220. R. M. Laine, R. G. Rinker, and P. C. Ford, *J. Am. Chem. Soc.* **99**, 252 (1977).
221. J. W. Reppe and E. Reindl, *Leibigs Ann. Chem.* **582**, 121 (1953).
222. H. Kang, C. H. Mauldin, T. Cole, W. Slegeir, K. Cann, and R. Pettit, *J. Am. Chem. Soc.* **99**, 8323 (1977).
223. P. Ford, R. G. Rinker, C. Ungermann, R. Laine, V. Landis, and S. Moya, *J. Am. Chem. Soc.* **100**, 4595 (1978).
224. R. B. King, C. C. Frazier, R. M. Hanes, and A. D. King, *J. Am. Chem. Soc.* **100**, 2925 (1978).
225. A. D. King, R. B. King, and D. B. Yang, *J. Am. Chem. Soc.* **102**, 1028 (1980).
226. R. G. Pearson and H. Mauermann, *J. Am. Chem. Soc.* **104**, 500 (1982).
227. K. R. Lane, R. F. Lee, L. Sallans, and R. R. Squires, *J. Am. Chem. Soc.* **106**, 5767 (1984).
228. K. R. Lane and R. R. Squires, *J. Am. Chem. Soc.* **108**, 7187 (1986).
229. J. Elzinga and H. Hogeveen, *J. Chem. Soc. Chem. Commun.* 705 (1977).
230a. G. D. Mercer, W. B. Beauliew, and D. M. Roundhill, *J. Am. Chem. Soc.* **99**, 6551 (1977).
230b. M. Kubota, F. S. Rosenberg, and M. J. Sailor, *J. Am. Chem. Soc.* **107**, 4558 (1985).
231. F. Rivetti and U. Romano, *J. Organomet. Chem.* **174**, 221 (1979).
232a. R. W. Johnson and R. G. Pearson, *Inorg. Chem.* **10**, 2091 (1971).
232b. T. F. Block, R. F. Fenske, and C. P. Casey, *J. Am. Chem. Soc.* **98**, 441 (1976).
233. C. H. Cheng, D. E. Hendriksen, and R. Eisenberg, *J. Am. Chem. Soc.* **99**, 2792 (1977).
234. H. W. Steinberg, R. Markby, and I. Wender, *J. Am. Chem. Soc.* **79**, 6116 (1957).
235a. D. M. Fenton, U.S. Patents 3490872 and 3539298 (1970).
235b. D. E. Morris and H. B. Tinker, *Chem. Technol.* 555 (1972).
236a. R. Laine, *in* "Aspects of Homogeneous Catalysis" (R. Ugo, Ed.), Vol. 5, pp. 217–240. Reidel, London, 1984. An updated version of this review is in press in that series. We thank Dr. R. M. Laine for a preprint of this manuscript.
236b. R. Eisenberg and D. Hendersen, *Adv. Catal.* **28**, 79 (1979).
236c. P. Escaffre, A. Thorez, P. Kalck *J. MOL. Catal.,* **33**, 87 (1985).
237. C. Ungermann, V. Landis, S. Moya, H. Cohen, H,. Walker, R. Pearson, R. Rinker, and P. C. Ford, *J. Am. Chem. Soc.* **101**, 5922 (1979).
238. D. M. Vandenberg, Ph.D. Dissertation, University of California, Santa Barbara, 1986.
239. D. M. Vandenberg, T. M. Suzuki, and P. C. Ford, *J. Organomet. Chem.* **272**, 309 (1984).
240. R. B. King, A. D. King, and D. B. Yang, *ACS Symp. Ser.* **152**, 107 (1981).
241. A. D. King, R. B. King, and D. B. Yang, *J. Am. Chem. Soc.* **1103**, 2699 (1981).
242. C. C. Frazier, R. Haines, A. D. King, Jr., and R. B. King, *ACS Adv. Chem. Ser.* **172**, 94 (1979).
243. P. C. Ford, C. Ungermann, V. Landis, S. A. Moya, R. G. Rinker, and R. M. Laine, *ACS Adv. Chem. Ser.* **173**, 81 (1979).
244. R. B. King, C. C. Frazier, R. M. Hanes, and A. D. King, *J. Am. Chem. Soc.* **100**, 2825 (1978).
245. Y. Doi and S. Tamura, *Inorg. Chim. Acta* **54**, L235 (1981).
246. A. D. King, R. B. King and D. B. Yang, *J. Chem. Soc. Chem. Commun.* 529 (1980).
247. H. Ishida, K. Tanaka, M. Morimoto and, T. Tanaka, *Organometallics* **5**, 724 (1986).
248. W. A. R. Slegeir, R. S. Sapienza, and B. Easterling, *ACS Symp. Ser.* **152**, 325 (1981).

249. K. Kaneda, H. Hiraki, K. Sano, T. Imanaka, and S. Teranishi, *J. Mol. Catal.* **9**, 227 (1980).
250. T. Yoshida, Y. Ueda, and S. Otsuka, *J. Am. Chem. Soc.* **100**, 3941 (1978).
251. T. Venalainen, E. Iiskola, J. Pursiainen, T. A. Pakkanen, and T. T. Pakkanen, *J. Mol. Catal.* **34**, 293 (1986).
252. T. Yoshida, T. Okano, Y. Ueda, and S. Otsuka, *J. Am. Chem. Soc.* **103**, 3411 (1981).
253. T. Venalainen, T. A. Pakkanen, T. T. Pakkanen, and E. Iiskola, *J. Organomet. Chem.* **314**, C49 (1986).
254a. P. A. Marnot, R. R. Ruppert, and J.-P. Sauvage, *Nouv. J. Chim.* **5**, 543 (1981).
254b. J. P. Collin, R. Ruppert, and J.-P. Sauvage, *Nouv. J. Chim.* **9**, 395 (1985).
255. A. Pardey, PhD. Dissertation, University of California, Santa Barbara, 1986.
256. R. G. Nuzzo, D. Feitler, and G. M. Whitesides, *J. Am. Chem. Soc.* **101**, 3683 (1979).
257. E. C. Baker, D. E. Hendricsen, an R. Eisenberg, *J. Am. Chem. Soc.* **102**, 1020 (1980).
258. T. C. Singleton, L. J. Park, J. L. Price, and D. Forster, *Prepr. Div. Petr. Chem. ACS* **24**, 329 (1979).
259. P. Yarrow, H. Cohen, C. Ungermann, D. Vandenberg, P. C. Ford, and R. G. Rinker, *J. Mol. Catal.* **22**, 239 (1984).
260a. C. P. Kubiak and R. Eisenberg, *J. Am. Chem. Soc.* **102** 3637 (1980).
260b. C. P. Kubiak, C. Woodcock, and R. Eisenberg, *Inorg. Chem.* **21**, 2119 (1982).
261. A. A. Frew, R. H. Hill, L. Manojlovic-Muir, K. W. Muir and R. J. Puddephatt, *J. Chem. Soc. Chem. Commun.* 198 (1982).
262a. C. H. Cheng and R. Eisenberg, *J. Am. Chem. Soc.* **100**, 5968 (1978).
262b. C. H. Cheng, L. Kuritzkes, and R. Eisenberg, *J. Organomet. Chem.* **190**, C21 (1980).
263. P. Giannoccaro, G. Vasapollo, and A. Sacco, *J. Chem. Soc. Chem. Commun.* 1136 (1980).
264. J. Kaspar, R. Spogliarich, G. Mestroni, and M. Graziani, *J. Organomet. Chem.* **208**, C15 (1981).
265. J. Kaspar, R. Spogliarich, A. Cernogoraz, and M. Graziani, *J. Organomet. Chem.* **255**, 371 (1983).
266. M. Pawlik, M. F. Hoq, and R. E. Shepherd, *J. Chem. Soc. Chem. Commun.* 1467 (1983).
267. P. Giannoccaro, E. Pannacciulli, and G. Vasapollo, *Inorg. Chim. Acta* **96**, 179 (1985).
268. P. Mahajan, C. Creutz, and N. Sutin, *Inorg. Chem.* **24**, 2063 (1985).
269. R. C. Ryan, G. M. Wiloman, M. P. Dalsanto, and C. U. Pittman, *J. Mol. Catal.* **5**, 319 (1979).
270. R. Ganzerla, M. Lenarda, F. Pinna and M. Graziani, *J. Organomet. Chem.* **208**, C43 (1981).
271. Y. Doi, A. Yakota, H. Miyake , and K. Soga, *J. Chem. Soc. Chem. Commun.* 394 (1984); *Inorg. Chim. Acta* **105**, 69 (1985).
272. J. C. Bricker, C. C. Nagel, and S. G. Shore, *J. Am. Chem. Soc.* **104** 1444 (1982).
273. D. C. Gross, D. Taube, and P. C. Ford, unpublished observations.
274. J. B. Keister, *J. Organomet. Chem.* **190**, C36 (1980).
275. J. C. Bricker, C. C. Nagel, A. A. Battacharyya, and S. G. Shore, *J. Am. Chem. Soc.* **107** 377 (1985).
276a. P. Ford., *Prepr. Dir. Petr. Chem, ACS* **2**, 267 (1984).
276b. D. J. Taube, Ph.D. Dissertation, University of California, Santa Barbara, 1985.
277. A. Delazarche, J. P. Hindermann, R. Kieffer, and A. Kiennemann, *Rev. Chem. Interm.* **6**, 255 (1985).
278. W. A. R. Slegeir, R. S. Sapienza, R. Rayford, and L. Lam, *Organometallics* **1**, 1728 (1982).

279. N. S. Imyanitov, B. E Kuvaev, and D. M. Rudkavskii, *Zh. Prikl. Khim.* **40**, 2831 (1967).
280. R. M. Laine, D. W. Thomas, L. W. Cary, and S. E. Buttrill, *J. Am. Chem. Soc.* **100**, 6527 (1978).
281. E. J. More, J. M. Sullivan, and J. R. Norton, *J. Am. Chem. Soc.* **108**, 2257 (1986).
282a. N. von Kutepow and H. Kindler, *Angew. Chem.* **72**, 802 (1960).
282b. G. Henrici-Olivé and S. Olivé, *Trans. Metal Chem.* **1**, 77 (1976).
283. R. M. Laine *J. Am. Chem. Soc.* **100**, 6451 (1978).
284. D. J. Taube and P. C. Ford, *Organometallics* **5**, 99 (1986).
285. G. Suss-Fink and J. Reiner, *J. Mol. Catal.* **16**, 231 (1982).
286. K. Murata, A. Matsuda, K. Bando, and Y. Sugi, *J. Chem. Soc. Chem. Commun.* 785 (1979).
287. K. Murata and A. Matsuda, *Bull. Chem. Soc. Jpn.* **54**, 245, 249, 2089 (1981).
288. T. Okano, T. Kobayashi, H. Konishi, and J. Kiji, *Bull. Chem. Soc. Jpn.* **54**, 3799 (1981).
289. K. Kaneda, Y. Yasumura, M. Hiraki, T. Imanaka, and T. Teranishi, *Chem. Lett,* 1763 (1981).
290a. J. Palagyi and L. Marko, *J. Organomet. Chem.* **236**, 343 (1982).
290b. E. Alessio, R. Vinzi, and G. Mestroni, *J. Mol. Catal.* **22**, 327 (1984).
290c. J. Kasper, R. Spogliarich, A. Cernogoraz, and M. Graziani, *J. Organomet. Chem.* **255**, 371 (1983).
291. K. Cann, T. Cole, W. Slegeir, and R. Pettit, *J. Am. Chem. Soc.* **100**, 3969 (1978).
292. T. Cole, R. Ramage, K. Cann, and R. Pettit, *J. Am. Chem. Soc.* **102**, 6184 (1980).
293a. K. Kaneda, I. Hiraki, T. Imanaka, and S. Teranishi, *J. Mol. Catal.* **12**, 385 (1981).
293b. T. Okano, K. Fujiwara, H. Konishi, and J. Kiji, *Chem. Lett.* 1083 (1981).
293c. Y. Watanabe, N. Suzuki, Y. Tsuji, S. C. Shin, and T. A. Mitsudo, *Bull. Chem. Soc. Jpn.* **55**, 1116 (1982).
293d. E. Alessio, G. Zassinovich, and G. Mestroni, *J. Mol. Catal.* **18**, 113 (1983).
293e. Y. Watanabe, Y. Tsuji, T. Ohsumi, and R. Takeuchi, *Tetrahedron Lett.* 4121 (1983).
293f. C. Crotti, S. Cenini, B. Rindone, S. Tallari, and F. DeMartin, *J. Chem. Soc. Chem. Commun.* 784 (1986).
294a. C. D. Meyer and R. Eisenberg, *J. Am. Chem. Soc.* **98**, 1364 (1976).
294b. M. Kubota, K. J. Evans, C. A. Koerntgen, and J. C. Marsters, *J. Am. Chem. Soc.* **100**, 342 (1978).
294c. W. P. Fang, C. H. Cheng, *J. Chem. Soc. Chemm. Commun.* 503 (1986).
295. W. J. Thomson and R. M. Laine, *ACS Symp. Ser.* **152**, 133 (1981).
296a. K. Kaneda, M. Yasumura, T. Imanaka, and S. Teranishi, *J. Chem. Soc. Chem. Commun.,* 935 (1982).
296b. L. Marko, M. A. Radhi, and T. Otvos, *J. Organomet. Chem.* **218**, 369 (1981).
297. R. Pettit, K. Cann, T. Cole, C. H. Mauldin, and W. Slegeir, *Ann. N. Y. Acad. Sci.* **333**, 101 (1980).
298a. R. H. Fish, A. D. Thormodsen, and G. A. Cremer, *J. Am. Chem. Soc.* **104**, 5234 (1982).
298b. T. J. Lynch, M. Banah, H. D. Kaesz, and C. R. Porter, *J. Org. Chem.* **49**, 1266 (1984).
299a. H. des Abbayes and H. Alper, *J. Am. Chem. Soc.* **99**, 98 (1977).
299b. H. Alper and S. Armaratunga, *Tetrahedron Lett.* **21**, 2603 (1980).
299c. H. Alper, *Adv. Organomet. Chem.* **19**, 183 (1981).
300. H. Alper and K. E. Hashem. *J. Am. Chem. Soc.* **103**, 6514 (1981).
301a. J. Falbe, "Organic Synthesis with Carbon Monoxide." Springer-Verlag, Berlin, 1970.

301b. J. Falbe, "New Synthesis with Carbon Monoxide." Springer-Verlag, Berlin, 1980.

301c. H. Alper, in "Organic Synthesis via Metal Carbonyls" I. Wender and P. Pino, Eds.) , Vol. 2, Ch. 7. Wiley, 1977.

301d. R. Ugo, in "Catalysis in C_1 Chemistry" (W. Keim, Ed.), Ch. 5. Reidel, Dordrecht, 1983.

302. K. L. Oliver, D. M. Fenton, and J. Biale, *Hydrocarb. Proc.* **51**, 95 (1972).

303a. D. M. Fenton, *J. Org. Chem.* **38**, 3192 (1973).

303b. D. E. James and J. K. Stille, *J. Am. Chem. Soc.* **98**, 1810 (1976).

303c. R. F. Heck, *J. Am. Chem. Soc.* **94**, 2712 (1972).

304. H. Alper and F. W. Hartstock, *J. Chem. Soc. Chem. Commun.* 1141 (1985).

305. Y. Tsuji, T. Ohsumi, T. Kando, and Y. Watanabe, *J. Organomet. Chem.* **309**, 333 (1986).

306. M. D. Curtis and K. R. Han, *Inorg. Chem.* **24**, 376 (1985).

307. M. Kubota, F. S. Rosenberg, and M. J. Sailor, *J. Am. Chem. Soc.* **107**, 4558 (1985).

308a. M. O. Albers and N. J. Coville, *Coord. Chem. Revs.* **53**, 227 (1984).

308b. J. Elzinga and H. Hogeveen, *J. Chem. Soc. Chem. Commun.* 705 (1977).

309. R. E. Stevens and W. L. Gladfelter, *Inorg. Chem.* **22**, 2034 (1983).

310. S. G. Davies, *J. Organomet. Chem.* **179**, C5 (1979).

311. J. A. Froelich and D. J. Darensbourg, *Inorg. Chem.* **16** 960 (1977).

312. F. Calderazzo and D. B. Dell'Amico, *Inorg. Chem.* **21**, 3639 (1982).

313a. Since the preparation of this article two reviews concerned with homogeneous catalysis of carbonylation reactions have been published (Refs. 313b and 313c). Notably, in each review many of the catalytic cycles proposed feature nucleophile carbonyl adducts as key intermediates.

313b. T. W. DeKleera and D. Forster, *Adv. Catal.* **34**, 81 (1987).

313c. F. J. Waller, *J. Mol. Catal.*, Review Issue, 1986, pp. 43–61.

ADVANCES IN ORGANOMETALLIC CHEMISTRY, VOL. 28

Organopalladium and Platinum Compounds with Pentahalophenyl Ligands[1]

RAFAEL USÓN AND JUAN FORNIÉS

Departamento de Química Inorgánica,
Instituto de Ciencia de Materiales de Aragón,
Universidad de Zaragoza-C.S.I.C.,
50009 Zaragoza, Spain

I

INTRODUCTION

The opening of the field reviewed in this chapter could not have occurred before the necessary precursors, i.e., the appropriate perhaloarylating reagents, had been synthesized. In the period 1957–1959 Grignard and organolithium derivatives C_6Cl_5MgX (*1*) and C_6F_5MgX (*2*) were described, and in 1962–1966 C_6F_5Li (*3*) and C_6Cl_5Li (*4*) were reported. In the mid-1960s Stone and co-workers prepared the first pentafluorophenyl derivatives of palladium and platinum, but these complexes were limited to the types $MRXL_2$ and MR_2L_2 (*5–7*). The first palladium and platinum complexes containing C_6Cl_5 group, mainly of the type $MRXL_2$, were described by Rausch annd co-workers (*8*). These authors attributed the

[1]Abbreviations: APPY, acethylmethylenetriphenylphosphorane; bibzim, 2,2'-bibenz-imidazolate; biim, 2,2'-biimidazolate; bipy, 2,2'-bipyridine; ^{i}Bu, ^{t}Bu, *iso-* or *tert*-butyl; Bz, benzyl; COD, 1,5-cyclooctadiene; Cy, cyclohexyl; dba, dibenzylidene acetone; DMF, dimethyl formamide; DMSO, dimethyl sulfoxide; dpae, 1,2-bis(diphenylarsino)ethane; dpam, bis(diphenylarsino)methane; dppa, bis(diphenylphosphino)amine; dppb, 1,4-bis(diphenylphosphino)butane; dppe, 1,2-bis(diphenylphosphino)ethane; dppm, bis-(diphenylphosphino)methane; dppp, 1,3-bis(diphenylphosphino)propane; Et, ethyl; en, ethylenediamine; Me, methyl; pdma, *o*-phenylenebis(dimethyl)arsine; Ph, phenyl; phen, 1,10-phenanthroline; pn, propylenediamine; py, pyridine; pyO, pyridine *N*-oxide; THF, tetrahydrofuran, OC_4H_8; tht, tetrahydrothiophene, SC_4H_8; tmbizim, 4,5,4',5'-tetra-methylbibenzimidazolate; tmeda, tetramethylethylenediamine; *p*-Tol, *p*-C_6H_4Me-4.

observed lack of reactivity toward further arylation to steric factors arising from the presence of the bulky C_6Cl_5 group and the two ancillary ligands. The enhanced thermal stability of the C_6F_5 derivatives in comparison with the analogous aryl complexes was noticed and was thought to be due to the greater electronegativity of the perfluoroaryl ligand and the resulting ionic–covalent resonance energy leading to stronger M–C bonds (9–11).

In the mid-1970s the role of the stabilizing π-acceptor ligands began to be questioned in our laboratory. The use of precursors without any Group Vb ligand, or even of completely ligand-free species, opened new routes for the preparation of novel types of complexes and their application in a variety of unprecedented reactions.

The purpose of this chapter is to summarize the existing synthetic methods, which permit introduction of up to four C_6X_5 groups in the coordination sphere of palladium or platinum, as well as to survey the reactivity of the different complexes obtained. We begin with the much more numerous compounds with the metals in oxidation state (II), since these provide the gateway for the preparation of other complexes involving oxidation state (I), (III), or (IV), whose chemistry is more reduced in scope.

All the complexes show infrared bands characteristic of the presence of the C_6X_5 groups, most readily observed for C_6F_5 compounds. Some of these internal absorptions are related to the skeletal symmetry of the molecule and to the oxidation state of the metal center and have therefore been used for structural diagnosis. The information obtained from IR data will be summarized in a final section.

II

COMPLEXES OF PALLADIUM(II) OR PLATINUM(II)

Many neutral, anionic, or cationic metal complexes, containing one to four C_6X_5 groups, mono- or polynuclear, as well as homo- or heterometallic, have been reported. In all of them the metal is in approximately a square planar environment. They can be synthesized by a variety of methods. Sometimes several methods lead to a specific type of complex; in other cases, a compound that is not accessible by the direct arylation of a precursor can be prepared by a suitable substitution reaction, starting from a substrate that contains the desired number of C_6X_5 groups.

A. *Direct Arylation of Palladium or Platinum Complexes*

The following arylating agents have been used: (1) organomagnesium and organolithium reagents, (2) organothallium and organomercury reagents, (3) perhaloaryl halide.

1. *Grignard and Organolithium Reagents (RMgX and RLi, R = C_6F_5, C_6Cl_5)*

These reagents are the most frequently used, the lithium derivative generally being the most reactive. Since the type of product formed is also influenced by the nature of the palladium or platinum precursor, it is useful to arrange this discussion according to the different starting compounds.

a. Arylation of MX_2L_2 (L = Group Vb Ligand). Mono- as well as bisarylation [Eq. (1)] has been observed. The Grignard reagent normally

$$MX_2L_2 \xrightarrow{\text{RMgX (RLi)}} MRXL_2 \xrightarrow{\text{RMgX (RLi)}} MR_2L_2 \quad (1)$$

leads to substitution of only one halide ligand X, while C_6F_5Li more readily substitutes both halide groups, affording the compounds MR_2L_2 when used in an excess. Monoaryl complexes can also be obtained by adjusting the molar ratio of starting complex to C_6F_5Li. Thus, Stone *et al*, prepared *cis*-Pt(C_6F_5)Cl(PEt_3)$_2$ (80%) from *cis*-PtCl$_2$(PEt_3)$_2$ and C_6F_5Li (1:2), and *cis*-Pt(C_6F_5)$_2$(PEt_3)$_2$ (92%) from the same reagents (in 1:4 ratio) (5). By using *trans*-PtCl$_2$(PEt_3)$_2$ and C_6F_5Li (1:1) a mixture of *trans*-Pt(C_6F_5)Cl(PEt_3)$_2$ (58%) and *trans*-Pt(C_6F_5)$_2$(PEt_3)$_2$ (1.5%) was obtained, while *cis*-PtCl$_2$(PPh_3)$_2$ reacts (1:2) with C_6F_5Li to give *cis*-Pt(C_6F_5)-Cl(PPh_3)$_2$ (44%) and *cis*-Pt(C_6F_5)$_2$(PPh_3)$_2$ (41%). The reaction between *trans*-PdCl$_2$(PEt_3)$_2$ and C_6F_5Li (1:2) gives a mixture of products: *trans*-Pd(C_6F_5)Cl(PEt_3)$_2$ (9%), *trans*-Pd(C_6F_5)$_2$(PEt_3)$_2$ (74%), and *cis*-Pd(C_6F_5)$_2$(PEt_3)$_2$ (8%) (6).

With C_6F_5MgBr, halide exchange takes place and thus arylation of MCl_2L_2 yields $M(C_6F_5)BrL_2$, often mixed with $M(C_6F_5)ClL_2$ (5,12). Therefore, it is advisable to use an excess of C_6F_5Li when the preparation of the bisaryl complexes is desired and to prepare the monoaryl derivatives by another procedure (seen Section II,A,1b).

Both C_6Cl_5MgBr and C_6Cl_5Li react with the compounds MCl_2L_2 (L = tertiary phosphine) to give monoaryl complexes, even if an excess of the arylating agent is used (8,13–16). It is possible to prepare the Grignard compound in the presence of $M_2Cl_2L_2$ [Eq. (2)], with benzyl chloride or 1,2-dibromoethane as initiator.

$$MCl_2L_2 + Mg + C_6Cl_6 \longrightarrow M(C_6Cl_5)ClL_2 + MgCl_2 \quad (2)$$

$$M= Pd, Pt; L= PR_3$$

The decreased arylating power of the pentachlorophenyl derivatives is probably due to kinetic factors. Thus *trans*-$PtCl_2(PEt_3)_2$ cannot be arylated with either C_6Cl_5MgCl or C_6Cl_5Li. However, sublimation of *cis*-$Pt(C_6Cl_5)Cl(PEt_3)_2$ (240°C), (0.005 mm Hg) gives *trans*-$Pt(C_6Cl_5)$-$Cl(PEt_3)_2$ (14). The compounds $M(C_6Cl_5)_2L_2$ (L = N, P, As donors) have been prepared by other methods.

Arylation of $PdCl_2L_2$ with C_6Cl_5Li affords bisaryl derivatives [Eq. (3)] in

$$\underline{trans}\text{-}PdCl_2L_2 + C_6Cl_5Li \ (excess) \longrightarrow \underline{trans}\text{-}Pd(C_6Cl_5)_2L_2$$

$$L= AsPh_3, py, bipy, tht$$

$$(3)$$

50–60% yield (8,17). The complexes with L = $AsPh_3$ or tht are suitable precursors for the synthesis of bis(pentachlorophenyl) derivatives through substitution [Eq. (4)] of the neutral ligands (17). Although complexes

$$\underline{trans}\text{-}Pd(C_6Cl_5)_2L_2 + 2L' \longrightarrow \underline{trans}\text{-}Pd(C_6Cl_5)_2L'_2 + 2L$$

$$for \ L= AsPh_3; \ L'= bipy, phen, dppm, dppb$$

$$for \ L= tht; \ L'= PEt_3, PPh_3, AsPh_3, BzNH_2, dpam$$

$$(4)$$

trans-$PdCl_2(PR_3)_2$ are relatively unreactive, the bis(aryl) derivatives can be obtained, and the substitution process [Eq. (4)] is a general one.

The influence of the metal center, of the ancillary ligands, and even of the configuration of the precursors can be illustrated by the following:

(i) *trans*-$Pd(C_6Cl_5)_2py_2$ does not react with tertiary phosphines, while *cis*-$Pd(C_6Cl_5)_2$(bipy) reacts with dppe or PEt_3 to afford *cis*-$Pd(C_6Cl_5)_2L_2$ (18).

(ii) Although tht is a weakly coordinating ligand with palladium [Eqn. (4)], *trans*-$Pt(C_6Cl_5)_2$(tht)$_2$ is inert toward substitution when refluxed with PPh_3 or PEt_3 in chloroform (19).

cis-$M(C_6X_5)_2L_2$ (M = Pd, Pt; L = N, P, As donor) can be prepared by adding the neutral ligand to the anionic binuclear species ($^nBu_4N)_2[M(\mu\text{-}X')(C_6X_5)_2]_2$ (M = Pd, Pt; X = F, Cl; X' = Cl, Br) [see Eq. (43)].

b. Arylation of the Compounds [Pd(μ-X)XL]₂. As mentioned earlier (Section II,A,1a), attempts to prepare monoaryl derivatives Pd(C_6F_5)-(X)L_2, even when controlling the ratio of $PdCl_2L_2$ to arylating agent, lead to mixtures of mono- and bis(pentafluorophenyl) complexes. Their separation and further purification results in low yields of the desired compounds. The binuclear complexes [Pd(μ-X)XL]₂ can be selectively arylated at the terminal sites when using C_6F_5MgBr (or Mg and C_6Cl_6) [Eq. (5)].

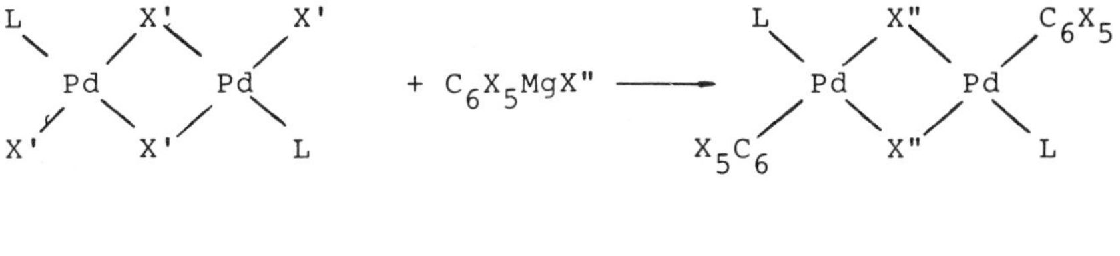

$$+ \; MgX'_2 \qquad (5)$$

X= F, X'= Cl; L= PPh_3, PEt_3, PBu_3; X"= Br

X= Cl; X'= Cl; L= PEt_3; X"= Cl

The Grignard reagent does not cleave the halogen bridge (*20*). Subsequent addition of a neutral ligand L or L' gives the monoarylated monomeric derivatives Pd(C_6X_5)X"L_2 or Pd(C_6X_5)X"LL', respectively. A similar synthesis has not yet been reported for platinum complexes.

c. Arylation of the Compounds MCl₂(tht)₂. Since the carbanions $C_6X_{15}{}^-$ are not strong nucleophiles they do not displace neutral Group Vb ligands in the starting compounds MCl_2L_2, and bisaryl derivatives M(C_6X_5)₂L_2 are obtained. Only in a few instances does displacement of phosphines by an organolithium occur, leading to anionic alkyl derivatives, e.g., [Pt(CH_3)₆]²⁻ (*21*).

Starting from precursors that contain no ancillary ligand or only weakly bound ones (such as tht), new types of perhaloaryl complexes of palladium and platinum, including anionic ones, can readily be prepared. Thus arylation (*22*) of *trans*-PdCl₂(tht)₂ with .C_6F_5MgBr (1:3.2 ratio) affords *trans*-Pd(C_6F_5)₂(tht)₂ (75% yield). Use of a lower ratio of C_6F_5MgBr (1:6) dereases the yield of the bisaryl complex (54%), but addition of Bu_4NBr to the solution allows isolation of the complex (Bu_4N)[Pd(C_6F_5)₃(tht)] (7%). The more inert platinum complex PtCl₂(tht)₂ reacts with C_6F_5MgBr (1:6) to give a mixture of *cis*- and *trans*-Pt(C_6F_5)₂(tht)₂ (overall yield 8%). Because of the expected lower reactivity, arylation with C_6Cl_5MgCl has not been attempted.

Arylation with C_6F_5Li gives mixtures of $M(C_6F_5)_2(tht)_2$, (Bu_4N)-$[M(C_6F_5)_3(tht)]$, and $(Bu_4N)_2[M(C_6F_5)_4]$. However, for M = Pt, the tetrakis (pentafluorophenyl) platinate(II) cannot be isolated because of the formation of uncrystallizable oils. The yields for the different species depend on the molar ratios; i.e., for $MCl_2(tht)_2/C_6F_5Li = 1:2.2$ the neutral complexes predominate (60% for *trans*-Pd; 56% for *cis*- *trans*-Pt), whereas for a 1:3 ratio the anionic $(Bu_4N)[M(C_6F_5)_3(tht)]$ (24% Pd; 56% Pt) compounds along with the neutral ones (36% Pd; 26% Pt) are obtained (*22*). With a 1:6 ratio, the complex $(NBu_4)_2[Pd(C_6F_5)_4]$ (70%) is the major product. The reactions proceed stepwise [Eq. (6)], displacement of halide ions preceding substitution of the neutral ligands.

$$MCl_2(tht)_2 \xrightarrow{RLi} MR_2(tht)_2 \xrightarrow{RLi} |MR_3(tht)|^- \xrightarrow{RLi} |MR_4|^{2-}$$

(6)

Arylation of $MCl_2(tht)_2$ with C_6Cl_5Li (ratio up to 1:6) yields only the neutral compounds *trans*-$[M(C_6Cl_5)_2(tht)_2]$ (*17,19*). This is not unexpected because of the lower reactivity of the pentachlorophenyl reagent.

As found for the neutral complexes [Eq. (4)], the tht ligand can be readily displaced in the salt $(Bu_4N)[M(C_6F_5)_3(tht)]$ to afford other complexes [Eq. (7)] (*22,23*). Halide ions do not displace tht, and only the pseudohalide CN reacts, giving $(Bu_4N)_2[M(C_6F_5)_3CN]$ as the product (*24*).

$$(Bu_4N)|M(C_6F_5)_3(tht)| + L \longrightarrow (Bu_4N)|M(C_6F_5)_3L| \quad (7)$$

M= Pd; L= PPh_3, $AsPh_3$

M= Pt; L= PPh_3, $AsPh_3$, $SbPh_3$, dppm, dppb

d. Arylation of $K_2[MCl_4]$. Attempts to arylate $K_2[PtCl_4]$ with C_6F_5Li, C_6F_5MgBr, or C_6Cl_5Li (1:8) in diethyl ether solution affords unreacted $K_2[PtCl_4]$ (*19,22*). Stirring a suspension of $K_2[PdCl_4]$ in a solution of C_6F_5Li in diethyl ether (1:8) (21 h, room temperature), with subsequent addition of Bu_4NBr, gives $(Bu_4N)_2[Pd(C_6F_5)_4]$, albeit in low yield (26%). The better thermal stability of the Grignard reagent, which can be refluxed in diethyl ether (1:8 ratio, 19 h), results in the same compound being obtained in substantially higher yield (54%). Increased molar ratio and/or decreased reaction time give oily products from which $(Bu_4N)_2[Pd(\mu-Br)(C_6F_5)_2]_2$ can be isolated in low yield. The compound $K_2[PtCl_4]$ is unreactive even on treatment with C_6Cl_5Li.

It has been claimed that $K_2[PdCl_4]$ reacts with C_6F_5MgHr in refluxing tetrahydrofuran and that subsequent addition of dioxane gives solutions from which (after removal precipitated magnesium salts) a bewildering array of neutral palladium complexes can be obtained (25). On the basis of their C and H analyses, the compounds "cis-Pd(C_6F_5)$_2$-(μ-dioxane)" (33% yield), "trans-Pd(C_6F_5)$_2$(dioxane)$_3$" (71%), "trans-Pd(C_6F_5)$_2$(dioxane)$_2$" (80%), and "{Pd(C_6F_5)$_2$(dioxane)(μ-Cl)}$_2$" (53%) have been described. In contrast, from K_2[PtCl$_4$] and C_6F_5MgBr in refluxing tetrahydrofuran a single compound, trans-Pt(C_6F_5)$_2$(dioxane)$_2$ (34%), can apparently be isolated.

Even if an excess of chloride ion is present, no anionic compounds have been characterized, despite the fact that reaction of "cis-Pd(C_6F_5)$_2$-(μ-dioxane)" with R$_4$PX gives halide-bridged binuclear anionic complexes (26) [Eq. (8)].

$$\underline{\text{cis}}\text{-Pd}(C_6F_5)_2(\mu\text{-dioxane}) \quad + \quad R_4PX \quad \longrightarrow$$

$$(R_4P)_2|(C_6F_5)_2Pd(\mu\text{-X})_2Pd(C_6F_5)_2| \quad (8)$$

$$X = Cl, \quad R_4P = PPh_3(CH_2Ph_2); \quad X = Br, I, \quad R_4P = PPhEt_3$$

The ready displacement of the neutral dioxane ligand has been exploited to prepare neutral cis- or trans- derivatives M(C_6F_5)$_2$L$_2$ (L = NH$_3$, amines, OPy, OPPh$_3$, AsPh$_3$) (25,27–29). The configuration of the end product is in every case determined by the starting complex. Other neutral complexes with O donors (DMF, DMSO, and ketones) have been described, and their formulation as trans-M(C_6F_5)$_2$(DMF)$_x$, trans-M(C_6F_5)$_2$(DMSO)$_x$ (x = 1.5), trans-Pt(C_6F_5)$_2$(RR'CO) (R = R' = CH$_3$; RR'CO = cyclohexanone;   R = Me,   R' = Et), trans-Pt(C_6F_5)$_2$-(RR'CO)$_{1.5}$ (R = Me, R' = ^{i}Bu, Ph), and trans-Pd(C_6F_5)$_2$(RR'CO)$_{0.5}$ (R = R' = ^{i}Bu; R = Me, R' = hexyl) has been reported (30).

It has also been claimed that Pt(C_6F_5)$_2$(acetone)$_x$ can be desolvated by heating; the dry air-stable white residue Pt(C_6F_5)$_2$ adds acetone, benzene, or toluene to give trans-Pt(C_6F_5)$_2$(acetone)$_x$ or trans-Pt(C_6F_5)$_2$(arene), respectively, even if a PPh$_3$ solution in the arene is used (31). Other complexes such as Pt(C_6F_5)$_2$ · H$_2$O, Pd(C_6F_5)$_2$ · OC$_4$H$_8$, and Pd(C_6F_5)$_2$(γ-butyrolactone)$_{1.5}$ have been reported (32). Telomeric structures with bridging C_6F_5 groups have been assigned to all the coordinatively unsaturated complexes. Since the structures of the proposed compounds have not been confirmed by X-ray crystallography, doubt about their true nature seems unavoidable. A recent paper reports the synthesis of cis-Pd(C_6F_5)$_2$(NCPh)$_2$ and describes its reaction to give the products Pd(C_6F_5)$_2$(L–L) (L–L = 1,5-cyclooctadiene, norbornadiene) (33).

e. Arylation of the Complex $(NBu_4)_2[M_2(\mu\text{-}Br)_2Br_4]$. This reaction takes place stepwise [Eq. (9)]. Thus for $M = Pd$ and C_6F_5MgBr (1:4.2) in

$$(NBu_4)_2 |M_2(\mu\text{-}Br)_2Br_4| \xrightarrow{\text{arylating agent}} (NBu_4)_2 |M_2(\mu\text{-}Br)_2R_4|$$

$$\xrightarrow{\text{arylating agent}} (NBu_4)_2 |MR_4| \qquad (9)$$

refluxing diethyl ether it affords $(NBu_4)_2[Pd_2(\mu\text{-}Br)_2(C_6F_5)_4]$ (81%), while with C_6F_5Li (1:20) the complex $(NBu_4)_2[Pd(C_6F_5)_4]$ (64%) is the end product. For $M = Pt$, no reaction takes place either with C_6F_5MgBr or with C_6F_5Li (22).

For $M = Pd$ and C_6Cl_5Li (1:14), $(NBu_4)_2[Pd(C_6Cl_5)_4]$ is the main product (64%), along with a small amount of $(NBu_4)_2[Pd_2(\mu\text{-}Br)_2\text{-}(C_6Cl_5)_4]$ (9%) (19). No reactions of $(NBu_4)_2[Pd(\mu\text{-}X)X_2]_2$ with C_6Cl_5MgCl, or of $(NBu_4)_2[Pt(\mu\text{-}X)X_2]_2$ with either C_6Cl_5MgCl or C_6Cl_5Li, have been reported so far.

f. Arylation of Halides MCl_2. Reactions of $PdCl_2$ or $PtCl_2$ with an excess of C_6F_5MgBr, followed by addition of NBu_4Br, give the binuclear derivatives $(NBu_4)_2[M(\mu\text{-}Br)(C_6F_5)_2]_2$ ($M = Pd, Pt$) (22). Since the yields (40%) are only moderate, the analogous reactions with C_6Cl_5MgCl have not been studied.

The arylations of $PdCl_2$ and $PtCl_2$ with C_6X_5Li ($X = F, Cl$) follow different pathways. The palladium chloride begins to react with C_6F_5Li or C_6Cl_5Li below room temperature (19,22). The reactions are very exothermic and within a few seconds lead to decomposition (metallic palladium). In contrast, with $PtCl_2$ the reactions are slow, and the resulting products vary with the $PtCl_2$/arylating agent ratio. Thus for C_6F_5Li, using a 1:2.3 ratio, $(NBu_4)_2[trans\text{-}PtCl_2(C_6F_5)_2]$ (60%) can be isolated after addition of NBu_4Br. For a molar ratio of 1:8 only 3% of the bisaryl derivative is formed, and the main product is $(NBu_4)_2[Pt(C_6Cl_5)_4]$ (66%) (22). Arylation of $PtCl_2$ with C_6Cl_5Li (1:10) gives a mixture of $(NBu_4)_2[Pt(C_6Cl_5)_4]$ (47%) and $(NBu_4)_2$ $[trans\text{-}PtCl_2(C_6Cl_5)_2]$ (9%) (19). The structure of $(NBu_4)_2[Pt(C_6Cl_5)_4]$ has been established by X-ray crystallography) (Fig. 1) (34,35). The four bulky C_6Cl_5 groups lie at angles of $\theta_{av} = 63.1°$ to the central PtC_4 plane.

2. Arylation with $[Tl(\mu\text{-}Br)(C_6F_5)_2]_2$ or $(C_6X_5)_2Hg$

The thallium reagent (36) is remarkable because it can transfer two C_6F_5 groups to a suitable precursor containing either a transition or a posttransi-

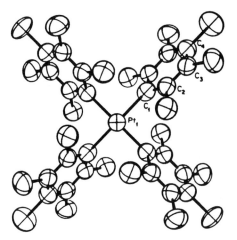

FIG. 1.   Structure of $[Pt(C_6Cl_5)_4]^{2-}$ (34,35).

tion element (37,38); e.g., Rh(I), Au(I), Ni(0), Pd(0), or Pt(0) compounds undergo oxidative addition [Eq. (10)].

$$2 \text{ M(PPh}_3)_4 + |\text{Tl}(\mu\text{-Br})(C_6F_5)_2|_2 \longrightarrow 2 \text{ } \underline{cis}\text{-M}(C_6F_5)_2(\text{PPh}_3)_2$$
$$+ 4 \text{ PPh}_3 + 2 \text{ TlBr} \qquad (10)$$

M= Pd, Pt

Oxidative addition is, however, not possible if the oxidation number of palladium or platinum in the precursor is 2+. Thus, trans-PdCl$_2$L$_2$ (L = PPh$_3$, PPh$_2$Me, AsPh$_3$) reacts (2:1) with [Tl(μ-Br)(C$_6$F$_5$)$_2$]$_2$ in refluxing benzene to give a mixture of [Pd(μ-Cl)(C$_6$F$_5$)L]$_2$ and trans-Pd(C$_6$F$_5$)ClL$_2$, which can be separated by fractional recrystallization (39). Oxidation to Pd(IV), as had initially been reported (37,38), does not take place. Neither trans-PtCl$_2$(PEt$_3$)$_2$ nor cis-PtCl$_2$(PPh$_3$)$_2$ undergoes oxidative addition with [Tl(μ-Cl)(C$_6$F$_5$)$_2$]$_2$ in refluxing benzene, and the heterobinuclear halogen-bridged M^{II}–Tl^{III} compounds are formed (40).

The synthesis of the slightly soluble compound [Tl(μ-Cl)(C$_6$Cl$_5$)$_2$]$_2$ and its use as an arylating agent have been described (41,42), but no reaction with either a palladium or a platinum complex has been reported.

Arylation with HgR$_2$ (R = C$_6$H$_5$, p-CH$_3$C$_6$H$_4$, 2-arylazoaryl) has been studied (43–45). Sales et al. (46) found that on heating a mixture of cis-PtCl$_2$(PPh$_3$)$_2$ and HgR$_2$ (R = C$_6$H$_{5-n}$Cl$_n$, n = 0–5) to near the melting point, one chloride ligand is exchanged with one chlorophenyl group and the monoaryl complex can be isolated by vacuum sublimation.

3. *Arylation with* C_6F_5X

Although these reagents do not arylate Pd(II) or Pt(II) precursors, they are active toward M(0) compounds and the finely powdered metals. Thus C_6F_5H adds oxidatively to $Pt(PCy_3)_2$ to give *trans*-$PtH(C_6F_5)(PCy_3)_2$ (*47*) and C_6F_5COX (X = Cl, Br) reacts with $Pd(PMePh_2)_4$ to give *trans*-$Pd(C_6F_5)X(PMePh_2)_2$, CO being released during the reaction (*48*).

Klabunde *et al.* (*49*) studied the cocondensation of palladium or platinum vapors with C_6F_5X. Acetone solutions of $[Pd(C_6F_5)Br]_n$ have been prepared from which an orange–brown solid (10.5% yield) can be isolated. Purity of the solvent and rapid workup of the solutions are critical factors. Otherwise, decomposition to Pd, $PdBr_2$, and $C_{12}F_{10}$ is observed. The isolated compound seems to be trimeric in acetone but monomeric in diethyl ether solution.

Addition of neutral ligands to solutions of $[Pd(C_6F_5)Br]_n$ give stable four-coordinated derivatives [Eq. (11)].

$$1/n \; |Pd(C_6F_5)Br|_n \; + \; 2L \; \longrightarrow \; Pd(C_6F_5)BrL_2 \quad (11)$$

$$L = SPh_2, SEt_2, SMe_2, NHEt_2, NMe_3, NH_3, py, PEt_3, bipy, AsPh_3, COD$$

Cocondensation of palladium and C_6F_5COCl and subsequent addition of PEt_3 (25°C) affords $Pd(C_6F_5)Cl(PEt_3)_2$ (23%) (*50*). With platinum atoms and C_6F_5Br, $[Pt(C_6F_5)Br]_n$ is initially formed. The latter can be trapped at -78°C with PEt_3 to give a mixture of *cis*- and *trans*-$Pt(C_6F_5)Br(PEt_3)_2$. Warming the original acetone solution without addition of PEt_3 causes disproportionation [Eq. (12)] while $[Pd(C_6F_5)Br]_n$, remarkably, is stable at 100°C (*51*).

$$|Pt(C_6F_5)Br|_n \; \xrightarrow{\text{acetone, } 25°C} \; Pt(C_6F_5)_2(\text{acetone})_2 \; + \; PtBr_2$$
$$(12)$$

Electrochemical oxidation of a palladium anode in a 1:1 mixture of $EtOH/C_6F_5Br$ and NEt_4Br as electrolyte gives $Pd(C_6F_5)_2L_2$ ($L_2 = 2PEt_3$, dppe, 2py) in good yields, a method which, however, does not work with platinum (*52*).

Reactive slurries of both Pd and Pt have been prepared by reduction of $MX_2(PEt_3)_2$ with metallic potassium in tetrahydrofuran or 1,2-dimethoxyethane. Addition of C_6F_5Br to the slurries gives *trans*-$Pd(C_6F_5)Br(PEt_3)_2$ (76%) or *trans*-$Pt(C_6F_5)Br(PEt_3)_2$ (40%), respectively. Zero-valent compounds such as $M(PEt)_4$ may also exist in these slurries (*53,54*). Reduction of PdI_2 with Li/naphthalene gives a reactive powder which, on addition of C_6F_5I, forms a solution that is thought to contain

$[Pd(C_6F_5)I]_n$ since, after addition of PEt_3 trans-$Pd(C_6F_5)I(PEt_3)_2$ (44%) can be isolated (54).

B. Reactivity of Perhalophenyl Complexes of Palladium(II) and Platinum(II)

1. Complexes with One C_6X_5 Group Bonded to each Metal Center.

Neutral or cationic as well as monomeric or polymeric complexes have been described.

a. *Neutral Complexes.* These are of the types $M(C_6X_5)X'L_2$ and $M_2\text{-}(\mu\text{-}X')_2(C_6X_5)_2L_2$ (M = Pd, Pt; X = F, Cl; X' = halide or pseudo-halide; L = neutral N, P, As, Sb, O, S donor). Although only the structure of the complex cis-$Pt(C_6F_5)Cl(PEt_3)_2$ (55) has been established by single-crystal X-ray crystallography, an approximately square planar environment for the metal is plausible in all cases.

Mononuclear Complexes $M(C_6X_5)X'L_2$. Both substitution of X' by other anionic ligands and displacement of L by other neutral ligands have been studied. Reaction of $M(C_6X_5)X'L_2$ with salts M'X'' in acetone or ethanol (at room or at reflux temperature) affords the substitution product [Eq. (13)] (yield over 60%) (13,14,16,18,20,56–59).

$$M(C_6X_5)X'L_2 \ + \ M'X'' \ \longrightarrow \ M(C_6X_5)X''L_2 \ + \ M'X' \qquad (13)$$

M= Pd,Pt; X= F,Cl; X'= Cl,Br; M'= Li,Na,K; X''= Br,I,SCN,

N_3,CN,CNO,CH_3-COO,NO_3,MeO

Occasionally, silver salts (in benzene, acetone, or chloroform) have been used when the alkali salts are unreactive [Eq. (14)] (13,14,16). With the

$$M(C_6Cl_5)X'L_2 \ + \ AgX'' \ \longrightarrow \ M(C_6Cl_5)X''L_2 \ + \ AgX' \qquad (14)$$

M= Pd,Pt; X'= Cl,Br; L= PEt_3; X''= ClO_4,NO_2,NO_3,CH_3-COO,CN

more inert pentachlorophenyl derivatives a two-step process leads to better results. Reaction of $Pd(C_6Cl_5)Br(PEt_3)_2$ with $AgClO_4$ gives the perchlorato complex $Pd(C_6Cl_5)(OClO_3)(PEt_3)_2$ [Eq. (13)], which can be isolated. The anion can subsequently be displaced by reaction with sodium or potassium salts of the anions Cl^-, $N_3{}^-$, and OC_6Cl_5 (13,16).

$$Pd(C_6X_5)ClL_2 + 2L' \longrightarrow Pd(C_6X_5)ClL'_2 + 2 L \quad (15)$$

X= F; L= AsPh$_3$; 2L'= 2 PPh$_3$,dppe,bipy,phen,tmeda

X= Cl; L$_2$= bipy; 2L'= 2 PEt$_3$, 2 PPh$_3$, dppe

The neutral ligands can also be substituted by others (*12,18*) [Eq. (15)].
Binuclear Complexes.

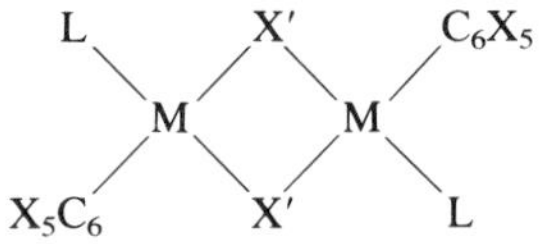

In addition to the preparative methods mentioned already, neutral binuclear double-bridged palladium complexes can be prepared (75–80%) by reactions between bisaryl derivatives and PdCl$_2$ (1:1) in refluxing acetone (*60*) [Eq. (16)]. This reaction does not proceed when L$_2$ is a

$$\underline{trans}\text{-}Pd(C_6X_5)_2L_2 + PdCl_2 \longrightarrow \qquad (16)$$

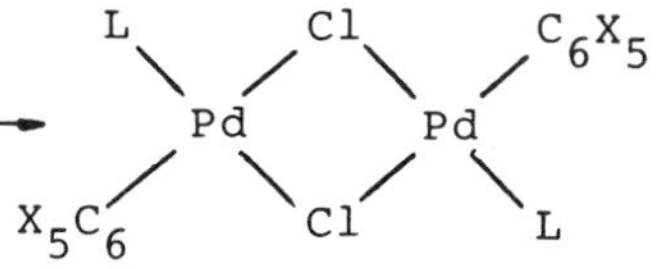

X= F,Cl; L= PPh$_3$,AsPh$_3$,SbPh$_3$,tht,py,BzNH$_2$

chelating ligand (L–L), which suggests dissociation of L as the rate-determining step. When L = SbPh$_3$ or tht, the reactions can be run at room temperature, whereas with L = py, or PPh$_3$ about 20 h refluxing acetone is required. The nature of the C$_6$X$_5$ group seems to play a minor role; only with L = BzNH$_2$ does a dramatic difference arise. Thus [Pd(μ-Cl)(C$_6$Cl$_5$)(BzNH$_2$)]$_2$ can be prepared, but the analogous C$_6$F$_5$ derivative has not been isolated, since no reaction takes place at room temperature and metallic palladium precipitates at reflux temperature. Nevertheless, this is the best method for synthesizing this type of palladium complex. It is noteworthy that M(C$_6$X$_5$)$_2$L$_2$ complexes act as arylating agents. The only analogous reaction reported for platinum, Pt(C$_6$F$_5$)$_2$(tht)$_2$ + PtCl$_2$ to give the binuclear derivative, requires more drastic conditions (refluxing toluene, 5 h) (*61*).

The structure assigned to these complexes is based on their IR spectra; no diffraction studies have been reported. The binuclear complexes undergo two classes of reactions: (a) substitution reactions at the chloride bridge to give new halide- or pseudohalide-bridged binuclear complexes and (b) cleavage with neutral ligands to give monomeric derivatives

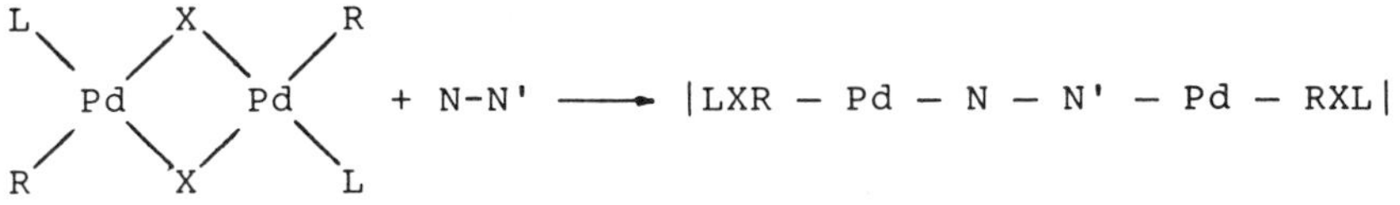

$$\underset{R}{\overset{L}{\diagdown}}Pd\diagup\overset{X}{\diagdown}\diagup Pd\diagdown\overset{R}{\diagup} \quad + \text{ N-N}' \longrightarrow |LXR - Pd - N - N' - Pd - RXL|$$

R= C_6F_5; L= PPh_3,py,$AsPh_3$; N-N'= 4,4'-bipy

R= C_6Cl_5; L= PPh_3, $SbPh_3$

$$(17)$$

PdRXLL′ or single-bridged binuclear derivatives when the neutral ligand is a bidentate nonchelating species as illustrated by the following case (62) [Eq. (17)].

In contrast, binuclear complexes containing the bridging ligands biim, bibzim, and tmbibzim, all of which contain a system of four N atoms suitably disposed to bridge two metal centers with a pair of N atoms chelating each of them, have been prepared (63) [Eqs. (18) and (19)]. Both

$$2 \; Pd(C_6X_5)(acac)(PPh_3) \; + \; H_2(N-N)_2 \longrightarrow 2 \; H(acac) \; +$$

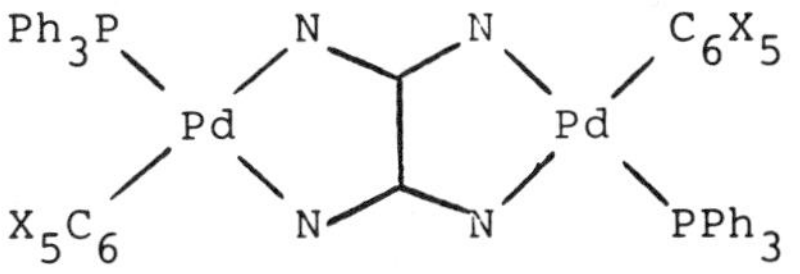

$$\underset{X_5C_6}{\overset{Ph_3P}{\diagdown}}Pd\diagup\overset{N}{\diagdown}\cdots\overset{N}{\diagup}Pd\diagdown\overset{C_6X_5}{\diagup}PPh_3 \qquad (18)$$

X= F,Cl; $(N-N)_2^{2-}$ = biim,bibzim,tmbibzim

$$Pd_2(\mu\text{-Cl})_2(C_6F_5)_2(PPh_3)_2 \; + \; Tl_2(N-N)_2 \longrightarrow 2 \; TlCl \; +$$

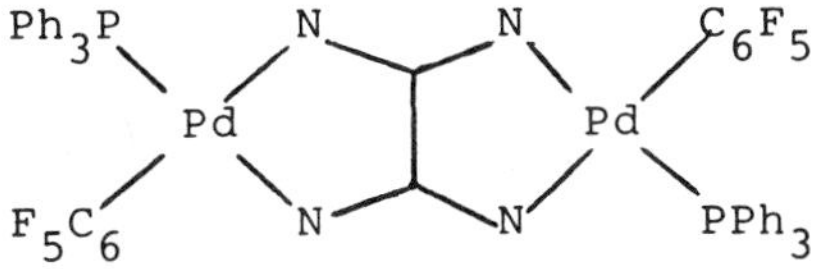

$$\underset{F_5C_6}{\overset{Ph_3P}{\diagdown}}Pd\diagup\overset{N}{\diagdown}\cdots\overset{N}{\diagup}Pd\diagdown\overset{C_6F_5}{\diagup}PPh_3 \qquad (19)$$

methods lead to the same complexes. However, the pure thallium salt of tmbibzim has not as yet been prepared; hence in this case only the first method [Eq. (18)] is used. Moreover, with C_6Cl_5 and bibzim or tmbibzim polynuclear species $[Pd(\mu\text{-N-N})_2]_x$ are obtained. Their molecular weights (isopiestic, in $CHCl_3$ solution) confirm the binuclear nature of the compounds. Tetranuclear complexes, as described for rhodium (64), are not observed.

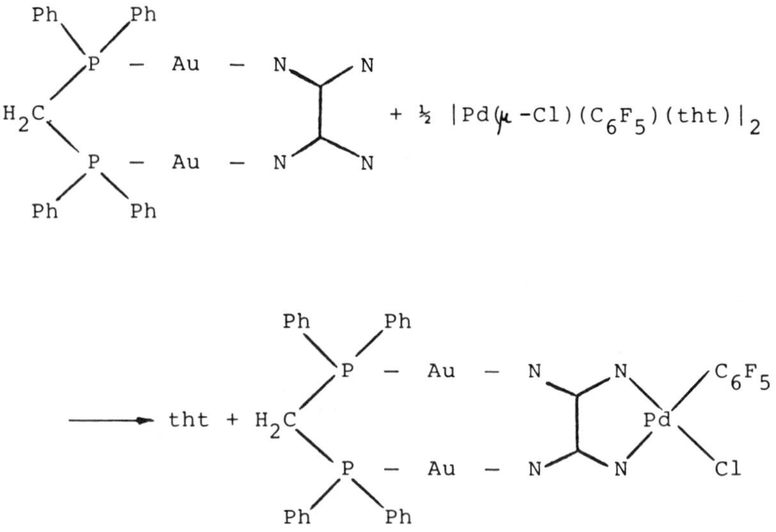

$$\text{(20)}$$

Reaction of the compound $(dppm)Au_2(\mu\text{-N-N})_2$ $(N\text{-N})_2$ = bibzim) with the binuclear species $[Pd(C_6F_5)Cl(tht)]_2$ in dichloromethane solution give a trinuclear complex (65) [Eq. (20)].

Finally, if the halide bridge in the binuclear complexes is replaced by CN groups (by using an excess of AgCN in CH_2Cl_2), tetranuclear cyclic complexes are obtained with two mutually cis CN linear groups each bridging two metal centers (66) [Eq. (21)]. This occurs because the CN ligand is not able to form bent bridges.

$$2\ Pd_2(\mu\text{-Cl})_2(C_6X_5)_2L_2 + 4\ AgCN \longrightarrow 4\ AgCl +$$

$$Pd_4(\mu\text{-CN})_4(C_6X_5)_4L_4 \quad \text{(21)}$$

$X= F;\ L= PEt_3,\ PPh_3,\ AsPh_3;\ X= Cl;\ L= PEt_3$

b. Cationic Compounds. As mentioned above [Eq. (14)], addition of $AgClO_4$ to acetone, dichloromethane, or benzene solutions of $M(C_6X_5)X'L_2$ complexes causes precipitation of silver halide and yields solutions of the perchlorato complexes $M(C_6X_5)(OClO_3)L_2$. Because the $OClO_3$ ligand is a very weak one, it can readily be displaced either by anionic ligands X'' or neutral ones L'. With the latter, cationic derivatives

$$M(C_6X_5)(OClO_3)L_2 + L' \longrightarrow |M(C_6X_5)L_2L'|ClO_4 \qquad (22)$$

M= Pd,Pt; X= F,Cl; L= N,P,As donor; L'= O,N,P,As,Sb donor

$$M(C_6F_5)(OClO_3)(AsPh_3)_2 + 3\ L \longrightarrow |M(C_6F_5)L_3|ClO_4 +$$
$$+\ 2\ AsPh_3$$

M= Pd; L= py, PEt_3, PPh_3

M= Pt; L= PPh_3, PEt_3

$$(23)$$

are formed (*13,16,57,67,68*) [Eq. (22)]. The reaction is a general one, and if L is a weakly coordinating ligand an excess of L' can also replace it (*57*) [Eq. (23)]. However, for M = Pt and L' = py, only one $AsPh_3$ group is substituted and $[Pt(C_6F_5)(AsPh_3)py_2]ClO_4$ is obtained, even if an excess of py is used (*58*).

The intermediate perchlorato complexes are excellent precursors for the synthesis of a variety of cationic complexes of new types that are inaccessible or difficult to prepare by other methods.

Synthesis of Cationic Palladium(II) Carbonyls. Bubbling CO at atmospheric pressure and room temperature through solutions or suspensions of the compounds $[Pd(C_6F_5)(OClO_3)L_2]$ or $[Pd(C_6F_5)(acetone)L_2]ClO_4$ in toluene or chloroform causes precipitation of cationic monocarbonyls [Eqs. (24) and (25)] (*69*). At room temperature, the complexes with

$$Pd(C_6F_5)(OClO_3)L_2 + CO \longrightarrow |Pd(C_6F_5)(CO)L_2|ClO_4 \qquad (24)$$

L_2= 2 PPh_3, 2 PPh_2Me, 2 $PPhEt_2$, 2 PEt_3, bipy

$$|Pd(C_6F_5)(acetone)L_2|ClO_4 + CO \longrightarrow |Pd(C_6F_5)(CO)L_2|ClO_4$$

L_2= phen, tmeda

$$(25)$$

monodentate ligands evolve CO, to give again the original perchlorato complexes. The $\nu(CO)$ band lies very high, at even higher energies than in the free CO (2.143 cm^{-1}) (see Table I) (*70*).

It is noteworthy that the complexes with higher $\nu(CO)$ stretches are more stable, indicating for the first time that π-backbonding is unimportant

TABLE I

RELEVANT DATA FOR $[Pd(C_6F_5)(CO)L_2]ClO_4$ COMPLEXES (69)

L_2	$\nu(CO)$ (cm^{-1})	Temperature at which complete decomposition is observed (°C)	Time for complete disappearance of $\nu(CO)$ at room temperature (days)
2PEt$_3$	2.132	32	15
2PPh$_3$	2.143	39	11
2PPhEt$_2$	2.146	65	170
2PPh$_2$Me	2.147	45	162
bipy	2.153	142	a
phen	2.160	133	a
tmeda	2.163	80	a

^{a} In 5 months the intensity of $\nu(CO)$ did not decrease.

for the stability of palladium(II) carbonyls, which are obtained under very mild conditions compared to the usual ones (71–76).

Synthesis of Acetylmethylenetriphenylphosphorane derivatives. The 1:1 reaction of $Pd(C_6F_5)(OClO_3)L_2$ (L = PPh$_3$, PBu$_3$, $\frac{1}{2}$ bipy) with APPY (acetylmethylenetriphenylphosphorane) in benzene solution causes precipitation of the cationic complexes $[Pd(C_6F_5)(APPY)L_2]ClO_4$ (77). The IR spectra of the complexes show a (CO) band at ~1520 cm^{-1} and therefore $\Delta\nu = \nu(CO)$ complex $- \nu(CO)$ free ylide < 0, implying O coordination of the ylide ligand. The ^{1}H and ^{31}P NMR spectra also confirm this result. The complex with $L_2 = $ bipy is necessarily the *cis* isomer, whereas the complexes with monodentate phosphines are the *trans* isomers. In deuterochloroform solutions the complex with L = PPh$_3$ has both the cisoid (A) and transoid (B) forms present (1:4 ratio). For L = PBu$_3$ both forms (1:1 ratio) are also present, while for $L_2 = $ bipy only the cisoid isomer has been detected.

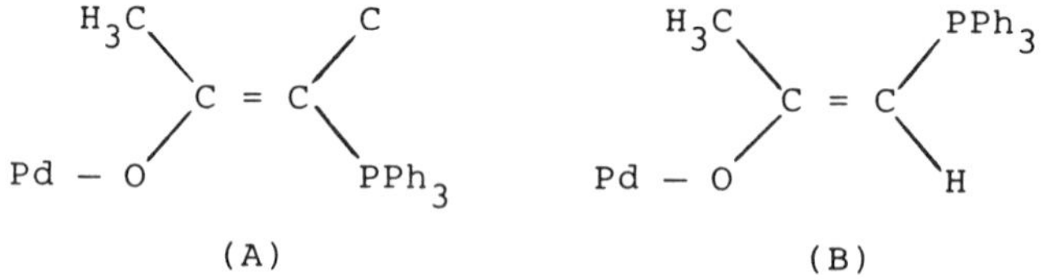

(A)   (B)

Treating $[Pd(\mu\text{-Cl})(C_6F_5)(tht)]_2$ with APPY results in the formation of neutral $[Pd(C_6F_5)Cl(APPY)(tht)]$, for which $\Delta\nu = +115$, suggesting C coordination of the APPY ligand, a result confirmed by the NMR data.

The other known palladium complexes with keto-stabilized ylides, both neutral (*78–83*) and cationic ones (*83*), are C-bonded.

Synthesis of Trialkylphosphonium Dithiocarboxylate Complexes. Perchlorato complexes of palladium and platinum $M(C_6F_5)(OClO_3)L_2$ react with $^-S_2C^{-+}PR_3$ to give cationic complexes containing unidentate S_2C-PR_3 [Eqs. (26) and (27) in 80–85% yield (*84*):

$$M(C_6F_5)(OClO_3)L_2 \ + \ ^-S_2C^{\overset{+}{-}}PR_3 \quad \xrightarrow{\ C_6H_6\ }$$

$$|M(C_6F_5)(\eta^1\text{-}S_2C\text{-}PR_3)L_2|ClO_4 \quad (26)$$

$$M= \text{Pd}; \ L= \text{PPh}_3, \ \text{PBu}_3, \ \text{PhEt}_2; \quad R_3= \text{Et}_3, \ \text{Bu}_3, \ \text{PhEt}_2$$

$$L= \text{PBu}_3; \quad\quad\quad\quad\quad\quad\quad\quad R_3= \text{Cy}_3$$

$$L= \text{PPh}_3; \quad\quad\quad\quad\quad\quad\quad\quad R_3= \text{Et}_3, \ \text{Cy}_3$$

$$M= \text{Pt}; \ L= \text{PEt}_3, \ \text{PBu}_3; \quad\quad\quad R_3= \text{Et}_3, \ \text{Bu}_3$$

$$M(C_6F_5)(OClO_3)(\text{bipy}) \ + \ ^-S_2C^{\overset{+}{-}}PR_3 \quad \xrightarrow[\text{or CH}_2\text{Cl}_2]{\ \text{OCMe}_2\ }$$

$$|M(C_6F_5)(\eta^1\text{-}S_2C\text{-}PR_3)(\text{bipy})|ClO_4 \quad (27)$$

$$M= \text{Pd, Pt}; \ R_3= \text{Et}_3, \ \text{Bu}_3, \ \text{Cy}_3$$

The complex $[Pd(C_6F_5)(\eta^1-S_2C-PEt_3)(PEt_3)_2]ClO_4$ can also be obtained (~60% yield) by insertion of CS_2 in the Pd-P bond, trans to the C_6F_5 group in $[Pd(C_6F_5)(PEt_3)_3]ClO_4$. The structure of this complex has been established by single-crystal X-ray crystallography (Fig. 2). The palladium atom

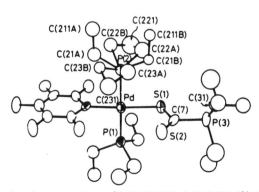

Fig. 2. Structure of *trans*-[Pd(C₆F₅)(PEt₃)₂(S₂C-PEt₃)]⁺ (*84*).

lies in an approximately square planar environment, the $S(1)-C(7)$ distance [1.699(16) Å] being a little greater than $S(2)-C(7)$ [1.632(15) Å], as a consequence of the coordination. In the crystal the $P(2)Et_3$ group is disordered.

Synthesis and Reactivity of Isonitrile Complexes. Reaction of $Pd(C_6F_5)$-ClL_2 (in acetone solution) with $NaClO_4$, in the presence of a stoichiometric amount of an isonitrile CNR, affords cationic complexes [Eq. (28)]

$$Pd(C_6F_5)ClL_2 + NaClO_4 + CNR \longrightarrow |Pd(C_6F_5)(CNR)L_2|ClO_4$$

$$L_2= 2\ PPh_3,\ 2\ PPh_2Me,\ 2\ PEt_3,\ bipy,\ tmeda,\ dppe$$

$$R=\ Cy,\ p\text{-}Tol,\ {}^tBu$$

(28)

(*85*) and because of the coordination $\nu(C{\equiv}N)$ is shifted (70–100 cm^{-1}) to higher energies (*86*). These complexes show marked differences in behavior toward amines and alcohols:

(i) When L = a unidentate ligand (PPh_3, PPh_2Me, or PEt_3) decomposition takes place.
(ii) When L_2 = bipy and R = ^tBu or Cy no reaction is observed.
(iii) When L_2 = a chelate ligand and R = *p*-tolylamine, the complex reacts with amines and alcohols in refluxing chloroform or acetone to give carbene complexes [Eq. (29)].

$$|Pd(C_6F_5)(chel)(CNTol)|ClO_4 + HX \longrightarrow$$

$$|Pd(C_6F_5)(chel)\{C(NHTol)X\}|ClO_4 \quad (29)$$

$$HX=\ p\text{-}TolNH_2,\ BzNH_2,\ tert\text{-}BuNH_2,\ MeOH,\ EtOH$$

$$chel=\ bipy,\ tmeda,\ dppe\ (only\ reacts\ with\ amine)$$

The carbene complexes can be deprotonated with KOH (1:1 molar ratio) to give neutral imidoyl complexes [Eq. (30)].

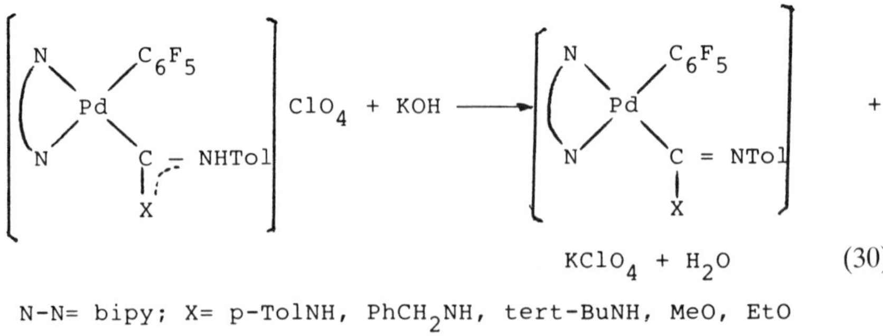

$$KClO_4 + H_2O \quad (30)$$

$$N\text{-}N=\ bipy;\ X=\ p\text{-}TolNH,\ PhCH_2NH,\ tert\text{-}BuNH,\ MeO,\ EtO$$

c. Single-Bridged Polynuclear Complexes. The poorly coordinating perchlorato ligand can also be displaced by the free end of a potentially bidentate ligand when it is acting as unidentate ligand in a complex (*87,88*). Thus, homo- or heteroatomic binuclear complexes can be prepared [Eq. (31)]:

$$
M(C_6F_5)XL_2 \; + \; M'(C_6F_5)(OClO_3)L'_2 \longrightarrow
$$

$$
\left[F_5C_6 - \underset{\underset{L}{|}}{\overset{\overset{L}{|}}{M}} - X - \underset{\underset{L'}{|}}{\overset{\overset{L'}{|}}{M'}} - C_6F_5 \right] ClO_4 \quad (31)
$$

M=M'= Pd; X= CN, SCN, N_3; L=L'= PPh_3

M= Pd; M'= Pt; X= CN; L= PPh_3, L'= PEt_3

The above process [Eq. (31)] does not take place when X = Cl, Br, or I, although the reaction between $Pd(C_6F_5)X(bipy)$ and $Pd(C_6F_5)$-$(OClO_3)bipy$ yields $\{[(\mu\text{-}X)Pd(C_6F_5)(bipy)]_2\}ClO_4$ (X = Cl, Br, SCN, CN).

This is the only reported method for the deliberate preparation of single-bridged complexes but rearrangement of ligands sometimes occurs, a mixture of products being the result [Eq. (32)]:

$$
2 \; Pd(C_6F_5)(NCS)(PPh_3)_2 \; + \; 2 \; Pd(C_6F_5)(OClO_3)(bipy) \longrightarrow
$$

$$
|Pd(\mu\text{-}SCN)(C_6F_5)(PPh_3)|_2 \; + \; 2 \; |Pd(C_6F_5)(PPh_3)(bipy)|ClO_4
$$

$$
(32)
$$

The same result (rearrangement, giving a mixture of products) occurs in the reaction between $Pd(C_6F_5)(OClO_3)(PPh_3)_2$ and $Pd(C_6F_5)(SCN)(bipy)$. If both reactants bear chelating ligands the rearrangement reaction is stopped [Eq. (33)]. With a starting complex containing two X groups, trinuclear

$$
Pd(C_6F_5)(SCN)(dppe) \; + \; Pd(C_6F_5)(OClO_3)(bipy) \longrightarrow
$$

$$
|(dppe)(C_6F_5)Pd(\mu\text{-}SCN)Pd(C_6F_5)(bipy)|ClO_4
$$

$$
(33)
$$

complexes can be prepared [Eq. (34)] when X = CN, but with more labile groups (X = SCN, N_3, I) redistribution reactions take place and mixtures of products are therefore obtained. Because of their inert behavior toward trans-$PdX_2(PPh_3)_2$ + 2 $Pd(C_6F_5)(OClO_3)(PPh_3)_2$ ⟶

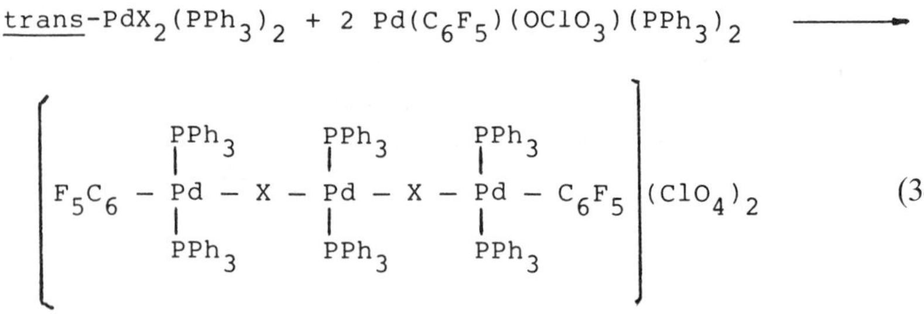

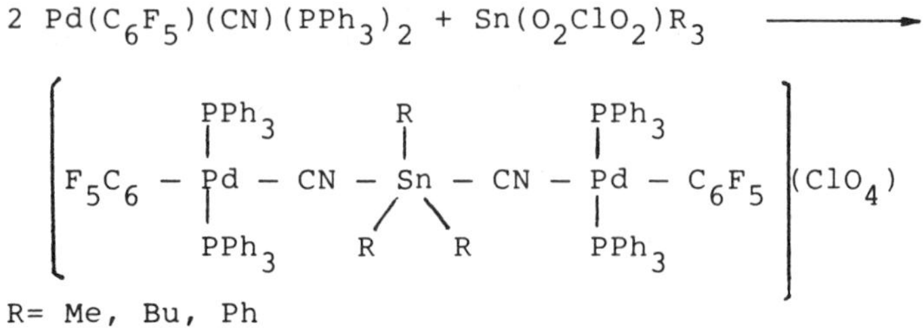

$$\left[F_5C_6 - \underset{\underset{PPh_3}{|}}{\overset{\overset{PPh_3}{|}}{Pd}} - X - \underset{\underset{PPh_3}{|}}{\overset{\overset{PPh_3}{|}}{Pd}} - X - \underset{\underset{PPh_3}{|}}{\overset{\overset{PPh_3}{|}}{Pd}} - C_6F_5 \right](ClO_4)_2 \qquad (34)$$

2 $Pd(C_6F_5)(CN)(PPh_3)_2$ + $Sn(O_2ClO_2)R_3$ ⟶

$$\left[F_5C_6 - \underset{\underset{PPh_3}{|}}{\overset{\overset{PPh_3}{|}}{Pd}} - CN - \underset{\underset{R}{\diagdown}\;\underset{R}{}}{\overset{\overset{R}{|}}{Sn}} - CN - \underset{\underset{PPh_3}{|}}{\overset{\overset{PPh_3}{|}}{Pd}} - C_6F_5 \right](ClO_4) \qquad (35)$$

R= Me, Bu, Ph

rearrangement, cyanide complexes have also been used for the synthesis of heteronuclear derivatives (89) [Eq. (35)].

When no rearrangement takes place the reactions occur with stereo-retention. Thus, $(NBu_4)_2[cis\text{-}Pd(C_6F_5)_2(CN)_2]$ reacts (1:2) with $Pd(C_6F_5)$-$(OClO_3)(PPh_3)_2$ to give (C) (24) while $(NBu_4)_2[trans\text{-}Pd(C_6F_5)_2(CN)_2]$ gives (D):

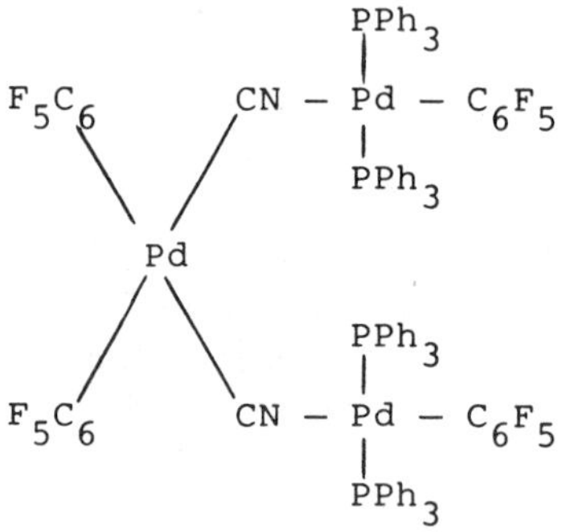

$$\underset{(C)}{F_5C_6 \underset{F_5C_6}{\diagup\diagdown}\overset{\diagdown\diagup}{Pd}\begin{matrix} CN - \underset{\underset{PPh_3}{|}}{\overset{\overset{PPh_3}{|}}{Pd}} - C_6F_5 \\ \\ CN - \underset{\underset{PPh_3}{|}}{\overset{\overset{PPh_3}{|}}{Pd}} - C_6F_5 \end{matrix}}$$

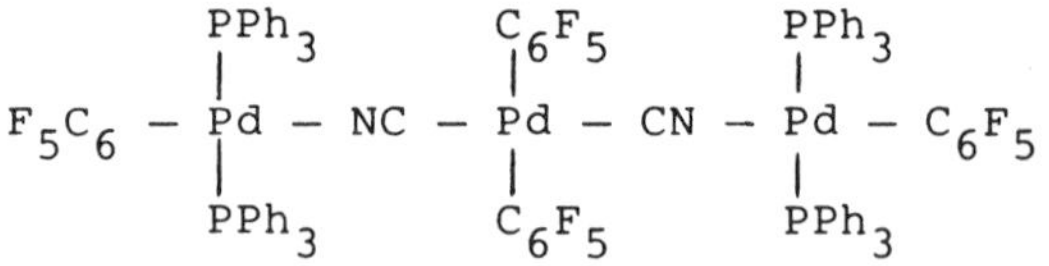

$$\underset{(D)}{F_5C_6 - \underset{\underset{PPh_3}{|}}{\overset{\overset{PPh_3}{|}}{Pd}} - NC - \underset{\underset{C_6F_5}{|}}{\overset{\overset{C_6F_5}{|}}{Pd}} - CN - \underset{\underset{PPh_3}{|}}{\overset{\overset{PPh_3}{|}}{Pd}} - C_6F_5}$$

With $(NBu_4)_2[Pd(C_6F_5)(CN)_3]$ only two of the three CN groups form bridges, even on treatment with an excess of $Pd(C_6F_5)(OClO_3)(PPh_3)_2$ (24). Notwithstanding this result, $(NBu_4)_2[Pd(CN)_4]$ reacts with an excess of the perchlorato complex to give the pentanuclear palladium compound $\{Pd[\mu\text{-}CN)Pd(C_6F_5)(PPh_3)_2]_4\}(ClO_4)_2(88)$.

Other neutral bidentate ligands, for instance, 4,4'-bipy (N–N') can also be used to prepare cationic binuclear complexes (*62*) [Eq. (36)]. When the starting complex has other weakly coordinating neutral ligands in addition to $-OClO_3$, cationic polymeric compounds are obtained [Eq. (37)].

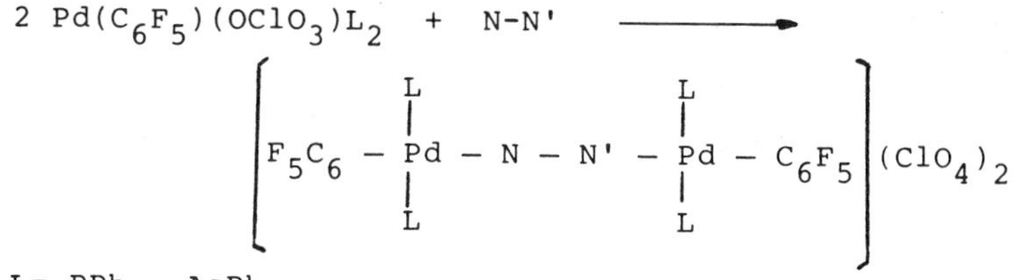

$$2 \ Pd(C_6F_5)(OClO_3)L_2 \ + \ N-N' \ \longrightarrow$$

$$\left[F_5C_6 - \overset{\overset{\displaystyle L}{|}}{\underset{\underset{\displaystyle L}{|}}{Pd}} - N - N' - \overset{\overset{\displaystyle L}{|}}{\underset{\underset{\displaystyle L}{|}}{Pd}} - C_6F_5 \right] (ClO_4)_2 \quad (36)$$

L= PPh$_3$, AsPh$_3$, py

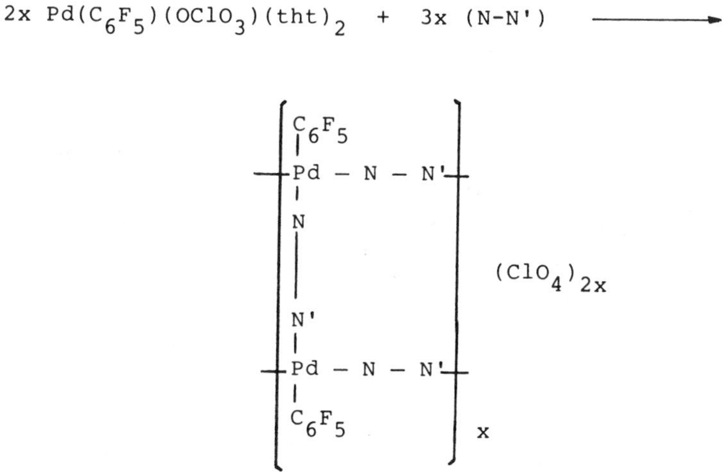

$$2x \ Pd(C_6F_5)(OClO_3)(tht)_2 \ + \ 3x \ (N-N') \ \longrightarrow$$

$$\left[\begin{array}{c} \overset{\displaystyle C_6F_5}{|} \\ -Pd - N - N'- \\ | \\ N \\ | \\ | \\ | \\ N' \\ | \\ -Pd - N - N'- \\ | \\ C_6F_5 \end{array} \right]_x (ClO_4)_{2x} \quad (37)$$

Finally, the versatility of this method can be illustrated by two more examples. Thus, binuclear palladium or platinum complexes containing a bridging CS_3^{2-} group can be obtained by treating a perchlorato complex with Tl_2CS_3 (*90*) [Eq. (38)]. The structure of the binuclear complex was established from the ^{31}P NMR data.

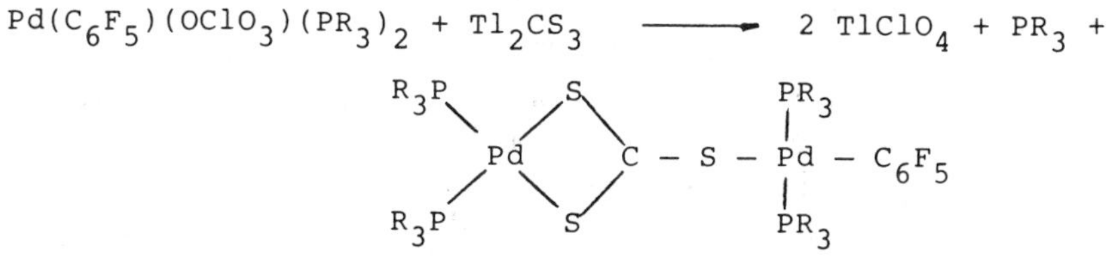

$$Pd(C_6F_5)(OClO_3)(PR_3)_2 + Tl_2CS_3 \longrightarrow 2 \ TlClO_4 + PR_3 +$$

$$\begin{array}{c} R_3P \diagdown \quad \diagup S \diagdown \\ \qquad Pd \qquad C - S - \overset{\overset{\displaystyle PR_3}{|}}{\underset{\underset{\displaystyle PR_3}{|}}{Pd}} - C_6F_5 \\ R_3P \diagup \quad \diagdown S \diagup \end{array}$$

PR$_3$= PEt$_3$, PEt$_2$Ph, PMePh$_2$, PPh$_3$

$$(38)$$

In contrast, displacement of the perchlorato ligand by the exo-S atom in mononuclear CS_3 complexes of palladium gives cationic homo- or heteronuclear derivatives [Eq. (39)]:

$$M(C_6F_5)(OClO_3)L_2 + Pd(\eta^2\text{-}S_2CS)L_2 \longrightarrow$$

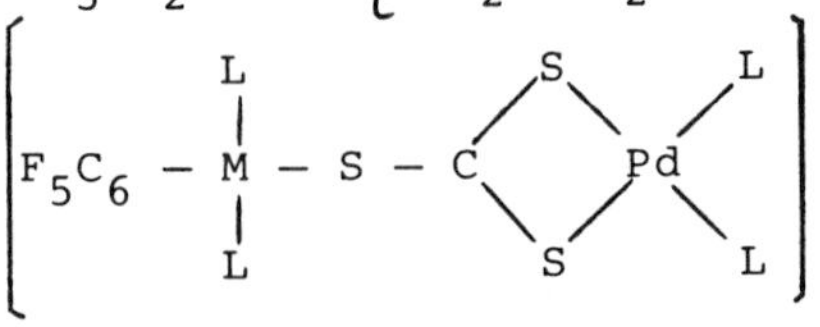

$$\left[\begin{array}{c} L \\ | \\ F_5C_6 - M - S - C \\ | \\ L \end{array} \begin{array}{c} S \\ \diagup \diagdown \\ \diagdown \diagup \\ S \end{array} \begin{array}{c} L \\ \diagup \\ Pd \\ \diagdown \\ L \end{array} \right] ClO_4 \qquad (39)$$

M= Pd; L= PEt_3, PEt_2Ph, $PMePh_2$, PPh_3

M= Pt; L= PEt_3, PEt_2Ph, $PMePh_2$

2. Complexes with Two C_6X_5 Groups Bonded to Each Metal Center

Arylation of the compounds MX_2L_2 (L = N,P, or As donor) with an excess of RLi in some cases [see Eq. (1)] leads to the complexes MR_2L_2. With M = Pd the *trans* isomer is commonly obtained, while for M =Pt the *cis*, the *trans*, or a mixture of both isomers can result.

Arylation of MCl_2 (M = Pd,Pt) or of $(NBu_4)_2[Pd(\mu\text{-}Br)Br_2]_2$ with C_6F_5MgBr gives the binuclear species $(NBu_4)_2[M(\mu\text{-}Br)R_2]_2$ (see Sections II,A,le and II, A,I,f). Moreover, other indirect preparative methods have also been developed.

a. Synthesis of the Complexes $(NBu_4)_2[M(\mu\text{-}X)(C_6X_5)_2]_2$. For M = Pd, arylation of $PdCl_2$ with $(NBu_4)_2[Pd(C_6X_5)_4]$ (1:1 ratio) in refluxing acetone gives both the pentafluorophenyl (*22*) (85%) and the pentachlorophenyl (50%) derivative, (*19*) [Eq. (40)]:

$$(NBu_4)_2|Pd(C_6X_5)_4| + PdCl_2 \longrightarrow (NBu_4)_2|Pd(\mu\text{-}Cl)(C_6X_5)_2|_2$$

X= F, Cl

$$(40)$$

Under these conditions $PtCl_2$ does not react, and if the reaction is carried out at higher temperatures (refluxing toluene) metallic platinum is formed.

A general method that gives the four possible complexes in yields over 85% involves addition of aqueous HCl to the corresponding tetrakis(pentahalophenyl) compound in methanolic solution (2:1 ratio) [Eq. (41)] albeit half of the pentahalophenyl groups are lost as C_6X_5H (*19,23*). The

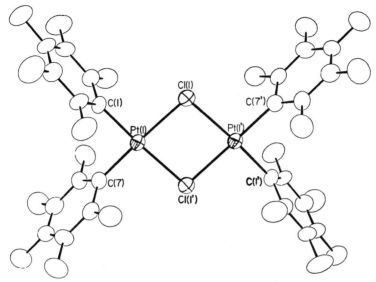

Fig. 3.   Structure of $[Pt_2(\mu\text{-}Cl)_2(C_6F_5)_4]^{2-}$ (106).

$$2 \ (NBu_4)_2|M(C_6X_5)_4| \ + \ 4 \ HCl \ \longrightarrow \ 4 \ C_6X_5H \ + \ 2 \ NBu_4Cl \ +$$

$$(NBu_4)_2|M(\mu\text{-}Cl)(C_6X_5)_2|_2$$

M= Pd, Pt; X= F, Cl

$$(41)$$

$$(NBu_4)_2|M(\mu\text{-}Cl)R_2|_2 \ + \ M'X' \ \longrightarrow \ M'Cl \ + \ (NBu_4)_2|M(\mu\text{-}X')R_2|_2$$

R= C_6F_5;   M= Pd; X'= Br, I, SCN

M= Pt; X'= I, SCN

R= C_6Cl_5;   M= Pd; X'= Br, I

M= Pt, X'= Br, I, SCN

$$(42)$$

structure of $(NBu_4)_2[Pt(\mu\text{-}Cl)(C_6F_5)_2]_2$ is shown in Fig. 3. Each platinum atom lies in a square planar environment and the dihedral angle of both planes is 175.10°, the whole anion also being practically planar.

By reaction with salts M'X' in refluxing acetone or ethanol, the chloride ions in the bridge can be replaced by other halides or pseudohalides without cleavage of the bridge (19,22) [Eq. (42)].

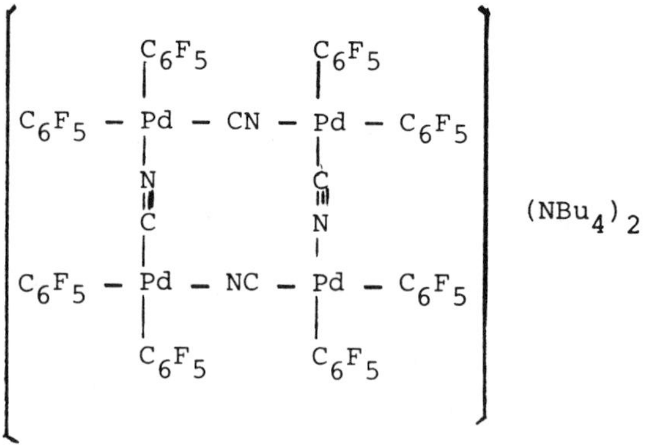

$$\left[\begin{array}{cc}
\overset{\displaystyle C_6F_5}{|} & \overset{\displaystyle C_6F_5}{|} \\[4pt]
C_6F_5 - Pd - CN - Pd - C_6F_5 \\
\underset{\displaystyle C}{\overset{\displaystyle N}{\|}} \qquad \underset{\displaystyle N}{\overset{\displaystyle C}{\|}} \\[4pt]
C_6F_5 - Pd - NC - Pd - C_6F_5 \\[4pt]
\underset{\displaystyle C_6F_5}{|} \qquad \underset{\displaystyle C_6F_5}{|}
\end{array}\right] (NBu_4)_2$$

FIG. 4.   Proposed structure for $(NBu_4)_2[Pd_4(\mu\text{-}CN)_4(C_6F_5)_8]$.

Although the complex $(NBu_4)_2[Pd(\mu\text{-}SCN)(C_6Cl_5)_2]_2$ cannot be prepared by this method, since decomposition occurs, it is accessible through a similar reaction between $(NBu_4)_2[Pd(\mu\text{-}I)(C_6Cl_5)_2]_2$ and AgSCN.

When silver cyanide reacts with $(NBu_4)_2[Pd(\mu\text{-}Br)(C_6F_5)_2]_2$ (in CH_2Cl_2) the tetranuclear palladium complex $(NBu_4)_4[Pd(\mu\text{-}CN)(C_6F_5)_2]_4$ is formed. The assigned structure (Fig. 4) with four cis-$Pd(C_6F_5)_2$ moieties linked by four linear CN groups is deduced from IR and conductivity data (66).

The thallium(I) salts of azolate ligands (N–N' = biim, bibzim) react with the binuclear compound $(NBu_4)_2[Pd(\mu\text{-}X')(C_6X_5)_2]_2$ (X = F, X' = Br; X = Cl, X' = Cl) with precipitation of thallium(I) halide and formation of the products $(NBu_4)_2[Pd(\mu\text{-}N\text{-}N')(C_6X_5)_2]_2$ (63).

Addition of neutral ligands to the anionic binuclear palladium complexes readily cleaves the bridges to afford anionic mononuclear complexes in a first step. Addition of more neutral ligand displaces the halide ligand and neutral complexes are obtained in a second step (19,91) [Eq. (43)]. This

$$(NBu_4)_2 |M(\mu\text{-}Cl)(C_6X_5)_2|_2 \xrightarrow{+2L} 2 \; (NBu_4)|\underline{cis}\text{-}M(C_6X_5)_2ClL|$$

$$\xrightarrow{+2L} 2 \; (NBu_4)Cl + 2 \; \underline{cis}\text{-}M(C_6X_5)_2L_2$$

M= Pd; X= F;   L= PPh_3, PPh_2Me, PEt_3, $P(OPh)_3$, $P(OMe)_3$, $BzNH_2$,

                py, $AsPh_3$, $SbPh_3$

        X= Cl;   L= PPh_3, PEt_3, PPh_2Me, py, $SbPh_3$

M= Pt; X= Cl,   L= PPh_3, PPh_2Me, py, $SbPh_3$

$$(43)$$

reaction takes place with stereoretention, i.e., always gives the cis isomers. For M = Pd, other preparative methods lead to the trans derivatives. For M = Pt, this process is the only general one for the synthesis of bis(pentachlorophenyl) derivatives.

Some neutral ligands exhibit a specific behavior: (a) py, $AsPh_3$, and $SbPh_3$ give the neutral complexes irrespective of the molar ratio used. The corresponding amount of the starting complex remains unreacted for ratios <1:4. (b) Even if used in excess, $BzNH_2$ does not displace the halide ligand, and the anionic mononuclear derivative remains the end product. (c) Bidentate chelating ligands, bipy, dpae, and dppe, give only the neutral complexes, as expected. (d) Some weakly coordinating ligands e.g., tetrahydrothiophene, tetrafluorophthalonitrile, and 1,5-cyclooctadiene, do not react because they are too weak to initiate cleavage of the bridge.

It is also possible to cleave the bridge in the binuclear complexes by using complexes with ligands containing free donor atoms (65) [Eq. (44)].

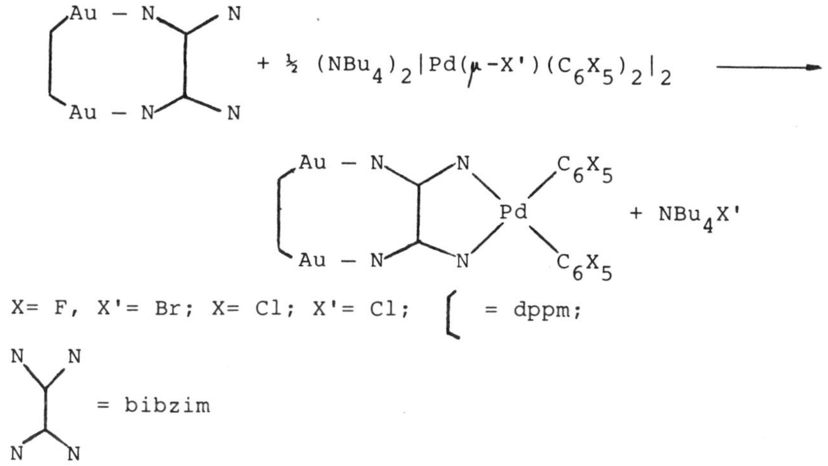

$$\text{[Au-N, N / Au-N, N]} + \tfrac{1}{2}\ (NBu_4)_2|Pd(\mu\text{-}X')(C_6X_5)_2|_2 \longrightarrow$$

$$\text{[Au-N, N / Au-N, N]Pd} \begin{smallmatrix} C_6X_5 \\ C_6X_5 \end{smallmatrix} + NBu_4X'$$

X= F, X'= Br; X= Cl; X'= Cl; $\big[\ $ = dppm;

$$\text{[N, N / N, N]} = \text{bibzim}$$

$$(44)$$

b. *Synthesis of the Complexes cis-M(C₆X₅)₂(OC₄H₈)₂.* The title compounds have interesting possibilities for a variety of syntheses. They are prepared [Eq. (45)] by room temperature (1:2) reactions between

$$(NBu_4)_2|M(\mu\text{-}X')(C_6X_5)_2|_2 + 2\ AgClO_4 \xrightarrow{\text{THF}}$$

$$2\ \underline{cis}\text{-}|M(C_6X_5)_2(OC_4H_8)_2| + 2\ AgX' + 2\ NBu_4ClO_4$$

M= Pd; X'= Br; X= F, Cl

M= Pt; X'= Cl; X= F, Cl

$$(45)$$

$(NBu_4)_2[M(\mu-X')(C_6X_4)_2]_2$ and $AgClO_4$ in tetrahydrofuran as a solvent and working under anhydrous conditions. After evaporation to dryness, the bisaryl complexes are extracted with diethyl ether while insoluble AgX' and NBu_4ClO_4 remain in the residue. From the ether solution the neutral products cis-$M(C_6X_5)_2(OC_4H_8)_2$ can be crystallized (92).

In the solid state the complexes are stable at room temperature, but they slowly decompose in dichloromethane solution. The facile displacement of the two OC_4H_8 ligands makes them excellent precursors for the synthesis of complexes containing the moieties "cis-$M(C_6X_5)_2$."

Bis(acetylene) Metal Complexes. Dichloromethane solutions of cis-$M(C_6F_5)_2(OC_4H_8)_2$ react with diphenylacetylene (dpa) to give bis-(acetylene) derivatives, which can be isolated by evaporation of the solutions *in vacuo* (94) [Eq. (46)]. Both complexes are stable solids. The

$$cis\text{-}M(C_6F_5)_2(OC_4H_8)_2 + 2\ dpa \longrightarrow 2\ cis\text{-}M(C_6X_5)_2(dpa)_2$$

$$+ 2\ OC_4H_8$$

M= Pd, Pt

(46)

molecular structure of the platinum complex has been established by X-ray crystallography (Fig. 5). The Pt atom is in an almost square planar environment of two C atoms (of the cis-C_6F_5 groups) and the midpoints of the C≡C bonds of the two acetylene molecules. It is noteworthy that the C–C distance (1.197 Å) is practically unchanged (1.19 Å) from that in the free alkyne, and the coordinated alkynes are almost linear, suggesting the absence of any significant π-backbonding.

Carbonyl Derivatives. Bubbling CO through dichloromethane solutions of cis-$M(C_6X_5)_2(OC_4H_8)_2$ precipitates the cis-dicarbonyl derivatives (92) [Eq. (47)]. The reactions take place at 1 CO pressure. The platinum

$$cis\text{-}M(C_6X_5)_2(OC_4H_8)_2 \xrightarrow{\ CO\ } cis\text{-}M(C_6X_5)_2(CO)_2 + 2\ OC_4H_8$$

M= Pd; X= F,Cl; M= Pt; X= F,Cl

(47)

complexes can be obtained (X = F, 79%; X = Cl, 91%) at room tempera-ture. For the palladium complexes, lower temperatures must be used (X = F, −78°C, 66%; X = Cl, −50°C, 50%).

The palladium complexes are the only monomeric dicarbonyls so far reported. The compounds [cis-$PtX_2(CO)_2$] (X = Cl, Br, I) have already been prepared by a high-pressure method, and $trans$-$PtCl_2(CO)_2$ has been

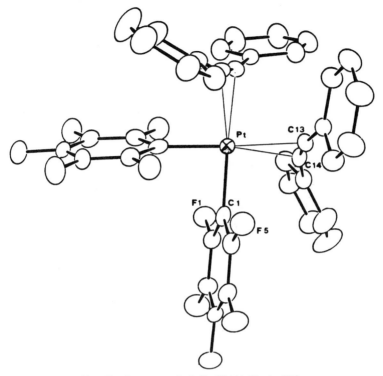

FIG. 5.   Structure of *cis*-Pt $(C_6F_5)_2$(dpa)$_2$ (*94*).

prepared using CO at atmospheric pressure and low temperature ($-80°C$) (*93*). They are moderately stable under a CO atmosphere. In contrast, the platinum complexes are stable at room temperature, even when exposed to moist air, and in solutions of nonpolar solvents *cis*-Pd$(C_6F_5)_2$(CO)$_2$ sublimes without decomposition at 205°C, while *cis*-Pt$(C_6Cl_5)_2$(CO)$_2$ decomposes at 208°C. The IR spectra confirm the cis configuration of the complexes; two ν(CO) bands appearing (Table II) at higher energies than that of the free CO ligand, suggesting negligible π-backbonding, as in the case of the bis(acetylene) complexes mentioned above.

A similar route has been used for the synthesis of *trans*-Pt$(C_6F_5)_2$(CO)$_2$. Addition of AgClO$_4$ to a tetrahydrofuran solution of (NBu$_4$)[*trans*-Pt$(C_6F_5)_2$Cl(CO)] (*95*), followed by evaporation to dryness and extraction with diethyl ether, leaves AgCl and NBu$_4$ClO$_4$ as the insoluble residue. Evaporation of the ether solution gives *trans*-Pt$(C_6F_5)_2$(CO)(OC$_4$H$_8$), which can be further carbonylated by bubbling CO at normal pressure

TABLE II

$\nu(CO)$ for Neutral and Anionic Bis (perhaloaryl)carbonyl–Palladium or -Platinum Complexes

Complex	$\nu(C\equiv O)$ (cm^{-1})
cis-Pd$(C_6F_5)_2(CO)_2$	2186, 2163
cis-Pd$(C_6Cl_5)_2(CO)_2$	2173, 2152
cis-Pt$(C_6F_5)_2(CO)_2$	2174, 2143
cis-Pt$(C_6Cl_5)_2(CO)_2$	2160, 2126
$trans$-Pt$(C_6F_5)_2(CO)_2$	2151
cis-Pt$(C_6F_5)_2(CO)(PPh_3)$	2112
cis-Pt$(C_6F_5)_2(CO)py$	2112
cis-Pt$(C_6Cl_5)_2(CO)(PPh_3)$	2100
$trans$-Pt$(C_6F_5)_2(CO)py$	2112
(PNP)[cis-Pt$(C_6F_5)_2Cl(CO)$]	2089
(PPh$_3$Et)[cis-Pt$(C_6F_5)_2Br(CO)$]	2090
(PPh$_2$Me$_2$)[cis-Pt$(C_6Cl_5)_2I(CO)$]	2084[a]
(NBu$_4$)[$trans$-Pt$(C_6F_5)_2Cl(CO)$]	2084
(PPh$_3$Et)[$trans$-Pt$(C_6F_5)_2Br(CO)$]	2112
(PPh$_2$Me$_2$)[$trans$-Pt$(C_6F_5)_2I(CO)$]	2082

[a] In CH$_2$Cl$_2$ solution; two bands (2080, 2052) are observed in Nujol mulls.

through a dichloromethane solution [Eq. (48)]:

$$\underline{trans}\text{-Pt}(C_6F_5)_2(CO)(OC_4H_8) \xrightarrow{\ CO\ } \underline{trans}\text{-Pt}(C_6F_5)_2(CO)_2 + OC_4H_8$$

$$(48)$$

Only a single band due to $\nu(CO)$ is found in the IR spectrum (Table II), in accordance with the trans structure.

In both the cis- and $trans$-dicarbonyl derivatives, one CO group can readily be displaced by one equivalent of either a neutral or an anionic ligand to give monocarbonyl complexes (92). This reaction occurs with stereoretention, as may be inferrred from the ^{19}F NMR data. The monocarbonyl derivatives are stable in the solid state as well as in solution. An excess of PPh$_3$ or py does not displace the second CO group. Isomerization from $trans$- to cis-Pt$(C_6F_5)_2(CO)py$ is observed (IR spectrum) in only trace amounts after 170 h in the presence of an excess of py in dichloromethane solution at room temperature.

Complexes Containing η^1- or η^2-Bonded TricyclohexylPhosphonium Dithioformate. Reaction in diethyl ether of $M(C_6X_5)_2(OC_4H_8)_2$ with $^-S_2C–PCy_3$ (1:0.95) allows isolation of complexes containing the ligand η^2-S_2C–PCy_3 (*96*) [Eq. (49)]:

$$M(C_6X_5)_2(OC_4H_8)_2 + {}^-S_2C{-}\overset{+}{P}Cy_3 \longrightarrow M(C_6X_5)_2(\eta^2\text{-}S_2C\text{-}PCy_3)\downarrow$$

$$+ \ 2 \ OC_4H_8$$

M= Pd, Pt; X= F, Cl

$$(49)$$

Figure 6 shows the molecular structure of the complex (M = Pd, X = F) with a square planar arrangement of two C and two S atoms around the metal center. The two S–C distances are 1.662(6) and 1.672(4) Å, respectively. Addition of neutral unidentate ligands gives the corresponding four-coordinated complexes with η^1-S_2C–PCy_3 [Eq. (50)]. Even though

$$(C_6X_5)_2M\overset{\displaystyle S}{\underset{\displaystyle S}{\diagup\diagdown}}C{-}PCy_3 + L \longrightarrow (C_6X_5)_2M(\eta^1\text{-}S_2C\text{-}PCy_3)L \quad (50)$$

M= Pd,Pt; X= F; L= PPh$_3$,py; M= Pt; X= F,Cl; L= CO

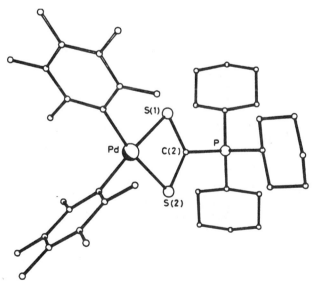

Fig. 6.   Structure of $[Pd(C_6F_5)_2(S_2CPCy_3)]$ (*96*).

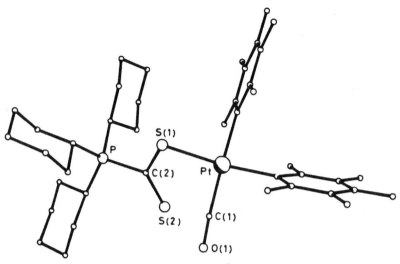

FIG. 7.   Structure of [Pt(C$_6$F$_5$)$_2$(η^1-SC(S)PCy$_3$)CO] (96).

the palladium complexes do not react with CO, the platinum derivatives Pt(C$_6$X$_5$)$_2$(η^1-S$_2$CPCy$_3$)(CO) can be obtained. Figure 7 shows the structure of the pentafluorophenyl complex. Both C$_6$F$_5$ groups are mutually cis, and the S$_2$C–PCy$_3$ ligand is unidentate. The square planar coordination at the

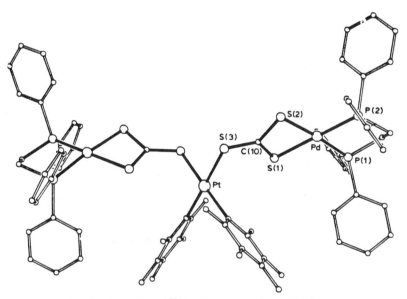

FIG. 8.   Structure of [{(dppe)Pd(μ-S$_2$CS)}$_2$Pt(C$_6$F$_5$)$_2$] (97).

metal center is also shown. The Pt–S(1) distance is 2.348(2) Å and the nonbonded Pt–S(2) distance is 3.589 Å. The C(2)–S(1) and C(2)–S(2) distances are 1.682(6) and 1.643(7) Å, respectively, only slightly different, as a consequence of electron delocalization in the S_2C moiety.

Polynuclear Complexes with Trithio- or Dithiocarbonate Bridges. Polynuclear derivatives containing the moieties cis-$M(C_6X_5)_2$, and including the groups S_2CS and S_2CO as bridging ligands, can also be prepared from $M(C_6X_5)_2(OC_4H_8)_2$ precursors.

From suitable trithiocarbonate complexes of palladium or platinum, trinuclear homo- or heterometallic derivatives (97) are obtained [Eq. (51)], since the exo-S atom in the $M(\eta^2\text{-}S_2CS)(L\text{-}L)$ complex readily

$$2 \; M(\eta^2\text{-}S_2CS)(L\text{-}L) + M'(C_6X_5)_2(OC_4H_8)_2 \longrightarrow$$

$$|\{(L\text{-}L)M(\mu\text{-}S_2CS)\}_2M'(C_6X_5)_2| + 2 \; OC_4H_8 \quad (51)$$

X= F;   M= Pd;   M'= Pd;   (L–L)= dppe, dppb

X= F;   M= Pd;   M'= Pt;   (L–L)= dppe, dppp, dppb

X= F;   M= Pt;   M'= Pd;   (L–L)= dppe

X= F;   M= Pt;   M'= Pt;   (L–L)= dppe

X= Cl;   M= Pd;   M'= Pd;   (L–L)= dppe

displaces the OC_4H_8 ligands. The molecular structure of the complex with M = Pd, M' = Pt, L–L = dppe has been determined and is shown in Fig. 8. The metal centers display square planar geometry; some distortion due to the chelating dppe is noticeable around the Pd atoms. The bridging S_2CS group supports the trimetallic unit.

In contrast, dithiocarbonate complexes of palladium or platinum react (1:1) with the bisaryl derivatives to give polymeric complexes [Eq. (52)].

$$M(S_2CO)(L\text{-}L) + M'(C_6F_5)_2(OC_4H_8)_2 \longrightarrow$$

$$|(L\text{-}L)M(\mu\text{-}S_2CO)M'(C_6F_5)_2|_n + 2 \; OC_4H_8 \quad (52)$$

M= Pd, Pt;   M'= Pd, Pt;   (L–L)= dppe

These complexes are relatively insoluble and somewhat unstable in solution, which precludes molecular weight determinations. The IR spectra confirm the presence of two mutually cis-C_6F_5 groups. The ^{31}P NMR spectra show only one signal (with Pt satellites when dppe is coordinated to

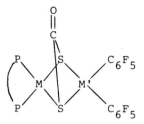

FIG. 9.   Binuclear dithiocarbonate complex.

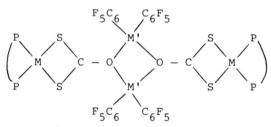

FIG. 10.   Tetranuclear dithiocarbonate complex.

the Pt center). These data are compatible with either a binuclear (Fig. 9) or a tetranuclear (Fig. 10) structure (*98*).

Polynuclear Halide-Bridged Complexes. The halide ligands in the complexes $M'X_2'(COD)$ (M' = Pd, Pt) also displace the two OC_4H_8 groups in the *cis*-$M(C_6X_5)_2(OC_4H_8)_2$ precursors and thereby afford asymmetric homo- or heterometallic complexes (*99,100*). The halide ligands act as bridges between metal atoms [Eq. (53)]. Molecular weights in chloroform

$$n \; \underline{cis}\text{-}M(C_6X_5)_2(OC_4H_8)_2 + n \; M'X'_2(COD) \longrightarrow$$

$$|(C_6F_5)_2M(\mu\text{-}X')_2M'(COD)|_n + 2n \; OC_4H_8 \quad (53)$$

M=M'= Pt; X= F; X'= Cl,Br,I; X= Cl, X'= I

M=M'= Pd; X= F, X'= Cl,Br;   X= Cl, X'= Cl

M= Pt; M'= Pd; X= F; X'= Cl,Br; X= Cl; X'= Br

M= Pd; M'= Pt; X= F; X'= Br;   X= Cl; X'= Br

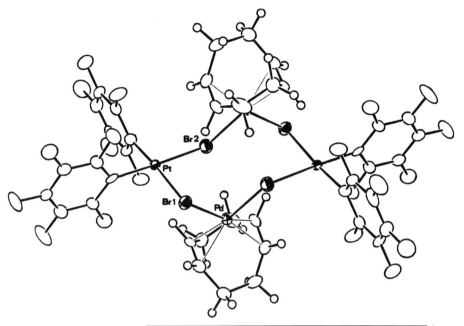

FIG. 11. Structure of [Pt(C$_6$F$_5$)$_2$(μ-Br)Pd(COD)(μ-Br)Pt(C$_6$F$_5$)$_2$(μ-Br)Pd(COD)(μ-Br)] (99,100).

(isopiestic method) indicate that in this solvent the complexes are binuclear. The molecular structure of the complex with M = Pt, X = F, X' = Br, and M' = Pd (Fig. 11) is, however, tetranuclear. It consists of an eight-membered ring formed by two palladium and two platinum atoms, with four bromide ligands singly bridging each pair of M-M' atoms. The ring has the form of a distorted chair, and each platinum atom has two cis-C$_6$F$_5$ groups and two bromide ions, each of which is connected to a different palladium atom. The two palladium atoms are also square planar, being coordinated to two bromide ions and to the midpoints of the two double bonds of the diolefin. Moreover, crystals of [(C$_6$F$_5$)$_2$Pt-(μ-Br)Pd(COD)(μ-Br)Pt(C$_6$F$_5$)$_2$(μ-Br)Pd(COD)(μ-Br)] are isomorphous with the palladium compound, and thus the two species are probably isostructural.

The single-bridged tetranuclear structure is unusual, as is the reversible change binuclear (solution) ⇌ tetranuclear (solid). In some cases, dramatic changes in color are observed; for instance, [(C$_6$F$_5$)$_2$-Pt(μ-X)$_2$Pd(COD)]$_n$ (X = Cl, Br) and [(C$_6$Cl$_5$)$_2$Pt(μ-Br)$_2$Pd(COD)]$_n$ are salmon- or garnet-colored as solids but green in dichloromethane solutions. Acetone solutions of the complexes are pale yellow or colorless, due to decomposition [Eq. (54)]:

$$|(C_6X_5)_2M(\mu-X')_2M'(COD)|_n \xrightleftharpoons[-OCMe_2]{OCMe_2} |\underline{cis}-M(C_6F_5)_2(OCMe_2)_x| +$$

$$|M'X'_2(COD)| \qquad (54)$$

but evaporation to dryness and subsequent heating of the residue (60°C) regenerates the original complexes.

Heteronuclear Complexes with Donor–Acceptor Bonds M → Pt (M = Co, Rh). Binuclear complexes with metal-to-Pt bonds have been prepared

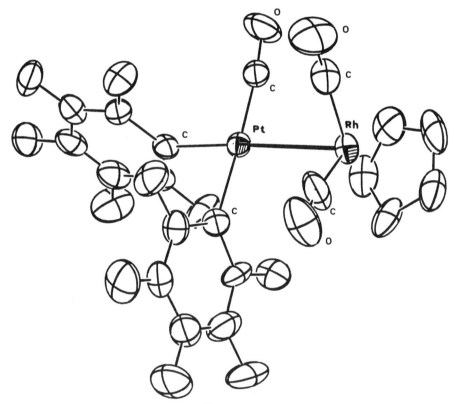

Fig. 12.   Structure of $(\eta^5-C_5H_5)(CO)_2Rh \to Pt(C_6F_5)_2(CO)$.

by treating dichloromethane solutions of cis-Pt$(C_6F_5)_2(CO)(OC_4H_8)$ with $(\eta^5$-$C_5R_5)M(CO)L$, whereby the metallic donor displaces the OC_4H_8 ligand (*101*) [Eq. (55)].

$$(\eta^5\text{-}C_5R_5)M(CO)L + \underline{cis}\text{-}Pt(C_6F_5)_2(CO)(OC_4H_8) \longrightarrow$$

$$(\eta^5\text{-}C_5R_5)(CO)LM \longrightarrow Pt(C_6F_5)_2CO + OC_4H_8 \quad (55)$$

M= Co; R= H; L= CO, PPh$_3$

M= Rh; R= H; L= CO, PPh$_3$

M= Rh; R= Me; L= CO

The molecular structure of $(\eta^5$-$C_5H_5)(CO)_2Rh \to Pt(C_6F_5)_2(CO)$ is shown in Fig. 12. As may be seen, there is an Rh–Pt bond (distance 2.749(1) Å) unsupported by any bridging ligand, and the platinum atom is square planar-coordinated. The structure in the solid state and the IR spectra of the complexes in solution reveal that the two C_6F_5 groups are mutually cis, but the ^{19}F NMR spectra of all these complexes (in CDCl$_3$) show the equivalence of both groups. Some mechanism such as the following [Eq. (56)] makes the two C_6F_5 groups equivalent. The activation energy must be low, since the equivalence persists at $-40°C$.

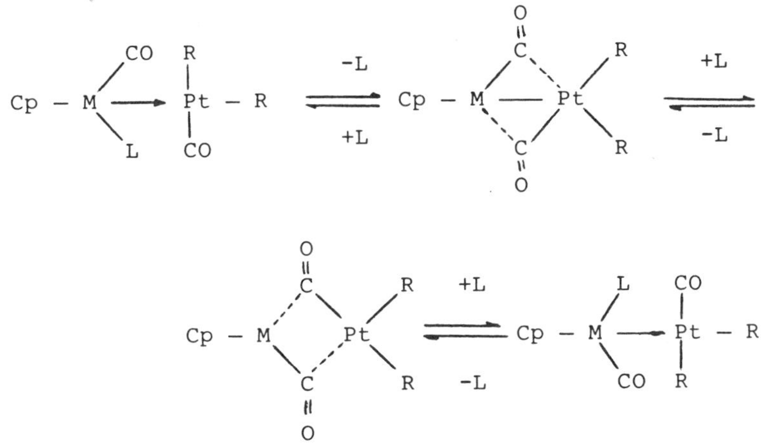

$$R = C_6F_5$$

As expected, addition of a stronger nucleophilic base (1:1 ratio) such as PPh$_3$ cleaves the donor–acceptor bond, leading to two mononuclear species [Eq. (57)]:

$$(\eta^5\text{-}C_5R_5)(CO)_2Rh \longrightarrow Pt(C_6F_5)_2(CO) + PPh_3 \longrightarrow$$
$$(\eta^5\text{-}C_5R_5)(CO)_2Rh + Pt(C_6F_5)_2(CO)(PPh_3) \qquad (57)$$

R= H, Me

c. Miscellaneous $M(C_6X_5)_2L_2$ Complexes. As mentioned earlier [see Eq. (4)], *trans*-Pd$(C_6X_5)_2$(tht)$_2$ reacts with neutral ligands L (N, P, As donors) to give *trans*-Pd$(C_6X_5)_2L_2$, a noteworthy method for the synthesis of pentachlorophenyl derivatives.

Moreover, the reactions of *trans*-Pd$(C_6F_5)_2$(tht)$_2$ with (NBu$_4$)-[XAu(C$_6$F$_5$)] or (NBu$_4$)[XAu(C$_6$F$_5$)$_3$] (X = SCN, CN) afford polynuclear heterometallic Pd–Au complexes since the free end of the bidentate SCN or CN readily displaces the weakly coordinating tht (*102*). If a 1:1 molar ratio is used, binuclear complexes result [Eq. (58)], while with a 1:2 molar ratio trinuclear complexes are obtained [Eq. (59)].

$$\underline{trans}\text{-PdR}_2(\text{tht})_2 + (\text{NBu}_4)|\text{XAuR}_n| \longrightarrow$$

$$(\text{NBu}_4)\left[\text{tht} - \overset{\overset{\textstyle R}{|}}{\underset{\underset{\textstyle R}{|}}{\text{Pd}}} - X - \text{AuR}_n\right] + \text{tht} \qquad (58)$$

$$\underline{trans}\text{-PdR}_2(\text{tht})_2 + 2\ (\text{NBu}_4)|\text{X-AuR}_n| \longrightarrow$$

$$(\text{NBu}_4)_2\left[\text{R}_n\text{Au} - X - \overset{\overset{\textstyle R}{|}}{\underset{\underset{\textstyle R}{|}}{\text{Pd}}} - X - \text{AuR}_n\right] \qquad (59)$$

R= C$_6$F$_5$, n= 1,3; X= CN, SCN

The IR spectra of the complexes show ν(CN) bands displaced toward higher energies, relative to the starting gold complexes. This is as expected from the changed position of these ligands from terminal to bridging sites.

Another class of polynuclear heterometallic palladium–gold complexes can be obtained from the reaction between PdRR'$(\eta^1$-dppm)$_2$ (R = R' = C$_6$F$_5$, C$_6$Cl$_5$ or R' = C$_6$F$_5$, R' = Cl) and neutral tht-containing gold

trans-PdRR'(η^1-dppm)$_2$ + XAu(tht) $\longrightarrow$

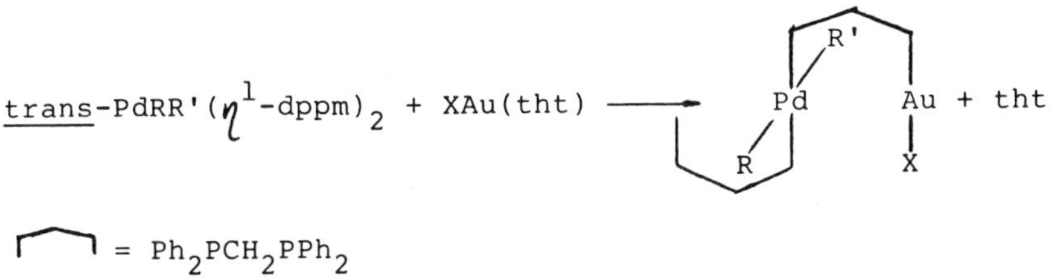

Au + tht

$\boxed{}$ = Ph$_2$PCH$_2$PPh$_2$

R=R'= C$_6$F$_5$, C$_6$Cl$_5$; R= C$_6$F$_5$, R'= Cl; X= C$_6$F$_5$, (C$_6$F$_5$)$_3$

$$(60)$$

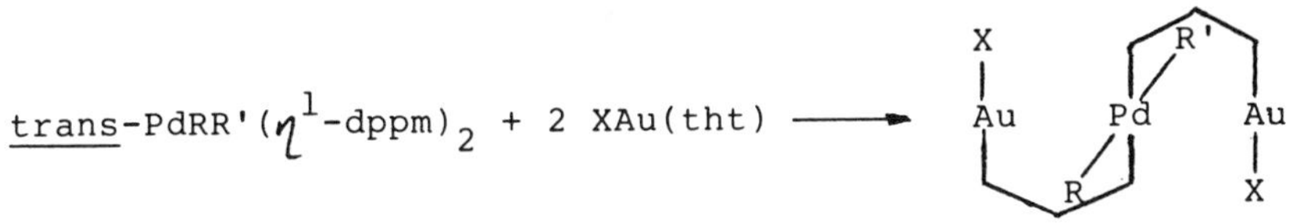

trans-PdRR'(η^1-dppm)$_2$ + 2 XAu(tht) $\longrightarrow$

R=R'= C$_6$F$_5$, C$_6$Cl$_5$; X= C$_6$F$_5$, (C$_6$F$_5$)$_3$

$$(61)$$

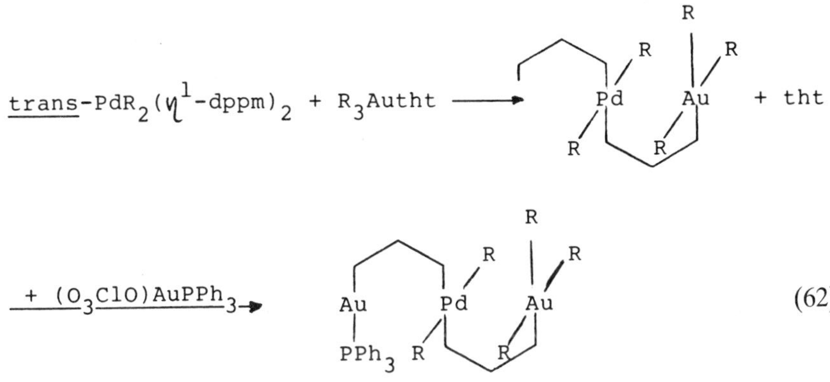

trans-PdR$_2$(η^1-dppm)$_2$ + R$_3$Autht $\longrightarrow$

+ tht

$\underline{+ (O_3ClO)AuPPh_3}\rightarrow$

$$(62)$$

R= C$_6$F$_5$

complexes, such as ClAu(tht), C$_6$F$_5$Au(tht), or (C$_6$F$_5$)$_3$Au(tht), since the uncoordinated end of the dppm ligand readily displaces tht (103). By appropriate control of the molar ratio, binuclear or trinuclear metal complexes are obtained [Eqs. (60) and (61)]. A heterotrinuclear complex can also be obtained, illustrating the synthetic potential of this method [Eq. (62)]. The structure of the trinuclear species RAu(dppm)-PdR$_2$(dppm)AuR (R = C$_6$F$_5$) is shown in Fig. 13, illustrating the approximately square planar environment around the Pd atom and the linear coordination at the gold centers.

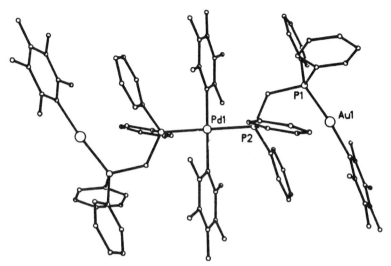

Fig. 13.   Structure of $(C_6F_5)Au(dppm)Pd(C_6F_5)_2(dppm)Au(C_6F_5)$ (*103*).

3. Complexes with Three C_6X_5 Groups

Only pentafluorophenyl derivatives have been reported, since arylation of $MCl_2(tht)_2$ with C_6Cl_5Li stops at the stage of the neutral bisaryl complexes $M(C_6Cl_5)_2(tht)_2$.

Displacement of the neutral ligand in $(NBu_4)[M(C_6F_5)_3(tht)]$ with other unidentate ligands ($L = PPh_3$, $AsPh_3$, $SbPh_3$) gives the complexes $(NBu_4)[M(C_6F_5)_3L]$ (*22*). With diphosphines, the end product is dependent on the phosphine employed. With dppb, the binuclear platinum complex $(NBu_4)_2[(C_6F_5)_3Pt(dppb)Pt(C_6F_5)_3]$ is obtained, while the mononuclear complex $(NBu_4)[Pt(C_6F_5)_3(dppm)]$, containing unidentate dppm, results even if an excess of the platinum precursor is used.

Halide ligands do not displace tht, but CN^- gives the complex $(NBu_4)_2[Pt(C_6F_5)_3(CN)]$ (*24*). Moreover, compounds containing terminal CN or SCN groups displace tht to give binuclear complexes such as $(NBu_4)_2[(C_6F_5)Au-X-Pd(C_6F_5)_3]$ ($X = CN$, SCN) (*102*).

An alternative method, leading to the synthesis of platinum complexes containing the moiety "$Pt(C_6F_5)_3$," is based on the (1:1) reaction between $[Pt(C_6F_5)_4]^{2-}$ and aqueous HCl in methanol solution. The complex $(NBu_4)_2[Pt(C_6F_5)_3Cl]$ can be isolated (*23*) in 76% yield, one C_6F_5 group being removed as C_6F_5H. So far, this is the only available method for the synthesis of halotriaryl platinate(II) species. In practice, similar reactions starting from $[Pd(C_6X_5)_4]^{2-}$ (X = F, Cl) or $[Pt(C_6Cl_5)_4]^{2-}$ yield the binuclear metal $[M(\mu\text{-}Cl)R_2]_2^{2-}$ species.

Addition of anionic X^- ligands to $(NBu_4)_2[Pt(C_6F_5)_3Cl]$ leads to substitution of the Cl ion by X = Br, I, CN, or SCN. Substitution by neutral ligands is also possible if a chloride abstractor such as $AgClO_4$ or $NaClO_4$ is simultaneously present [Eq. (63)].

$$(NBu_4)_2|Pt(C_6F_5)_3Cl| \; + \; MClO_4 \; + \; L \longrightarrow (NBu_4)ClO_4 \; + \; MCl$$

$$+ \; (NBu_4)|Pt(C_6F_5)_3L| \qquad (63)$$

M= Ag, L= C_2Ph_2

M= Na, L= ½ dppm, ½ dppe, dppm, dppe

The bidentate ligands dppm and dppe give bridged binuclear platinum complexes $(NBu_4)_2[(C_6F_5)_3Pt(P\text{-}P)Pt(C_6F_5)_3]$ for a Pt/P–P = 2:1 molar ratio. If a 1:1 ratio is used, mononuclear metal complexes (NBu_4)-$[Pt(C_6F_5)_3(P\text{-}P)]$ (the diphosphine acting unidentate) are obtained. The complex with dppm has been used for the synthesis of neutral or anionic polynuclear complexes (23) [Eqs. (64)–(66)].

$$(NBu_4)|Pt(C_6F_5)_3(\eta^1\text{-}dppm)| \; + \; Pd(C_6F_5)(OClO_3)(PPh_3)_2 \longrightarrow$$

$$|(C_6F_5)_3Pd(dppm)Pd(C_6F_5)(PPh_3)_2| \; + \; NBu_4ClO_4 \qquad (64)$$

$$2 \; (NBu_4)|Pt(C_6F_5)_3(\eta^1\text{-}dppm)| \; + \; \underline{trans}\text{-}MCl_2(tht)_2 \longrightarrow$$

$$2 \; tht \; + \; (NBu_4)_2 \qquad (65)$$

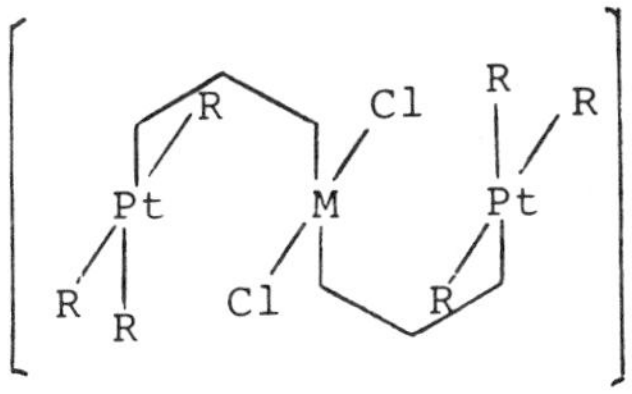

M= Pd, Pt

$$2 \; (NBu_4)|Pt(C_6F_5)_3(\eta^1\text{-}dppm)| \; + \; AgClO_4 \longrightarrow NBu_4ClO_4 \; +$$

$$(NBu_4) \qquad (66)$$

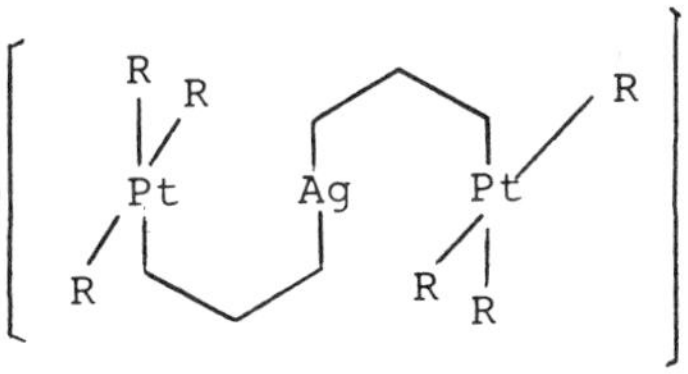

⌒ = dppm

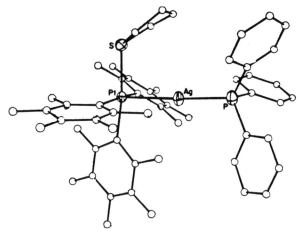

FIG. 14.   Structure of $[(SC_4H_8)(C_6F_5)_3Pt \rightarrow AgPPh_3]$ *(104,105)*.

a. Basic Behavior of Anionic Perhaloaryl Platinum Complexes. Some recent research has focused on the basic properties of certain perhaloaryl platinum derivatives that react with Lewis acids, namely Ag^+ or $(AgL)^+$, to give clusters often possessing Pt–Ag bonds. Our present knowledge can be summarized as follows.

The Pt Center as the Only Basic Site. The compound $(NBu_4)[Pt(C_6F_5)_3L]$ reacts (1:1) with $(O_3ClO)AgL'$ in dichloromethane solution to give binuclear complexes [Eq. (67)], which can be isolated by evaporation to

$$(NBu_4)|Pt(C_6F_5)_3L| + (O_3ClO)AgL' \longrightarrow NBu_4ClO_4 + $$

$$|L(C_6F_5)_3Pt \rightarrow AgL'| \qquad (67)$$

$$L= SC_4H_8,\ PPh_3,\ PEt_3,\ PPh_2Et,\ NC_5H_5;\quad L'= PPh_3,\ PEt_3$$

dryness of the CH_2Cl_2 solution followed by extraction of the solid with diethyl ether. Insoluble $(NBu_4)ClO_4$ remains as a residue *(104,105)*.

The molecular structure of a member of this family of compounds has been established by single-crystal X-ray crystallography (Fig. 14). As may be seen, the platinum environment is square pyramidal, with three C atoms (of the C_6F_5 groups) and the S atom (of the SC_4H_8 ligand) on the base and the Ag atom at the apical site. There is also a direct Pt–Ag bond, unsupported by any bridging ligand and roughly perpendicular (Pt–Ag–P angle = 174.3(1)°) to the PtC_3S plane. The Pt–Ag distance [2.637(1) Å] is shorter than any previously known Pt–Ag bond.

It is a distinctive feature of this structure that one ortho fluorine atom of each C_6F_5 group makes a close contact with the silver atom [2.757(7), 2.791(7), and 2.763(8) Å, respectively]. This could be an additional factor contributing to the stability of the molecule. One C_6F_5 ring is tilted (39°) from the Pt–Ag–C(1) plane, since by eliminating the twist of this ring a series of unacceptable contacts would occur (104).

The donor–acceptor Pt–Ag bond is readily cleaved by neutral ligands L to give ionic complexes, with L coordinated to the silver center (105) [Eq. (68)]. A similar reaction can take place with acetone, since the solutions behave as 1:1 electrolytes.

$$|(H_8C_4S)(C_6F_5)_3Pt \rightarrow Ag(PPh_3)| \xrightarrow{\quad 2PPh_3 \quad} SC_4H_8 \ + $$

$$|Ag(PPh_3)_2||Pt(C_6F_5)_3(PPh_3)| \xrightarrow{\quad PPh_3 \quad} \qquad (68)$$

$$|Ag(PPh_3)_3||Pt(C_6F_5)_3(PPh_3)|$$

The anionic binuclear complexes of the type $(NBu_4)_2[Pt_2(\mu\text{-}X)_2(C_6F_5)_4]$ (X = Cl, Br) react with $AgClO_4$ in dichloromethane solution to give trinuclear metal complexes $(NBu_4)[Pt_2AgX_2(C_6F_5)_4 \cdot OEt_2]$ [Eq. (69)],

$$(NBu_4)_2|Pt_2(\mu\text{-}X)_2(C_6F_5)_4| + AgClO_4 \longrightarrow NBu_4ClO_4 \ +$$

$$+ (NBu_4)|Pt_2AgX_2(C_6F_5)_4 \cdot O(C_2H_5)_2| \qquad (69)$$

X= Cl, Br

which can be isolated by extraction with Et_2O (106). It is noteworthy that this reaction occurs without precipitation of AgX, as is the case if the same reaction is run in tetrahydrofuran solution [see Eq. (45)].

The structure of the trinuclear anion is shown in Fig. 15. The almost planar binuclear fragment "$Pt_2(\mu\text{-}Cl)_2(C_6F_5)_4$" has lost its planarity on coordination to the "$AgO(C_2H_5)_2$" group. The long Pt(1)–Pt(2) distance [3.263(1) Å] excludes any Pt–Pt bond. There are two Pt–Ag bonds [at distances of 2.782(1) and 2.759(1) Å, respectively] unsupported by any bridging ligand. These donor–acceptor Pt–Ag bonds are stronger than those present in the compounds $L(C_6F_5)_3Pt \rightarrow AgL'$, since addition of PPh_3 to the trinuclear complex only replaces the OEt_2 molecule, the Pt–Ag bonds remaining unaltered [Eq. (70)].

Not surprisingly, similar compounds can be obtained if the anionic binuclear platinum(II) complexes are treated with $Ag(OClO_3)L$ in CH_2Cl_2 solution. Trinuclear complexes $(NBu_4)[Pt_2Ag(\mu\text{-}Cl)_2(C_6F_5)_4L]$ are formed, for which an analogous structure can be assumed.

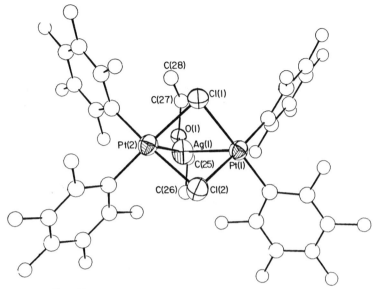

FIG. 15.   Structure of [Pt$_2$AgCl$_2$(C$_6$F$_5$)$_4$O(C$_2$H$_5$)$_2$]$^-$ *(106)*.

$$(NBu_4)\,|\,Pt_2Ag\,(\mu-Cl)_2\,(C_6F_5)_4O(C_2H_5)_2\,| \xrightarrow{\;PPh_3\;} O(C_2H_5)_2 \;+$$

$$+ \;(NBu_4)\,|\,Pt_2Ag\,(\mu-Cl)_2\,(C_6F_5)_4\,(PPh_3)\,| \qquad (70)$$

b. The Platinum Center and Another Atom Act as Bases.   The reaction between (NBu$_4$)$_2$[*trans*-PtCl$_2$(C$_6$F$_5$)$_2$] and AgClO$_4$ (or AgNO$_3$) (1:1), in methanol or acetone, yields the tetranuclear metal cluster (NBu$_4$)$_2$-[Pt$_2$Ag$_2$Cl$_4$(C$_6$F$_5$)$_4$] (90% yield) *(95,107)*. The structure of the anion is shown in Fig. 16. Two square planar *trans*-PtCl$_2$(C$_6$F$_5$)$_2$ units are linked to a central Ag$_2$ group [Ag–Ag distance 2.994(6) Å]. The essentially planar Ag$_2$Pt$_2$Cl$_2$ central unit is shown in Fig. 17. Each silver atom has a close contact with one platinum atom [2.772(3) Å], as well as a longer contact [3.063(3) Å] with the other Pt atom. Each Ag atom also has contacts with Cl atoms at a distances of 2.408(8) and 2.724(8) Å, respectively. Again, a remarkable feature of this structure is the close approach of two ortho fluorine atoms to each Ag atom [Ag$\cdots$F distances 2.60(1) and 2.69(1) Å]. The nature of this interaction is still obscure, although it may contribute to the stability of the cluster. Seemingly, not only the platinum atoms but also the chlorine ligands (and perhaps the ortho

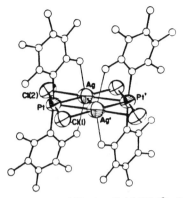

FIG. 16.   Structure of $[Pt_2Ag_2Cl_4(C_6F_5)_4]^{2-}$ (95,107).

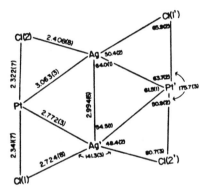

FIG. 17.   Drawing of the Central $Ag_2Pt_2Cl_4$ plane (95,107).

fluorine ones also) are acting as Lewis bases toward the acceptor Ag^+ ions in this cluster.

Addition of neutral ligands L to the tetranuclear cluster compound gives different results, depending on the nature of the neutral ligand (95). Thus, addition of $AsPh_3$ or $SbPh_3$ causes instantaneous precipitation of AgCl and two platinum complexes, an anionic and a neutral species [Eq. (71)].

$$(NBu_4)_2|Pt_2Ag_2Cl_4(C_6F_5)_4| + 2L \longrightarrow 2\ AgCl +$$

$$\underline{trans}\text{-}Pt(C_6F_5)_2L_2 + (NBu_4)_2|\underline{trans}\text{-}PtCl_2(C_6F_5)_2|$$

$$L= AsPh_3,\ SbPh_3$$

(71)

Bubbling of CO through a CH_2Cl_2 suspension of the cluster compound also gives AgCl, and $(NBu_4)[trans\text{-}Pt(C_6F_5)_2Cl(CO)]$ (71%) can be isolated from the solution. The phosphine PPh_3 behaves similarly to $AsPh_3$ and $SbPh_3$ if longer reaction times are allowed. If, after addition of PPh_3, the reaction mixture is worked up rapidly, a binuclear metal derivative is obtained [Eq. (72)].

$$(NBu_4)_2|Pt_2Ag_2Cl_4(C_6F_5)_4| + 2\ PPh_3 \longrightarrow$$

$$2\ (NBu_4)|PtAgCl_2(C_6F_5)_2(PPh_3)| \quad (72)$$

The structure of the complex anion is shown in Fig. 18. It consists of an almost square planar $[trans\text{-}PtCl_2(C_6F_5)_2]^{2-}$ unit linked to the cationic part $[Ag(PPh_3)]^+$ by both a Pt–Ag bond and a chloride bridge. Within the square planar unit the bond lengths are normal. The short Pt–Ag distance [2.796(2) Å] points to a bond of considerable strength. The structure of this complex is not simply that of half of the tetranuclear anion. Cleavage of this precursor is accompanied by changes in the strengths of the platinum–silver and silver–chlorine bonds (95). The binuclear anion does not exhibit any Ag $\cdots$ o-fluorine distances less than 3.1 Å. Seemingly, the

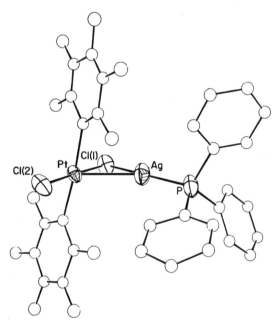

FIG. 18.   Structure of $[PtAgCl_2(C_6F_5)_2PPh_3]^-$ (95).

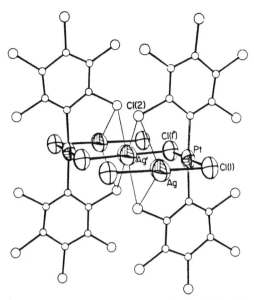

Fig. 19. Structure of a fragment of the polymeric $[Pt(C_6Cl_5)_2(\mu\text{-}Cl)_2Ag]_n^{n+}$ (108,109).

coordination to PPh_3 increases the electron density around the silver atom, making it less electrophilic.

c. *Polymeric* $(NBu_4)_n[(C_6Cl_5)_2Pt(\mu\text{-}Cl)_2Ag]_n$ *and Its Derivatives.* Although the reaction between $(NBu_4)_2[trans\text{-}PtCl_2(C_6Cl_5)_2]$ and $AgNO_3$ (or $AgClO_4$) in methanol–acetone gives (85%) a complex of stoichiometry similar to that obtained from the pentafluorophenyl derivative (108,109), the structure of the pentachlorophenyl complex is very different and consists of NBu_4^+ cations and polymeric anions (a representative segment of the latter is shown in Fig. 19). Each Pt atom is in a planar *trans*-$PtCl_2(C_6Cl_5)_2$ unit and each Ag atom in a linear Cl–Ag–Cl group. Significant Pt–Ag or Ag–Ag bonding seems unlikely in view of the long distances. There is also no Pt → Ag donor–acceptor bond in this case. One ortho chlorine of each C_6Cl_5 group is positioned so as to bridge adjacent Ag atoms, thus making the environment of each Ag atom a rhombically distorted octahedron of Cl atoms. The ortho chlorine atoms play a role reminiscent of that played by the ortho fluorine atoms in some of the compounds discussed above. Perhaps the greater size of the Cl atoms is the reason for the different structures of the pentafluoro- and pentachlorophenyl complexes.

Dichloromethane suspensions of the polynuclear complex $(NBu_4)_n$-$[Pd(C_6Cl_5)_2(\mu\text{-}Cl)_2Ag]_n$ react with tertiary phosphines, $AsPh_3$ or $SbPh_3$ (but not with CO, at normal pressure), to give bi- or trinuclear metal complexes, depending on the neutral ligand (*109*) [Eqs. (73) and (74)]. The

$$(NBu_4)_n |Pt(C_6Cl_5)_2(\mu\text{-}Cl)_2Ag|_n + nL \longrightarrow$$

$$n\ (NBu_4)\ |PtAgCl_2(C_6Cl_5)_2L| \qquad (73)$$

$$L = PPh_3,\ PEt_3,\ AsPh_3,\ SbPh_3$$

$$(NBu_4)_n |Pt(C_6Cl_5)_2(\mu\text{-}Cl)_2Ag|_n + nL \longrightarrow$$

$$n/2\ (NBu_4)_2 |\underline{trans}\text{-}PtCl_2(C_6Cl_5)_2| \ +$$

$$+\ n/2\ |Pt(C_6Cl_5)_2(\mu\text{-}Cl)_2Ag_2L_2| \qquad (74)$$

$$L = PPh_2Me,\ PPh_2Et,\ PPhMe_2$$

structure of each type of complex has been established by X-ray crystallography. Figure 20 shows that of the anion $[PtAgCl_2(C_6Cl_5)_2PPh_3]^-$ formed by the two moieties "*trans*-$PtCl_2(C_6Cl_5)_2$" and "$Ag(PPh_3)$" linked

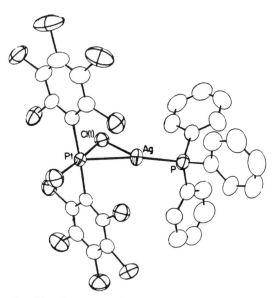

Fig. 20. Structure of $[PtAgCl_2(C_6Cl_5)_2PPh_3]^-$ (*109*).

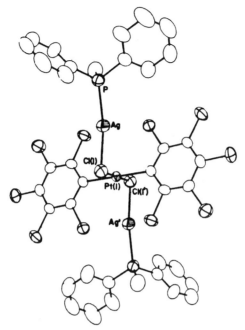

Fig. 21.   Structure of $Pt(C_6Cl_5)_2(\mu\text{-}Cl)_2Ag_2(PPh_2Me)_2]$ (109).

by a strong Pt–Ag bond [2.782(1) Å]. One chlorine atom of the first moiety also bridges both metal centers [Cl–Pt distance 2.667(3) Å]. The silver atom is also bonded to the P atom in PPh_3 [Ag–P, 2.395(2) Å], the bond angle Pt–Ag–P [163.42(8)°] being nearly linear. Moreover, the Ag atom has two contacts [2.951(3) and 3.012(3) Å] with two ortho chlorine atoms of the C_6Cl_5 groups located above and below the Pt–Cl–Ag plane.

A comparison of distances and angles observed with $[PtAgCl_2\text{-}(C_6Cl_5)_2PPh_3]^-$ with those found in similar pentafluorophenyl derivative (Fig. 18) shows only small differences around the platinum center, but differences are significant around the silver atom. Perhaps in the C_6Cl_5 complex the two Cl $\cdots$ Ag contacts, which do not have any F $\cdots$ Ag counterpart, are related to the longer Ag–Cl and Ag–P distances as well as to the opening of the Pt–Ag–PPh_3 angle. If some electron density is derived from the o-chloride–Ag contacts, then the bonds to Cl and P should be weakened.

Figure 21 shows the structure of the trinuclear metal complex $[Pt(C_6Cl_5)_2Cl_2Ag_2(PPh_2Me)_2]$. The Pt atom displays a trans square planar environment; the two Ag atoms are in an almost linear one. Both Cl atoms bonded to platinum act as monobridges between the Pt and the two Ag atoms. Finally, the silver atoms are each also bonded to a P atom. The absence of any other kind of interactions (Pt$\cdots$Ag bond or Cl$\cdots$Ag contacts) requires the existence of strong Cl–Ag [2.418(2) Å] and Ag–P [2.370(2) Å] bonds.

d. Basic Behavior of Platinum with Environmental Change. The complex $(NBu_4)_2[Pt(C_6F_5)_3Cl]$ reacts with an excess of $AgClO_4$, in CH_2Cl_2 solution, affording a trinuclear cluster *(104)* [Eq. (75)]:

$$2 \ (NBu_4)_2|Pt(C_6F_5)_3Cl| \ + \ 3 \ AgClO_4 \ \longrightarrow \ 2 \ AgCl \ +$$

$$3 \ NBu_4ClO_4 \ + \ (NBu_4)|Pt_2Ag(C_6F_5)_6O(C_2H_5)_2|\cdot O(C_2H_5)_2$$

$$(75)$$

which can be isolated after recrystallization from an ether–hexane mixture. The structure of the trinuclear anion (Fig. 22) shows the unit $[(C_6F_5)_2Pt(\mu\text{-}C_6F_5)_2Pt(C_6F_5)_2]$, with two bridging and four terminal C_6F_5 groups. The two Pt atoms are only 2.698(15) Å apart, which implies either that a Pt–Pt bond exists or that this distance is due to the steric demands of the bridging C_6F_5 groups (both Pt–C–Pt angles are 74.1°). This is the first known example in which a C_6F_5 group has acted as a bridging ligand. The silver atom is directly bonded to the two platinum atoms [Pt–Ag = 2.827(2) and 2.815(2) Å] unsupported by any bridging ligand.

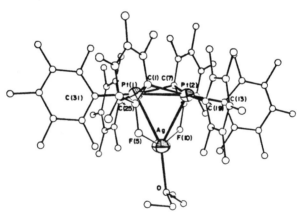

FIG. 22.   Structure of $[Pt_2Ag(C_6F_5)_6(OEt_2)]^-$ *(104)*.

4. *Isocyanide Insertion Reactions into Pd–C_6X_5 Bonds*

In most of the chemistry of compounds with M–C_6X_5 bonds surveyed so far, these bonds behave as if they were very strong, since these complexes enter into a variety of reactions in which the M–C bonds remain unmodified. Only HCl seems able to cleave M–C bonds in the anionic complexes [see Eq. (41)]. A few examples of transfer of the C_6X_5 ligand from one metal center to another have been observed [see Eq. (40)]. Moreover, it has been reported (*86,110–112*) that M–C_6X_5 bonds are remarkably inert toward insertion of small molecules.

Nevertheless, insertion reactions of isocyanides have been described. The complexes $Pd(C_6F_5)_2(CNR)_2$ (R = Me, p-Tol) react (1:1) with $Pd_2Cl_2(NCPh)_2$ in refluxing benzene to give red polymeric imidoyl derivatives with both halide and imidoyl bridges, as a result of the insertion of CNR in the Pd–C_6F_5 bonds (*113,114*) [Eq. (76)]. If the same reaction is

$$Pd(C_6F_5)_2(CNR)_2 + PdCl_2(NCPh)_2 \xrightarrow{\text{benzene}}$$

R= Me, p-Tol

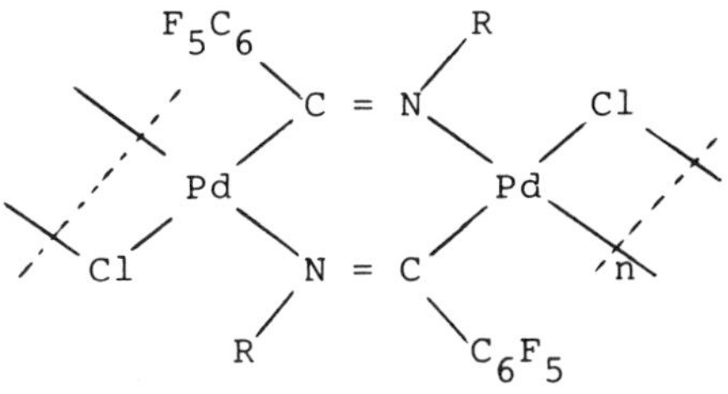

(76)

$$Pd(C_6F_5)_2(CNR)_2 + PdCl_2(NCPh_2) \xrightarrow{\text{acetone}}$$

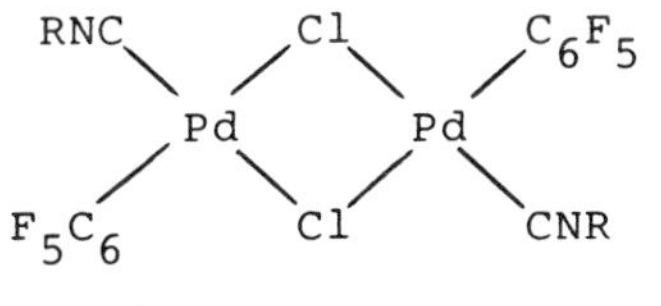

(77)

R= Me, p-Tol, tert-Bu, Cy

carried out with acetone as the solvent, no insertion takes place, and binuclear metal halide-bridged complexes are obtained [Eq. (77)] in a process similar to that discussed in Section II,B,l,a. Even in refluxing benzene, neither *t*-BuNC nor CyNC gives insertion products [Eq. (76)]. Refluxing benzene solutions of the binuclear palladium compound $Pd_2(\mu\text{-}Cl)_2(C_6F_5)_2(CNR)_2$ (R = Me, *p*-Tol) for short periods gives the imidoyl derivatives. Molecular weight measurements on $cis\{Pd(\mu\text{-}Cl)_2[\mu\text{-}C(C_6F_5)\!=\!N(p\text{-}Tol)]_2\}_n$ (115) show that it is tetranuclear ($n = 2$). It should also have a closed structure with four Pd atoms alternatively linked via chlorine and imidoyl bridges. Unfortunately, suitable crystals for X-ray studies have not been obtained.

The chloride bridges can be substituted by other halide or carboxylate groups. The latter can be accomplished stepwise with silver salts (AgOOCR′) in acetone solution, and mixed halide–carboxylate and carboxylate derivatives are produced (115) [Eq. (78)]. Both the mixed

$$\{|Pd_2\{\mu\text{-}C(C_6F_5)\!=\!NR\}_2\,|\,(\mu\text{-}Cl)_2\}_2 + 2\ AgOOCR' \longrightarrow 2\ AgCl +$$

$$+ \{\,|Pd_2(\mu\text{-}C(C_6F_5)\!=\!NR\}_2\,|\,(\mu\text{-}Cl)(\mu\text{-}OOCR')\}_2 \xrightarrow{\ 2AgOOCR'\ }$$

$$2\ AgCl + \{|Pd_2\{\mu\text{-}C(C_6F_5)\!=\!NR\}_2\,|\,(\mu\text{-}OOCR')_2\}_2 \qquad (78)$$

R= Me, p-Tol; R′= Me, CF$_3$

halide–carboxylate-bridged and the bis(carboxylate)-bridged derivatives are tetranuclear. Fortunately, suitable crystals for X-ray studies could be obtained and the structure of $\{[Pd_2\{\mu\text{-}C(C_6F_5)\!=\!NMe_2\}](\mu\text{-}Cl)(\mu\text{-}OOCMe)\}_2$ is shown in Fig. 23 (115). The complex has a crown structure with two bridging Cl atoms above and two carboxylate groups below the plane containing the four palladium atoms. Each palladium atom lies in a

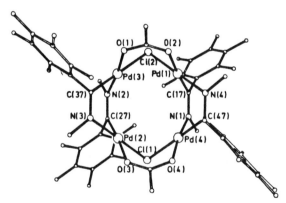

FIG. 23.   Structure of $\{[Pd_2\{\mu\text{-}C(C_6F_5)\!=\!NMe_2](\mu\text{-}Cl)(\mu\text{-}MeCOO)\}_2$ (115).

square planar environment. There are two pairs of palladium atoms, each pair with a different coordination. With Pd(1) and Pd(2) the O atom of the OOCMe group is trans to the C atom of the imidoyl ligand and the Cl atom is trans to the N atom of the imidoyl. The atoms Pd(3) and Pd(4) show the opposite coordination.

It is possible to selectively eliminate the chloride bridge in the tetranuclear complexes by reaction with Tl(acac) in dichloromethane solution, giving binuclear metal species (molecular weight) (113,114) [Eq. (79)].

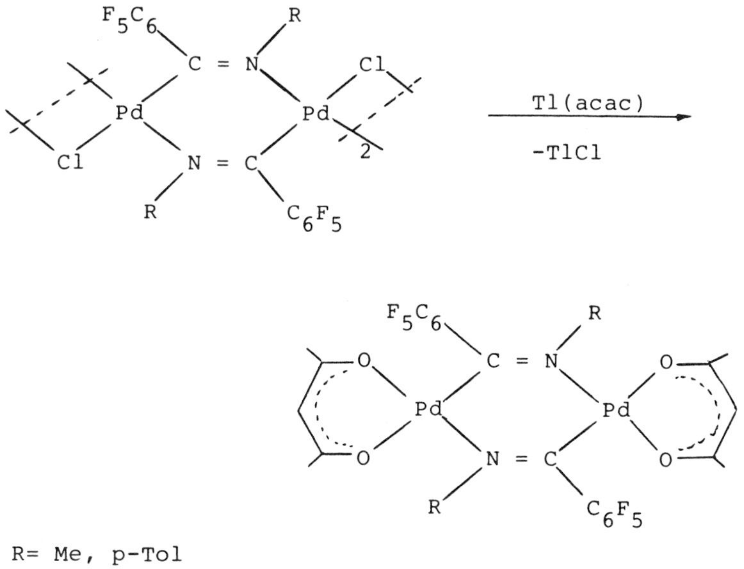

R= Me, p-Tol

$$(79)$$

The chloride bridge can also be selectively cleaved by neutral ligands (Pd/L = 1:1 ratio) to form binuclear imidoyl-bridged complexes [Eq. (80)], whereby the isomer (A) or (B), or mixtures of both, can result.

The structure of these complexes has been assigned from IR and from ^1H, ^{13}C, and ^{19}F NMR data. Moreover, the molecular structure of $Pd_2\{\mu\text{-}C(C_6F_5)\text{=}NMe\}_2Cl_2(SC_4H_8)_2$ has been established by X-ray crystallography (Fig. 24). The two palladium atoms are in an approximately square planar environment, bridged by two imidoyl groups. Two Cl atoms are trans to the C atom, with the S atom of the two SC_4H_8 ligands trans to the N atom of each imidoyl group [type A structure, Eq. (80)].

By addition of N–N chelating ligands (Pd/N-N = 1:1 ratio) the tetranuclear metal complexes undergo asymmetric cleavage of the chloride bridge system to give zwitterionic binuclear imidoyl-bridged complexes

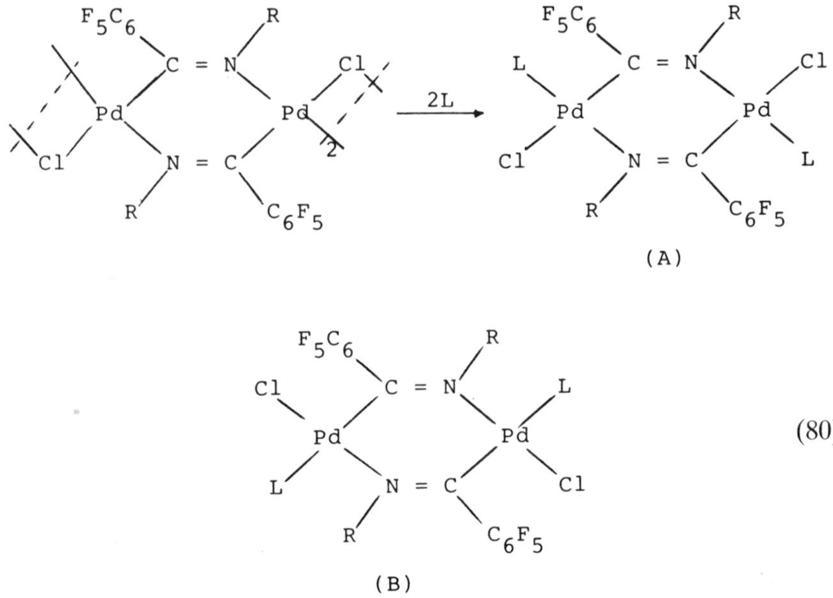

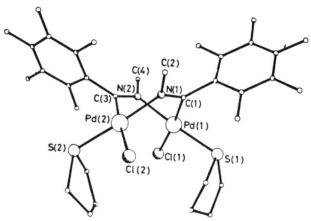

(A)

(80)

(B)

R= Me, L= PPh$_3$,SC$_4$H$_8$,   CNMe   (A);   L=   NMe$_3$ (B);   L=   py,

4-Mepy, 4-tert-Bupy [(A) 20-25% + (B) 80-75%].

R= p-Tol, L= SC$_4$H$_8$ [(A) 94% + (B) 6%|; py [(A) 8% + (B)

92%]; 4-Mepy [(A) 10% + (B) 90%].

F_IG_. 24.   Structure of Pd$_2${μ-C(C$_6$F$_5$)=N(Me)}$_2$Cl$_2$(SC$_4$H$_8$)$_2$ (*113*).

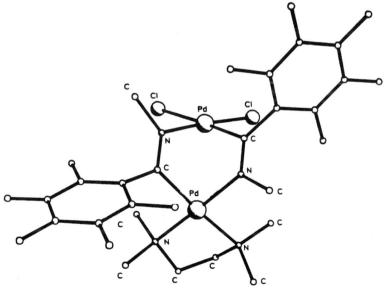

$$N-N= \text{bipy, tmeda}$$

$$(81)$$

(*116*) [Eq. (81)]. The structure of the complex with N–N = tmeda is shown in Fig. 25. Both palladium atoms display a square planar coordination. One of them is coordinated to two Cl atoms and the other to the two N donor atoms of the neutral ligand, thus confirming the zwitterionic nature of the complex.

FIG. 25.　Structure of $[Cl_2Pd\{\mu\text{-}C(C_6F_5)\text{=}NMe\}_2Pd(tmeda)]$ (*116*).

The[19]F NMR spectra of the imidoyl-bridged complexes described in this section exhibit five signals showing that rotation of the C_6F_5 groups is hindered by the syn conformation of the bridging ligand.

Addition of PPh_2H to benzene solutions of the tetranuclear metal complex with R = Me ($Pd/PPh_2H = 1:1$) symmetrically cleaves the Cl bridges, as is the case with other unidentate ligands, to afford the binuclear palladium compound $\{Pd_2[\mu\text{-}C(C_6F_5)\text{=}NMe]_2Cl_2(PPh_2H)_2\}$ (117). However, the same reaction in acetone solution, with addition of alcoholic KOH ($Pd/KOH = 1:1$), gives a complex [Eq. (82)] in which a new C–P

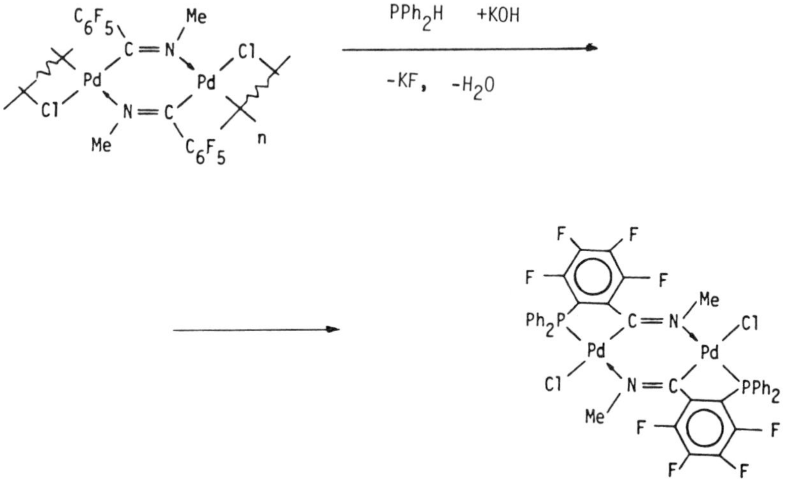

(82)

bond is created as a consequence of elimination of an ortho F atom (as KF). The complex (Fig. 26) has a boat structure, and both platinum atoms are in a very distorted square planar environment. The Cl ligands now terminal can be readily substituted by other halides. Addition of neutral unidentate or bidentate ligands (in this case, with $NaClO_4$ as chloride abstractor) gives mononuclear neutral or cationic complexes, respectively [Eq. (83)].

Some mononuclear derivatives with terminal imidoyl ligands have been prepared (118) by addition of isocyanides ($Pd/CNR' = 1:1$) to dichloromethane solutions of the complexes $[Pd_2\{\mu\text{-}C(C_6F_5)\text{=}NMe\}_2Cl_2(CNR')_2]$ or $[Pd(CNR')Cl(\mu\text{-}C(C_6F_5)\text{=}N(p\text{-}Tol))]_n$ [Eq. (84)].

The ^{19}F spectra of the mononuclear complexes show that the two F_o and the two F_m are isochronous, suggesting that the terminal imidoyl has an anti configuration. Free rotation of the C_6F_5 groups is thereby

PPh3
bipy
NaClO4
Ph2P
Pd
C=NMe
Cl
PPh3
ClO4
Me
PPh2

(83)

F5C6
Me
R'NC
Pd
Cl
CNR'
C6F5
PTol
n
R'= Me, p-Tol, tert-Bu

(84)

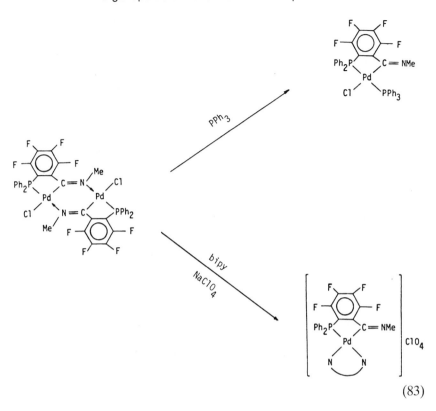

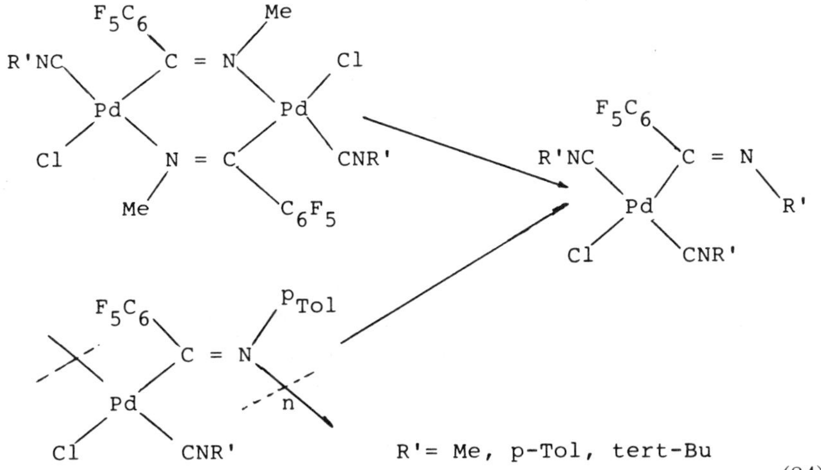

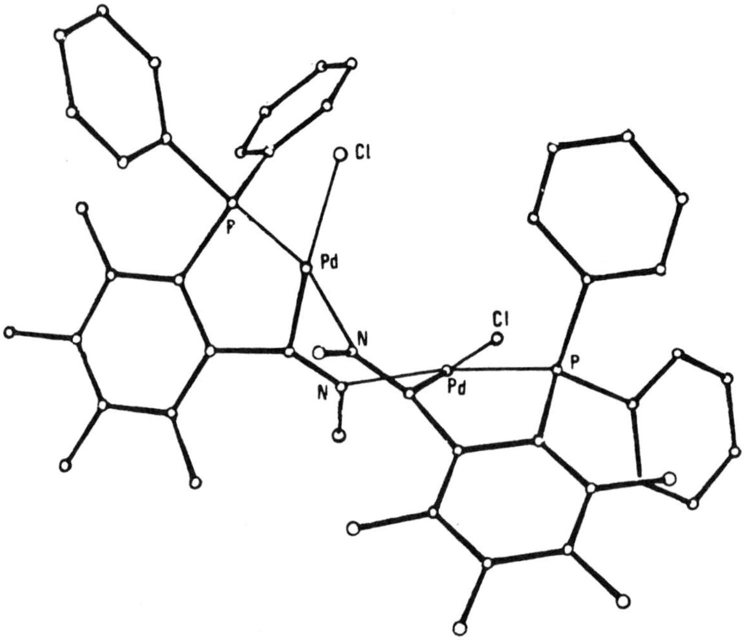

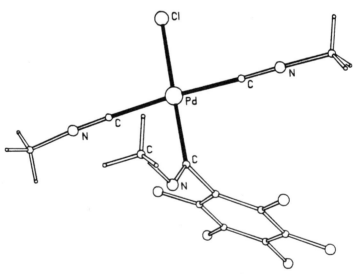

FIG. 26.   Structure of $\{Pd[C(C_6F_4PPh_2\text{-}o)N(Me)Cl\}_2$ (*117*).

FIG. 27.   Structure of $Pd\{C(C_6F_5)\!=\!N(Me)\}Cl(CNMe)_2$ (*118*).

achieved. From ir and ^{1}H nmr data, a trans structure can be assigned, which was confirmed by X-ray studies in case of the complex $Pd\{C(C_6F_5)=NMe\}Cl(CNMe)_2$ (Fig. 27). The Pd atom displays the usual square planar coordination, the long Pd–Cl distance (2.404 Å) being noteworthy. The anti configuration of the imidoyl ligand is also patent.

By adding CNR' to acetone or dichloromethane solutions of the mononuclear imidoyl complexes in the presence of $NaClO_4$ (or $NaBPh_4$), cationic complexes are obtained [Eq. (85)].

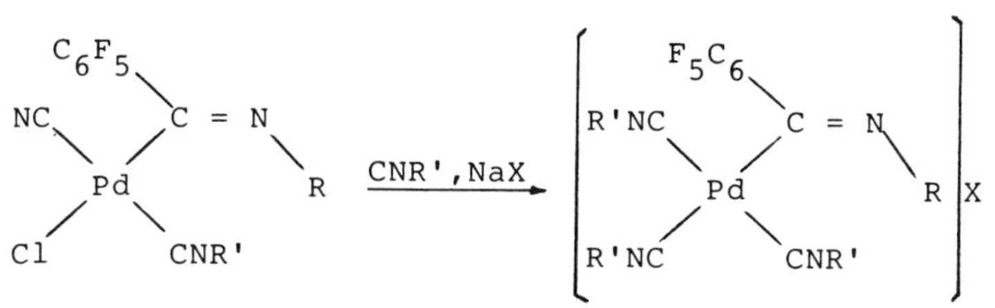

$$\underset{\substack{\text{Cl}}}{\overset{\substack{C_6F_5}}{\underset{\substack{\text{Pd}}}{\text{NC}}}}\overset{\substack{C\ =\ N}}{\underset{\substack{\text{CNR'}}}{}}\overset{R}{\xrightarrow{\text{CNR',NaX}}}\left[\underset{\substack{R'NC}}{\overset{\substack{F_5C_6}}{\underset{\substack{\text{Pd}}}{R'NC}}}\overset{\substack{C\ =\ N}}{\underset{\substack{\text{CNR'}}}{}}R\right]X \quad (85)$$

$= ClO_4,\ BPh_4;\ R=\ Me,\ p\text{-Tol};\ R'=\ Me,\ p\text{-Tol, tert-Bu}$

An unprecedented reaction in the chemistry of palladium takes place when PPh_3 is added to the binuclear compound $[Pd_2\{\mu\text{-}C(C_6F_5)=NMe\}_2Cl_2(PPh_3)_2]$ with the object of preparing mononuclear complexes with the terminal imidoyls. Instead, imidoyl elimination is observed, the mononuclear complex possibly being an undetected intermediate [Eq. (86)].

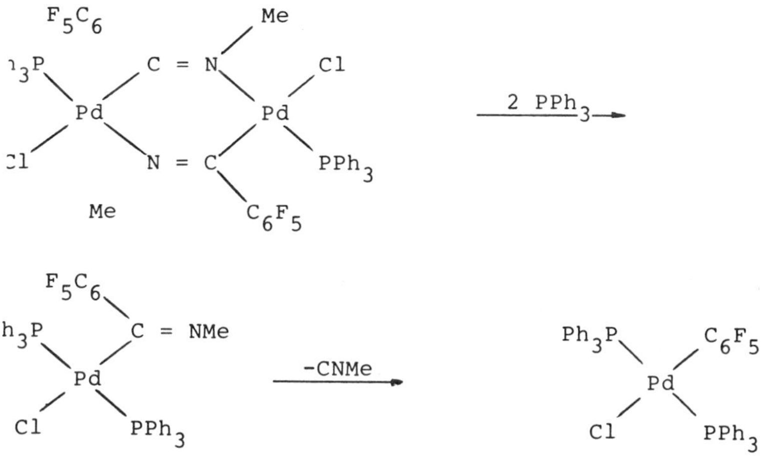

$$\underset{\substack{\text{Cl}}}{\overset{\substack{F_5C_6}}{\underset{\substack{\text{Pd}}}{\overset{\substack{3P}}{}}}}\overset{\substack{C\ =\ N}}{\underset{\substack{N\ =\ C}}{\overset{Me}{}}}\overset{\substack{Me}}{\underset{\substack{C_6F_5}}{\overset{\substack{Cl}}{\underset{\substack{\text{Pd}}}{}}}}\overset{Cl}{\underset{\substack{PPh_3}}{}}\xrightarrow{2\ PPh_3}$$

$$\underset{\substack{Cl}}{\overset{\substack{F_5C_6}}{\underset{\substack{\text{Pd}}}{\overset{\substack{h_3P}}{}}}}\overset{\substack{C\ =\ NMe}}{\underset{\substack{PPh_3}}{}}\xrightarrow{-CNMe}\underset{\substack{Cl}}{\overset{\substack{Ph_3P}}{\underset{\substack{\text{Pd}}}{}}}\overset{\substack{C_6F_5}}{\underset{\substack{PPh_3}}{}}$$

$$(86)$$

Though less well documented, some isocyanide insertion reactions into Pd–C_6Cl_5 bonds have also been reported (119). Refluxing benzene solutions of trans-$Pd(C_6Cl_5)X(CNMe)_2$ results in insertion of a terminal isocyanide ligand into the Pd–C_6Cl_5 bond [Eq. (87)]. The terminal Cl

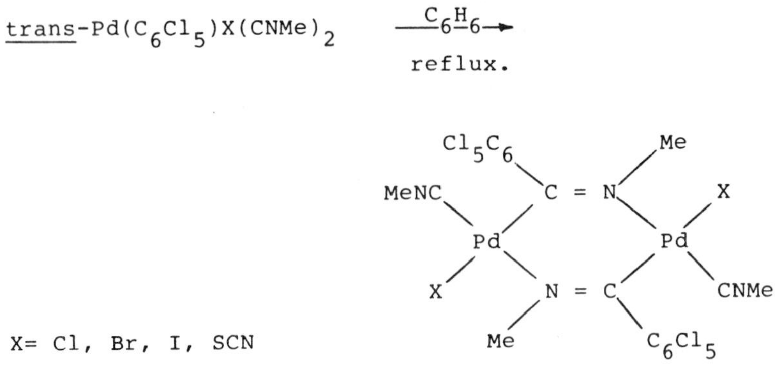

$$\text{trans-Pd(C}_6\text{Cl}_5\text{)X(CNMe)}_2 \quad \xrightarrow[\text{reflux.}]{\text{C}_6\text{H}_6}$$

X= Cl, Br, I, SCN

(87)

ligands in the binuclear complex $\{Pd_2[\mu\text{-}C(C_6Cl_5){=}NMe]_2Cl_2(CNMe)_2\}$ can be replaced in refluxing acetone by halide or pseudonhalide ligands, employing the corresponding alkaline salts.

Addition of CNMe (excess) to diethyl ether solutions of the binuclear metal complexes $\{Pd_2[\mu\text{-}C(C_6Cl_5){=}NMe]_2X_2(CNMe)_2\}$ (X = Cl, Br, I) affords mononuclear complexes with a terminal imidoyl ligand [Eq. (88)].

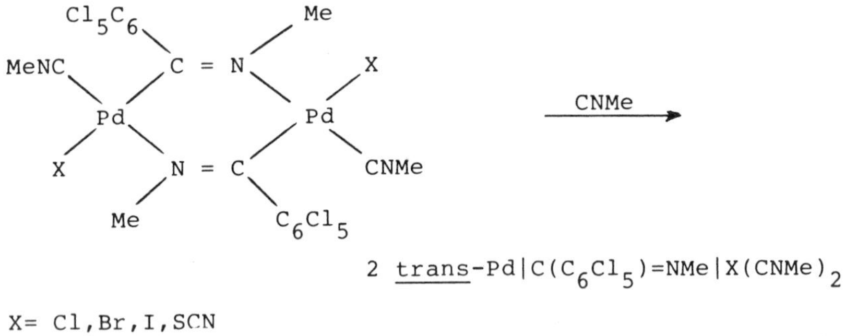

$$\xrightarrow{\text{CNMe}}$$

2 trans-Pd|C(C$_6$Cl$_5$)=NMe|X(CNMe)$_2$

X= Cl,Br,I,SCN

(88)

These complexes are also accessible by metathetical reactions of the mononuclear derivative with alkaline salts MX in the presence of free CNMe to avoid the formation of the binuclear complexes with a bridging imidoyl. The iodine derivatives give $PdI_2(CNMe)_2$ under these conditions.

In contrast, reactions between the complexes $Pd(C_6X_5)_2(CNR)_2$ and

PdCl$_2$ (which with X = F lead to binuclear or tetranuclear complexes) lead to metallic palladium when X = Cl.

III

PLATINUM(III) COMPLEXES

The only fully characterized mononuclear platinum(III) complex is the pentachlorophenyl derivative (NBu$_4$)[Pt(C$_6$Cl$_5$)$_4$]. It can be obtained by oxidation of (NBu$_4$)$_2$[Pt(C$_6$Cl$_5$)$_4$] in dichloromethane solutions with the stoichiometric amount of Cl$_2$ or Br$_2$ in carbon tetrachloride solution (*34,35*) [Eq. (89)]. Evaporation of the deep blue solutions permits isolation of the compound (90–92% yield). Use of I$_2$ or TlCl$_3$ as an oxidizing agent gives lower yields. Also electrochemical oxidation is possible at 0.54 V, through a reversible one-electron process.

$$(NBu_4)_2|Pt(C_6Cl_5)_4| + \tfrac{1}{2} X_2 \longrightarrow (NBu_4)|Pt(C_6Cl_5)_4| +$$
$$+ NBu_4X \qquad (89)$$

The structure of the complex anion is shown in Fig. 28. The Pt(III) ion lies in a square planar environment and is 9.7 Å distant from the nearest Pt atom, excluding any interaction or stacking. The structure is almost undistinguishable from that adopted by the [Pt(C$_6$Cl$_5$)$_4$]$^{2-}$ anion (Fig. 1), since the average Pt–C, C–C, and C–Cl distances and even the twist angle between the C$_6$Cl$_5$ rings and the central Pt–C$_4$ plane are basically identical

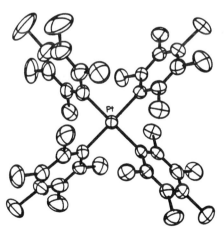

FIG. 28.   Structure of [Pt(C$_6$Cl$_5$)$_4$]$^-$ (*34,35*).

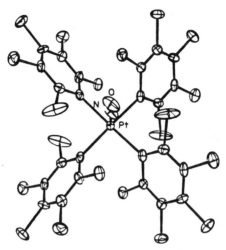

FIG. 29.   Structure $[Pt(C_6Cl_5)_4NO]^-$ (35).

in both anions. Determination of the magnetic susceptibility gives $\mu_{eff} = 2.57$ B.M. (260 K) and 2.42 B.M. = (80 K), consistent with spin 1/2, with a strong orbital contribution arising from the population of excited levels, as has been observed for other square planar complexes of d^7 ions (120).

The Pt(III) complex is air- and moisture-stable. An excess of chlorine in refluxing CH_2Cl_2 has no effect. Only in refluxing 1,1,2,2-tetrachloroethane (146°C) with an excess of Cl_2 does cleavage of the $Pt-C_6Cl_5$ bonds occur, with C_6Cl_6 and $(NBu_4)_2[PtCl_6]$ as the reaction products. A methanolic HCl solution (1:2 ratio) does not yield a binuclear Pt(III) complex. Instead, a mixture of C_6Cl_5H, $(NBu_4)_2[Pt(C_6Cl_5)_4]$, and $(NBu_4)_2[Pt_2(\mu\text{-}Cl)_2(C_6Cl_5)_4]$ is obtained.

Reaction with NO leads to $(NBu_4)[Pt(C_6Cl_5)_4NO]$, whose structure (35) is noteworthy (Fig. 29). The anion is a distorted square pyramid with the $Pt-C_4$ unit deviating only slightly from planarity. There is a linear $Pt-N-O$ unit.

IV

COMPLEXES OF Pd(IV) AND Pt(IV)

Only pentafluorophenyl derivatives have been reported, and all of them have been synthesized by oxidative addition of halogen to Pd(II) and Pt(II) precursors.

A. *Palladium Derivatives*

Bubbling of Cl_2 gas through dichloromethane or benzene solutions of $Pd(C_6F_5)_2(N-N)$ (*121*) or $Pd(C_6F_5)X(N-N)$ (*122*) affords Pd(IV) derivatives [Eqs. (90) and (91)].

$$Pd(C_6F_5)_2(N-N) + Cl_2 \longrightarrow Pd(C_6F_5)_2Cl_2(N-N) \quad (90)$$

N-N= bipy, phen, en, pn

$$Pd(C_6F_5)Cl(N-N) + Cl_2 \longrightarrow Pd(C_6F_5)Cl_3(N-N) \quad (91)$$

N-N= bipy, phen, tmeda

An excess of chlorine cleaves the Pd—C bonds with formation of C_6F_5Cl. Orange-colored and insoluble products $PdCl_4(N-N)$ are obtained.

When heated, these Pd(IV) complexes evolve Cl_2, and the deep yellow dichloromethane solutions of $Pd(C_6F_5)_2Cl_2(N-N)$ turn pale within a few hours. Dichloromethane solutions of the Pd(IV) complexes react with (colorless) iodide to give (brown) iodine solutions, a reaction that has not been observed for Pd(II) compounds. Moreover, some ir internal absorptions of the pentafluorophenyl group are shifted toward higher energies after oxidation of the Pd(II) precursors.

The existence of the previously described compounds $Pd(C_6F_5)_2Cl_2$-$(PPh_3)_2$ (*37,38*) and $Pd(C_6F_5)Cl_3(PPh_3)_2$ (*11*) could not be confirmed (*39*). Attempts to oxidize pentafluorophenyl palladium(II) complexes containing phosphine ligands have failed (*123*). The presence of the chelating N—N donor ligand seems to be a decisively stabilizing factor and the Pd(IV) complexes described above remain the only unambiguously* characterized organometallic Pd(IV) complexes (*124*).

B. *Platinum Derivatives*

These are more numerous than the Pd(IV) complexes. Reactions of *cis*-$Pt(C_6F_5)_2L_2$ with Cl_2, Br_2 or I_2 have been studied in detail (*40,61,125*)

---

* The synthesis and structure of *fac*-[PdMe$_3$(2,2'-bipyridyl) I] have been reported recently (*124b*).

[Eq. (92)]. The precursor used, the neutral ligand, the halogen, and the reaction conditions can influence the results. Thus, when L_2 = bipy, PBu_3

$$\underline{cis}\text{-}Pt(C_6F_5)_2L_2 + X_2 \longrightarrow Pt(C_6F_5)_2X_2L_2$$

$$X= Cl, Br; \quad L_2= bipy,phen,dpam,pdma,2\ PBu_3,2\ PEt_3 \quad (92)$$

$$X= I; \quad L_2= pdma$$

or PEt_3, even the use of excess halogen does not preclude isolation of the Pt(IV) complex (40,125). However, careful use of stoichiometric amount of X_2 is necessary when L_2 = dpam or phen, to avoid the formation of C_6F_5Cl. When L = py, a mixture of products is obtained in every case. With trans-$Pt(C_6F_5)_2(PPh_3)_2$ or with a mixture of cis- and trans-$Pt(C_6F_5)_2(AsPh_3)_2$, no reaction takes place (125).

The use of iodine as the oxidizing agent ($CHCl_3$, room temperature) is effective only when L_2 = o-phenylenebis(dimethyl)arsine (61). Even in refluxing toluene, iodine does not oxidize $Pt(C_6F_5)_2L_2$ complexes. However, isomerization (both trans–cis and cis–trans) and anion exchange [to $Pt(C_6F_5)IL_2$] have been observed. The compound cis-$Pt(C_6F_5)I(PEt_3)_2$ reacts with I_2 (toluene, 90°C) to yield trans-$PtI_4(PEt_3)_2$ (40). Complexes of the type $Pt(C_6F_5)X_3L_2$ have also been reported (L = PEt_3, X = Cl, Br; L_2 = pdma, X = Cl, Br, I).

The structure of the Pt(IV) complexes has been assigned from dipole moments (40) and from nmr spectra (126).

V

PALLADIUM(I) AND PLATINUM(I) COMPLEXES

A. Binuclear Metal Complexes Containing Bridging Ligands

These compounds have been prepared by condensation redox reactions between M(II) species $M(C_6X_5)X'(\eta^1\text{-dpppm})_2$ and zero-valent complexes [$Pd_2(dba)_3 \cdot CHCl_3$, Pt (COD)$_2$, Pt(PPh$_3$)$_4$]. Suitable choice of the reaction partners leads to homo- (127–129) or heterometallic (130) derivatives [Eqs. (93–95)]. Similarly [Eq. (93)], $Pd(C_6F_5)_2(dppa)_2$ gives the compound $(F_5C_6)Pd(\eta^1\text{-dppa})_2Pd(C_6F_5)$ (131).

The reactions are carried out at room temperature in oxygen-free dichloromethane or benzene. Yields are commonly 65–90%. The complexes are colored (deep yellow to orange–brown), and after isolation

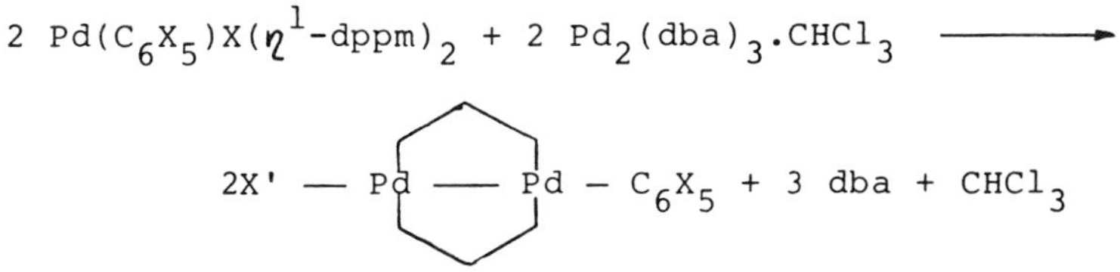

$$2 \ Pd(C_6X_5)X(\eta^1\text{-dppm})_2 + 2 \ Pd_2(dba)_3 \cdot CHCl_3 \longrightarrow$$

$$2X' - \ Pd \text{——} Pd - C_6X_5 + 3 \ dba + CHCl_3 \quad (93)$$

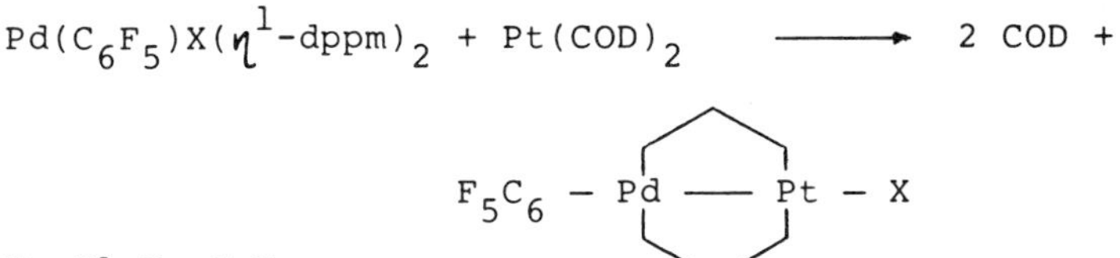

$$X = F; \ X' = Cl, Br, I, OCN, C_6F_5; \ X = Cl; \ X' = Cl, C_6Cl_5$$

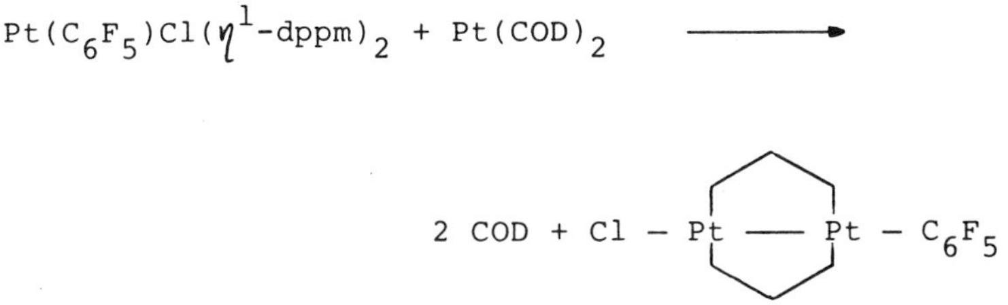

$$Pt(C_6F_5)Cl(\eta^1\text{-dppm})_2 + Pt(COD)_2 \longrightarrow$$

$$2 \ COD + Cl - Pt \text{——} Pt - C_6F_5 \quad (94)$$

$$Pd(C_6F_5)X(\eta^1\text{-dppm})_2 + Pt(COD)_2 \longrightarrow 2 \ COD +$$

$$F_5C_6 - Pd \text{——} Pt - X \quad (95)$$

$$X = Cl, Br, C_6F_5$$

$$X' - M \text{——} M' - C_6X_5 + M''X'' \longrightarrow M''X' +$$

$$X'' - M \text{——} M' - C_6X_5 \quad (96)$$

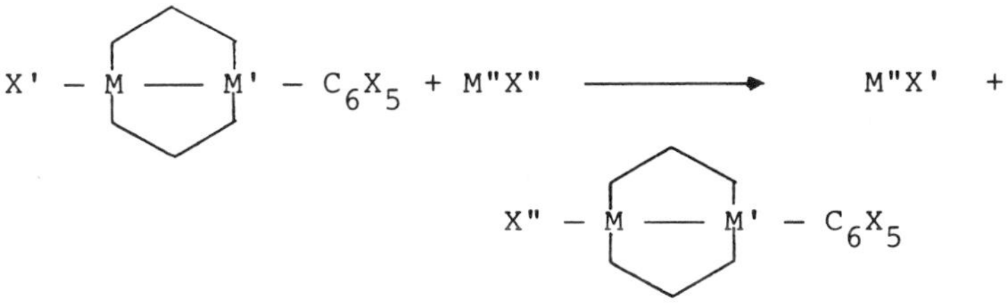

$$M = M' = Pd; \ X = Cl; \ X' = Cl; \ X'' = Br, I, SCN, CNO$$

$$M = M' = Pt; \ X = F; \ X' = Cl; \ X'' = Br, SCN$$

$$M = Pt, \ M' = Pd; \ X = F; \ X' = Cl; \ X'' = SCN$$

they are air-stable at room temperature. The structures of $(Cl_5C_6)Pd(\mu\text{-dppm})_2Pd(C_6Cl_5)$ and $ClPt(\mu\text{-dppm})_2Pd(C_6F_5)$ have been determined (Figs. 30 and 31) and show a slightly distorted square planar coordination at each metal atom. The Pd–Pd and Pd–Pt distances are 2.6704(21) and 2.643(1) Å, respectively. In the heterodimetallic complex, the structure and the ^{19}F nmr spectrum reveal that the C_6F_5 group is linked to the Pd

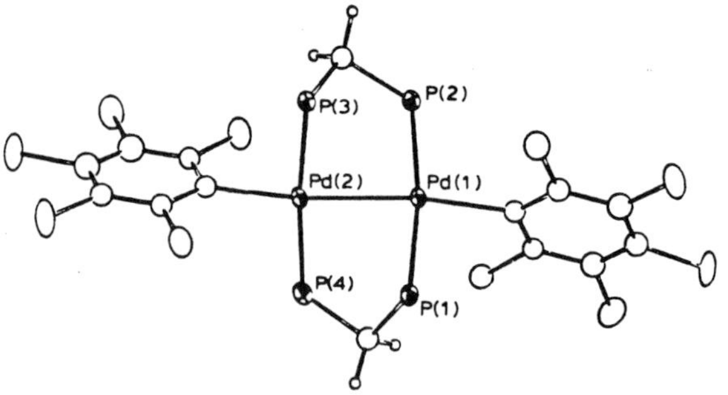

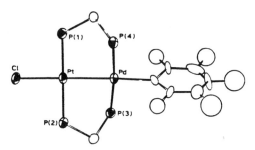

FIG. 30.   Structure of Pd$_2$ (μ-dppm)$_2$(C$_6$Cl$_5$)$_2$. Central section with Ph rings removed.

FIG. 31.   Structure of ClPt(μ-dppm)$_2$Pd(C$_6$F$_5$). Central section with Ph rings removed.

atom, thus confirming that the redox condensation takes place through migration of the X group from the Pd(II) to the Pt(0) center.

The binuclear metal complexes undergo two classes of reactions:

1. Substitution Reaction at the M–X Bond

The halide ligand X′ can be substituted by other halides or pseudo-halides (X′$^-$) in methanol or acetone solutions (127–130) [Eq. (96)]. Methanol suspensions of X′M(μ-dppm)$_2$M′(C$_6$F$_5$) react with equivalent amounts of a neutral ligand and NaBPh$_4$ to give cationic derivatives [Eq. (97)]. Normally, the Cl derivative can be used at room temperature. However, the halo complexes (Cl, Br, I) are unreactive even when refluxed if C$_6$X$_5$ = C$_6$Cl$_5$ (128). In this case, the cyanate complex must be used.

Finally, insertion of SnCl$_2$ can occur in the Cl–M bond (but not in the M–M′ bond) [Eq. (98)].

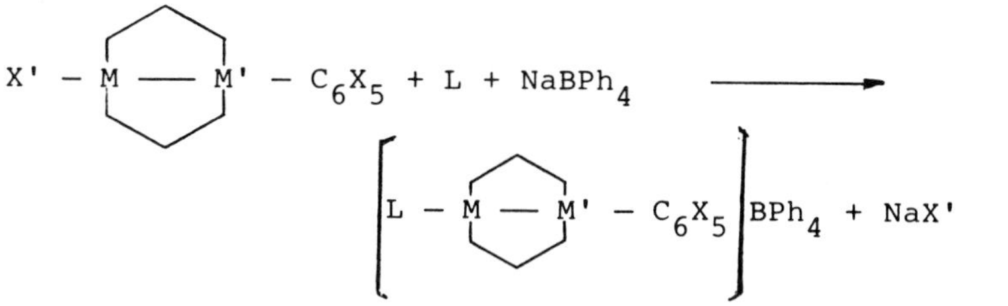

$$X' - M \underset{\square}{\overline{}} M' - C_6X_5 + L + NaBPh_4 \longrightarrow$$

$$\left[L - M \underset{\square}{\overline{}} M' - C_6X_5 \right] BPh_4 + NaX' \qquad (97)$$

M=M'= Pd; X= F; X'= Cl; L= PPh$_3$,P(OPh)$_3$,py,tht

M=M'= Pd; X= Cl; X'= CNO; L= PPh$_3$,P(OPh)$_3$,AsPh$_3$,SbPh$_3$,tht

M=M'= Pt; X= F; X'= Cl; L= PPh$_3$,AsPh$_3$,py

M= Pt, M'= Pd; X= F; X'= Cl; L= PPh$_3$,py

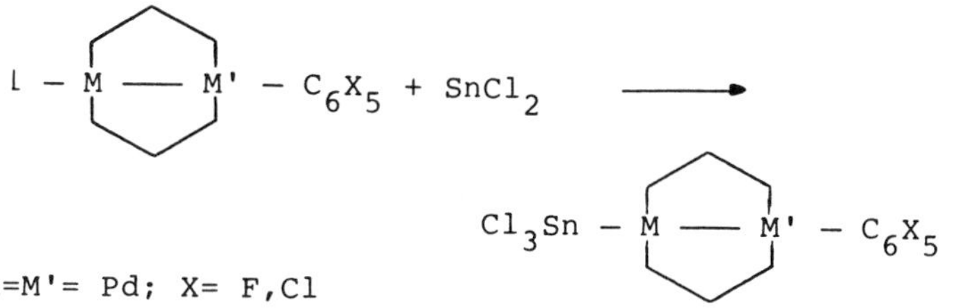

$$L - M \underset{\square}{\overline{}} M' - C_6X_5 + SnCl_2 \longrightarrow$$

$$Cl_3Sn - M \underset{\square}{\overline{}} M' - C_6X_5 \qquad (98)$$

=M'= Pd; X= F,Cl

l=M'= Pt;, X= F

l= Pt, M'= Pd; X= F

2. Reactions Affecting the M–M' Bond

Since the M–M' bond is electron-rich it readily undergoes insertion reactions with electrophiles such as SO_2, diazonium salts, acetylenes, CO, and isocyanide to give the so-called A-frame complexes (Scheme 1).

The cationic complex $[ClPt(\mu\text{-dppm})_2(\mu\text{-H})Pt(C_6F_5)]ClO_4$, prepared by addition of $HClO_4$ to a Pt–Pt precursor, is fluxional with a very low activation energy, as found from the ^1H and ^{19}F nmr spectra [Eq. (99)].

Insertion of CO does not take place when both metal atoms are linked to a C_6X_5 group. If only one metal atom is linked to C_6X_5 and the other to a halogen atom, carbonylation in solution is observed [M = M' = Pd, $\nu(CO), 1715 \text{ cm}^{-1}$ for C_6F_5, 1708 cm^{-1} for C_6Cl_5]. However, on evaporation of the dichloromethane solutions, only the binuclear starting complex is recovered.

Bubbling CO through a dichloromethane solution of $ClPt(\mu\text{-dppm})_2Pd(C_6F_5)$ for 15 min gives a solution showing *two* ir absorptions at

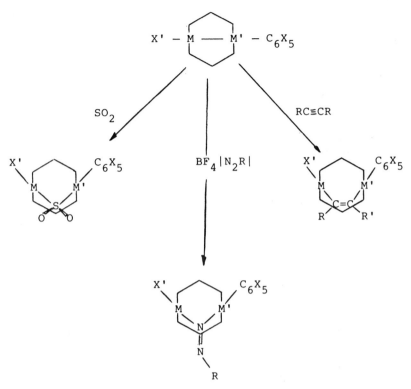

M= Pd; M'= Pt; X= F, Cl; X'= Cl; R= p-Tol, o-NO$_2$C$_6$H$_4$;

R'= -COOMe

SCHEME 1

(99)

2055(w) and 1710 cm^{-1} (130). After 60 min only the band at 2055 cm^{-1} is present. Seemingly, carbonylation gives primarily the insertion product (A) which later isomerizes to (B) with a terminal CO group [Eq. (100)]. For M = M' = Pt, only the coordinated complex can be detected (even in solution).

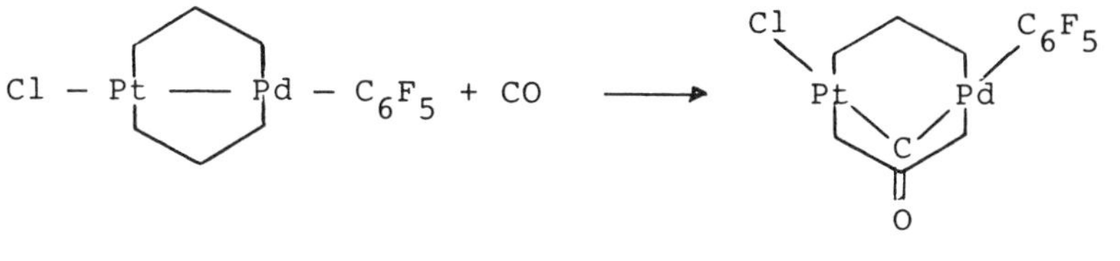

$$Cl - Pt \overline{\qquad} Pd - C_6F_5 + CO \longrightarrow$$

(A)

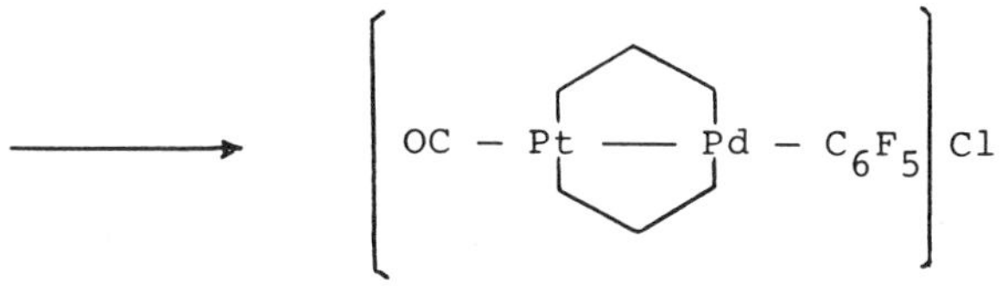

$$\longrightarrow \left[OC - Pt \overline{\qquad} Pd - C_6F_5 \right] Cl \qquad (100)$$

(B)

The cationic species can be isolated by addition of NaBPh$_4$. The complex [(OC)Pt(μ-dppm)$_2$Pt(C$_6$F$_5$)]BPh$_4$ has also been prepared, with BPh$_4$ as counterion. The presence of C$_6$X$_5$ groups apparently hinders the insertion of CO into the M–M' bond, and even the formation of a noninserted cationic species is possible only if a platinum centre is present.

In reactions of the binuclear metal complexes with isocyanides, three different pathways are observed: (a) insertion, (b) displacement of X to give an uninserted cationic derivative, and (c) insertion and coordination of the isocyanide with a excess of isocyanide [Eq. (101)].

The results of a study of the reactions of the compounds X'M(μ-dppm)$_2$M'(C$_6$X$_5$) (M = M' = Pd, X' = Cl, X = F; M = M' = Pd, X' = C$_6$F$_5$, X = F; M = M' = Pd, X' = C$_6$Cl$_5$, X = Cl; M = M' = Pt, X' = Cl, X = F; M = M' = Pt, X' = C$_6$F$_5$, X = F, M = Pt, M' = Pd, X' = Cl, X = F; M = Pt, M' = Pd, X' = C$_6$F$_5$, X = F) with the reagents CN(p-Tol), CN(t-Bu), and CNCy can be summarized as follows:

(a) Displacement of a C$_6$X$_5$ group has not been observed in any instance.

(b) The ease of insertion decreases with increasing number of C$_6$X$_5$ groups.

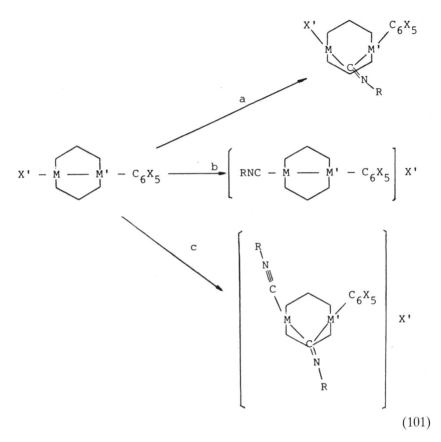

$$(101)$$

$$\underline{cis}\text{-}Pt(C_6X_5)_2L_2 + Pt(\eta^2\text{-}C_2H_4)(PPh_3)_2 \xrightarrow{-C_2H_4}$$

$$Pt_2(C_6X_5)_2L_2(PPh_2)_2 \qquad (102)$$

X= F, L= CO, CN(p-Tol), CNCy, CN(tert-Bu)

X= Cl, L= CO, CN(p-Tol)

(c) The tendency for insertion decreases in the sequence CN(p-Tol) > CNCy > CN(t-Bu), the same as the order observed for insertion into M–C bonds, in both alkyl and aryl derivatives (114,132). CN(t-Bu) does not insert into the M–M′ bonds.

(d) In polar solvents (CH$_2$Cl$_2$, OCMe$_2$, CH$_3$CN) coordination is favored, while in nonpolar ones (C$_6$H$_6$) insertion is favored. Infrared spectroscopy is a useful tool for structural diagnosis; with inserted CNR $\nu(C=N) = 1550-1650$ cm^{-1}, and with CNR as a terminal ligand $\nu(C\equiv N) = 2100-2200$ cm^{-1}.

(e) Cationic complexes $[(RNC)M(\mu\text{-dppm})_2(\mu\text{-CNR})M'(C_6F_5)]X'$ (X' = Cl, BPh$_4$, ClO$_4$) have been isolated only in cases where the neutral insertion complex is also known.

(f) The ir spectrum of dichloromethane solutions of ClPd(μ-dppm)$_2$(μ-CNCy)Pd(C$_6$F$_5$) reveals the isocyanide coordinated as a terminal ligand. However, the solid obtained on evaporation and addition of *n*-hexane contains inserted isocyanide. This reversible interconversion of two linkage isonitrile isomers is unprecedented in palladium chemistry.

(g) Equimolar amounts of ClPd(μ-dppm)$_2$Pd(C$_6$Cl$_5$) and CN(*p*-tol) react in CH$_2$Cl$_2$ solution, and the ir spectrum shows the presence of an equilibrium mixture of inserted isocyanide and the cationic complex with terminal isocyanide. On evaporation, only the insertion complex can be isolated (*128*). Acetone solutions show a high value of Λ_M (molar conductivity), again due to the presence of the ionic species.

(h) Addition of NaBPh$_4$ to the complexes containing inserted CNR gives the corresponding cationic complexes.

B. *Binuclear Metal Complexes without Any Bridging Ligand*

Platinum(II) complexes *cis*-Pt(C$_6$X$_5$)$_2$L$_2$ undergo redox condensation with Pt(η^2-C$_2$H$_4$)(PPh$_3$)$_2$ in oxygen-free refluxing tetrahydrofuran to give binuclear Pt(I) complexes (*133*) [Eq. (102)]. However, the method is not general, since when X = F and L = PPh$_3$, THF or when X = Cl and L = PPh$_3$, CNCy, CN (*t*-Bu) no reaction takes place and the starting products are recovered.

Molecular weight determinations (isopiestic method) suggest formation of binuclear species. The structure of Pt$_2$(C$_6$F$_5$)$_2$(CO)$_2$(PPh$_3$)$_2$ is shown in Fig. 32. Each Pt atom is in a square planar environment. Two Pt(C$_6$F$_5$)(CO)(PPh$_3$) fragments are connected by a direct Pt–Pt bond (2.5971 Å) unsupported by any bridging ligand. The two terminal CO groups [$\nu(CO) = 2054, 2036$ cm^{-1}] and the two C$_6$F$_5$ groups are in cisoid positions with respect to the Pt–Pt bond. The ^{31}P NMR spectra of all these complexes show similar patterns, pointing to analogous structures.

An initial study of the reactivity of this new class of Pt(I) complexes reveals departures from the behavior exhibited by the complexes discussed in Section V,a. No insertion reactions in the Pt–Pt bond have been observed, even with CN(p-Tol). The reagent SnCl$_2$ is also unable to insert into the Pt–Pt bond, although it does react with [Pt$_2$Cl$_4$(CO)]$_2$]$^{2-}$ to give

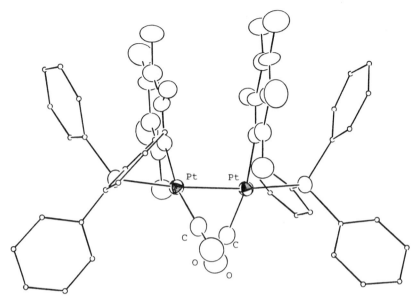

FIG. 32.   Structure of $Pt_2(C_6F_5)_2(CO)_2(PPh_3)_2$ (*133*).

$[PtX_2(CO)_2SnCl_2]^{2-}$ (*134*). Oxidative addition of I_2 to give mononuclear complexes of Pt(II) has also been observed [Eq. (103)].

$$Pt_2(C_6X_5)_2L_2(PPh_3)_2 + I_2 \longrightarrow 2\ Pt(C_6X_5)IL(PPh_3) \quad (103)$$

X= F,Cl; L= CO, CN(p-Tol)

VI
APPENDIX. INFRARED SPECTRA OF THE
PENTAHALOPHENYL METAL COMPOUNDS

A. *Pentafluorophenyl Derivatives*

The C_6F_5 groups show strong absorptions in the 1650, 1500, 1300, 1050, 950, and 800 cm^{-1} regions. The band at ~800 cm^{-1} has been assigned to an "X-sensitive mode that has mainly $\nu(M\text{–}C)$ character (*135*), and is related to the skeletal symmetry of the molecule." In practice, this absorption gives structural information and has often been used to elucidate the structure of new compounds. Table III lists some data for penta-

TABLE III

Absorptions Assigned to the X-Sensitive Mode of the C_6F_5 Group

Complex	Point group	Symmetry of X-sensitive modes	Number of IR-active bands	(X-sensitive)^a (cm^{-1})	Ref.
(NBu$_4$)$_2$[Pd(C$_6$F$_5$)$_4$]	D_{4h}	E_u	1	752(s)	137
(NBu$_4$)$_2$[Pt(C$_6$F$_5$)$_4$]	D_{4h}	E_u	1	765(s)	137
(NBu$_4$)$_2$[Pt(C$_6$F$_5$)$_3$Cl]	C_{2v}	$2A_1 + B_1$	3	800(s), 780(s), 765(s)	23
(NBu$_4$)$_2$[Pt(C$_6$F$_5$)$_3$CN]	C_{2v}	$2A_1 + B_1$	3	778(s), 765(sh), 760(s)	24
(NBu$_4$)[Pd(C$_6$F$_5$)$_3$(tht)]	C_{2v}	$2A_1 + B_1$	3	782(s), 770(m), 760(s)	137
(NBu$_4$)[Pt(C$_6$F$_5$)$_3$(tht)]	C_{2v}	$2A_1 + B_1$	3	798(s), 782(s), 770(s)	137
(NBu$_4$)$_2$[Pd$_2$(μ-Cl)$_2$(C$_6$F$_5$)$_4$]	D_{2h}	$A_g + B_{2g} + B_{2u} + B_{3u}$	2	795(s), 781(s)	137
(NBu$_4$)$_2$[Pt$_2$(μ-Br)$_2$(C$_6$F$_5$)$_4$]	D_{2h}	$A_g + B_{2g} + B_{2u} + B_{3u}$	2	807(s), 798(s)	137
cis-Pd(C$_6$F$_5$)$_2$py$_2$	C_{2v}	$A_1 + B_1$	2	790(s), 780(s)	91
cis-Pd(C$_6$F$_5$)$_2$(CO)$_2$	C_{2v}	$A_1 + B_1$	2	798(s), 786(s)	92
cis-Pt(C$_6$F$_5$)$_2$(tht)$_2$	C_{2v}	$A_1 + B_1$	2	802(s), 790(s)	137
trans-Pd(C$_6$F$_5$)$_2$py$_2$	D_{2h}	$B_{2u} + A_g$	1	768(s)	91
trans-Pt(C$_6$F$_5$)$_2$(tht)$_2$	D_{2h}	$B_{2u} + A_g$	1	778(s)	137
trans-Pt(C$_6$F$_5$)$_2$(CO)$_2$	D_{2h}	$B_{2u} + A_g$	1	790(s)	93
(NBu$_4$)[trans-PtCl$_2$(C$_6$F$_5$)$_2$]	D_{2h}	$B_{2u} + A_g$	1	764(s)	137
(NBu$_4$)[cis-PdCl(C$_6$F$_5$)$_2$(PPh$_3$)]	C_s	$2A'$	2	788(s), 769(s)	91
(NBu$_4$)[cis-PtCl(C$_6$F$_5$)$_2$(tht)]	C_s	$2A'$	2	802(s), 788(s)	137
(NBu$_4$)[cis-PtCl(C$_6$F$_5$)$_2$(CO)]	C_s	$2A'$	2	803(s), 798(s)	93
cis-Pt(C$_6$F$_5$)$_2$py(CO)	C_s	$2A'$	2	806(s), 799(s)	93
(NBu$_4$)[trans-PtCl(C$_6$F$_5$)$_2$(CO)]	C_{2v}	$A_1 + B_1$	2	797(vw), 779(s)	93
trans-Pt(C$_6$F$_5$)$_2$py(CO)	C_{2v}	$A_1 + B_1$	2	808(s)	93

^{a} S, strong; m, medium; vw, very weak; sh, shoulder.

TABLE IV

Absorptions Assigned to $\nu(C-F)$ Related to Formal Oxidation State of the
Metal for $M(II)$ and $M(IV)$ Pentafluorophenyl Complexes

M(II) complexes	$\nu(C-F)$ (cm^{-1})	M(IV) comples	$\nu(C-F)$ (cm^{-1})	Ref.
Pd(C$_6$F$_5$)$_2$bipy	953	PdCl$_2$(C$_6$F$_5$)$_2$bipy	970	138
Pt(C$_6$F$_5$)$_2$bipy	961, 959	PtCl$_2$(C$_6$F$_5$)$_2$bipy	971, 965	139
		PtBr$_2$(C$_6$F$_5$)$_2$bipy	971, 965	139
Pd(C$_6$F$_5$)$_2$phen	955	PdCl$_2$(C$_6$F$_5$)$_2$phen	970	138
Pt(C$_6$F$_5$)$_2$phen	959, 952	PtCl$_2$(C$_6$F$_5$)$_2$phen	970, 968	139
		PtBr$_2$(C$_6$F$_5$)$_2$phen	970, 968	139
Pd(C$_6$F$_5$)$_2$en	950	PdCl$_2$(C$_6$F$_5$)$_2$en	960	138
Pd(C$_6$F$_5$)$_2$pn	950	PdCl$_2$(C$_6$F$_5$)$_2$pn	966	138
PdCl(C$_6$F$_5$)bipy	950	PdCl$_3$(C$_6$F$_5$)bipy(C$_6$H$_6$)	970	138
PdCl(C$_6$F$_5$)phen	950	PdCl$_3$(C$_6$F$_5$)phen(C$_6$H$_6$)	968	138
cis-Pt(C$_6$F$_5$)$_2$(PBu$_3$)$_2$	955	PtCl$_2$(C$_6$F$_5$)$_2$(PBu$_3$)$_2$	970	139
		PtBr$_2$(C$_6$F$_5$)$_2$(PBu$_3$)$_2$	970	139
Pt(C$_6$F$_5$)$_2$(dpae)	953	PtCl$_2$(C$_6$F$_5$)$_2$(dpae)	965	139
		PtBr$_2$(C$_6$F$_5$)$_2$(dpae)	965	139

fluorophenyl complexes of palladium and platinum of different stoichio-metries and symmetries. Their normal vibration modes $\nu(M-C)$ (the IR-active modes are underlined), predicted absorptions arising from the symmetry, and the positions of the absorptions assigned to the X-sensitive modes of the C$_6$F$_5$ group are given.

As may be seen (Table III), whenever internal absorptions due to the presence of other ligands do not interfere in the 800 cm^{-1} region, analysis of the X-sensitive modes does permit structural assignments. There is only one exception. In complexes of the type M(C$_6$F$_5$)$_2$LL', in the cis(C_s, $2A'$) as well as the trans-isomers ($C_2\nu$, $A_1 + B_1$), the two normal vibration modes $\nu(M-C_F)$ are IR-active and the assignment is not an unequivocal one.

In practice, the cis isomers show two strong bands, whereas the trans isomers show only one strong band, or one strong and one weak absorption. Therefore, structural diagnosis is possible for these species.

Moreover, all the C$_6$F$_5$ derivatives of palladium and platinum synthe-sized so far show one absorption (sometimes broad or double) in the ~950 cm^{-1} region (135). The position of this absorption is related to the formal oxidation number of the central atom. In the Pd(IV) and Pt(IV) derivatives this band is shifted ~10–15 cm^{-1} toward higher energies relative to the M(II) precursors (Table IV) and ~10cm^{-1} toward lower energies in the M(I) complexes (Table V).

TABLE V

Absorptions Assigned to ν(C–F) Related to Formal Oxidation State of the Metal for M(II) and M(I) Pentafluorophenyl Complexes

M(II) complexes	ν(C–F) (cm^{-1})	M(I) complexes	ν(C–F) (cm^{-1})	Ref.
PdCl(C_6F_5)(dppm)$_2$	952	ClPd(μ-dppm)$_2$Pd(C_6F_5)	941	127
		ClPt(μ-dppm)$_2$Pd(C_6F_5)	945	130
PdBr(C_6F_5)(dppm)$_2$	951	BrPd(μ-dppm)$_2$Pd(C_6F_5)	940	127
Pd(C_6F_5)$_2$(dppm)$_2$	949	(C_6F_5)Pd(μ-dppm)$_2$Pd(C_6F_5)	943	127
		(C_6F_5)Pt(μ-dppm)$_2$Pd(C_6F_5)	945, 940	130
PtCl(C_6F_5)(dppm)$_2$	958	ClPt(μ-dppm)$_2$Pt(C_6F_5)	947	129
Pt(C_6F_5)$_2$(dppm)$_2$	959	(C_6F_5)Pt(μ-dppm)$_2$Pt(C_6F_5)	949	140
cis-Pt(C_6F_5)$_2$(CO)$_2$	961	[Pt(C_6F_5)(CO)(PPh$_3$)$_2$]	951	140
cis-Pt(C_6F_5)$_2$(CNTol)$_2$	956	[Pt(C_6F_5)(CNTol)(PPh$_3$)]$_2$	948	140
cis-Pt(C_6F_5)$_2$(CNCy)$_2$	958	[Pt(C_6F_5)(CNCy)(PPh$_3$)]$_2$	950	140
cis-Pt(C_6F_5)$_2$(CN^{t}Bu)$_2$	958	[Pt(C_6F_5)(CN^{t}Bu)(PPh$_3$)]$_2$	948	140

TABLE VI

Absorptions Assigned to X-Sensitive Mode or to ν(M–C) for the Pentachlorophenyl Complexes

Complexes	Point group	Symmetry of X-sensitive or ν(M–C) mode	Number of IR-active bands	(X-sensitive) (cm^{-1})	ν(M–C) (cm^{-1})[a]	Ref.
(NBu$_4$)$_2$[Pd(C$_6$Cl$_5$)$_4$]	D_{4h}	E_u	1	815(s)	595(s), (585(m))	19
(NBu$_4$)$_2$[Pt(C$_6$Cl$_5$)$_4$]	D_{4h}	E_u	1	820(m)	590(m)	19
(NBu$_4$)$_2$[trans-PtCl$_2$(C$_6$Cl$_5$)$_2$]	D_{2h}	$B_{2u} + A_g$	1	820(m)	590(m)	19
(NBu$_4$)$_2$[Pd(μ-Cl)(C$_6$Cl$_5$)$_2$]$_2$	D_{2h}	$A_g + B_{2g} + B_{2u} + B_{3u}$	2	835, 825(m)	620, 610(m)	19
(NBu$_4$)$_2$[Pt(μ-Br)(C$_6$Cl$_5$)$_2$]$_2$	D_{2h}	$A_g + B_{2g} + B_{2u} + B_{3u}$	2	845, 835(w)	633, 625(m)	19
(NBu$_4$)[cis-PdCl(C$_6$Cl$_5$)$_2$PPh$_3$]	C_s	$2A'$	2	833, 825(m)	613, 605(m)	19
(NBu$_4$)[cis-PtCl(C$_6$Cl$_5$)$_2$py]	C_s	$2A'$	2	845, 835(w)	630, 625(m)	19
cis-Pd(C$_6$Cl$_5$)$_2$(PEt$_3$)$_2$	C_{2v}	$A_1 + B_1$	2	830, 825(m)	605, 600(m)	19
cis-Pt(C$_6$Cl$_5$)$_2$py$_2$	C_{2v}	$A_1 + B_1$	2	847, 840(m)	630, 623(m)	19
trans-Pt(C$_6$Cl$_5$)$_2$(tht)$_2$	D_{2h}	$B_{2u} + A_g$	1	833(m)	605(m)	19

[a] s, strong; m, medium; w, weak.

B. *Pentachlorophenyl Derivatives*

The C_6Cl_5 group shows two diagnostic absorptions from a structural viewpoint (*135*): that above 800 cm^{-1} has been assigned to the X-sensitive mode; the second one ($\sim$600 cm^{-1}) has been assigned to the $\nu(M-C)$ mode (*136*). Table VI lists relevant data for C_6Cl_5 derivatives.

ACKNOWLEDGMENTS

We thank the Comisión Asesora de Investigación Científica y Técnica (CAICYT) for support. We acknowledge permission to reproduce Figs. 1, 2, 5–8, 11, 13, 14, and 16–31, granted respectively by *J. Chem. Soc. Dalton Trans.* (Figs. 2, 6, 7, 23, 24), *J. Chem. Soc. Chem. Commun.* (Figs. 1 and 28), *J. Am. Chem. Soc.* (Figs. 14, 16, 17 and 22), *Inorg. Chem.* (Figs. 16, 17, 18, 19, 20, and 21), *Organometallics (Figs. 1, 28, and 29), Pure Appl. Chem.* (Fig. 11), *Polyhedron* (Fig. 19), and *J. Organomet. Chem.* (Figs. 5, 8, 13, 25, 26, 27, 30, and 31).

REFERENCES

1. E. Nield, R. Stephens, and J. C. Tatlow, *J. Chem. Soc.* 166 (1959).
2. H. E. Ramsden, A. E. Balint, W. R. Whitford, J. J. Walburn, and R. Cserr, *J. Org. Chem.* **22,** 1202 (1957).
3. P. L. Coe R. Stephens, and J. C. Tatlow, *J. Chem. Soc. A* 3227 (1962).
4. M. D. Rausch, F. E. Tibbetts, and H. B. Gordon, *J. Organomet. Chem.* **5,** 493 (1966).
5. D. T. Rosevear and F. G. A. Stone. *J. Chem. Soc.* 5275 (1965).
6. F. J. Hopton, A. J. Rest, D T. Rosevear, and F. G. A. Stone, *J. Chem. Soc. A* 1326 (1966).
7. A. J. Rest, D. T. Rosevear, and F. G. A. Stone, *J. Chem. Soc. A* 66 (1967).
8. M. D. Rausch and F. E. Tibbetts, *J. Organomet. Chem.* **21,** 487 (1970).
9. P. M. Treichel and F. G. A. Stone, *Adv. Organomet. Chem.* **1,** 143 (1964).
10. F. G. A. Stone, *Endeavour* **25,** 33 (1966).
11. R. S. Nyholm, *Q. Rev. Chem. Soc.* **24,** 1 (1970).
12. R. Usón, J. Forniés, S. Gonzalo, F. Martinez, and R. Navarro, *Rev. Acad. Ci. Zaragoza* **32,** 75 (1977).
13. J. M. Coronas, C. Polo, and J. Sales, *Synth. React. Inorg. Met. Org. Chem.* **10,** 53 (1980).
14. J. M. Coronas, C. Peruyero, and J. Sales, *J. Organomet. Chem.* **128,** 291 (1977).
15. J. M. Coronas and J. Sales, *J. Organomet. Chem.* **94,** 107 (1975).
16. J. M. Coronas, G. Müller, and J. Sales, *Synth. React. Inorg. Met. Org. Chem.* **6,** 217 (1976).
17. R. Usón, J. Forniés, R. Navarro, M. P. Garcia, and B. Bergareche, *Inorg. Chim. Acta* **25,** 269 (1977).
18. R. Ceder, J. Granell, G. Müller, O Rosell, and J. Sales, *J. Organomet. Chem.* **174,** 115 (1979).
19. R. Usón, J. Forniés, F. Martinez, M. Tomás, and I. Reoyo, *Organometallics* **2,** 1386 (1983).
20. R. Usón, J. Forniés, and F. Martinez, *J. Organomet. Chem.* **132,** 429 (1977).
21. G. W. Rice and R. S. Tobias, *J. Chem. Soc. Chem. Commun.* 994 (1975); *J. Am. Chem. Soc.* **99,** 2141 (1977).

22. R. Usón, J. Forniés, F. Martinez, and M. Tomás, *J. Chem. Soc. Dalton Trans.* 888 (1980).
23. R. Usón, J. Forniés, M. Tomás, and R. Fandos, *J. Organomet. Chem.* **263**, 253 (1984).
24. R. Usón, J. Forniés, P. Espinet, and A. Arribas, *J. Organomet. Chem.* **199**, 111 (1980).
25. G. Garcia and G. López, *Inorg. Chim. Acta* **52**, 87 (1981).
26. G. Garcia and G. López, *An. Univ. Murcia* **41**, 91 (1983).
27. G. López, G. Garcia, C. de Haro, G. Sanchez, and M. C. Vallejo, *J. Organomet. Chem.* **263**, 247 (1984).
28. G. López, G. Garcia, N. Cutillas, and J. Ruiz, *J. Organomet. Chem.* **241**, 269 (1984).
29. G. Garcia, G. López, and M. D. Santana, *An. Quim.* **79B**, 214 (1983).
30. G. Garcia, G. López, J. Ruiz, and G. Sanchez, *An. Univ. Murcia* **43**, 13 (1984).
31. G. López, G. Garcia, J. Gálvez, and N. Cutillas, *J. Organomet. Chem.* **258**, 123 (1983).
32. G. López, G. Garcia, J. Ruiz, and N. Cutillas, *J. Organomet. Chem.* **246**, C83 (1983).
33. C. de Haro, G. Garcia, G. Sánchez, and G. López, *J. Chem. Res. (S)* 119 (1986).
34. R. Usón, J. Forniés, M. Tomás, B. Menjón, K. Süinkel, and R. Bau, *J. Chem. Soc. Chem. Commun.* 751 (1984).
35. R. Usón, J. Forniés, M. Tomás, B. Menjón, R. Bau, K. Sünkel, and E. Kuwabara, *Organometallics* **5**, 1576 (1986).
36. G. B. Deacon, J. H. S. Green, and R. S. Nyholm, *J. Chem. Soc.* 3411 (1965).
37. R. S. Nyholm and P. Royo, *J. Chem. Soc. Chem. Commun.* 421 (1969).
38. P. Royo, *Rev. Acad. Cien. Zaragoza* **27**, 235 (1972).
39. R. Usón, P. Royo, J. Forniés, and F. Martinez, *J. Organomet. Chem.* **90**, 367 (1975).
40. R. Usón, P. Royo, and J. Gimeno, *Rev. Acad. Cien. Zaragora* **28**, 355 (1973).
41. P. Royo and R. Serrano, *J. Organomet. Chem.* **136**, 309 (1977).
42. P. Royo and R. Serrano, *J. Organomet. Chem.* **144**, 33 (1978).
43. V. I. Sokolov, V. V. Bashilov, L. M. Anischenko, and O. A. Reutov, *J. Organomet. Chem.* **71**, C41 (1974).
44. V. I. Sokolov, L. L. Troitskaya, and O. A. Reutov, *J. Organomet. Chem.* **93**, C11 (1975).
45. V. I. Sokolov, V. V. Bashilov, and O. A. Reutov, *J. Organomet. Chem.* **97**, 299 (1975).
46. O. Rossell, J. Sales, and M. Seco, *J. Organomet. Chem.* **205**, 133 (1981).
47. J. Forniés, M. Green, J. L. Spencer, and F. G. A. Stone, *J. Chem. Soc. Dalton Trans.* 1006 (1977).
48. A. J. Mukhedkar, M. Green, and F. G. A. Stone, *J. Chem. Soc. (A)* 3023 (1969).
49. K. J. Klabunde, B. B. Anderson, and K. Neuenschwander, *Inorg. Chem.* **19**, 3719 (1980).
50. K. J. Klabunde and J. Y. Low, *J. Am. Chem. Soc.* **96**, 7674 (1974).
51. S. T. Lin and K. J. Klabunde, *Inorg. Chem.* **24**, 1961 (1985).
52. J. J. Habeb and D. G. Tuck, *J. Organomet. Chem.* **139**, C17 (1977).
53. R. D. Rieke, W. J. Wolf, N. Kujundzic, and A. V. Kavaliunas, *J. Am. Chem. Soc.* **99**, 4159 (1977).
54. R. D. Rieke and A. V. Kavaliunas, *J. Org. Chem.* **44**, 3069 (1979).
55. N. Bresciani-Pahor, M. Plazzotta, L. Randaccio, G. Bruno, V. Ricevuto, R. Romeo, and U. Belluco, *Inorg. Chim. Acta* **31**, 171 (1978).
56. R. Usón, P. Royo, and J. Forniés, *Rev. Acad. Cien. Zaragoza* **28**, 349 (1973).
57. R. Usón, J. Forniés, and S. Gonzalo, *J. Organomet. Chem.* **104**, 253 (1976).
58. R. Usón, J. Forniés, P. Espinet, and M. P. Garcia, *Rev. Acad. Cien. Zaragoza* **32**, 85 (1977).
59. T. Yoshida, T. Okano, and S. Otsuka, *J. Chem. Soc. Dalton Trans.* 993 (1976).
60. R. Usón, J. Forniés, R. Navarro, and M. P. Garcia, *Inorg. Chim. Acta* **33**, 69 (1979).

61. R. Usón, J. Forniés, P. Espinet, and G. Alfranca, *Synth. React. Inorg. Met. Org. Chem.* **10**, 579 (1980).
62. R. Usón, J. Forniés, R. Navarro, and A. Gallo, *Trans. Met. Chem.* **5**, 284 (1980).
63. R. Usón, J. Gimeno, J. Forniés, and F. Martinez, *Inorg. Chim. Acta* **50**, 173 (1981).
64. S. W. Kaiser, R. B. Saillant, W. N. Buttler, and P. G. Rasmussen, *Inorg. Chem.* **17**, 2688 (1976).
65. R. Usón, J. Gimeno, J. Forniés, F. Martinez, and C. Fernández, *Inorg. Chim. Acta* **63**, 91 (1982).
66. R. Usón, J. Forniés, and P. Espinet, *Synth. React. Inorg. Metal. Org. Chem.* **13**, 513 (1983).
67. R. Usón, P. Royo, and J. Forniés, *Synth. React. Inorg. Met. Org. Chem.* **4**, 157 (1974).
68. R. Usón, P. Royo, and J. Gimeno, *J. Organomet. Chem.* **72**, 299 (1974).
69. R. Usón, J. Forniés, and F. Martinez, *J. Organomet. Chem.* **112**, 105 (1976).
70. P. S. Braterman, "Metal Carbonyl Spectra," p. 177. Academic Press, New York, 1975.
71. J. Browning, P. L. Goggin, R. J. Goodfellow, M. G. Morton, A. J. M. Rattray, B. F. Taylor, and J. Mink, *J. Chem. Soc. Dalton Trans.* 2061 (1977).
72. T. Majima and H. Kurosawa, *J. Organomet. Chem.* **134**, C45 (1977).
73. F. Calderazzo and D. B. Dell'Amico, *Inorg. Chem.* **20**, 1310 (1981).
74. J. Powell and B. L. Shaw, *J. Chem. Soc. (A)* 1839 (1967).
75. H. Onoue, K. Nakagawa, and I. Moritani, *J. Organomet. Chem.* **35**, 217 (1972).
76. S. A. Dias, A. W. Downs, and W. R. McWhinnie, *J. Chem. Soc. Dalton Trans.* 162 (1975).
77. R. Usón, J. Forniés, R. Navarro, P. Espinet, and C. Mendivil, *J. Organomet. Chem.* **290**, 125 (1985).
78. E. T. Weleski, Jr.,J. L. Silver, M. D. Jansson, and J. L. Burmeister, *J. Organomet. Chem.* **102**, 365 (1972).
79. P. Bravo, G. Fronza, and C. Ticcozi, *J. Organomet. Chem.* **111**, 361 (1976).
80. P. A. Arnup and M. C. Baird, *Inorg. Nucl. Chem. Lett.* **5**, 65 (1965).
81. H. Koezuka, G. Matsubayashi, and T. Tanaka, *Inorg. Chem.* **15**, 417 (1976).
82. Y. Oosawa, H. Urabe, T. Saito, and Y. Sasaki, *J. Organomet. Chem.* **122**, 113 (1976).
83. M. Onishi, Y. Ohama, K. Hiraki, and H. Shintam, *Polyhedron* 1, 539 (1982).
84. R. Usón, J. Forniés, R. Navarro, M. A. Usón, M. P. Garcia, and A. J. Welch, *J. Chem. Soc. Dalton Trans.* 345 (1984).
85. R. Usón, J. Forniés, P. Espinet, A. Garcia, and A. Sanaú, *Trans. Met. Chem.* **8**, 11 (1983).
86. R. Usón, J. Forniés, P. Espinet and E. Lalinde, *J. Organomet. Chem.* **220**, 393 (1981).
87. R. Usón, J. Forniés, P. Espinet, R. Navarro, and M. A. Usón, *Inorg. Chim. Acta* **33**, L103 (1979).
88. R. Usón, J. Forniés, P. Espinet, and R. Navarro, *Inorg. Chim. Acta* **82**, 215 (1984).
89. R. Usón, J. Forniés, M. A. Usón, and E. Lalinde, *J. Organomet. Chem.* **185**, 359 (1980).
90. R. Usón, J. Forniés, M. A. Usón, and M. A. Orta, *Inorg. Chim. Acta* **89**, 175 (1984).
91. R. Usón, J. Forniés, P. Espinet, F. Martinez, and M. Tomás, *J. Chem. Soc. Dalton Trans.* 463 (1981).
92a. R. Usón, J. Forniés, M. Tomas, and B. Menjón, *Organometallics* 4, 1912 (1985).
92b. R. Usón, J. Forniés, M. Tomás, and B. Menjón, *Organometallics* 5, 1581 (1986).
93a. L. Malatesta and L. Naldini, *Gazz. Chim. Ital.* **90**, 1505 (1960).
93b. D. B. Dell'Amico and F. Calderazzo, *Gazz. Chim. Ital.* **109**, 99 (1979).
93c. D. B. Dell'Amico, F. Calderazzo, C. A. Veracini, and N. Zandona, *Inorg. Chem.* **23**, 3030 (1984).

94. R. Usón, J. Forniés, M. Tomás, B. Menjón, and A. J. Welch, *J. Organomet. Chem.* **304**, C24 (1986).
95. R. Usón, J. Forniés, B. Menjón, F. A. Cotton, L. R. Falvello, and M. Tomás, *Inorg. Chem.* **24**, 4651 (1985).
96. R. Usón, J. Forniés, M. A. Usón, J. F. Yagüe, P. G. Jones, and K. Meyer-Bäse, *J. Chem. Soc. Dalton Trans.* 947 (1986).
97. J. Forniés, M. A. Usón, J. I. Gil, and P. G. Jones, *J. Organomet. Chem.* **311**, 243 (1986).
98. G. Thiele, G. Liehr, and E. Lindner, *J. Organomet. Chem.* **70**, 427 (1970).
99. R. Usón, *Pure Appl. Chem.* **58**, 647 (1986).
100. R. Usón, J. Forniés, M. Tomás, B. Menjón, and A. J. Welch, unpublished results.
101. R. Usón, J. Forniés, P. Espinet, M. Tomás, C. Fortuño, and A. J. Welch, unpublished results.
102. R. Usón, J. Forniés, A. Laguna, and J. I. Valenzuela, *Synth. React. Inorg. Met. Org. Chem.* **12**, 935 (1982).
103. R. Usón, A. Laguna, J. Forniés, I. Valenzuela, P. G. Jones, and G. M. Sheldrick, *J. Organomet. Chem.* **273**, 129 (1984).
104. R. Usón, J. Forniés, M. Tomás, J. M. Casas, F. A. Cotton, and L. R. Falvello, *J. Am. Chem. Soc.* **107**, 2556 (1985).
105. F. A. Cotton, L. R. Falvello, R. Usón, J. Forniés, M. Tomás, J. M. Casas, and I. Ara, *Inorg. Chem.* **26**, 1366 (1987).
106. R. Usón, J. Forniés, M. Tomás, J. M. Casas, F. A. Cotton, and L. R. Falvello, *Inorg. Chem.*, in press.
107. R. Usón, J. Forniés, M. Tomás, F. A. Cotton, and L. R. Falvello, *J. Am. Chem. Soc.* **106**, 2482 (1984).
108. R. Usón, J. Forniés, M. Tomás, J. M. Casas, F. A. Cotton, and L. R. Falvello, *Polyhedron* **5**, 901 (1986).
109. R. Usón, J. Forniés, M. Tomás, J. M. Casas, F. A. Cotton, and L. R. Falvello, *Inorg. Chem.* **25**, 4519 (1986).
110. P. M. Treichel, Adv. Organomet. Chem. **11**, 21 (1973).
111. P. M. Treichel and R. W. Hess, *J. Am. Chem. Soc.* **92**, 4731 (1970).
112. H. D. Empsall, M. Green, S. K. Shaksooki, and F. G. A. Stone, *J. Chem. Soc. (A)* 3472 (1971).
113. R. Usón, J. Forniés, P. Espinet, E. Lalinde, P. G. Jones, and G. M. Sheldrick, *J. Chem. Soc. Dalton Trans.* 2389 (1982).
114. R. Usón, J. Forniés, P. Espinet, and E. Lalinde, *J. Organomet. Chem.* **254**, 371 (1983).
115. R. Usón, J. Forniés, P. Espinet, E. Lalinde, A. Garcia, P. G. Jones, K. Meyer-Bäse, and G. M. Sheldrick, *J. Chem. Soc. Dalton Trans* 259 (1986).
116. R. Usón, J. Forniés, P. Espinet, E. Lalinde, P. G. Jones, and G. M. Sheldrick, *J. Organomet. Chem.* **253**, C47 (1983).
117. R. Usón, J. Forniés, P. Espinet. A. Garcia, C. Foces-Foces, and F. H. Cano, *J. Organomet. Chem.* **282**, C35 (1985).
118. R. Usón, J. Fornies, P. Espinet, E. Lalinde, P. G. Jones, and G. M. Sheldrick, *J. Organomet. Chem.* **288**, 249 (1985).
119. R. Usón, J. Forniés, P. Espinet, L. Pueyo, and E. Lalinde, *J. Organomet. Chem.* **229**, 251 (1986).
120. A. T. Casey and S. Mifra, "Theory and Applications of Molecular Paramagnetism" (E. A. Boudreaux and L. N. Mulay, Eds.), p. 220. Wiley, New York, 1976.
121. R. Usón, J. Forniés, and R. Navarro, *J. Organomet. Chem.* **96**, 307 (1975).
122. R. Usón, J. Forniés, and R. Navarro, *Synth. React. Inorg. Met. Org. Chem.* **7**, 235 (1977).

123. R. Usón and J. Forniés, unpublished results.
124a. P. M. Maitlis, P. Espinet, and M. J. H. Russell, "Comprehensive Organometallic Chemistry" (G. Wilkinson, F. G. A. Stone, and E. W. Abel, Eds.), Vol. 6, p. 340. Pergamon, New York, 1982.
124b. P. K. Byers, A. J. Canty, B. W. Skelton, and A. H. White, *J. Chem. Soc. Chem. Commun.* 1722 (1986).
125. R. Usón, J. Forniés, and P. Espinet, *J. Organomet. Chem.* **116**, 353 (1976).
126. C. Crocker, R. J. Goodfellow, J. Gimeno, and R. Usón, *J. Chem. Soc. Dalton Trans.* 1448 (1977).
127. R. Usón, J. Forniés, P. Espinet, F. Martinez, C. Fortuño, and B. Menjón, *J. Organomet. Chem.* **256**, 365 (1983).
128. P. Espinet, J. Forniés, C. Fortuño, G. Hidalgo, F. Martinez, M. Tomás, and A. J. Welch, *J. Organomet. Chem.* **317**, 105 (1986).
129. R. Usón, J. Forniés, P. Espinet, and C. Fortuño, *J. Chem. Soc. Dalton Trans.* 1849 (1986).
130. J. Forniés, F. Martinez, R. Navarro, A. Redondo, M. Tomás, and A. J. Welch, *J. Organomet. Chem.* **316**, 351 (1986).
131. R. Usón, J. Forniés, R. Navarro, and J. I. Cebollada, *J. Organomet. Chem.* **304**, 381 (1986).
132. S. Otsuka, A. Nakamura, and T. Yoshida, *J. Am. Chem. Soc.* **91**, 7196 (1969).
133. R. Usón, J. Forniés, P. Espinet, C. Fortuño, M. Tomás, and A. J. Welch, unpublished results.
134. R. J. Goodfellow and I. R. Herbert, *Inorg. Chim. Acta* **65**, L161 (1982).
135. E. Maslowsky, Jr., "Vibrational Spectra of Organometallic Compounds," p. 437. Wiley, New York, 1977.
136. J. Casabó, J. M. Coronas, and J. Sales, *Inorg. Chim. Acta* **11**, 5 (1974).
137. M. Tomás, Ph.D. Thesis, University of Zaragoza, 1979.
138. R. Navarro, Ph.D. Thesis, University of Zaragoza, 1975.
139. P. Espinet, Ph.D. Thesis, University of Zaragoza, 1975.
140. C. Fortuño, Ph.D. Thesis, University of Zaragoza, 1986.

ADVANCES IN ORGANOMETALLIC CHEMISTRY, VOL. 28

H–H, C–H, and Related Sigma-Bonded Groups as Ligands

ROBERT H. CRABTREE
and
DOUGLAS G. HAMILTON

Department of Chemistry
Yale University
New Haven, Connecticut 06511

I
INTRODUCTION

Ligands that act by donation of lone pairs (e.g., NH_3) or of π-bonding electrons (e.g., C_2H_4) are common in transition metal chemistry. Only recently has it become clear that σ-bonding electron pairs can also ligate to metals. The resulting structures are isolobal (*1*) with H_3^+ (**1**), a triangular species well known in the gas phase (*2*). From the point of view of the coordination chemist, H_3^+ is a dihydrogen complex of the proton (**2**).

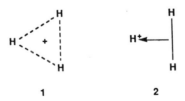

1   **2**

Replacing H^+ with the the isolobal fragments CH_3^+ and L_nM gives the structure of the gas-phase CH_5^+ cation **3** and of the transition metal dihydrogen complex **4**, a species only generally recognized since 1984.

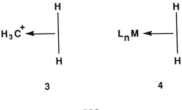

3   **4**

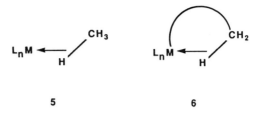

5	**6**

Replacing an H in **4** with CH_3 gives a methane complex **5**, no isolable examples of which are yet known. Complexes analogous to **5**, in which a C–H bond is chelated to the metal (e.g., **6**) are known in considerable number and termed "agostic" species (4). Also related to **1–6** is **7**, in which a main group or transition metal hydride binds to a metal to give a bridging hydride (5).

$$L_nM \nearrow \overset{H}{} \searrow ML_n$$

7

H^+ and CH_3^+ are powerful Lewis acids, and so the bonding (6) in **1–3** is best described in terms of donation of the X–Y σ-bonding electrons to the Lewis acid (X–Y being the ligand). This weakens but does not break the X–Y bond, because the resulting three center molecular orbital (MO) is bonding over all three centers (Fig. 1c). In considering **4–7**, we also have to take into account the fact that a transition metal often has $d\pi$ electrons available for backbonding. This interaction can be described in terms of

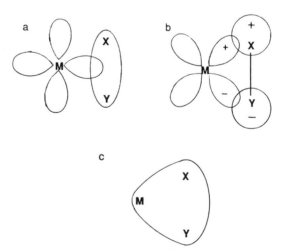

FIG. 1.   Bonding of an X–Y σ-bonded fragment to a low-valent transition metal.

electron donation from the $M(d_\pi)$ orbital to the X–Y σ^* orbital. This back donation component, if strong enough, *can* break the X–Y bond because the X–Y σ^* orbital is being filled. When X–Y bond breaking takes place, e.g., Eq. (1), we regard **10** as the product of an oxidative addition, with conventional 2*e*, two-center bonds.

$$L_nM \longleftarrow \overset{\displaystyle X}{\underset{\displaystyle Y}{\big|}} \quad \longrightarrow \quad L_nM{\big<}{\overset{\displaystyle X}{\underset{\displaystyle Y}{}}} \tag{1}$$

10

Two-electron, three-center bonds seem to form most readily when at least one of the three centers is H. This is probably a result of the small size of the hydrogen atom, which allows close approach of the X–Y (Y = H) group to the metal, and the fact that the hydrogen atom has no lone pairs, which would otherwise bind instead.

On the other hand, the stablest complexes of such weak donors as the H–H and C–H bonds are formed with transition metals. This suggests that the back donation component of the binding (only efficient with non-d^0 transiton metals) stabilizes the L_nM–(XY) binding to a significant degree. Species such as CH_5^+ are very strong proton acids, as is consistent with a substantial $H_2(\sigma)$-to-CH_3^+ electron transfer depleting the H_2 molecule of electron density. Transition metals increase the proton acidity of a C–H or H–H bond to a much lesser degree. Not only is the transition metal a better π base than CH_3^+, but also it is no doubt a weaker Lewis acid.

As we shall see, M–(XY) species are plausible intermediates not only in the oxidative addition of X–Y but also in other X–Y bond-breaking pathways, such as heterolytic activation. They may also be relevant to binding of molecules such as H_2 and CH_4 to metalloenzymes. These structures are therefore not just interesting in themselves; they cast a new light on a number of chemical problems of wider interest.

II

DIHYDROGEN COMPLEXES

Dihydrogen complexes are the simplest transition metal species isolobal with H_3^+, but they have been recognized only in the past three years. The pioneering work of Kubas *et al.* (*7*) demonstrated the existence of an H–H bond in $[M(H_2)(CO)_3(PR_3)_2]$ (M = Mo, W; R = *i*-Pr or Cy) by a number of methods including neutron diffraction. A similar arrangement was subsequently found in species such as $Cr(CO)_5(H_2)$ (8), which are stable in a matrix at low temperature or in liquid xenon. More recently, H_2 ligands have been recognized in such well-known hydrides as $ReH_7(PPh_3)_2$ and $FeH_4(PEt_2Ph)_3$ (9). Although known for some 20 years, these species had always been considered to have a classical structure with terminal M–H bonds only. In one 1976 report, Ashworth and Singleton (10) speculated that $RuH_4(PPh_3)_3$ might "be effectively considered as a ruthenium(II) complex containing neutral dihydrogen." They did not offer any spectroscopic or structural evidence, however, and the nonclassical structure has only been proved very recently (9). Kubas (11) indicated the possibility that the bonding of H_2 in $[WH_2(CO)_3(PCy_3)_2]$ "may be novel" in his original 1980 report on these complexes (11).

It might be thought that crystallographic evidence would offer the most direct proof of structure for a dihydrogen complex. X-ray methods suffer from the drawback that hydrogen atoms are the least easily located atoms by this technique, especially in the vicinity of a heavy-metal atom. Morris et al. (12) have used X-ray methods, but in conjunction with conclusive spectroscopic evidence to characterize $[Fe(H_2)H(dppe)_2]^+$ (dppe = $Ph_2PCH_2CH_2PPh_2$). Neutron diffraction is more reliable, but growing the larger crystals required for this technique is often difficult , and disorder problems have been encountered for polyhydrides. William's group used neutron diffraction to help characterize $[W(H_2)(CO)_3(PCy_3)_2]$. There is evidence that the H_2 ligand may be able to rotate easily about the $M–(H_2)$ axis. This is reasonable in the bonding model we have discussed, because only the $M(d_\pi)$-to-$H_2(\sigma^*)$ backbonding can offer any resistance to such a rotation, and we have seen that we expect this backbonding to be limited in extent. In the presence of rotational disorder, the observed H–H distance as determined by neutron diffraction is expected to be shorter than the real internuclear distance because the true banana-shaped atomic positional distribution may not be properly modeled by the simple ellipsoid normally assumed in the refinement.

The structural work shows that the H–H vector is colinear with the P–W–P rather than the CO–W–CO axis. The likely reason is that the d orbital in the P–W–P axis (we will call it the d_{xz}) is less stabilized than

$$W(CO)_3L_2 \xrightarrow{\ H_2\ } \quad (2)$$

the one in the CO–W–CO axis (d_{yz}), because the phosphines back bond less well. This means that the d_{xz} orbital will give the more effective bonding to the $H_2(\sigma^*)$ and so fix its orientation.

Infrared spectroscopy is a valuable method for matrix work with $M(CO)_n$ fragments because the symmetry of the fragment can be deduced and in addition the $\nu(H_2)$ vibration at 2300–2700 cm^{-1} can often be observed. Although it might be thought that H–H stretching should not be IR-active, in fact there seems to be enough mixing with allowed $\nu(CO)$ vibrations to give it adequate intensity. Isotopic studies confirm the $\nu(H_2)$ character of the band (see Table I). The IR bands are often broad, typically ca. 150 cm^{-1} at half-height, which may be related to the ease of rotation of the H_2 molecule about the $M–(H_2)$ axis.

Oxidative addition of the H_2 to the $Cr(CO)_5$ fragment to give a classical dihydride $H_2Cr(CO)_5$ can be ruled out becaust the A_1 mode of the $Cr(CO)_5$ fragment shifts only 3 cm^{-1} on binding of H_2. Where oxidative addition does occur, e.g., in $H_2Fe(CO)_4$ (13), a shift of 42 cm^{-1} to higher energy is found (8).

In noncarbonyl complexes, IR spectroscopy seems to be a very much less effective method, because the $\nu(H_2)$ bands are often too weak to be observed. Counterexamples exist, however; as early as 1971, Aresta et al. (14) noted an IR band at 2380–2400 cm^{-1} in the spectrum of [FeH$_4$(PEtPh$_2$)$_3$] which they were not able to assign. Only with the realization (9) that this complex has the nonclassical structure [Fe(H$_2$)H$_2$(PEtPh$_2$)$_3$] did the true identification of this band with the $\nu(H_2)$ stretch become clear. In contrast, the isostructural Ru analogs show no $\nu(H_2)$ band.

NMR spectroscopy has been one of the most useful methods for the identification of dihydrogen complexes. A number of different cases can be distinguished. In the simplest type, the only hydrides present are those of the H_2 ligand, e.g., [CpRu(H$_2$)(PPh$_3$)(CNt-Bu)]$^+$ (15a). In such a case,

TABLE I

IR DATA FOR KUBAS' COMPLEX

Assignment	W(CO)$_3$(PCy$_3$)$_2$			W(CO)$_3$\{P(i-Pr)$_3$\}$_2$		
	H$_2$	HD	D$_2$	H$_2$	HD	D$_2$
$\nu(H_2)$, cm^{-1}	2690	2360	1900 (Raman)	2695	2360	—
$\nu(MH_2)$, cm^{-1}	1574	1350	1140	1567	1350	1140
$\nu(MH_2)$, cm^{-1}	953	791	703	953	793	704
$\delta(MH_2)$, cm^{-1}	—	—	442	—	—	444
δMH_2, cm^{-1}	462	—	319	465	—	312

TABLE II

Physical Data on Some Dihydrogen Complexes

Complex	$\delta(H_2)$ (ppm)	$\omega(1/2)$ (Hz)	T_1^a (ms)	r (Å)	$^1J(H, D)$ (Hz)	$\nu(H_2)$ cm^{-1}	Remarks
W(CO)$_3${P(i-Pr)$_3$}$_2$(H$_2$)	-4.2	24[b] 8(HD)	4	0.84[c]	33.5	2695	
[W(CO)$_3$(PCy$_3$)$_2$(H$_2$)] Cr(CO)$_5$H$_2$	-3.2[d] -3.85	— 200	1700[d,e]			2690 3030 2241(D$_2$) —	Stable in liq. Xe
[IrH(H$_2$)bq(PPh$_3$)$_2$]BF$_4$	-2.9[f] -15.2[g]	160[f]	30[f] 350[g]	0.80[i]	29.5		
[IrH$_2$(H$_2$)$_2$(PCy$_3$)$_2$](PhC(SO$_2$CF$_3$)$_2$)	-5.05[f] -15.2[g]	175	48[f] 73[g,h]				
Mo(CO)(dppe)$_2$(H$_2$) [CpRu(PPh$_3$)(CNt-Bu)(H$_2$)]$^+$	-5 -8	300 15 3(HD)			34 28.6	— —	
[FeH(H$_2$)(dppe)$_2$]$^+$	-8[c] -12.9[g]	— —	7 146	0.89[i]		—	
[RuH(H$_2$)(dppe)$_2$]$^+$	-4.6[f] -10[g]		11 280		32		

Complex						
Fe(CO)(NO)₂(H₂)					2973	Matrix
Co(CO)₂(NO)(H₂)					2195(D₂); 3100, 2976	Matrix
Os(H₂)H₃(PPH₃)₃⁺	5.7	10(300k)	35(220k)	0.87[j]	2228(D₂)	J(P,H) not obs.
Mo(CO)₅(H₂)					3080	Matrix
W(CO)₅(H₂)					2711	Matrix
Fe(H₂)H₂(PEtPh₂)₃	−11.5	2	24		2380	
Ru(H₂)H₂(PPh₃)₂	−7	20	38	0.87[j]		
Re(H₂)H₅(PPh₃)₂	−4.2	2	78	0.90[j]		
[Os(H₂)H(depe)₂]BF₄	−10	25	52			
	−9.7		260			

[a] At −80°C in CD₂Cl₂ unless stated.
[b] At 35°C, rises to 40 Hz at −50°C.
[c] By neutron diffraction.
[d] Classical tautomer.
[e] In toluene.
[f] For M–(H₂).
[g] For M–H.
[h] Short because of exchange with M(H₂).
[i] By X-ray.
[j] By T₁ (see text).

the proton resonance for the H_2 appears to high field of TMS as a broad feature [$\omega(1/2)$ = 3–300 Hz; see Table II]. On substituting HD for H_2, ^1J(H, D) coupling of 22–32 Hz can usually be seen in the spectrum. Free HD exhibits a ^1J(H, D) coupling of 43 Hz, so the H–D bond order appears to be somewhat reduced in the complex. Chinn and Heinkey's (*15b*) [CpRu(dmpe)(HD)]$^+$ shows a ^1J(H, D) as low as 22 Hz.

There is as yet no general agreement on the maximum value of ^2J(H, H′) to be expected for a classical hydride. Values up to 10 Hz are common, but exceptional cases are known in which ^1J(H, H′) is larger than 10 Hz but not as large as 180 Hz (the ^1J(H, H′) value that corresponds to a ^1J(H, D) of 30 Hz). For example, Gilbert and Bergman (*16*) find a value of ^1J(H, H′) of 56 Hz in [Cp*Ir(PMe$_3$)H$_3$]BF$_4$. This would correspond to a ^1J(H, H′) of ca. 9 Hz , and so a nonclassical formulation appears to be unjustified for the present, but further work is required. Until we understand the significance of these intermediate values of ^1J(H, H′), it is unsafe to use them as a criterion of structural type.

A second possible case is represented by [W(H_2)(CO)$_3$(P*i*-Pr$_3$)$_2$] (*8*), for which there is an equilibrium between the classical and nonclassical tautomers in solution [Eq. (3)]. Separate resonances for each tautomer are only observable by NMR at low temperature; a similar equilibrium for [CpRu(dmpe)(H_2)]$^+$ is slow at room temperature. It is interesting that both tautomers of these complexes should have similar energies, rather than a single ground state structure with an intermediate H–H distance. This may mean that there will prove be two distinct classes of hydride, rather than a smooth gradation of structures.

$$L_2(CO)_3W \xleftarrow{} \begin{matrix} H \\ | \\ H \end{matrix} \rightleftarrows L_2(CO)_3W \begin{matrix} H \\ \diagdown \\ H \end{matrix} \qquad (3)$$

85%   15%

A slightly more complex situation is represented by [FeH(H_2)-(dppe)$_2$]BF$_4$ (*12*) or [IrH(H_2)bq(PPh$_3$)$_2$]BF$_4$ (bq = 7,8-benzoquinolinate) (*17a*). At sufficiently low temperatures (e.g., −60°C) two resonances are observed corresponding to the two types of proton, but on warming, protons exchange between the two sites and the two peaks coalesce. At the low-temperature limit the ^1J(H, D) may be detected in the HD analog, but on warming the coupling can no longer be observed as the exchange becomes faster. In some cases, ^1J(H, D) cannot be resolved even at the lowest accessible temperatures, e.g., [IrH$_2$(H_2)$_2$(PCy$_3$)$_2$]$^+$ (*17b*).

Finally, the complex may be so fluxional that a single proton NMR resonance is observed for all the hydrides in the molecule at all accessible temperatures, e.g., Re(H_2)H$_5$(PPh$_3$)$_2$. The ^1J(H, D) coupling has not been seen in such cases and so provides no useful structural criterion (*18*). Where the proton the NMR resonances are broad (see Table I), it is often difficult to distinguish coupling to other nuclei, such as the metal or the phosphine ligand. Morris *et al.* (*19*) have carefully simulated the temperature dependence of the proton NMR spectra of $[MH(H_2)(depe)_2]BF_4$ (M = Os, Ru, and Fe; depe = $Et_2PCH_2CH_2PEt_2$), which enable them to assign the following $^2J(H, P)$ values for the dihydrogen resonances: Fe, ~5 Hz; Ru, ~1 Hz.; Os, 5.8 Hz.

A T_1 criterion for the presence of dihydrogen ligands in a given complex was developed in response to the problem of characterizing $[IrH_2(H_2)_2(PCy_3)_2]^+$ (*20*). This complex could not be crystallized, gives no useful IR bands, and is too fluxional to allow $^1J(H, D)$ to be resolved.

The T_1 of any resonance can be measured by a standard inversion–recovery pulse sequence ($180°$-t-$90°$). This reveals how rapidly the inverted spins recover their equilibrium magnetization in the direction of the magnetic field. The dipole–dipole mechanism is the major contributor to this relaxation. By this mechanism a given dipolar nucleus, say a proton, is relaxed by other nearby dipolar nuclei. As the molecule rotates under the influence of collisions with the solvent molecules (Brownian motion), the magnetic field felt by the first proton fluctuates as the second dipolar nucleus rotates around the first. If the fluctuations happen to include components at the Larmor frequency, then a transition may occur. This has the effect of dispersing the excess energy of the inverted spins to the thermal sink of the solvent and so reestablishing the normal thermal spin equilibrium that was originally present before the $180°$ inversion pulse was applied. This mechanism becomes more efficient with the inverse sixth power of the distance between the dipolar nuclei. The H–H distance in H_2 being the shortest in chemistry, the corresponding relaxation rate is expected to be very large, and essentially the whole of the relaxation will arise from the dipole–dipole mechanism under the conditions of the experiment. The equation governing the rate of relaxation $R(DD)$ is shown in Eq. (4) (*21*). The characteristic relaxation time $T_1(DD)$ is simply the inverse of $R(DD)$ and is of the order of 1 s for protons in organic molecules or classical metal hydrides in solution. For dihydrogen complexes, on the other hand, the T_1 often lies in the range of 4–100 ms.

The simplest way to apply this method is to measure T_1 for a given hydride in solution and compare the value with those found for a range of hydrides having known T_1 values (see Table II). Since the T_1 depends on temperature, a suitable temperature must be used; $-80°C$ was chosen for the earliest work. A nonclassical structure should have an H–H distance of 0.8–1.0 Å as opposed to a classical hydride, which would have a non-bonded H $\cdots$ H distance of at least 1.5, if not more. The relaxation rates in

TABLE III
SOME T_1 VALUES FOR METAL HYDRIDES[a]

Complex	T_1 (ms)	Conclusion
$H_2Fe(CO)_3$	3000	Classical
$IrH_5(PCy_3)_2$	820*	Classical
$ReH_5(PPh_3)_3$	540	Classical
$WH_6(PMe_2Ph)_3$	166[b]	?
$WH_5(PMe_2Ph)_4^+$	148*[b]	?
$MoH_3(H_2)(PMePh_2)_4^+$	44*	Nonclassical
$ReH_5(H_2)(PPh_3)_2$	78	Nonclassical

[a] Toluene ($-70°C$) except * = $CdCl_2$, $-80°C$.
[b] Classical by variable temperature T_1 data.

the two cases should differ by several orders of magnitude. One of the most striking examples is the Kubas complex itself, in which the $T_1(-80°C$, toluene) of 4 ms for the nonclassical tautomer is very different from the T_1 of 1700 ms for the classical tautomer.

The method was applied to a number of molecules (Table III), and although clear-cut distinctions were possible in many cases, doubt remained in others because T_1 values were intermediate between the 4–100 ms usually observed for nonclassical species and values of >350 ms found for some classical hydrides. The problem seemed to be that the Brownian motion of a molecule varies in a way that depends largely on its moment of inertia, the temperature, and the viscosity of the solvent. For example, the T_1 for free H_2 in solution in toluene at $-80°C$ is 1600 ms (9). The reason for the long T_1 is the rapid rotation of the H_2 molecule. It turns out that the magnetic field fluctuations tend to be more rapid than the Larmor frequency, and relaxation is inefficient. (The long T_1 for free H_2 also rules out rapid exchange with free H_2 as the source of the short T_1 values observed for nonclassical hydrides.)

Observations made on molecules having different moments of inertia and in different solvents are therefore not comparable. The key to making the T_1 criterion quantitative is the study of its temperature dependence. Equation (4) predicts that the T_1 will go through a minimum when the Brownian motion (measured by a rotational correlation time τ_C) is best matched with the Larmor frequency ω. This should happen when $\tau_C = 0.63/\omega$, and so if we can observe the minimum experimentally, we will know τ_C at that temperature, and therefore we can calculate the H–H distance r from Eq. (4). Table IV shows the H–H distances calculated

$$R(DD) = \{T_1(DD)\}^{-1}$$
$$= 0.3\gamma^4 h^2 r^{-6} \{\tau_C/(1 + \omega^2\tau_C^2) + 4\tau_C/(1 + 4\omega^2\tau_C^2)\} \quad (4)$$

TABLE IV

TEMPERATURE DEPENDENCE OF T_1

Compound	Temperature at minimum (K)	$T_1{}^a$ (ms)	Corresponding H–H distance (Å)	Conclusion
$MoH_4(PMePh_2)_4$	250	165	1.27	Classical
$ReH_8(PPh_3)^-$	185	245	1.35	Classical
$Re(H_2)H_5(PPh_3)_2$	185	21.7	0.90	Nonclassical
$Ru(H_2)H_2(PPh_3)_3$	267	17.6	0.87	Nonclassical

^{a} Having allowed for statistical factors using the formulation shown in the first column, with $T_1(c) = 200$ ms.

in this way for a number of hydrides. In a fluxional system, such as $Re(H_2)H_5(PPh_3)_2$, we have to allow for the statistical effects due to exchange between the different sites; for the case of the Re compound Eq. (5) describes the situation.

$$7/T_1(obs) = 5/T_1(c) + 2/T_1(n) \qquad (5)$$

$T_1(obs) = $ observed T_1; $T_1(c)$ and $T_1(n)$ are the T_1 values for classical and nonclassical sites, respectively.

$T_1(obs)$ is dominated by $T_1(n)$ and so the value chosen for $T_1(c)$ is not critical [we usually use $T_1(c) = 200$ ms]. The choice of the number of dihydrogen ligands to assign to a given complex [i.e., $Re(H_2)H_5(PPh_3)_2$ rather than $Re(H_2)_2H_3(PPh_3)_2$ is guided by the available chemical evidence [e.g., Eq. (6)].

$$Re(H_2)H_5(PPh_3)_2 + PPh_3 = ReH_5(PPh_3)_3 \qquad (6)$$

The H–H distance obtained in this way assumes that the whole of the relaxation is a result of dipole–dipole interactions in the H_2 ligand. This is clearly a great oversimplification and the r values obtained are correspondingly approximate and constitute a lower limit for the true H–H distance. As can be seen, the nonclassical cases give r values of ca. 0.8 Å and the classical species give $r > 1.1$ Å. Since the H–H bond will be vibrating in the H_2 complex, the H–H distance calculated from Eq. (4) will be slightly shorter than the real internuclear distance, but comparative data among different complexes should not be much affected by this consideration.

It is not yet clear why the proton resonances for η^2-H_2 complexes tend to be broad. Counterexamples do exist: both $Re(H_2)H_5(PPh_3)_2$ and $FeH_2(H_2)(PEtPh_2)_3$ give narrow resonances at room temperature. As we lower the temperature, broadening is to be expected for all nonclassical hydrides because T_2, which controls the linewidth by Eq. (7), is expected to continue to fall as T_1 goes through its minimum and starts to rise again (Fig. 1).

$$\text{linewidth} = 1/\pi T_2 \tag{7}$$

Above $T_1(\text{min})$, T_1 usually equals T_2, so we would expect to find linewidths of $1/\pi T_1$ at room temperature. We know T_1 by direct measurement and so we can be sure that in the majority of cases there is an additional, and as yet unknown, factor that is broadening the hydride resonances of nonclassical hydrides. As a rule, the identification of broad lines with a dihydrogen complex may be useful *prima facie* evidence, but until we understand its origin, it cannot be relied on. The coupling of a quadrupolar nucleus to a proton can sometimes broaden proton signals, but anomalous broadening of nonclassical hydride resonances occurs even with dipolar nuclei. Quadrupolar effects should not alter T_1 values for a coupled proton. The most likely origin for this anomalous broadening is some exchange process: either of bound H_2 with free H_2 or of proton exchange with adventitious base with the H_2 complex acting as a proton acid.

The case of $[ReH_9]^{2-}$ is a particularly interesting one. It has a slightly broad hydride resonance at room temperature, but its T_1 in methanol at $-80°C$ is 2400 ms. This does not rule out a nonclassical structure because of the exceptionally low moment of inertia of the molecule. Variable-temperature studies gave no evidence that a minimum T_1 had been reached at any accessible temperature. A band in the IR spectrum at 2700 cm^{-1} has never been satisfactorily explained. $Re(H_2)H_5(PPh_3)_2$ is definitely nonclassical, but $ReH_8(PPh_3)^-$ is clearly classical by the $T_1(\text{min})$ evidence (18). The changeover may be due to the replacement of a good donor, PPh_3, by a better one, H^-. Simple extrapolation might suggest a classical structure for $[ReH_9]^{2-}$, as was suggested on the basis of neutron diffraction (27), but the case remains unproved. We do not believe neutron diffraction necessarily provides definitive evidence because of the rotational and positional disorder that might be present. $[ReH_9]^{2-}$ certainly provides one of the most challenging structural problems in this area. Table V lists this and other candidate species worthy of further study.

Solid-state NMR spectroscopy has provided further insights into the properties of H_2 complexes. Under the usual conditions, the resonances from the phosphine ligands obscure the resonance for the H_2. Zilm *et al.* (29) have shown that the ligand resonances in $[W(H_2)(CO)_3(PCy_3)_2]$ can be largely removed by selective decoupling; because the ligand protons are extensively mutually coupled, the H_2 resonance only has a "hole" burned in it. From the form of the H_2 resonance, Zilm *et al.*, were able to deduce the value of the dipole–dipole coupling and therefore the H–H distance: 0.89_0 Å (cf. 0.82 Å by neutron diffraction). It can also be shown that the

TABLE V
Candidate Nonclassical Hydrides

Entry	Ref.	Original formulation	Possible new formulation	^{1}H NMR	Isoelectronic species and remarks
1	22	$RuH_6(PCy_3)_2$	$Ru(H_2)_2H_2L_2$	-7.48 br	$Ru(H_2)_2H_2L_2^+$
2	22	$RuH_4(PCy_3)_3$	$Ru(H_2)H_2L_2$	-9.1 br	$Ru(H_2)H_2(PPh_3)_2$
3	23	L'_2PtH_2[a]	Nonclassical tautomer in equilibrium		Inequivalent H exchanging by NMR
4	24	$IrHCl_2(Pi\text{-}Pr_3)_3 + H_2$	$Ir(H_2)HCl_2L_3$		Unusual NMR behavior
5	25	$IrH_4(PMe_2Ph)_3^+$	$Ir(H_2)H_2L_3^+$	$+8.4$ br	No $J(H, P)$ observed; N_2 displaces H_2
6	26	Cp_2NbH_3	$Cp_2NbH(H_2)$		Unusual temperature and field dependence of NMR signals; absence of diastereotopy for $CpCp'NbH_3$
7	27	ReH_9^{2-}	?		Broad NMR resonance; IR band at 2700 cm^{-1}; T_1 long, but see text

[a] $L' = Cy_2P_2(CH_2)_n$ ($n = 2, 3$); br = broad.

H_2 is librating by ca. 45° about its equilibrium position. This motion would, of course, cloud the interpretation of the H–H distance determined by neutron diffraction as discussed above.

Now that we can confidently distinguish the two classes of complex, we can try to understand why some complexes adopt one structure, and others, the other. Based on the bonding model discussed earlier, we would expect that poor π-donor metal sites would favor dihydrogen and good π-donor sites favor dihydride structures. Morris *et al.* (30) have quantified this approach in octahedral d^6 systems by using $E_{1/2}(ox)$, $\nu(CO)$, and $\nu(N_2)$ of the corresponding CO or N_2 complexes to probe the electronic character of the site. They find that H_2 will normally bind as a dihydrogen ligand where $\nu(N_2)$ of the corresponding N_2 complex is greater than 2060 cm^{-1}, while a dihydride structure is adopted if $\nu(N_2) < 2050$ cm^{-1}. This supports the view that increased back donation tends to split the H–H bond.

Dihydrogen ligands are generally found trans to a high trans effect ligand, such as H or CO. By weakening the trans M–(H_2) bond, this sterochemistry probably favors an η^2-H_2 structure. A cationic charge on the complex would be expected to favor a nonclassical structure by weakening the back donation from the metal; it is notable how many of the complexes in Table I are indeed cationic. In such cases, a noncoordinating anion such as BF$_4$ seems to be required to prevent displacement of the H_2. Third-row elements, as stronger π-bases, would be expected to favor the classical structure, as illustrated by the series [Fe(H_2)H$_2$(PEtPh$_2$)$_3$], [Ru(H_2)H$_2$(PPH$_3$)$_3$], [OsH$_4$(Po-tolyl$_3$)$_3$] (9).

d^8 fragments seem to favor binding H_2 in the dihydride form, H_2Fe(CO)$_4$, for example, being classical (13). This is perhaps because oxidative addition leads to an octahedral d^6 structure having a particuarly strong ligand field stabilization. Even a cationic charge and the presence of electron-withdrawing ligands are apparently insufficient to prevent oxidative addition, although the evidence for the classical structure of [(cod)$_2$IrH$_2$]$^+$ is not yet unequivocal (31).

$$(cod)_2Ir^+ \underset{}{\overset{H_2}{\rightleftharpoons}} (cod)_2IrH_2{}^+ \tag{8}$$

d^{10} fragments seem to prefer oxidative addition, e.g., (R$_3$P)$_2$PtH$_2$.

Almost all the nonclassical hydrides so far established can be described as H_2 adducts of d^6 metals; only Re(H_2)H$_5$(PPh$_3$)$_2$ is a verified H_2 adduct of a d^2 fragment, but Cp$_2$NbH$_3$ is a d^2 candidate. Fe(CO)(NO)$_2$(H_2) and Co(CO)$_2$(NO)(H_2), detected in liquid Xe (8d), are formally d^{10} species; Ni(CO)$_3$(H_2) has also been seen in a matrix (8c). It will be interesting to see how far H_2 complexes extend beyond d^6. The d^0 case is an interesting one in that several main group fragments such as H^+ and CH_3^+ form nonclassical, albeit rather unstable, H_2 adducts. The failure of $16e$, d^0 complexes of the Cp_2MX_2 (M = Ti, Zr, or Hf) type to give isolable H_2 adducts may mean that a certain amount of backbonding is *required* to make the H_2 adduct stable. Too much, of course, and a classical dihydride is formed, so the balance is a delicate one. $16e$ d^8 species seem to form classical dihydrides only, perhaps because the empty $d_{x^2-y^2}$ orbital lies in the ML_4 plane ($62j$).

The dihydrogen ligand seems to show a reactivity pattern that is quite distinct from that of a classical dihydride. In one case [Eq. (3)], the two tautomers appear to be in equilibrium, in which case the complex should be able to show the reactivity pattern of each type. It is not yet clear how general this tautomerism will be; in other cases it is likely that the hydride will normally show the reactivity pattern of the ground state structure. Purely classical complexes presumably have a nonclassical tautomer, which may be unstable by several or several tens of kilocalories per mole with respect to the ground state. It is still an open question whether the nonclassical tautomer may play a part in the chemistry of classical hydrides, for example, in their fluxionality, photochemistry, and subsitution chemistry.

The simplest reaction of an H_2 complex is dissociation. In the cases for which this has been studied, the $M–(H_2)$ dissociation energy is 3–10 kcal/mol, but measurements have been made only on the most labile complexes. Once the H_2 departs, the resulting $2e$ vacant site is filled by a solvent molecule or by an agostic C–H bond of a ligand, so the observed values do not correspond to the formation of the presumed $16e$ intermediate ($17,20$). Coordinated H_2 can usually be displaced irreversibly by CO and in many cases also reversibly by N_2. In the cationic iridium species, displacement by N_2 is not found, the site being too weakly π-basic, but reversible displacement with water is observed.

$$L_2(bq)HIr(H_2)^+ \overset{H_2O}{\rightleftharpoons} L_2(bq)HIr(H_2O)^+ \qquad (9)$$

There is evidence that the reaction of Eq. (9) provides for the selective transport of H_2 across a liquid phase consisting of the complex dissolved in an organic solvent ($32b$).

Coordinated H_2 can be more acidic than free hydrogen, probably because the $H_2(\sigma)$-to-$M(d\sigma)$ electron transfer leads to depletion of the charge on the H_2 (20). One of the most acidic hydrides of this sort is $[IrH(H_2)bq(PPh_3)_2]^+$ and it was possible to show by a labeling study that the base does indeed deprotonate the H_2 molecule and not the terminal hydride.

It has been suggested that both nitrogenase and the Ni-containing H_2-uptake hydrogenases bind H_2 in an η^2-H_2 fashion *(33)*. H_2 is a reversible inhibitor of N_2 reduction in nitrogenase, and coordinated H_2 undergoes isotopic exchange with the solvent in the hydrogenases.

The fact that binding in an η^2-H_2 fashion can turn H_2 into a proton acid suggests that heterolytic H_2 activation may go via an H_2 complex as intermediate. Schwartz *(34)* suggested a polarization of the type shown in **12**, but a dihydrogen complex **13** now seems to have all the necessary properties. Binding would be weak in a d^0 species, but the lack of back donation would enhance the acidity of the coordinated H_2.

$$M\text{–}R + H_2 = 12 \text{ or } 13 = M\text{–}H + R\text{–}H \tag{10}$$

12   **13**

d^0 alkyls can undergo rapid hydrogenolysis [Eq. (11) *(35)*], and several d^0 systems are active hydrogenation catalysts, e.g., $Cp_2TiCl_2/Li[Al(OR)_2H_2]$ *(36)*.

$$WMe_6 \xrightarrow[-78° \text{ to RT}]{H_2,\ PMe_3} WH_2(PMe_3)_5 \tag{11}$$

There is clear evidence in many cases that H_2 ligands can undergo rapid exchange with terminal M–H groups *(12,20)*. What is not yet clear is whether this goes via the classical tautomer **(14)** or proton transfer [Eq. (12)]. The former looks more likely, since an M–H bond may not be sufficiently basic to deprotonate the H_2 ligand.

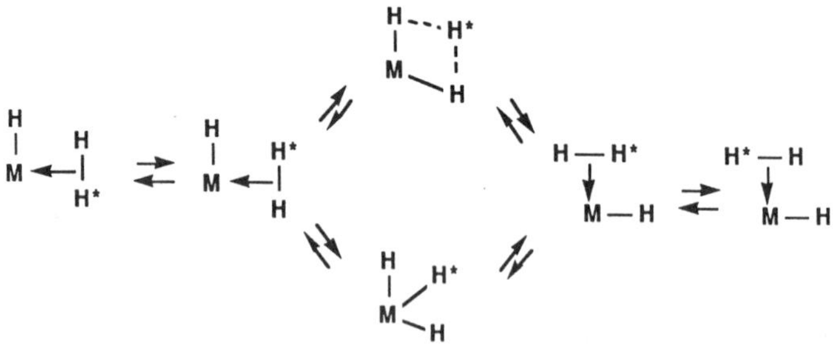

14

Both classical and nonclassical polyhydrides show very low barriers for fluxionality *(37)*. Whichever tautomer is the ground state structure, the other tautomer might be accessible thermally and so account for the facile fluxionality of both classes of polyhydride. The accessibility of the nonclassical tautomer might explain the rapid ($\Delta G^{\ddagger}$ = 19 kcal/mol) intramolecular scrambling of H_a and H_b sites in **15**, as observed by Stobart *et al.* (*38*) (see also Table V, entry 3).

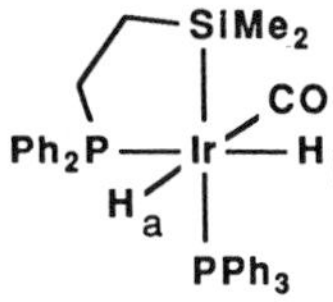

$$
\begin{array}{c}
\text{SiMe}_2 \\
\text{Ph}_2\text{P} - \text{Ir} \overset{\text{CO}}{\underset{\underset{\text{PPh}_3}{|}}{\overset{|}{\diagdown}}} \text{H}_b \\
\underset{H_a}{\diagup}
\end{array}
$$

15

Kubas *et al.* (*7*) have observed exchange between H_2 and D_2 catalyzed by $[W(CO)_3(P\{i\text{-Pr}\}_3)_2]$ even in the solid state. It is difficult to see how a $16e$ fragment could do this and so it is probable that the phosphine dissociates to generate an additional site for binding H_2, or perhaps the coordinated H_2 exchanges with trace water in the system.

The unusually rapid isotopic scrambling observed by Brown *et al.* (*39*) in the hydrogenation of **16** by $[Ir(cod)(PCy_3)py]BF_4$ led them to undertake detailed kinetic studies. These suggested that exchange takes place at the alkyl hydride stage, and they postulate that a dihydrogen complex is the most reasonable means of accomplishing this, especially in light of the known tendency of the cationic iridium system to form H_2 complexes.

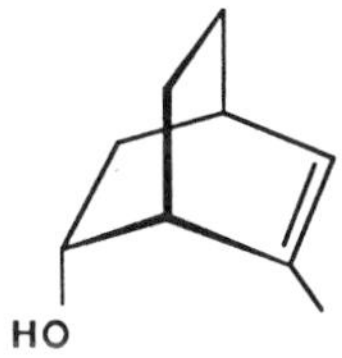

HO

16   (13)

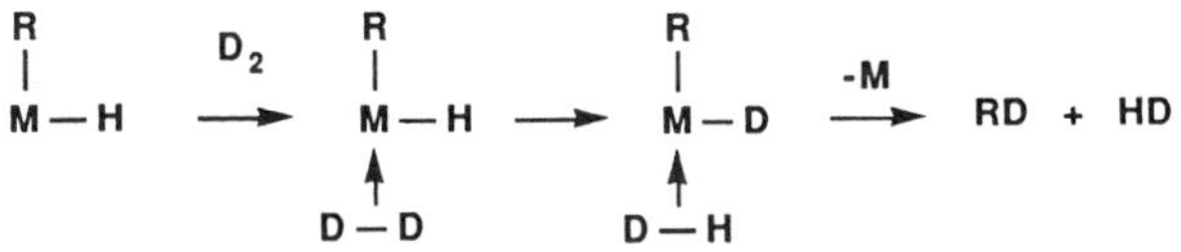

$$
\begin{array}{ccccc}
\overset{\text{R}}{\underset{|}{\text{M}-\text{H}}} & \xrightarrow{\text{D}_2} & \overset{\text{R}}{\underset{|}{\text{M}-\text{H}}} & \xrightarrow{} & \overset{\text{R}}{\underset{|}{\text{M}-\text{D}}} & \xrightarrow{-\text{M}} & \text{RD} + \text{HD} \\
& & \uparrow & & \uparrow \\
& & \text{D}-\text{D} & & \text{D}-\text{H}
\end{array}
$$

There is disagreement about the factors that control the direction of H_2 addition d^8 square planar complexes. Eisenberg *et al.* (*40a*) observed that H was always trans to CO in the kinetic products (**17**) of H_2 to complexes of the Vaska type [Eq. (14)]. They proposed that a favorable interaction between the CO π^* and the H_2 in a reagentlike transition state was responsible for the stereochemistry of addition. Rearrangement to the thermodynamic product (**18**) was also observed. Burk, Crabtree, and Uriarte (*40b*) found examples in which the kinetic product of addition has the opposite stereochemistry and proposed that the direction of addition is decided by the values of K, k_1, and k_2 for the proposed intermediate dihydrogen adducts, as shown in Eq. (15).

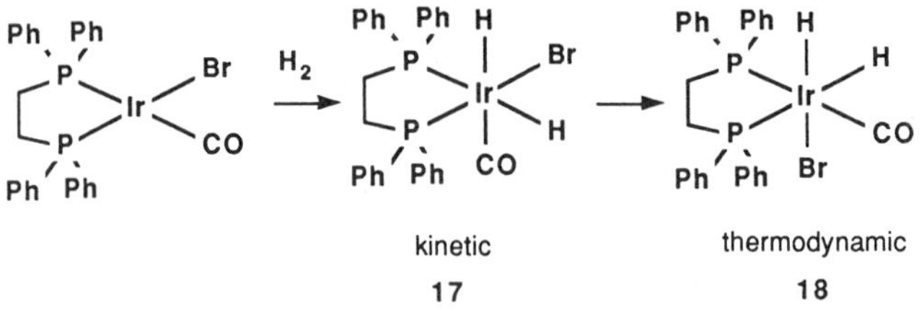

$$(14)$$

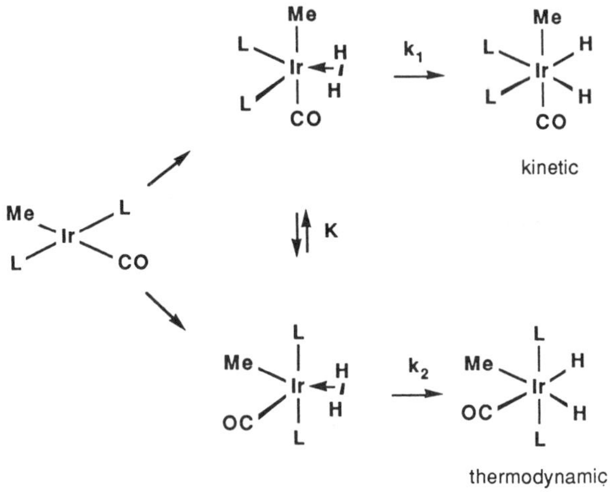

kinetic   thermodynamic
17   18

$$(15)$$

Ozin *et al.* (*41*) studied the interaction between H_2 and Pd atoms in Xe or Kr matrices at 10–12 K. In contrast to the oxidative addition that is observed with other transition metals, they find that Pd gives an H_2 adduct. They propose that an η^2-H_2 adduct is present in the Xe matrix, but both η^1-H_2 and η^2-H_2 adducts are present in the Kr matrix. Theoretical studies support the formation of η^2-H_2 (*42,43*) and η^1-H_2 (*42*) adducts with Pd atoms. It will be interesting to see if η^1-H_2 adducts can be detected in isolable complexes. The bonding model we have discussed suggests that the η^2-H_2 form will usually be more stable because only in this form is backbonding possible. On the other hand, in d^0 systems this would not matter.

Siegbahn *et al.* (*44*) looked at the potential energy surface for NiH_2 by theoretical methods and identified a state that is bound by 9.5 kcal/mol with respect to Ni + H_2 and has $H \cdots H = 1.41$ Å; this may be an η^2-H_2 complex with an unusually long H–H distance. Intracavity Ni^+ ions in zeolites form labile complexes with H_2 that may involve an η^2-H_2 structure (45).

There is an intriguing possibility that H_3^+ may be able to bind to transition metals to form stable η^3-H_3 complexes. It is even possible that some of the nonclassical polyhydrides identified in the past year or so may contain this η^3-H_3 group. Finding definitive proof for such a structure constitutes a formidable experimental problem.

III

C–H BONDS AS LIGANDS

Interactions between a C–H bond and coordinatively unsaturated metal fragments (**19** and **20**) were first observed in the 1960s by Ibers (46) and by Maitlis (47). Only in the past few years have these structures achieved greater prominence with a rise in interest in the problem of C–H activation (48a). Brookhart and Green reviewed the area recently and used the term agostic to describe the chelation of a C–H bond. In this review we will look

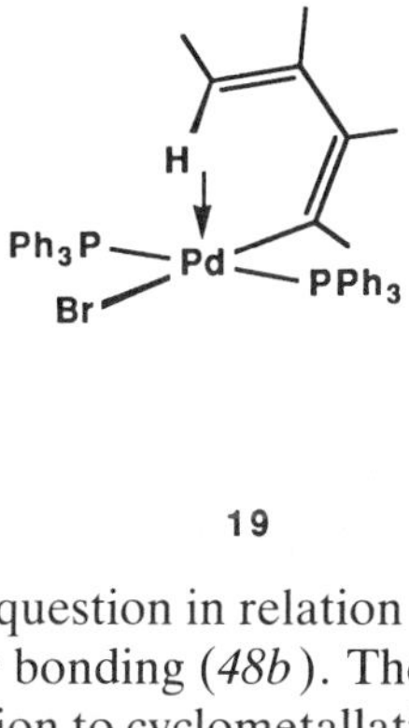

19

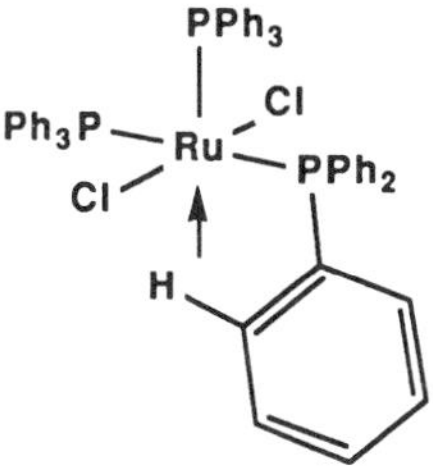

20

at the same question in relation to the more general phenomenon of $2e(\sigma)$, three-center bonding (48b). The special importance of agostic interactions is their relation to cyclometallation. Just as η^2-H_2 complexes are thought to be intermediates in the oxidative addition of H_2, so agostic systems may play that role in cyclometallation and alkane complexes in the oxidative addition of alkanes.

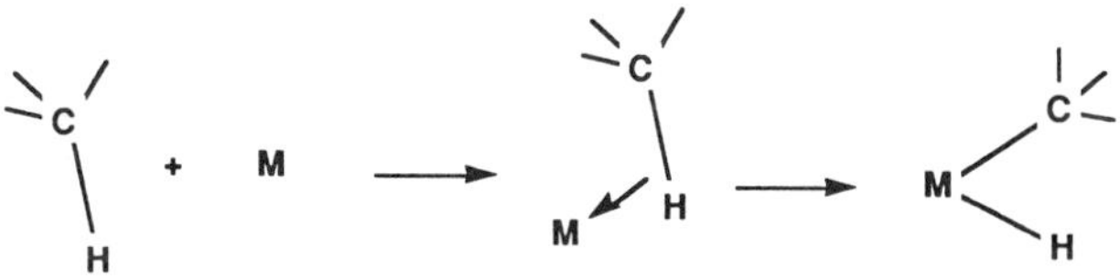

$$(16)$$

Agostic interactions are best recognised by crystallography. Even if the C–H hydrogen cannot be detected with certainty in an X-ray study, its position can often be deduced with considerable reliability from the known positions of the other substituents about the agostic carbon. This technique was useful in both Ibers' and Maitlis' complexes. Neutron diffraction studies have been carried out in a number of cases, and of course these give the most reliable results, especially with regard to the C–H distance. Instead of being strictly side-on as in the case of η^2-H_2 complexes, the C–H bond is usually skewed in such a way that the H atom is closer to the metal. A Burgi–Dunitz (50) analysis of the crystallographic data has led to a proposal for a kinetic trajectory (32) for reaction (16). The approach to the transition state may involve a pivoting motion of the C–H bond such as that shown in Fig. 2. This approach could be opposed by two factors: (i) the presence of other substituents on carbon (a steric effect) and (ii) conformational restrictions on the way the C–H bond can approach, which might be particularly severe in cyclometallation. Indeed, alkane activation by oxidative does seem to be strongly retarded by steric effects in both the metal fragment and the substrate. The successful alkane activation systems are all relatively unhindered. Methylcyclohexane, for example, is preferentially attacked at the unhindered methyl group by various alkane dehydrogenation catalysts based on Ir and Re (51). Jones et al. (52) have shown that in a stoichiometric C–H activation [Eq. (17)] the cyc-

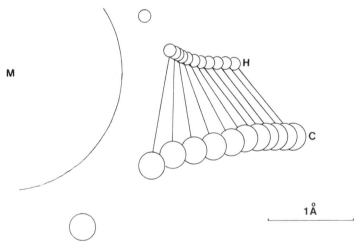

Fig. 2. Burgi–Dunitz trajectory for the reaction M + C–H = C–M–H. The metal is shown on the left and the approaching C–H bond on the right. The isolated circles show the final C and H positions. (We thank the American Chemical Society for permission to reproduce this figure.)

lometallation product is thermodynamically but not kinetically favored. The additional kinetic barrier for cyclometallation could be conformational in origin. Discussions of this problem are presented in a number of recent papers (*53*).

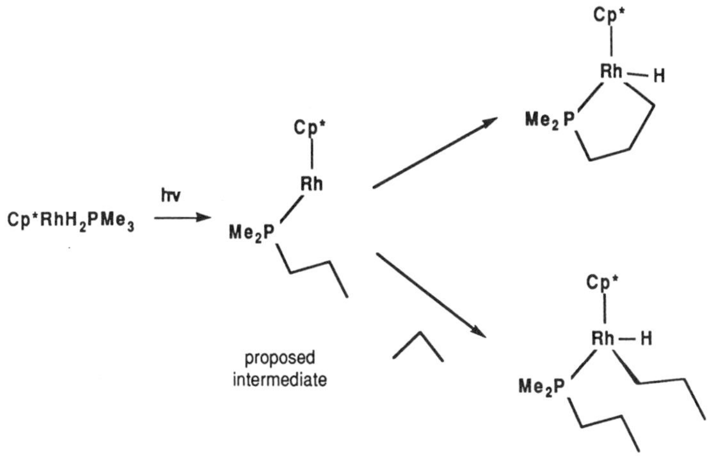

Cp*RhH2PMe3
hν
Cp*
Rh
Me2P
proposed
intermediate
Cp*
Rh—H
Me2P
Cp*
Rh—H
Me2P

(17)

Another interesting feature of the crystallographic data is the relatively small lengthening of the C–H bond as a result of the agostic interaction. Although the picture is as yet incomplete, and strictly comparable systems are rare, it looks as if η^2-H_2 binding can lead to greater lengthening (up to 0.2 Å) of the X–H bond (X = C or H) than does η^2-C–H binding (ca. 0.1 Å). Since we expect M(d_π) to X–H(σ^*) backbonding to be the chief factor affecting the X–H bond lengthening, the explanation may be that the C–H σ^* is less accessible than the corresponding H–H σ^*. In η^2-H_2 complexes, the side-on structure and the lack of other substituents make for maximum accessibility of both σ^* lobes. In contrast, the more skewed structure of the η^2-C–H complex, together with the presence of substituents at carbon, makes the C–H σ^* less accessible for back donation from the metal (**21** versus **22**).

M
H
H
M
C
H

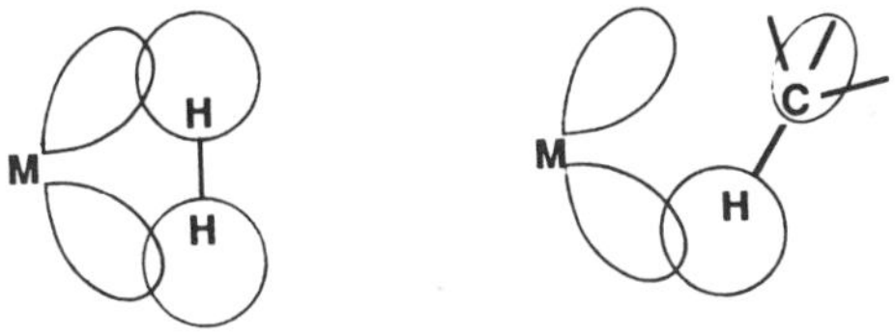

21 **22**

Two of the most interesting agostic structures are **23** and **24**; these were characterized largely by X-ray methods (*54*). More recently, a neutron diffraction structure has been determined for **24**, and the Ti–C–H angle quoted in the diagram comes from the neutron data (*54b*). As d^0 systems, the metals have no nonbonding electrons with which to populate the C–H σ^* and so break the C–H bond. The stability of these species suggests that $d^0 \ \eta^2$-H$_2$ complexes may be capable of existence. Another d^0 system with an unusual agostic interaction is Cp$_2$*Th(CH$_2t$-Bu)$_2$, in which *both* the α-C–H bonds of a single neopentyl group are agostic. The resulting Th–C–H angles are 84.4° and 87.1°, compared to angles of 101.1° and 104.2° for the classical neopentyl group (*55*).

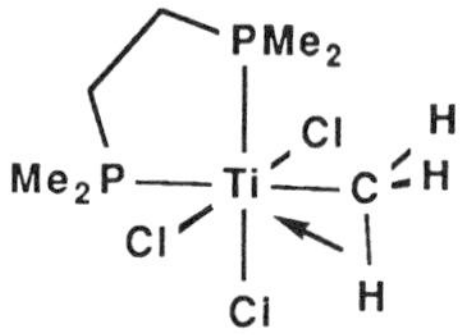

Ti‐C‐‐H = 93.7°   Ti‐‐C‐‐C = 85.9°

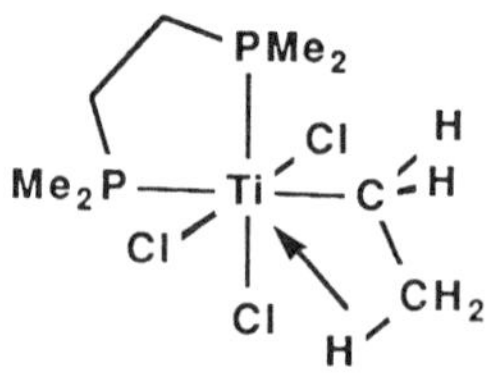

23   **24**

IR data have often been used to confirm the presence of an agostic interaction. The stretching frequency in such a case can usually be observed at 100–300 cm^{-1} to lower energies compared to that of an unbound C–H bond.

NMR observations can be a powerful method for detecting agostic C–H groups. The coordination shift of the C–H proton in the ^{1}H NMR (up to 13 ppm to high field) and the reduced ^1J(C, H) coupling (values as low as 50–90 Hz, compared with 125 Hz for an unperturbed C–H bond, are not uncommon) as observed in the ^{13}C NMR have been most widely used, but there appear to be cases of authentic agostic systems in which there is little, if any, lowering of ^1J(C, H) (*56*). Any decrease on binding is not simply a reflection of C–H bond weakening. The agostic carbon tends to rehybridize so that the agostic C–H bond acquires more *p* character. At the same time any terminal C–H bonds at the agostic carbon usually acquire more *s* character, and so their coupling constants tend to rise. This means that if we have a fluxional agostic methyl group in which each proton spends some time in the bridging position [Eq. (18)], then the averaged ^1J(C, H) that results may be indistinguishable from that of a classical methyl group.

$$M \xleftarrow{} \overset{\diagdown}{C} \diagdown \overset{-H}{\underset{H^* \quad H}{\diagup \diagdown}} \quad \rightleftarrows \quad M \xleftarrow{} \overset{\diagdown}{C} \diagdown \overset{-H}{\underset{H \quad H^*}{\diagup \diagdown}} \quad (18)$$

It is not unusual to find an agostic methyl group that shows a singlet at all accessible temperatures, and so the agostic interaction cannot be characterized from this experiment alone. A valuable technique in such a case is isotopic perturbation of resonance (IPR) (57). The methyl in question has to be partially deuterated so that the d_0, d_1, and d_2 isotopomers are present in solution. If the methyl group is not agostic then the isotopomers should show a negligible chemical shift separation. On the other hand, in an agostic methyl, the d_1 and d_2 isotopomers will show a shift with respect to the d_0 resonance which will be temperature-dependent and increase with cooling.

The origin of the effect lies in the smaller difference between the zero point energies of C–H versus C–D in the bridging C–H bond, C–H$_b$, compared to the terminal C–H bonds, C–H$_t$. The zero point energies will be smaller in the weaker C–H$_b$ bond, and so protons will tend to accumulate in this position and deuterons in the C–H$_t$ bond. If ΔE is the energy difference between the two situations in Eq. (19), then Eq. (20) will govern the observed average chemical shift, δ_{av}.

$$M \xleftarrow{} \overset{\diagdown}{C} \diagdown \overset{-H}{\underset{D \quad H}{\diagup \diagdown}} \quad \rightleftarrows \quad M \xleftarrow{} \overset{\diagdown}{C} \diagdown \overset{-H}{\underset{H \quad D}{\diagup \diagdown}} \quad (19)$$

$$\delta_{av} = [\delta_b + \delta_t + \exp(-\Delta E/RT)]/[2 \exp(-\Delta E/RT) + 1] \quad (20)$$

The protons will therefore report, by means of their chemical shift, an average chemical environment that differs in the isotopomers. Quantitative analysis of the data allows ΔE and δ_b and δ_t, the chemical shifts of the C–H$_b$ and C–H$_t$ sites, to be calculated. This is a very useful method for verifying that an agostic interaction seen in the solid state is also present in solution. Care must be taken with the technique, however, because any perturbation of the methyl group, such as rapid interconversion between a free methyl and the corresponding oxidative addition adduct, will give an IPR.

Schrock et al. (58) looked at a series of agostic neopentyl complexes of tantalum and found that as the interaction with the α-C–H group increases, $\delta(^{13}C)$ for the agostic carbon moves to higher frequency and $^1J(C, H)$ falls markedly (Table VI). Jolly et al. (59) found shifts in the ^1H and ^{13}C NMR of $[(\eta^3\text{-allyl})_3Mo(i\text{-Pr})]$, in which one of the isopropyl

TABLE VI

NMR Data for Some Agostic Neopentyls

Complex	$\delta C\alpha$ (ppm)	$^1J(^{13}C, H)$ (Hz)
$TaNp_2Cl_3L$	96	121
$TaNp_2EtCl_2$	112	117
$TaNp_3Cl_2$	115	119
$TaNp_2Cl_3$	123	116
$TaNpCl_4$	132	110
$Ta(C_2H_4)NpCl_2L_2$	139	98

β-C–H groups is agostic. At $-120°C$ the agostic H is frozen out and shows a shift of -8.43 δ, but at $-50°C$ only one signal is seen for all three β-C–H protons on one methyl of the i-Pr group. The agostic carbon shows a coordination shift of 20 ppm to low frequency from the uncoordinated one; this appears to be a shift in the opposite direction to that observed by Schrock *et al.*, perhaps because Jolly's complex is d^2 and Schrock's is d^0. The $^1J(C, H)$ of 88 Hz in the Mo ethyl complex suggests that the interaction is quite strong. The Jolly system is also interesting in that an agostic t-butyl complex was also prepared; t-butyl complexes of transition metals are very rare.

Other NMR criteria have been applied to agostic systems. Green *et al.* (*60*) found that $MeTiCl_3$ has the agostic structure **25** in which all three C–H bonds are bending toward the metal to retain the C_{3v} symmetry. Electron diffraction suggested that the C–H distances were unusually long (1.16 Å), but the authors naturally wanted to obtain spectroscopic evidence for the interaction. The methyl carbon appears at the unusually high frequency of 118.2 δ (CH_2Cl_2, $-73°C$) in the ^{13}C NMR, and the $^2J(H, H)$ coupling of $+11.27$ Hz, although normal in magnitude, has a reversed sign from those found in normal methyl groups (*61*).

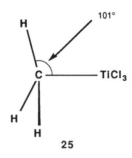

25

Useful theoretical work on C–H$\cdots$M systems has appeared (*62*). Goddard *et al.* (*62a*) traced the orbital interactions involved in the distortions observed by Schrock *et al.* in their alkylidene complexes and speculated on the distortions that a methyl group might undergo. Eisenstein and Jean (*62b*) treated the question of methyl distortion in detail and concluded that the methyl group will tilt in an octahedral but not a tetrahedral environment (i.e., in **23** but not in **25**). They regarded $MeTiCl_3$ as only slightly flattened and not agostic, but the spectroscopic evidence obtained by Green *et al.* suggests that there is a significant $C-H \cdots M$ interaction. Distortions from an octahedral disposition of the ligands about the metal in a d^0 complex were also discussed (*62c*).

Koga *et al.* (*62d,e*) have pointed out that agostic interactions may be more important in stabilizing reactive intermediates and transition states than they are in ground state structures. They found by calculation that the fluoro substituents in $Pd(CH_2CHF_2)H(PH_3)$ were able to suppress the agostic interaction found for the unsubstituted ethyl complex, because the F substituents make the C–H bond a poorer donor. Katz (*62f*) has postulated that agostic interactions may play a role in promoting the high selectivity often seen in alkyne polymerization.

Lichtenberger and Kellog (*62g*) have looked at the photoelectron spectrum of $(CO)_3Mn(\eta^3\text{-}C_6H_9)$, which contains an agostic C–H. They conclude that $C-H(\sigma)$-to-metal direct donation is more important than metal-to-$C-H(\sigma^*)$ back donation, a result entirely in harmony with the general picture that is emerging for $2e(\sigma)$, three-center interactions in general.

A particularly significant series of agostic complexes is $Cp^*Co(PR_3)$-(alkyl) (**26**) (*56,63*). These undergo alkyl chain growth with ethylene to give polyethylene. Most transition metals at best oligomerize ethylene, because β-elimination is usually fast compared to insertion. It is likely that the stability of the agostic form points to the unwillingness of the system to β-eliminate, and it may also be that the relatively Lewis acid site speeds up insertion as well. Studies on the PMe_2Ph ethylcobalt species show that there are three fluxional processes. In the first (coalescence, $-40°C$, $\Delta G^{\ddagger} = 40$ kJ/mol) the agostic C–H dissociates and the CH_3 protons of

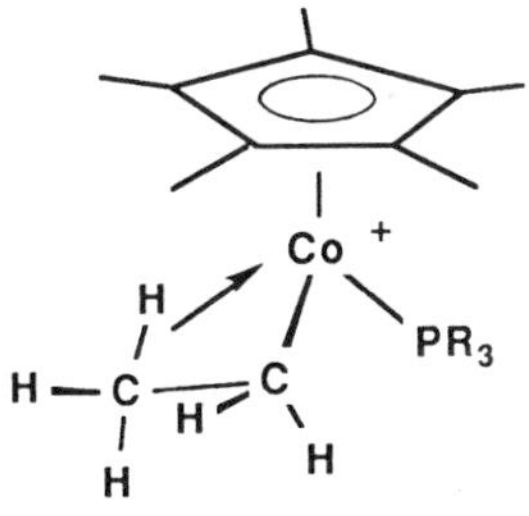

26

the ethyl group become equivalent. At a slightly higher temperature (coalescence, $-10°C$, $\Delta G^{\ddagger} = 56$ kJ/mol), inversion of chirality takes place at the metal. The third process, β-elimination, is so slow that it can be detected only by spin saturation transfer.

Another interesting complex is **27** (*64a*), in which the methyl group appears to prefer to form a normal covalence with Rh and a 2e, three-center bond with Ti, the better π-acid, rather than the reverse (**28**).

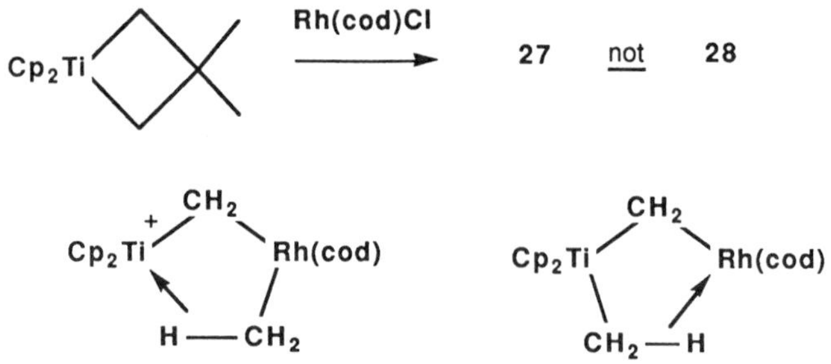

$$\text{(21)}$$

Orpen *et al.* (*64b*) found a molybdenum complex with a structure that corresponds to an arrested alkene-to-allyl-hydride rearrangement. C_1–C_2 has shortened to 1.471 Å in "anticipation" of the C–H cleavage.

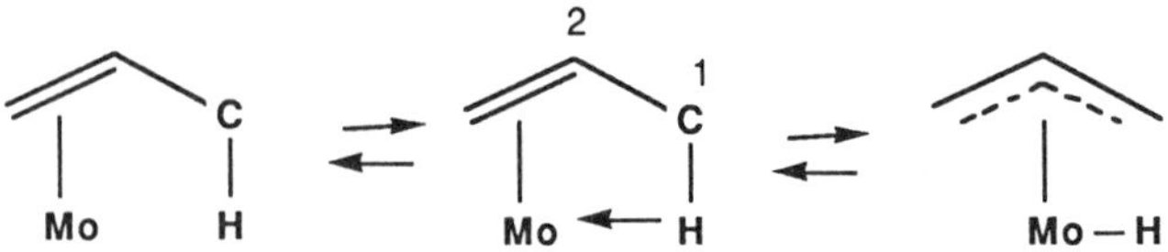

$$\text{(22)}$$

Other recently discovered agostic structures are listed in Table VII.

Although isolable alkane complexes of the type $L_nM(H–R)$ have not yet been reported, they are likely intermediates in alkane oxidative addition. Some evidence for interaction between an alkane and an unsaturated metal fragment has come from studies with matrices (*65*). Bergman *et al.* (*66a*) postulated an alkane complex to account for the isotopic exchange process

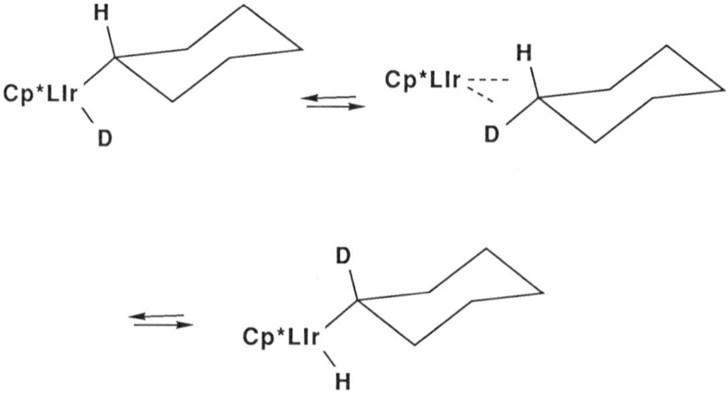

$$\text{(23)}$$

TABLE VII
Some Agostic and Related Systems

Complex	X-ray structure	(H) (ppm)	$^1J(^{13}C, H)$ (Hz)	Remarks[a]	Ref.
Mo(butadiene)(methylallyl)(PMe$_3$)$_2$[b]	Yes	-9.16	84	^1J(C, H terminal): -144 Hz	86
Mo(dtc)CO(PMe$_3$)$_2$(COCH$_3$)	Yes	—	—	η^2-binding via the acyl oxygen is usually preferred	87
Cp$_2$Ru$_2$(dpm)(μ-CO)(μ-CH$_3$)	No	-2.1	—		88
Cp$_2$ClZr–CH=CH-ZrClCp$_2$	Yes	—	"Reduced" (not quoted)		89
Cl(PMe$_3$)$_3$W=CH(t-Bu)	No	-8.3	45		90
Cl(PMe$_3$)$_3$W=CH$_2$	No	-8.0	119	IPR observed	90
IrH(PPh$_3$)$_2$(methylallyl)$^+$	No	-2.2(av)	—	IPR observed	91
RhH(PPh$_3$)$_3$	Yes	—	—	Rh$\cdots$H, 2.87Å	92
Me$_2$Si(C$_5$Me$_4$)$_2$Nd–CH(TMS)$_2$	Yes	—	—	Nd–H, 2.5Å; Nd–C–H, 76°	93
(CO)$_3$Mn(methylallyl)[c]	Yes	-11.8	—		94
Cp(CO)$_2$Mo(PR$_2$BH$_2$–H)	Yes	—	—		95
Cp*Ti(CH$_2$Ph)$_3$	Yes	—	—	Ti$\cdots$H, 2.32Å	96
P(OR)$_3$(CO)$_2$Cr(cycloheptadiene)	Yes	-10.7	—	Cr$\cdots$H, 1.85Å	97
(nbdl)Rh(C$_2$B$_9$H$_9$Me$_2$)	Yes	—	—	Rh–H, 1.9Å	98
(codl)Rh(C$_2$B$_9$H$_9$Me$_2$)	No	-3.94	—		99
[Fe$_3$H(CO)$_9$(CH$_2$)]	No	-10.1	67		100
W(CO)$_3$(PCy$_3$)$_2$	Yes	—	—	W–H, 2.27Å	7

[a] IPR, Isotopic perturbation of resonance.

[b] Agostic group underlined.

[c] Substituted methylallyls were studied; dtc, dithiocarbamate; dpm, bis(diphenylphosphino) methane; Cy, cyclohexyl; nbdl, 1,2,3-norbornadienyl; codl, 1,2,5-cyclooctadienyl.

observed in Eq. (23). Phosphine dissociation was eliminated by using labeled PMe_3. The alternative pathway involving α-elimination with a slipped Cp was thought to be unlikely. Complexation of methane and alkanes in general might be somewhat difficult to detect if the interaction is weak and fast exchange is occurring, but T_1 methods might be useful (67).

Avery (66b) has postulated the presence of an alkane complex to account for the C–H stretch observed at 2690 cm^{-1} in the electron energy loss spectrum of cyclopentane adsorbed on a Pt(111) surface. Suggs et al. (66c) have suggested that the alkane complex is a probable intermediate in the reductive elimination of an alkane from a metal dialkyl.

So far, alkanes such as CH_4 have not displaced H_2 from dihydrogen complexes to give an alkane complex, although it may be that insufficient examples have been studied. Methane is probably an intrinsically poorer ligand than H_2, both because of the greater steric hindrance of the alkane and because only one lobe of the C–H(σ^*) is accessible. Low and Goddard (68) calculate that the transition state for H_2 loss from PdH_2 is only 5 kcal/mol unstable with respect to Pd + H_2, while that for CH_4 elimination from MePdH is 30 kcal/mol unstable with respect to Pd + CH_4. If the trends found for these transition states can be taken to represent the trends that would be found for analogous dihydrogen and methane complexes, then the methane complex may be substantially less stable than the H_2 analog.

There are a number of cases in which essentially planar methyl groups bridge between two metals [e.g., 29 (69) and 30 (70)]. Although these could be considered agostic, it is not clear whether the C–H$\cdots$M interaction is dominant. They may perhaps be better thought of as examples of trigonal bipyramidal carbon with $2e$, three-center M–C–M bonding. There are many other bridging methyl groups, particularly in main group structures, in which C–H$\cdots$M interactions are believed to play a negligible part in the bonding. In Al_2Me_6, for example, the methyl group appears to be essentially sp^3-hybridized. Once again we have an M–C–M $2e$, three-center interaction, but the geometry is different from that found in 29 and 30. Where the metal acts only as a Lewis acid (main group and d^0 systems), M–C–M interactions seem to be preferred. The

29   30

availability of the C–H(σ^*) for backbonding may be the factor that makes C–H $\cdots$ M the preferred bridging mode for electron-rich transition metals.

Just as H_2 complexes tend to act as proton acids, so agostic C–H systems also tend to be acidic. Siegbhan's (71) theoretical study of the transition state for CH_4 elimination from MeNiH (**31**) gives us some indication of the charge distribution to be expected in an agostic system. The diagram shows the calculated charge distributions (electrons) and relative overlap populations.

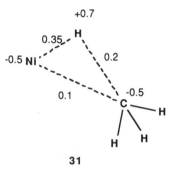

31

More direct evidene for the acidity of agostic C–H bonds comes from the work of Legg *et al.* (72) on the facile deprotonation of an agostic C–H in a Co(III) complex. An unstable methane complex is also a reasonable intermediate for Watson's (73a) heterolytic methane activation by Cp_2^*LuMe, a complex that is strongly Lewis acidic but a negligible π-base.

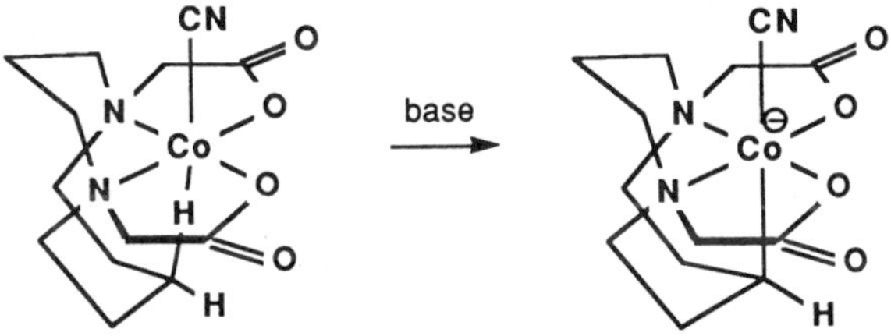

$$(24)$$

Agostic systems have also been recognized in which the C–H bond in question reversibly oxidatively adds to the metal. Isotopic exchange studies by Crabtree, Lavin, and Holt (32) led them to propose that this is the case for **32**. The dihydride group of **32** probably becomes a dihydrogen ligand in the oxidative addition product **33**, in view of the known structure of **34**, a species in which oxidative addition of a 7,8-benzoquinoline C–H bond has taken place. The stronger M–C bond in an aryl than a benzyliridium complex probably accounts for the difference in structure betwen **32** and **34**.

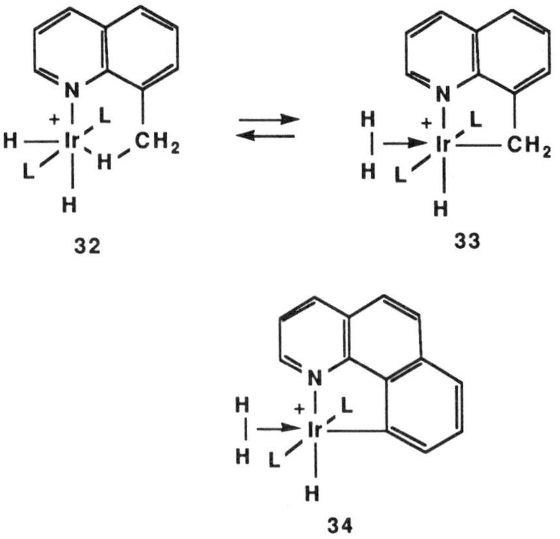

(25)

Shapley *et al.* (*73b*) have shown that the ethyl group in [HOs$_3$-(CO)$_{10}$(Et)] has an agostic α-C–H bond and that the complex is in equilibrium with the corresponding ethylidene hydride by an α-elimination. Interestingly, the β-elimination to give ethylene and H$_2$Os(CO)$_{10}$ is ca. 100 times slower. The reason may be that the α-C–H bond of the ethyl in an H–C–M$_1$–M$_2$ cluster bears the same geometric relationship to the metal that a β-C–H bond does in a mononuclear complex H–C–C–M.

Although the last two examples discussed contain agostic interactions with third-row elements, these remain rarer than cases involving second- and especially first row elements. For example, [(η^6-C$_6$H$_6$)Re(η^5-C$_8$H$_{11}$)(H)] shows no tendency to adopt the alternative agostic structures that are so common for related manganese species (*73c*). The probable reason is that metal–ligand bond strengths are higher in the third row and so the product of oxidative addition is usually thermodynamically favored. Legzdins *et al.* (*73d*) have cited several cases of 16e species in which *no* agostic interaction is apparent from their X-ray structures [e.g., CpM(NO)(CH$_2$SiMe$_3$) (M = Mo, W) and CpW(NO)(CH$_2$Me$_3$)].

All this evidence suggests that η^2-H$_2$ and η^2-C–H systems may be common intermediates for both the oxidative addition and heterolytic activation reactions [Eq. (26)].

The word agostic is probably better reserved for cases of C–H bond chelation, rather than 2e, two-center CF $\cdots$ M and CCl $\cdots$ M interactions, which are electronically quite distinct (*73e*).

A number of cases (*74*) have been reported in which the product of oxidative addition of a silane has shown unusual structural and spectro-

$$\begin{array}{c} \overset{\displaystyle R}{\underset{\displaystyle |}{}} \\ M \end{array} \quad + \quad H-X \quad \longrightarrow \quad \begin{array}{c} \overset{\displaystyle R}{\underset{\displaystyle |}{}} \\ M \end{array}\overset{\displaystyle H}{\underset{\displaystyle X}{}}$$

heterolytic
activation

oxidative
addition

(26)

$$\begin{array}{c} R-H \\ + \\ M-X \end{array} \qquad \begin{array}{c} \overset{\displaystyle R}{\underset{\displaystyle |}{}} \\ M-H \\ | \\ X \end{array}$$

(X = CR$_3$ or H)

scopic behavior. This has led several groups to postulate that there is an attractive Si $\cdots$ H interaction in the adduct. For example, the presence of an H $\cdots$ Si interaction in Cp(CO)$_2$Mn(SiPh$_3$)H (**35**) was first proposed by Graham and Bennett (*74d,74h*). Kaesz *et al.* (*74a*) observed a Raman band at 1355 cm^{-1}, consistent with the presence of such an interaction. Similarly, Re$_2$(CO)$_8$H$_2$(SiPh$_2$) shows no splitting of the bridging hydrogen IR absorptions [a splitting is found for H$_2$Re$_2$(CO)$_8$], suggesting a similar structure. Corriu *et al.* (*74g*) showed that the ^1J(Si, H) for the Si $\cdots$ H interaction is 65 Hz, intermediate between those characteristic of classical silyl hydrides ($\sim$6 Hz) and those of free silanes ($\sim$200 Hz). Finally, a neutron diffraction study of [(MeCp)Mn(CO)$_2$(SiPh$_2$F)H] (**35a**) by Schubert *et al.* (*74e*) gave Mn–H–Si, 88.2(2)°; Mn–H, 1.569(4) Å; and Si–H, 1.802(5) Å. One possible interpretation of these results is that these species are not true oxidative addition products but have 2*e*, three-center bonds. Bennett *et al.* (*74c*) challenged this view on the grounds that Re$_2$(CO)$_8$H$_2$(SiPh$_2$) has the same M–Si distance as found in the unbridged complex W$_2$(CO)$_8$(SiPh$_2$).

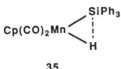

35

An alternative explanation, which we propose, is that the M–H bond is interacting with the Si–R σ^* orbital. This implies that the system should only be slightly distorted from the expected structure of the oxidative adduct and the M–Si bond should be normal, and it explains why analogous structures have not yet been observed for alkyl hydrides. The σ^*

orbital of an X–R (X = Si) bond would be expected to have a larger X character and to be lower in energy than an X–R σ^* (X = C), Si being more electropositive than C. This picture also explains why the H in **35a** is trans to F rather than P; the more electronegative the substituent at Si, the greater the Si character of the corresponding σ^* orbital. Consistent with this picture, the H in **35a** is trans to the F rather than the Ph substituent on Si; the σ^* orbital will have greater Si character and therefore overlap better with the M–H σ bond, the more electronegative the substituent.

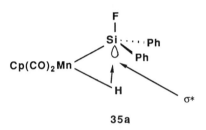

35a

Alkyl and aryl hydrides are worthy of closer structural study. Some may be significantly distorted; others may be so distorted as to be better described as alkane complexes. An X-ray study of [IrHCl(Ph)(P(*i*-Pr)$_3$)$_2$] (**36**) shows a C–Ir–H angle of 77.9° (*75*). In addition, there is no free rotation about the Ir–Ph bond and all six carbons of this ligand are distinct in the room temperature ^{13}C NMR. Possibly, the C$_1$–C$_2$ σ^* could be an acceptor orbital for the interaction with the Ir–H bond.

This picture predicts that only the rotamer shown in **36** would give such an interaction and that other groups containing relatively electropositive atoms X or relatively electronegative substituents R might behave in the same way (TeR, AsR$_3$, CF$_3$, PF$_3$, etc.).

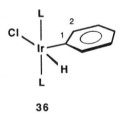

36

Genuinely coordinated Si–H bonds have recently been reported, for example, in the complexes **37** and **38** (*76a*). Other X–H groups can also chelate via 2e, three-center interactions; Roundhill *et al.* (*76b*) have found an example involving an amide N–H bond in a Pt(II) complex. The Pt $\cdots$ H distance is 2.318 Å and the ν(NH) frequency in the IR is lowered by 200 cm^{-1}.

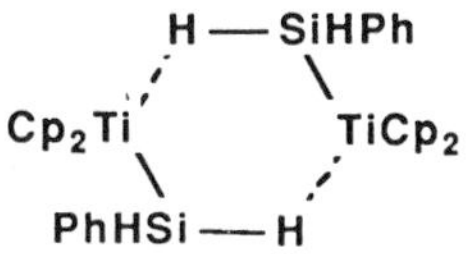

H — SiHPh

Cp$_2$Ti

TiCp$_2$

PhHSi — H

37

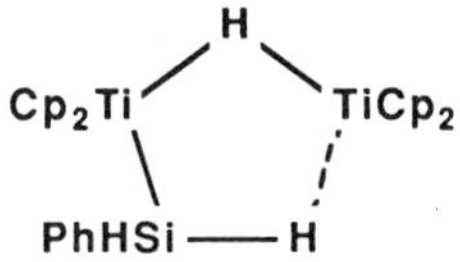

H

Cp$_2$Ti

TiCp$_2$

PhHSi —— H

38

The structure of **39** is interesting in that the gold hydride (*77a*) can be seen as isolobal with H$_2$. As mentioned in Table V, the analogous IrH$_4$(PR$_3$)$_3$ may be nonclassical (*25*) and the related [IrH$_2$(H$_2$)$_2$(PR$_3$)$_2$]$^+$ certainly is an H$_2$ complex (*20*). **39** can also be seen as a stepping stone from M–H–M′ interactions where M′ is a main group element and M–H–M′ structures in which both M and M′ are transition elements. In the latter case the bridging of an H atom via a 2*e*, three-center bond has been well recognized for many years.

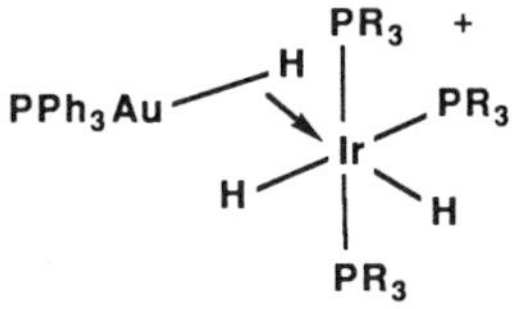

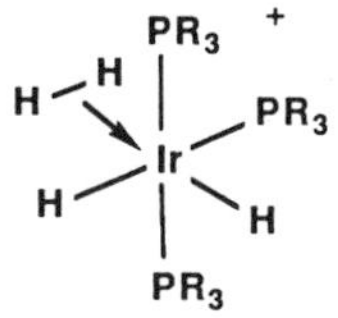

PR$_3$ +

PPh$_3$Au   H

PR$_3$

H   Ir   H

PR$_3$

39

PR$_3$ +

H   H

PR$_3$

H   Ir   H

PR$_3$

Stone's group (*77b*) has pointed out the formal isolobal relationship between an agostic methyl and the W–Au cluster shown below, although in this case the gold does have other available electrons and so may bond via a lone pair (**40b**), rather than via the pair of electrons in the C–Au bond (**40a**).

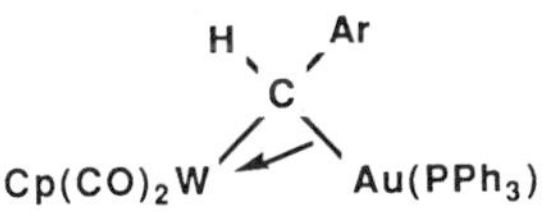

H   Ar

C

Cp(CO)$_2$W   Au(PPh$_3$)

40a

H   Ar

C

Cp(CO)$_2$W   :Au(PPh$_3$)

40b

An important analogy between M–H–M and M–H–X structures (X = H, C, or a main group element) is that both tend to be bent. This aspect of transition metal M–H–M structures has been stressed by Bau (*78*). The main driving force for the X–H or M–H bond going partially or fully side-on in its binding to M is presumably the contribution to the stabilization of the complex that is made by backbonding from M(d_π) to the H–X or the H–M σ^* orbital. This overlap is zero in an η^1 or end-on structure. Steric effects would be expected to oppose the formation of a strictly side-on structure except in the case of H_2 itself, so a compromise appears to be reached between the two extremes.

We might anticipate that bridging hydrides would be stronger proton acids than terminal ones, but no data seem to be available to test this hypothesis.

In a recent development, Bau et al. (79) have shown that the $[HW_2(CO)_{10}]^-$ anion has a bent W–H–W bridge; previous data had been collected on disordered salts and so were less reliable.

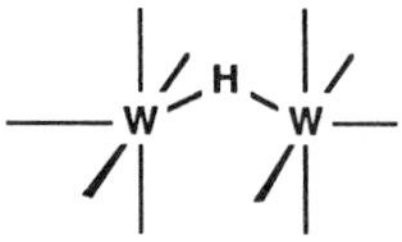

Bau (80) has also studied M–H–B bridges, also bent. One example is $FeH(\eta^1\text{-}H\text{-}BH_3)(dmpe)_2$, which is isolobal with Morris' H_2 complex $[FeH(H_2)(dppe)_2]^+$. To complete the series, Bau has also found that $[B_2H_7]^-$ contains a B–H–B bridge, again bent (81,82).

Rare exceptions to the bent M–H–M "rule" have been found; for example, Schumann et al. (83) reported a linear bridge in Cp_3Nd–H–$NdCp_3$-, but only on the basis of an X-ray study. The bonding model discussed above is able to rationalize this result because neodymium, as a d^0 metal, is not capable of significant backbonding. In contrast, the BH_3 fragment is capable of limited backbonding by hyperconjugation, which could account for the bent B–H–B unit in $[B_2H_7]^-$. Me_3Al–H–$AlMe_3$, on the other hand, is linear (84) because the $AlMe_3$ fragment, unlike BH_3-, is not capable of significant backbonding. Also in agreement with this picture is the recent discovery by Evans, Atwood, et al. (85) of a samarium(III) complex having one of the shortest $Ln \cdots HC$ agostic interactions known (2.29 Å). Such a strong interaction would normally imply a strongly side-on structure (32a), but in fact the C–H is almost exactly end-on. It is among the d^0 metals that the search for stable $\eta^1\text{-}H_2$ complexes might most usefully be carried out.

$2e$, 3-center interactions involving X–Y bonds in which neither X nor Y is H are very rare. There is some evidence, however, for an agostic C–C bond in $[(\eta^4\text{-}C_5H_4Me_2)IrL_2]^+$ (48a,101).

IV

ADDENDUM

Conroy-Lewis and Simpson (102) have reported on $[CpRu(Ph_2P(CH_2)_nPPh_2)H_2]^+$ ($n = 1$–3), formed in the protonation of the corresponding neutral monohydrides. For $n = 1$, the complex is classical. For $n = 3$, the species is nonclassical. For $n = 2$, both tautomers can be detected by ^1H NMR, and very different T_1 values are found for the two isomers: 30 ms and 1.3 s at 298 K. For $n = 3$, $^1J(H,D)$ is 21.9 Hz , the lowest value yet reported for a dihydrogen complex. Chinn and Heinekey (15b) have prepared $[CpRu(Me_2P(CH_2)_2PMe_2)H_2]^+$, which is also a tautomeric mixture. The nonclassical (major) isomer, which crystallizes from the mixture, is the better proton acid (pK $= 17.6$ in CH_2Cl_2).

The expected IPR effects are not always present for H_2 complexes and agostic alkyls. They are not seen for partially deuterated nonclassical polyhydrides such as RuH_4L_3 (103), for example. A possible reason is the small chemical shift difference between the MH and $M(H_2)$ sites, as suggested by the small differences found in the chemical shifts of dihydrogen and dihydride tautomers in cases (8, 15b, 102) where both can be seen independently in the NMR: $WH_2(CO)_3(Pi\text{-}Pr_3)_2$, -4.2 and -3.2 δ, $[CpRu(Me_2P(CH_2)_2PMe_2)H_2]^+$, -9 and -8.6 δ.

In agostic systems, there can be changes in $^1J(C-H)$ on deuterium substitution due to IPR effects. Unfortunately, there are cases of unambiguously agostic systems, such as $TiR(dmpe)Cl_3$ (R $=$ Me or Et), in which little, if any shift of δ_H or $^1J(C,H)$ is seen on isotopic substitution (104) Bercaw et al. (105) have reported on Cp_2^*ScR (R $=$ Me, Et, and n-Pr). The ethyl is almost certainly β-agostic from IR data, even though IPR effects were very small. In the methyl case, essentially no IPR is seen and the structure is probably classical; unfortunately, the complex was disordered in the crystal. The propyl derivative appears to be classical probably for steric reasons. The lack of back donation in d^0 agostic systems may be the factor that contributes to the lack of IPR, because the perturbation of the agostic C-H bond is expected to be small in the absence of back donation.

ACKNOWLEDGMENTS

We thank G. J. Kubas, K. Zilm, S. J. Simpson, D. M. Heinekey, and R. H. Morris for preprints and O. Eisenstein for discussions and the Department of Defence, Army Research Office, for supporting our work in this area.

REFERENCES

1. R. Hoffmann, *Angew. Chem. Int. Ed.* **21**, 711 (1982).
2. G. L. Wendt and R. S. Landauer, *J. Am. Chem. Soc.* **42**, 930 (1920); G. D. Carney and R. N. Porter, *J. Chem. Phys.* **60**, 4251 (1974); T. Oka, *Phys. Rev. Lett.* **45**, 531 (1980).
3. G. A. Olah, *Curr. Top. Chem.* **80**, 19 (1979).
4. M. Brookhart and M. L. H. Green, *J. Organomet.Chem.* **250**, 395 (1983).
5. R. G. Teller and R. Bau, *Struct. Bonding (Berlin)* **44**, 1 (1981).
6a. P. J. Hay, *J. Phys. Chem.*, **103**, 466 (1984).

6b. Y. Jean, O. Eisenstein, F. Volatron, B. Maouche, and F. Sefta, *J. Am. Chem. Soc.* **108**, 6587 (1986).

7. G. J. Kubas, R. R. Ryan, P. J. Vergamini, and H. J. Wasserman, *J. Am. Chem. Soc.* **106**, 451 (1984); H. J. Wasserman, G. J. Kubas, and R. R. Ryan, *ibid.* **108**, 2294; G. J. Kubas, R. R. Ryan, and D. Wrobleski, *ibid.* **108**, 1239; G. J. Kubas, C. J. Unkefer, B. I. Swanson, and E. Fukushima, *ibid.* **108**, 7000 (1986).

8a. S. P. Church, F. W. Grevels, H. Herrman, and K. Schaffner, *Chem. Commun.* 30 (1985).

8b. R. K. Upmacis, G. E. Gadd, M. Poliakoff, M. B. Simpson, J. J. Turner, R. Whyman, and A. F. Simpson, *Chem. Commun.* 27 (1985).

8c. R. L. Sweaney, *J. Am. Chem. Soc.* **107**, 2734 (1985); *Organometallics* **6**, 387 (1986).

8d. R. K. Upmacis, M. Poliakoff, and J. J. Turner, *J. Am. Chem. Soc.* **108**, 2547 and 3645 (1986).

9. R. H. Crabtree and D. G. Hamilton, *J. Am. Chem. Soc.*, **108**, 3124 (1986).

10. T. V. Ashworth and E. Singleton, *Chem. Commun.* 705 (1976).

11. G. J. Kubas, *Chem. Commun.* 61 (1980).

12. R. H. Morris, J. F. Sawyer, M. Shiralian, and J. P. Zubkowski, *J. Am. Chem. Soc.* **107**, 5581 (1985).

13. The classical formulation is supported by the T_1^1 of 3 sec at $-80°C$; R. H. Crabtree and D. G. Hamilton, unpublished data, 1986.

14. M. Aresta, P. Gioannoccaro, M. Rossi, and A. Sacco, *Inorg. Chim. Acta*, **5**, 115 (1971).

15a. F. M. Conroy-Lewis and S. Simpson, *Chem. Commun.* 506 (1986).

15b. M. Chinn and M. Heinekey, personal communication (May, 1987).

16. T. M. Gilbert and R. G. Bergman, *J. Am. Chem. Soc.* **107**, 3502 (1985).

17a. R. H. Crabtree and M. Lavin, *Chem. Commun.* 794; (1985).

17b. Ibid 1660 (1985).

18. R. H. Crabtree and D. G. Hamilton, unpublished data, 1986.

19. M. Bautista, K. A. Earl, R. H. Morris, and A. Sella, personal communication, 1986.

20. R. H. Crabtree, M. Lavin, and L. Bonneviot, *J. Am. Chem. Soc.* **108**, 4032 (1986).

21. N. Blombergen, E. M. Purcell, and R. V. Pound, *Phys. Rev.,* **73**, 679 (1948); J. A. Pople, W. G. Schmeider, and H. J. Bernstein "High Resolution NMR." McGraw-Hill, New York 1959.

22. B. Chaudret and R. Poilblanc, *Organometallics,* **4**, 1722 (1985).

23. H. C. Clark and M. J. Hampden-Smith, *J. Am. Chem. Soc.* **108**, 3829 (1986).

24. P. Mura and A. Segre, *Angew. Chem. Int. Ed.* **25**, 460 (1986).

25. L. F. Rhodes and K. G. Caulton, *J. Am. Chem. Soc.* **107**, 259 (1985).

26a. F. N. Tebbe, *J. Am. Chem. Soc.* **95**, 5412 (1973); M. D. Curtis, L. G. Bell, and W. M. Butler, *Organometallics* **4**, 701 (1985); J. F. Reynoud, J. C. Leblanc, and C. Moise, *Tr. Met. Chem.* **10**, 291 (1985).

26b. F. N. Tebbe (26c) has found a $^1J(H, H')$ of 100 Hz in $Cp_2NbH_3·AlEt_3$.

26c. F. N. Tebbe cited by J. A. Labinger in Chapter 25 of "Comprehensive Organometallic Chemistry" (G. Wilkinson, ed.). Pergamon, Oxford, 1982.

27. A. P. Ginsberg, J. M. Miller, and E. Loubek, *J. Am. Chem. Soc.* **83**, 4909 (1961); A. P. Ginsberg and C. R. Sprinkle, *Inorg. Chem.* **8**, 2212 (1969).

28. R. H. Crabtree, B. E. Segmuller, and R. J. Uriarte, *Inorg. Chem.* **24**, 1949 (1985).

29. K. W. Zilm, M. W. Kummer, R. A. Merrill, and G. J. Kubas, *J. Am. Chem. Soc.* **108**, 7837 (1986).

30. R. H. Morris, K. A. Earl, R. L. Luck, N. J. Lazarowyck, and A. Sella, *Inorg. Chem.* in press, and personal communication (1986).

31. R. H. Crabtree, H. Felkin, T. Fillebeen-Khan, and G. E. Morris, *J. Organomet. Chem.* **168,** 183 (1979).

32a. R. H. Crabtree, E. M. Holt, M. E. Lavin, and S. M. Morehouse, *Inorg. Chem.* **24,** 1986 (1985).

32b. M. Clague and R. H. Crabtree, unpublished data, 1986.

33. R. H. Crabtree, *Inorg. Chem Acta* **125,** L7 (1986).

34. K. I. Gill and J. Schwartz, *J. Am. Chem. Soc.* **100,** 3246 (1978).

35. K. W. Chiu, R. A. Jones, G. Wilkinson, A. M. R. Galas, M. B. Hursthouse, and K. M. A. Malik, *J. Chem. Soc. Dalton* 1204 (1981).

36. R. Stern and L. Sajus, *Tetrahedron Lett.* 6313 (1968).

37. G. G. Hlatky and R. H. Crabtree, *Coord. Chem. Rev.* **65,** 1 (1985).

38. M. J. Auburn and S. R. Stobart, *Chem. Commun.* 281 (1985).

39. J. M. Brown, A. E. Derome, and S. A. Hall, *Tetrahedron Lett.* **41,** 4647 (1985).

40a. C. E. Johnson, B. J. Fisher, and R. Eisenberg, *J. Am. Chem. Soc.* **105,** 7772 (1983).

40b. C. E. Johnson and R. Eisenberg, *J. Am. Chem. Soc.* **107,** 3148 (1985); M. J. Burk and R. H. Crabtree, unpublished data 1986.

41. G. A. Ozin and J. Gracia Prieto, *J. Am. Chem. Soc.* **108,** 3099 (1986).

42. C. Jargue and O. Novaro, *J. Am. Chem. Soc.* **108,** 3507 (1986).

43. J. J. Low and W. A. Goddard, *J. Am. Chem. Soc.* **106,** 8321 (1984).

44. M. R. A. Blomberg and P. E. M. Sieghahn, *J. Chem. Phys.* **78,** 986 (1983).

45. D. Olivier, M. Richard, M. Che, F. Bozon-Verdura, and R. B. Clarkson, *J. Phys. Chem.* **84,** 420 (1980). J. Michalik, M. Narayama, and L. Kevan, *J. Phys. Chem.* **88,** 5236 (1984); L. Bonneviot, personal communications, 1986.

46. J. J. La Placa and J. A. Ibers, *Inorg. Chem.* **4,** 778 (1965).

47. D. M. Roe, P. M. Bailey, K. Moseley, and P. M. Maitlis, *Chem. Commun.* 1273 (1972).

48a. R. H. Crabtree, *Chem. Rev.* **85,** 245 (1985).

48b. We use the designation $2e(\sigma)$ to distinguish the $2e$, three-center bonding in η^2-alkene complexes.

49. M. Brookhart and M. L. H. Green, *J. Organomet. Chem.* **250,** 395, (1983).

50. H. B. Burgi and J. D. Dunitz, *Acc. Chem. Res.* **16,** 153 (1983).

51. D. Baudry, M. Ephritikine, H. Felkin, and J. Zakrzewski, *Chem. Commun.* 1243 (1980), *Tetrahedron Lett.* 1283 (1984); M. J. Burk and R. H. Crabtree, *Chem. Commun.* 606 (1982); 1829 (1985).

52. W. D. Jones and F. J. Feher, *J. Am. Chem. Soc.* **104,** 4240 (1982).

53. H. H. Karsch, *Chem. Ber.* **117,** 3123 (1984); J. Halpern, *Inorg. Chim. Acta.* **100,** 41 (1985).

54a. Z. Dawoodi, M. L. H. Green, V. S. B. Mtetwa, and K. Prout. *Chem. Commun.* 1410 (1982).

54b. J. M. Williams, cited as ref. 8 of O. Eisenstein and Y. Jean, *J. Am. Chem. Soc.* **107,** 1177 (1985).

55. J. W. Bruno, G. M. Smith, T. J. Marks, C. K. Fair, A. J. Schultz, and J. M. Williams, *J. Am. Chem. Soc.* **108,** 40 (1986).

56. R. B. Cracknell, A. G. Orpen, and J. L. Spencer, *Chem. Commun.* 326 (1984).

57. M. Saunders, M. H. Jaffe, and P. Vogel, *J. Am. Chem. Soc.* **93,** 2558 (1971); R. B. Calvert and J. R. Shapley, *J. Am. Chem. Soc.* **100,** 7726 (1978).

58. J. D. Fellman, R. R. Schrock, and D. D. Traficante, *et al., Organometallics* **1,** 481 (1982).

59. R. Benn, S. Holle, P. Jolly, R. Mynott, and C. Ramao, *Angew. Chem. Int. Ed.* **25,** 555 (1986).

60. A. Berry, Z. Dawoodi, A. E. Derome, J. M. Dickinson, A. J. Downs, J. Green, M. L. H. Green, P. M. Hare, M. P. Hare, M. P. Payne, D. W. H. Rankin, and H. E. Robertson, *Chem. Commun.* 520 (1986).

61. J. D. Duncan, J. C. Green, M. L. H. Green, and K. A. McLauchlan, *Discuss. Faraday Soc.* **47**, 178 (1969).

62a. R. J. Goddard, R. Hoffman, and E. D. Jemmis, *J. Am. Chem. Soc.* **102**, 7667 (1980).

62b. O. Eisenstein and Y. Jean, *J. Am. Chem. Soc.* **107**, 1177 (1985).

62c. A. Demollions, Y. Jean and O. Eisenstein, *Organometallics* **5**, 1457 (1986).

62d. N. Kaga, S. Obara, and K. Morokuma, *J. Am. Chem. Soc.* **106**, 4625 (1984); *J. Organomet. Chem.* **270**, C33 (1984).

62e. N. Koga, S. Obara, K. Kitaura, and R. Morokuma, *J. Am. Chem. Soc.* **107**, 7109 (1985).

62f. T. J. Katz and T. M. Sivavec, *J. Am. Chem. Soc.* **107**, 737 (2985).

62g. D. L. Lichtenberger and G. E. Kellog, *J. Am. Chem. Soc.* **108**, 2560 (1986).

62h. J. Y. Saillard and R. Hoffman, *J. Am. Chem. Soc.* **106**, 2006 (1984).

62i. P. J. Hay, *Chem. Phys. Lett.* **103**, 466 (1984).

62j. J. K. Burdett, J. R. Philips, M. R. Pourian, M. Poliakoff, R. R. Turner, and R. Upmacis, *Inorg. Chem.*, submitted.

63a. G. F. Schmidt and M. Brookhart, *J. Am. Chem. Soc.* **107**, 1443 (1985).

63b. R. B. Cracknell, A. G. Orpen, and J. L. Spencer, *Chem. Commun.* 1005 (1986).

64a. J. W. Park, P. B. Mackenzie, W. P. Schafer and R. H. Grubbs, *J. Am. Chem. Soc.* **108**, 6402 (1986).

64b. D. G. Bourner, L. Brammer, M. Green, G. Moran, A. G. Orpen, C. Reeve, and C. J. Schaverien, *Chem. Commun.* 1409 (1985).

65. J. J. Turner, J. K. Burdett, R. N. Perutz, and M. Poliakoff, *Pure Appl. Chem.* **49**, 271, (1977).

66a. R. G. Bergman, *J. Am. Chem. Soc.* **108**, 1537 (1986).

66b. N. R. Avery, *Surf. Sci.* **163**, 357 (1985).

66c. J. W. Suggs, M. J. Workulich, and S. D. Cox, *Organometallics* **4**, 1101 (1985).

67. The rotational correlation times (and therefore the T_1 values) of free and bound CH_4 should differ substantially.

68. J. J. Low and W. A. Goddard, *J. Am. Chem. Soc.* **106**, 8321 (1984).

69. R. B. Waymouth, B. D. Santarsiero, and R. H. Grubbs, *J. Am. Chem. Soc.* **106**, 4050 (1984).

70. P. L. Watson, *J. Am. Chem. Soc.* **105**, 6491 (1983).

71. M. R. A. Blomberg, U. Brandemark and P. E. M. Siegbahn, *J. Am. Chem. Soc.* **105**, 5557 (1983).

72. K. Kanamori, W. E. Broaderick, R. F. Jordan, R. D. Willett and J. I. Legg, *J. Am. Chem. Soc.* **108**, 7122 (1986).

73a. P. L. Watson, *J. Am. Chem. Soc.* **105**, 6491 (1983).

73b. M. Cre-Uchima, J. R. Shapley, and G. M. St. George, *J. Am. Chem. Soc.* **108**, 1316 (1986).

73c. A. E. Derome, M. L. H. Green, and D. O'Hare, *Chem. Commun.* 343 (1986).

73d. L. Carlton, J. C. Davidson, P. Ewing, L. M. Muir, and K. W. Muir, *Chem. Commun.* 1474 (1985).

73e. H. H. Murray, J. P. Fackler, and D. A. Tocher, *Chem. Commun.* 1278 (1985).

74a. M. A. Andrews, S. W. Kirtley, and H. D. Kaesz, *Adv. Chem. Ser.* **167**, 215 (1978).

74b. A. J. Hart-Davis and W. A. G. Graham, *J. Am. Chem. Soc.* **94**, 4388 (1971).

74c. M. Cowie and M. J. Bennett, *Inorg. Chem.* **16**, 2321 and 2325 (1977).

74d. W. A. G. Graham and M. J. Bennett, *Chem. Eng. News* **48**(24), 75 (1970).

74e. U. Schubert, K. Ackermann, and B. Worle, *J. Am. Chem. Soc.* **104,** 7378 (1982).
74f. R. A. Smith and M. J. Bennett, *Acta Crystallogr. B.* **B33,** 1113 (1977).
74g. K. Colomer, R. J. P. Corrier, C. Manzin, and A. Vioux, *Inorg. Chem.* **21,** 368 (1982).
74h. W. A. G. Graham, *J. Organometal. Chem.* **300,** 81 (1986).
75. H. Werner, A. Hohn and M. Dzallias, *Angew. Chem. Int. Ed.* **25,** 1090 (1986).
76a. U. Schubert, G. Scholz, J. Muller, K. Ackermann, B. Worle, and R. Stansfield, *J. Organomet. Chem.* **306,** 303 (1986).
76c. O. Hedden, D. M. Roundhill, W. C. Fultz, and A. L. Rheingold, *Organometallics* **5,** 336 (1986).
77a. H. Lehner, B. Matt, P. S. Pregosin, L. M. Venanzi and A. Albinati, *J. Am. Chem. Soc.* **104,** 6825 (1982).
77b. G. A. Carriedo, J. A. R. Howard, F. G. A. Stone, and M. J. Went, *J. Am. Chem. Soc. Dalton* 2545 (1984).
78. R. Bau, *Acc. Chem. Res.* **12,** 176 (1979).
79. D. W. Hart, R. Bau, and T. F. Koetzle, *Organometallics* **4,** 1590 (1985).
80. R. Bau, H. S. H. Yuan, M. V. Baker, and L. D. Field, *Inorg. Chim. Acta* **114,** L27 (1986).
81. S. G. Shore, S. H. Lawrence, M. I. Watkins, and R. Bau, *J. Am. Chem. Soc.* **104,** 7669 (1982).
82. S. G. Shore, S. H. Lawrence, M. I. Watkins and R. Bau, *J. Chem. Soc. Dalton* 1743 (1986).
83. H. Schumann, W. Genthe, E. Hahn, M. B. Hossain, and D. van der Hehn, *J. Organomet. Chem.* **299,** 67 (1986).
84. J. L. Atwood, D. C. Hrcnir, R. D. Rogers, and J. A. R. Howard, *J. Am. Chem. Soc.* **103,** 6787 (1981).
85. W. J. Evans, D. K. Drummond, S. G. Bott, and J. L. Atwood, *Organometallics* **5,** 2389 (1986).
86. M. Brookhart, K. Cox, G. N. Cloke, J. C. Green, M. L. H. Green, D. M. Hare, J. Bashkin, A. E. Derome, and P. D. Grebnik, *J. Chem. Soc. Dalton* 423 (1985).
87. E. Carmona, L. Sanchez, J. M. Marin, M. L. Poveda, J. L. Atwood, R. D. Priester, and R. D. Rogers, *J. Am. Chem. Soc.* **106,** 3214 (1984).
88. D. L. Davies, B. P. Gracey, V. Guerchais, S. A. R. Knox, and A. G. Orpen, *Chem. Commun.* 841 (1984).
89. G. Erker, W. Fromberg, K. Angermund, and C. Kruger, *Chem. Commun.* 372 (1986).
90. S. J. Holmes, D. N. Clark, H. W. Turner, and R. R. Schrock, *J. Am. Chem. Soc.* **104,** 6322 (1982).
91. O. W. Howarth, C. H. McAteer, P. Moore, and G. E. Morris, *J. Chem. Soc. Dalton* 1171 (1984).
92. T. P. Hanusa and W. J. Evans, *J. Coord. Chem.* **14,** 223 (1986).
93. G. Jeske, L. E. Schlock, P. N. Swepston, H. Schumann, and T. J. Marks, *J. Am. Chem. Soc.* **107,** 8103 (1985).
94. C. G. Kreiter, M. Leyendecker, and W. S. Sheldrick, *J. Organomet. Chem.* **302,** 217 (1986); see also F. Timmers and M. Brookhart, *Organometallics* **4,** 1365 (1985).
95. W. F. McNamara, E. N. Duesler, and R. T. Paine, *Organometallics* **5,** 380 (1986).
96. G. Michael, J. Kaub, and C. G. Kreiter, *Chem. Ber.* **118,** 3944 (1985); see also *Angew. Chem. Int. Ed.* **24,** 502 (1985).
97. G. Michael, J. Kaub, and C. G. Kreiter, *Chem. Ber.* **118,** 3944 (1985); see also G. Micheal, J. Kaub, and C. G. Kreiter, *Angew. Chem. Int. Ed.* **24,** 502 (1985).
98. D. M. Speckman, C. B. Knobler, and M. F. Hawthorne, *Organometallics* **4,** 1692 (1985).

99. D. M. Speckman, C. B. Knobler, and M. F. Hawthorne, *Organometallics* **4,** 426 (1985).
100. J. C. Vites, G. Jacobsen, T. K. Dutta, and T. P. Fehlner, *J. Am. Chem. Soc.* **107,** 5563 (1985).
101. R. H. Crabtree, R. P. Dion, D. J. Gibboni, D. V. McGrath, and E. M. Holt, *J. Am. Chem. Soc.* **108,** 7222 (1986).
102. F. M. Conroy-Lewis and S. J. Simpson, *Chem. Commun.*, in press, (1987).
103. D. G. Hamilton and R. H. Crabtree, unpublished data, 1987; J. Halpern, personal communication, 1987.
104. M. L. H. Green, personal communication, 1987.
105. M. E. Thompson, S. M. Baxter, A. R. Bulls, B. J. Burger, M. C. Nolan, B. D. Santarsiero, W. P. Schaefer, and J. E. Bercaw, *J. Am. Chem. Soc.* **109,** 203 (1987).

ADVANCES IN ORGANOMETALLIC CHEMISTRY, VOL. 28

Organometallic Compounds Containing Oxygen Atoms

FRANK BOTTOMLEY and LORI SUTIN

Department of Chemistry
University of New Brunswick
Fredericton, New Brunswick, Canada E3B 6E2

I
INTRODUCTION

The area of chemistry covered by this review is both old and new. The first reported investigations of the oxidation of organometallic complexes with dioxygen appear to have been conducted by Fischer and co-workers and resulted in the preparation of $CpVCl_2(O)$ (*1*) and $[CpCrO]_4$ (*2*), though the true structure of the latter as $[CpCr(\mu_3-O)]_4$ (Cp = $\eta^5-C_5H_5$) was only realized much later (*3*) (Fig. 1). Oxidation of carbonyls and alkyls may have been attempted much earlier, but the results were not reported because no characterizable products were obtained. After the early work

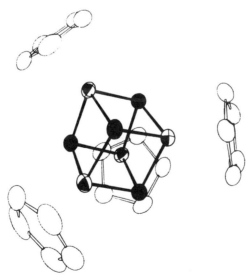

Fig. 1.   Structure of $[CpCr(\mu_3-O)]_4$.

339

of Fischer and the preparation of cis-$[CpMo(O)]_2(\mu\text{-}O)_2$ by Green (4), $[CpTiCl_2]_2(\mu\text{-}O)$ by Corradini and Allegra (5), and $(\eta^1\text{-}C_6H_5)VCl_2(O)$ by Reichle and Carrick (6), there was a period during which organometallic oxo compounds were obtained mainly serendipitously (by inadvertent admission of air into a reaction mixture) or as the unexpected product of a reaction between an organometallic compound and a reagent containing oxygen. These discoveries were not followed up. Only recently have a reasonably large number of different types of organometallic oxo compounds been prepared, and even then it is better to describe synthetic methods for several types as being rational exploitation of serendipitous or unexpected discoveries than to pretend that the compounds were obtained by a planned synthesis. The progress achieved by this exploitation is nonetheless spectacular, as shown by the work of Herrmann on $Cp^5Re(O)_3$ (7) $[Cp^5 = \eta^5\text{-}C_5(CH_3)_5]$, Bottomley on $[CpM]_m(\mu_3\text{-}O)_n$ clusters (8–10), and Thewalt on the hydrolysis products of Group 4 cyclopentadienyl derivatives (11–13).

The present article is restricted to compounds of the transition elements which have a metal–carbon and a metal–oxygen bond, with the same metal being involved in both bonds, and which have been isolated (i.e., are not reaction intermediates). We deal only with atomic oxygen as a ligand, thus precluding compounds containing O_2, alkoxides (OR), and other organic ligands with oxygen as the donor atom(s). Cyanides and isocyanides are excluded as organic ligands. These restrictions are arbitrary, necessitated by space rather than chemical sense. The reader should recognize that there is no clear distinction between $ReCl(O)_3$ and $ReMe(O)_3$ or $Cp^5Re(O)_3$, or between $[CpTiCl_2]_2(\mu\text{-}O)$ and $CpTiCl_2(OEt)$ (Me = CH_3, Et = C_2H_5).

At present the range of organic ligands found in organometallic oxo compounds is quite restricted, the majority being either $\eta\text{-}C_5R_5$ or alkyls or aryls having no β-hydrogen atom. Carbonyls are fairly common in clusters containing oxygen atoms (the cluster $[Os(CO)_3(\mu_3\text{-}O)]_4$ was the first cluster of this type to be characterized (14,14a), but the first monomeric oxocarbonyl with a terminal [M=O] group, $WCl_2(PMePh_2)_2(CO)(O)$, and its ethylene analog, $WCl_2(PMePh_2)_2(\eta^2\text{-}C_2H_4)(O)$, (Ph = C_6H_5), were discovered only recently (15). Interesting oxocarbonyls such as $[Cr(CO)_3(O)_2]^-$ (16) have been obtained in the gas phase and others such as $Mo(CO)_4(O)_2$ (17) by matrix isolation techniques. The metals involved in organometallic oxo compounds are also restricted, being confined at present to the oxophilic elements of Groups 4 to 7 (Group 7 being confined to rhenium) with the addition of a few examples from the lanthanides and actinides and some clusters containing metals from Groups 8 and 9. Only one monomeric complex,

$Os(CH_2SiMe_3)_4(O)$ (*18*), from Group 8 or 9 is known. The discussion of the noncluster complexes is therefore divided according to the ligand types; organometallic clusters containing oxygen atoms are discussed separately. The oxo ligand occurs in five different modes of bonding: terminal [M=O], doubly bridging [$M(\mu_2$-O$)_nM$], triply bridging [$M_3(\mu_3$-O)], and in interstitial quadruply and quintuply bridging forms. In this respect organometallic oxo complexes are no different from their purely inorganic analogs (*19*).

A large and growing area of research concerns the properties of organometallic complexes that are supported on a variety of materials based on silica or alumina. The complexes can be used as catalysts. This area will be excluded from this review since the coordination sphere of the supported complex has the effective formula $RM(OR')_n$, and the oxygen atoms are therefore bound to carbon. There is a small area of recent research devoted to the attachment of organometallic complexes to polyoxometallates. These compounds qualify within the definition of organometallic oxo complexes which has been adopted.

Metal oxides are used extensively as catalysts for organic reactions and as oxidizing agents for organic compounds (*20*). One of the most important of such reactions is the catalytic oxidation of olefins by oxo complexes such as $Os(O)_4$, $CrCl_2(O)_2$, or $Mo(O_2)_2(O)$ [Eq. (1)].

$$R_2C{=}CR_2 + 2[O] + 2H^+ \xrightarrow{\ Os(O)_4\ } R_2C(OH)CR_2(OH) \tag{1}$$

Two mechanisms have been proposed for this reaction. In the first mechanism direct attack of an olefin at the oxo ligand is proposed [Eq. (2)] (*21*). In the second mechanism a metallo-oxocyclobutane is formed [Eq. (3)] (*22*). In either case it is most likely that the olefin initially attacks the metal [Eq. (4)] (*22,23*). Both the intermediate olefin complex and the metallo-oxocyclobutane qualify as organometallic oxo complexes.

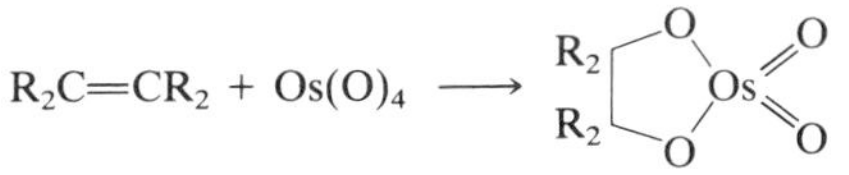

$$R_2C{=}CR_2 + Os(O)_4 \longrightarrow \begin{array}{c} R_2 \\ R_2 \end{array}\!\!\! \underset{O}{\overset{O}{\big\langle}} Os \underset{O}{\overset{O}{\big\rangle}} \tag{2}$$

$$R_2C{=}CR_2 + Os(O)_4 \longrightarrow \underset{R_2\ \ R_2}{O{-}\overset{\overset{O\ \ O}{\diagdown\!\!\parallel}}{Os}{-}O} \tag{3}$$

$$R_2C{=}CR_2 + Os(O)_4 \longrightarrow \begin{array}{c} R_2 \\ C \\ \parallel \\ C \\ R_2 \end{array} \longrightarrow Os(O)_4 \tag{4}$$

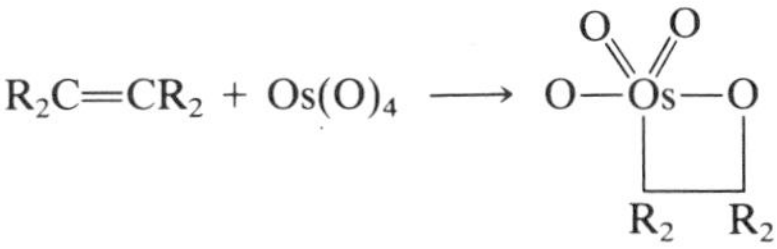

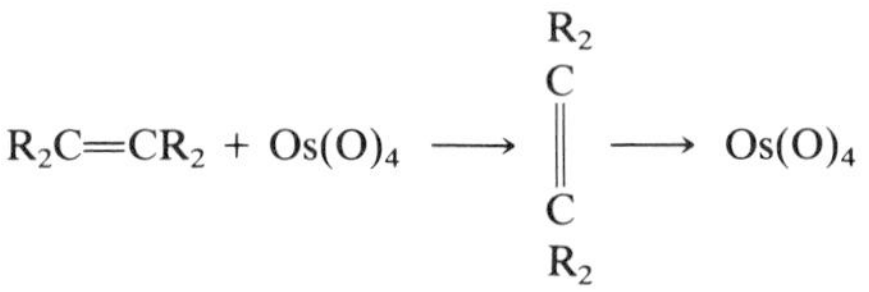

However, the mechanism of the reaction is still uncertain $(22–24)$, and the proposed organometallic intermediates have not been isolated $(23,25)$. Therefore these and other catalytic reactions are not discussed further in this review.

There is one reaction for which an [M=O] group appears to be essential to the catalyst although it is only a spectator in the reaction. This is olefin metathesis, exemplified by Eq. (5) (26).

$$CH_2{=}CH_2 + R_2C{=}CR_2 \xrightarrow[\text{Al(C}_2\text{H}_5)\text{Cl}_2]{\text{WCl}_4(\text{O})} 2CH_2{=}CR_2 \qquad (5)$$

In this case organometallic oxo complexes that either are themselves catalysts for reaction (5) or are the immediate precursors of such catalysts have been isolated. They are invariably oxoalkylidene complexes containing the [M=O] and [M=CR^{1}R^2] groups. An example is W(PEt$_3$)$_2$Cl$_2$-(CHCMe$_3$)(O) $(27,28)$. In discussing such compounds a problem with assigning oxidation states arises. An oxo ligand is usually considered as being O^{2-} attached to M^{2+}. An alkylidene or carbene ligand is regarded, on the other hand, as neutral CR^{1}R^2 attached to M^0. This assignment of oxidation states obscures the relationship between the oxo and alkylidene ligands that is obvious in the L$_n$M(=O)(=CR^{1}R^2) description. A similar problem arises in oxoolefin and oxoacetylene complexes. No description will please everyone and in any case is only formal; here we adopt the convention that the oxo ligand is O^{2-} and the alkylidene, olefin, acetylene, and carbonyl ligands are neutral.

As noted above, many organometallic compounds containing oxygen have been obtained seredipitously or unexpectedly. However, recent work has resulted in some general synthetic methods and we describe these first.

II

PREPARATION OF ORGANOMETALLIC OXO COMPOUNDS

Two basic routes to organometallic oxo compounds may be envisaged: addition of an organic ligand to an inorganic oxo complex, or addition of oxygen to an organometallic compound. The first approach requires that the inorganic oxo complex be soluble in an organic solvent. This requirement is fulfilled by Os(O)$_4$ and some oxyhalides and their derivatives, and addition of the appropriate ligand to these compounds gives organometallic oxo complexes. Some examples are given in Eqs. (6)–(11).

$$3[\text{HfCl}_2(\text{O})]_n + 6n\text{NaC}_5\text{H}_5 \rightarrow n[\text{Cp}_2\text{Hf}(\mu\text{-O})]_3 + 6n\text{NaCl} \quad (29) \qquad (6)$$

$$\text{WCl}_4(\text{O}) + \text{Me}_2\text{Mg} \rightarrow \text{WCl}_3(\text{OEt}_2)\text{Me}(\text{O}) \quad (26) \qquad (7)$$

$$ReI_3(PPh_3)_2(O) + MeC\equiv CMe \rightarrow ReI(\eta^2\text{-}C_2Me_2)_2(O) \quad (30) \qquad (8)$$

$$ReCl_4(O) + 4MeLi \rightarrow ReMe_4(O) + 4LiCl \quad (31) \qquad (9)$$

$$Os(O)_4 + 2(Me_3SiCH_2)_2Mg \rightarrow Os(CH_2SiMe_3)_4(O) \quad (18) \qquad (10)$$

$$Os(O)_4 + CO \rightarrow [Os(CO)_3(\mu_3\text{-}O)]_4 \quad (14) \qquad (11)$$

This general method of preparation is likely to increase in importance as oxo compounds with labile ligands such as phosphines, olefins, acetylenes, and CO become more common. These labile complexes, which are also soluble in organic solvents, can then be used for exchange reactions. Rational syntheses of monomeric and dimeric organometallic oxo compounds are more likely to be achieved by this approach than by the oxidation of an organometallic compound. The latter approach dominates in numbers at present, mainly because the inadvertent admission of air to an organometallic reaction belongs in this category.

The most useful oxidizing agents for the preparation of organometallic oxo compounds have been nitrogen oxides. Examples of such oxidations are given in Eqs. (12)–(21).

$$Cp_2^5Sm(thf)_2 + NO(N_2O) \rightarrow [Cp_2^5Sm]_2(\mu\text{-}O) \quad (32) \qquad (12)$$

$$[Cp_2Ti(\mu\text{-}Cl)]_2 + 2NO \rightarrow [Cp_2TiCl]_2(\mu\text{-}O) + N_2O \quad (33) \qquad (13)$$

$$[Cp_2Ti(\mu\text{-}Cl)]_2 + 4PhNO \rightarrow [Cp_2TiCl]_2(\mu\text{-}O) + Ph_2N_2 \\ + PhN(O){=}NPh \quad (34) \qquad (14)$$

$$Cp_2^5Ti + N_2O \rightarrow \\ [Cp^5Ti]_2(\mu\text{-}\eta^1 : \eta^5\text{-}C_5Me_4CH_2)(\mu\text{-}O)_2 \quad (35) \qquad (15)$$

$$Cp_2V + C_5H_5NO \rightarrow [CpV(\mu_3\text{-}O)]_4 + [CpV]_6(\mu_3\text{-}O)_8 \\ + Cp_5(O)V_6(\mu_3\text{-}O)_8 \quad (10) \qquad (16)$$

$$V(CH_2SiMe_3)_4 + NO \rightarrow V(CH_2SiMe_3)_3(O) \quad (36) \qquad (17)$$

$$Cp_2Cr + Me_3NO \rightarrow \\ [CpCr(\mu_3\text{-}O)]_4 + [CpCr]_4(\mu_3\text{-}\eta^2\text{-}C_5H_4)(\mu_3\text{-}O)_3 \quad (37) \qquad (18)$$

$$[CpMo(CO)_2]_2 + MeC_6H_4NO_2 \rightarrow \\ [CpMo(O)]_2(\mu\text{-}NC_6H_4Me)(\mu\text{-}O) \quad (38) \qquad (19)$$

$$CpW(NO)(CH_2SiMe_3)_2 \rightarrow \\ CpW(CHSiMe_3)(CH_2SiMe_3)(O) \quad (39) \qquad (20)$$

$$\overline{CpW(CO)_2(CH{=}CHCMe({=}O))} + NO \rightarrow \\ CpW(Me)(\eta^2\text{-}C_2H_4)(O) + CpW(C(O)Me)(\eta^2\text{-}C_2H_4)(O) \quad (40,40a) \qquad (21)$$

The structures of $[Cp^5Ti]_2(\mu\text{-}\eta^1 : \eta^5\text{-}C_5Me_4CH_2)(\mu\text{-}O)_2$, $Cp_5(O)V_6(\mu_3\text{-}O)_8$ and $[CpCr]_4(\mu_3\text{-}O)_3(\mu_3\text{-}\eta^2\text{-}C_5H_4)$ are shown in Figs. 2–4, respectively.

Reactions of this type are seldom quantitative and little is known about

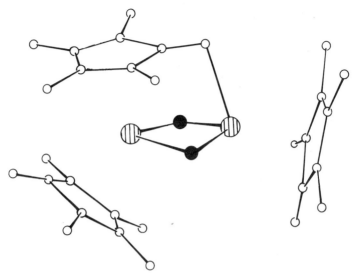

FIG. 2. Structure of $[Cp^5Ti]_2(\mu-\eta^1 : \eta^5-C_5Me_4CH_2)(\mu-O)_2$.

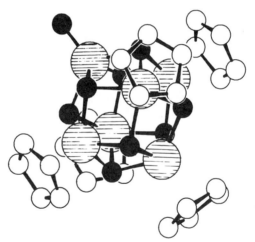

FIG. 3. Structure of $Cp_5(O)V_6(\mu_3-O)_8$.

the other products or the mechanisms. These reactions will not be suitable for the development of rational syntheses until more mechanistic information, which will allow the reactions to be controlled, is available. However, the products are among the most interesting of organometallic oxo compounds, and these products are not obtainable by other methods. In particular, this method gives compounds in which the metal is in a relatively low oxidation state, for example, Cr(III) in $[CpCr(\mu_3-O)]_4$ and

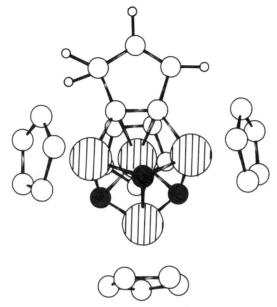

FIG. 4. Structure of $[CpCr]_4(\mu_3\text{-}\eta^2\text{-}C_5H_4)(\mu_3\text{-}O)_3$.

W(IV) in $CpW(Me)(\eta^2\text{-}C_2H_4)(O)$. Therefore the oxidation of organometallic complexes with nitrogen oxides will continue to be a source of oxo compounds.

Exhaustive decarbonylation with concomitant oxidation of a cyclopentadienyl metal carbonyl has proved to be a useful preparative route to cyclopentadienyl metal oxo compounds having no other ligands. Examples are given in Eqs. (22)–(26). This method has so far been used only when

$$3Cp_2Zr(CO)_2 + 3CO_2 \longrightarrow [Cp_2Zr(\mu\text{-}O)]_3 + 9CO \quad (41) \qquad (22)$$

$$Cp^5Cr(CO)_2(NO) + O_2 \longrightarrow [Cp^5Cr(O)(\mu\text{-}O)]_2 \quad (42) \qquad (23)$$

$$[CpMo(CO)_3]_2 + O_2 \xrightarrow{h\nu} [CpMo(O)(\mu\text{-}O)]_2 \\ + [CpMo(O)_2]_2(\mu\text{-}O) \quad (4) \qquad (24)$$

$$Cp^5Re(CO)_3 + O_2 \xrightarrow{h\nu} Cp^5Re(O)_3 \quad (43,44) \qquad (25)$$

$$Cp^5Re(CO)_3 + H_2O_2 \longrightarrow Cp^5Re(O)_3 \quad (45) \qquad (26)$$

cyclopentadienyl is the organic ligand. Similar reactions with suitable alkyl complexes appear to be viable. An ingenious and potentially useful extension of the decarbonylation reaction is the oxygen transfer shown in Eq. (27).

$$W(S_2CNR_2)_2(\eta^2\text{-}C_2H_2)(CO) + [Mo\{S_2P(OEt)_2\}_2(O)]_2(\mu\text{-}O) \rightarrow$$

$$W(S_2CNR_2)_2(\eta^2\text{-}C_2H_2)(O) + 2[Mo\{S_2P(OEt)_2\}_2(O)] \quad (46) \qquad (27)$$

An alternative to oxidation is hydrolysis, and this is the most important preparative method for Group 4 derivatives. Examples are given in Eqs. (28)–(33).

$$CpTiCl_3 + H_2O \rightarrow [CpTiCl_2]_2(\mu\text{-}O)$$
$$+ [CpTiCl(\mu\text{-}O)]_4 \quad (5,47\text{--}50) \quad (28)$$

$$Cp_2TiX_2 + H_2O \rightarrow [\{Cp_2TiX\}_2(\mu\text{-}O)]^{n+} \quad (11,13,51) \quad (29)$$

(X = Cl, Br, I, NO_3 for $n = 0$, X = H_2O for $n = 2$)

$$2Cp_2Ti(CO)_2 + C_2(CO_2Me)_2 + H_2O \rightarrow$$
$$[Cp_2Ti\{\eta^1\text{-}C(CO_2Me)=CH(CO_2Me)\}]_2(\mu\text{-}O) \quad (52) \quad (30)$$

$$6Cp_2Ti(CO)_2 + 8H_2O \rightarrow [CpTi]_6(\mu_3\text{-}O)_8 + 12CO + 5H_2$$
$$+ 6C_5H_6 \quad (53) \quad (31)$$

$$[Cp^5Zr(N_2)]_2(\mu\text{-}N_2) + H_2O \rightarrow [Cp_2^5Zr(H)]_2(\mu\text{-}O) + 3N_2 \quad (54) \quad (32)$$

$$Cp_2HfMe_2 + H_2O \rightarrow [Cp_2HfMe]_2(\mu\text{-}O) \quad (55) \quad (33)$$

The structure of $[CpTi]_6(\mu_3\text{-}O)_8$ is shown in Fig. 5.

Cyclopentadienyl oxo compounds of Group 6 can also be prepared by hydrolysis, and such reactions have been extended recently to tungsten alkylidyne complexes [Eqs. (34)–(39)]. Hydrolysis appears to be a generally applicable preparative method when the oxo complex formed is in a high enough oxidation state to force release of both protons from H_2O and when the organic ligands can survive the reaction conditions. The method lends itself to rational syntheses in such cases.

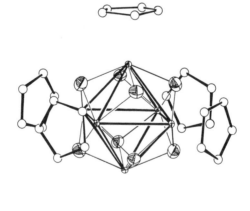

FIG. 5. Structure of $[CpTi]_6(\mu_3\text{-}O)_8$.

$$CpMo(CO)_3BF_4 + H_2O \rightarrow \{[CpMo(CO)_2]_3(\mu_3\text{-}O)\}BF_4 \quad (56) \quad (34)$$

$$Cp_2MoCl_2 + OH^- \rightarrow Cp_2Mo(O) \quad (57,58) \quad (35)$$

$$CpMoBr_4 + H_2O \rightarrow [CpMo(O)(\mu_3\text{-}O)]_4 \quad (59) \quad (36)$$

$$[W(CCMe_3)Cl_4]^- + H_2O + 2Et_3P + 2Et_3N \rightarrow W(CHCMe)(PEt_3)_2Cl_2(O)$$
$$+ Et_3NHCl + Et_4NCl \quad (60) \quad (37)$$

$$W(CCMe_3)(CH_2CMe_3)_3 + H_2O \rightarrow [W(CH_2CMe_3)_3(O)]_2(\mu\text{-}O) \quad (61) \quad (38)$$

$$W(CCMe_3)(OCMe_3)_3 + H_2O \rightarrow [W(CH_2CMe_3)(O)_3]^- \quad (62) \quad (39)$$

III

METAL OXOALKYL AND -ARYL COMPOUNDS

Organometallic oxo complexes containing σ-bonded alkyl or aryl groups are known for all three members of Groups 4 and 5, for the heavier members of Groups 6 and 7, and a very few examples for Group 8. A complete list of compounds is given in Table I. The lighter members of Groups 6 and 7 are very powerful oxidizing agents in the high oxidation

TABLE I

METAL OXOALKYLS AND -ARYLS

Compound	Ref.
$[(PhCH_2)_3Ti]_2(\mu\text{-}O)$	63
$[Cp_2Ti(\eta^1\text{-}CR^1{=}CR^2H)]_2(\mu\text{-}O)$	
$\quad R^1 = R^2 = CO_2Me$	
$\quad R^1 = R^2 = CO_2Et$	52
$\quad R^1 = R^2 = Ph$	64
$\quad R^1 = Ph, R^2 = H$	52
$\quad R^1 = R^2 = CF_3$	65
$[Cp^5Ti]_2(\mu\text{-}\eta^1{:}\eta^5\text{-}C_5Me_4CH_2)(\mu\text{-}O)_2$	35,66
$[Cp^5TiMe(\mu\text{-}O)]_3$	66a
$[Cp_2ZrMe]_2(\mu\text{-}O)$	67
$[\{Me_3Si\}_2N\}_2ZrMe]_2(\mu\text{-}O)$	68
$[Cp_2HfMe]_2(\mu\text{-}O)$	55
Group 5	
$VCl_2Ph(O)$	6,69,70
$VClPh(OC_3H_7{}^i)(O)$	71
$VPh(OC_3H_7{}^i)_2(O)$	71
$VMe(OC_3H_7{}^i)_2(O)$	72,73
$V(OR)_2Me(O)$	73
$\quad R = CHMeEt$	
$\quad R = CMe_3$	

(continued)

TABLE I (*continued*)

Compound	Ref.
$V(CH_2SiMe_3)_3(O)$	*36,74,75*
$Cp_2Nb(\eta^1-C_7H_5(CF_3)_2)(O)$	*76*
$Cp_2NbMe(O)$	*74*
$NbCl_2Me(O) \cdot 2L$	*77*
$L = OSMe_2, OPMe_3,$	
$OP(NMe_2)_3, OP(OMe)_3,$	
$OP(OMe)_2(NMe_2), OP(OMe)(NMe_2)_2,$	
$OAsPh_3, PPh_3, ONMe_3, ONC_5H_5,$	
$O\{OP(NMe_2)_2\}_2$	
$NbBr_2Me(O) \cdot 2OPPh_3$	*77*
$TaCl_2Me(O) \cdot 2OPPh_3$	*77*
$[Cp_2Nb(n-C_4H_9)]_2(\mu-O)$	*78*
$Cp_2NbR(O)$	*79*
$R = n-C_4H_9, \alpha-CH_2C_5H_4N, \alpha-C_4H_3S$	
$[Ta(CH_2CMe_3)_3(\mu-O)]_n$	*80*
$[(\eta^5-C_5Me_4Et)TaCl_2H-$	
$[(\eta^5-C_5Me_4Et)TaCl_2](\mu-CHPMe_3)(\mu-O)$	*81,82*
$Cp_2^5TaMe(O)$	*83*
<div align="center">Group 6</div>	
$Mo(\eta^1-1,3,5-C_6H_2Me_3)_3(O)$	*84*
$Mo(\eta^1-1,3,5-C_6H_2Me_3)_2(O)_2$	*85*
$Mo(\eta^1-1,3,5-C_6H_2Me_3)\{CPBu_3^n(1,3,5-C_6H_2Me_3)\}(O)_2$	*86*
$Mo(\eta^1-1,3,5-C_6H_2Me_3)_2(CH_2PBu_3)(O)_2$	*86a*
$Mo(bpy)(CH_2CMe_3)_2(O)_2$	*87*
bpy = 2,2'-bipyridine	
$Mo(bpy)BrR(O)_2$	*88,89*
$R = Me, Et, CH_2CMe_3, n-C_3H_7, i-C_3H_7, CMe_3$	
$Mo(bpy)R_2(O)_2$	
$R = Me$	*90*
$R = PhCH_2$	*91*
$R = C_2H_5, C_3H_7, C_4H_9, C_5H_9, C_6H_{11}$	*91a*
$[CpMo(CO)][CpMo(O)](\mu-CO)(\mu-\eta^1 : \eta^3-CH=CHCHCMe_2)$	*92*
$\overline{M(\eta^3-S_2COPr^i)}(\eta^2-S_2CPMe_3)(OPr^i)(O)$	*93*
M = Mo, W	
$MX(CH_2CMe_3)_3(O)$	
M = Mo, W; X = Cl	*61,95,96*
M = Mo, W; X = $OCMe_3$	*95*
$[W(CH_2CMe_3)_3(O)]_2(\mu-O)$	*61,96*
$[W(CH_2AMe_3)(O)_3]^-$	*62,96*
A = C, Si	
$[WCl_4(1-norbornyl)(O)]^{2-}$	*96a*
$W(\eta^2-OCMe_2CMe_2O)(CH_2CMe_3)(O)_2$	*62*
$WCl_3Me(O) \cdot OEt_2$	*26*
$WCl_3Me(O) \cdot L$	
$L = (NMe_2)_3PO, Ph_3PO, Ph_3AsO, Me_2SO$	*98*

(*continued*)

TABLE I *(continued)*

Compound	Ref.
WClMe(O)$_2 \cdot$2L	
L = (NMe$_2$)$_3$PO, Ph$_3$PO, Ph$_3$AsO, Me$_2$SO	97,98
W(DIPP)$_3$(η^1-CCMe$_3$=CR^{1}R^2)(O)	99
R^1 = R^2 = H, Me; R^1 = H, R^2 = Ph, EtO, Me$_2$N	
DIPP = 2,6-diisopropylphenoxide	
W(DIPP)$_3$(η^1-CEt=CR^{1}R^2)(O)	99
R^1 = R^2 = Me; R^1 = H, R^2 = Ph, EtO, Me$_2$N	
Cp'WMe(η^2-C$_2$H$_2$)(O)	40,40a
Cp' = Cp, Cp^1, Cp^5	
Cp^{1}W(C(O)Me)(η^2-C$_2$H$_2$)(O)	40,40a
CpW(CH$_2$SiMe$_3$)(O)$_2$	39
CpWPh(η^2-Ph$_2$C$_2$)(O)	100
{W[(2-CH$_2$C$_6$H$_4$)$_2$]$_2$(μ-O)}$_2$Mg(thf)$_4$	101
{W[2-CH$_2$C$_6$H$_4$CH$_2$]$_2$(μ-O)}$_2$Mg(thf)$_4$	102
WX(Fc)$_3$(O)	
X = Cl, OMe, OFc, OBu^n	102a
CpW(Fc)(O)$_2$; Fc = CpFeC$_5$H$_4$	263
[W(NPh)(PMe$_3$)Me$_2$(μ–O)]$_3$	266
Group 7	
ReMe$_4$(O)	31,103–108
cis-[ReMe$_3$(O)$_2$]	104,107
Re(CH$_2$SiMe$_3$)$_4$(O)	103,106,108
[Re(CH$_2$SiMe$_3$)$_3$(O)]$_2$(μ-O)	103,111
ReMe(O)$_3$	109
[Re(CH$_2$CMe$_2$R)$_2$(O)]$_2$(μ-O)$_2$	110
R = Me, Ph	
[ReMe$_2$(η^2-MeNO)(O)]$_2$(μ-O)	112
Re(NO)(η^2-ONCH$_2$SiMe$_3$)(CH$_2$SiMe$_3$)$_2$(O)	112
[Re(NO)(CH$_2$SiMe$_3$)$_2$(thf)(O)]$_3$	112
[Re(η^1-1,3,5-C$_6$H$_2$Me$_3$)$_2$(O)$_2$]$_2$Mg(thf)$_2$	112a
Re(η^1-1,3,5-C$_6$H$_2$Me$_3$)$_2$(O)$_2$	112a
Re(η^1-1,3,5-C$_6$H$_2$Me$_3$)$_4$(O)	112a
[ReR$_4$(μ-O)]$_2$[Mg(thf)$_4$]	111
R = Me, CH$_2$SiMe$_3$	
Cp^{5}ReMe$_2$(O)	113
Group 8	
[Ru(CH$_2$SiMe$_3$)$_3$]$_2$(μ-O)$_2$	111a
Os(CH$_2$SiMe$_3$)$_4$(O)	18
Os(η^1-1,3,5-C$_6$H$_2$Me$_3$)$_2$(O)$_2$	112a

states found in the known oxoalkyls and -aryls, making such compounds unstable. Metal alkyls can also decompose by a β-hydrogen mechanism and are unstable when this process is possible. The majority of known oxoalkyls and -aryls have no β-hydrogen, but rather contain methyl (Me), neopentyl (CH_2CMe_3), or its silicon analog CH_2SiMe_3. One of the few exceptions is $Mo(bipy)Br(C_2H_5)(O)_2$, which is remarkably stable (88,89). In most compounds of Groups 4, 5, and 6 the metal is in the highest possible oxidation state and has zero d electrons. Exceptions are the Mo(V) complex $Mo(\eta^1\text{-}1,3,5\text{-}C_6H_2Me_3)_3(O)$ (84) and the W(IV) compounds $CpW(\eta^2\text{-}C_2R_2^1)R^2(O)$ [R^1 = H for R^2 = Me, C(O)Me (40,40a) and $R^1 = R^2 = $ Ph (100)]. In compounds of the metals of Group 7 and 8 oxidation states lower than the maximum are the rule, for example, Re(VI) in $ReMe_4(O)$ (31,103–8), Re(V) in $Cp^5ReMe_2(O)$ (113), and Os(VI) in $Os(CH_2SiMe_3)_4(O)$ (18). Since the metal has zero or at the maximum two d electrons, there are effectively no electronic restrictions on the coordination number of metal oxoalkyls or -aryls, and values from four [e.g., $VCl_2Ph(O)$ (6,69,70)] to eight [e.g., $[Cp_2ZrMe]_2(\mu\text{-O})$ (67)] are known. The number of alkyl or aryl groups appears to be likewise restricted only by the maximum oxidation state attainable and values from one to four are observed. Examples of each are $WCl_3(OEt_2)Me(O)$ (26,97,98), $Mo(bipy)(CH_2Ph)_2(O)_2$ (91), $CpW(CH_2SiMe_3)_3(O)$ (39), and $ReMe_4(O)$ (31,103–106). The number of oxo ligands increases from left to right in the periodic table, i.e., with the maximum oxidation state. Examples are 0.5 in $[Cp_2Ti(\eta^1\text{-CR=CHR})]_2(\mu\text{-O})$ [R = CF_3 (65), CO_2Me (52)], 1 in $V(CH_2SiMe_3)_3(O)$ (36), 2 in $ReMe_3(O)_2$ (104,107), and finally 3 in $ReMe(O)_3$ (109). It is clear from the ranges in coordination numbers, from the relative numbers of alkyl or aryl compared to oxo ligands, and from the oxidation states of the metal atoms that the alkyl and oxo ligands are similar in nature, both being "hard" ligands.

The known oxoalkyls and -aryls cannot decompose by β-elimination of hydrogen to form an olefin and are in general remarkably stable toward dioxygen or water and also thermally stable. Only $ReMe_4(O)$ is sensitive to dioxygen (103); the oxoalkyls of vanadium $VX_2R(O)$ (X = Cl, OR) are all very water-sensitive and decompose thermally at room temperature or below to yield R_2 (6,69–73). However, the analogous niobium and tantalum compounds, along with all other oxoalkyls, are thermally stable. The most thorough investigation of thermal decomposition has been of $Mo(bipy)(CH_2CMe_3)_2(O)_2$, which thermolyzes slowly above 180° by hydrogen abstraction from the solvent, giving Me_4C (87). Oxoalkyls are also stable toward potential ligands such as CO, phosphines, or H_2.

Some oxoalkyls can be further alkylated with loss of the oxo group, for example, Eq. (40), and the oxo group can be wholly or partially removed by acids such as HCl [Eqs. (41) and (42)]. When hydrolysis is observed, it is only at very high pH [Eqs. (43)–(45)].

$$\text{ReMe}_4(\text{O}) + \text{AlMe}_3 \rightarrow \text{ReMe}_6 \quad (104) \quad (40)$$

$$[\text{W}(\text{CH}_2\text{CMe}_3)_3(\text{O})]_2(\mu\text{-O}) + 2\text{HCl} \rightleftharpoons$$
$$2[\text{WCl}(\text{CH}_2\text{CMe}_3)_3(\text{O})] + \text{H}_2\text{O} \quad (61) \quad (41)$$

$$[\text{Cp}^5\text{Ti}]_2(\mu\text{-}\eta^5\!:\eta^1\text{-C}_5\text{Me}_4\text{CH}_2)(\mu\text{-O})_2 + \text{HCl} \rightarrow \text{Cp}_2^5\text{TiCl}_2 + \text{Cp}^5\text{TiCl}_3 \quad (35) \quad (42)$$

$$[\text{W}(\text{CH}_2\text{CMe}_3)(\text{O})_3]^- + \text{OH}^- \rightarrow [\text{W}(\text{O})_4]^{2-} + \text{Me}_4\text{C} \quad (62) \quad (42a)$$

$$\text{MoBr(bipy)Me}(\text{O})_2 + \text{OH}^- \rightarrow [\text{MoMe}(\text{O})_3]^- \quad (89) \quad (43)$$

$$[\text{MoMe}(\text{O})_3]^- + \text{H}_2\text{O} \rightarrow [\text{Mo}(\text{O})_4\text{H}]^- + \text{CH}_4 \quad (89) \quad (44)$$

As noted, oxo complexes of alkyls having β-hydrogen atoms are very rare because of the extremely facile transfer of this hydrogen [Eq. (45)].

$$[\text{M}(\text{C}_2\text{H}_5)] \rightarrow \text{M(H)} + \text{C}_2\text{H}_4 \quad (45)$$

The possibility of α-hydrogen transfer [Eq. (46)] also exists and would be a convenient preparation of alkylidene complexes. Only one example of such a transfer has been observed so far, the thermal reaction (47).

$$\text{M(CHR}_2) \rightarrow \text{HM}(\!=\!\text{CR}_2) \quad (46)$$

$$\text{CpW}(\text{CH}_2\text{SiMe}_3)_3(\text{O}) \xrightarrow{\Delta} \text{CpW}(\text{CH}_2\text{SiMe}_3)(\text{CHSiMe}_3)(\text{O}) + \text{Me}_4\text{Si} \quad (39) \quad (47)$$

IV

OXO ALKYLIDENE, ALKYLIDYNE, OLEFIN, ACETYLENE, AND CARBONYL COMPOUNDS

We discuss oxo compounds with multiple bonds between the metal and the organic ligand under this heading and confine the discussion to monomeric complexes. There are several M_3 (triangular) or M_4 (tetrahedral) clusters containing oxygen atoms and carbonyl ligands. These will be considered in Section VI (cluster compounds). One carbonyl dimer, $[\text{Cp}^5\text{Re(CO)}_2]_2(\mu\text{-O})$ has been prepared $(7,114)$; trans-$[\text{M(CO)}_4(\text{O})_2]$ (M = Mo, W) $(17,115)$, $[\text{M(CO)}_2(\text{O})_2]$ (M = Cr, Mo, W) $(115,116)$, and $[\text{Au(CO)(O)}]$ (117) have been identified by matrix isolation techniques at low temperatures; and some carbonyl oxide anions of chromium, molybdenum, and iron have been observed in the gas phase (16). Monomeric oxo compounds with multiple bonding between the metal and the organic ligand are presently confined to the metals tantalum molybdenum, tungsten, rhenium, and an unusual diene complex of iridium, $[\text{IrCl(COD)}]_2(\mu\text{-OH})_2(\mu\text{-O})$ (COD = 1,5-cyclooctadiene) (118).

The rarest of the multiply bonded ligands to be found in combination with an oxo group is an olefin. Ricard and Weiss obtained the Mo(IV) compound $Mo(\eta^2\text{-}S_2CNPr_2^n)_2(\eta^2\text{-}C_2(CN)_4)(O)$ in 1972 (*119,120*), but only recently was a second oxoolefin, and one with ethylene itself, discovered. This was of W(IV), namely $WCl_2(PMePh_2)_2(\eta^2\text{-}C_2H_4)(O)$ (*15*), and was obtained via reaction (48).

$$WCl_2(PMePh_2)_3(O) + CH_2{=}CHR \rightarrow WCl_2(PMePh_2)_2(\eta^2\text{-}CH_2CHR)(O) \quad (15) \quad (48)$$

$$R = H, CH_3, CH{=}CH_2$$

In the ethylene complex the oxo and ethylene ligands are cis to one another and the C–C axis of the ethylene is perpendicular to the M–O bond, a configuration that maximizes π bonding between the W(IV) (d^2) and the ethylene (*15*). A reaction analogous to (48) with CO affords the only known monomeric oxocarbonyl, $WCl_2(PMePh_2)_2(CO)(O)$ (*15*).

Oxoacetylene complexes are more common than oxoolefins. The acetylene bonds to the metal in the same manner as an olefin; i.e., only one of the two available π bonds in the acetylene is used. The M(IV) complexes $CpMX(\eta^2\text{-}C_2R_2)(O)$ (X = Cl, Me, Ph, or SC_6F_5; R = H, Ph, CF_3) are well established for molybdenum and tungsten (*40,40a,100,122*). A variant of this type is the triangular $[Fe(CO)_3][CpW(CO)_2][CpW(O)]\{\mu_3\text{-}\eta^2\text{-}C_2(C_6H_4Me)_2\}$ (Fig. 6) (*122a*). A second general type of molybdenum or tungsten(IV) complex is the dithiocarbonate $M(S_2CNR_2)_2(\eta^2\text{-}R^1CCR^2)(O)$ (R = Me, Et; R^2, R^2 = H, Me, Ph) (*46,120,124–126*). Such complexes can be obtained for molybdenum by simple reversible addition

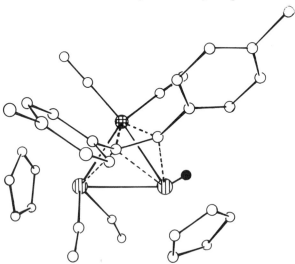

FIG. 6.   Structure of $[Fe(CO)_3][CpW(O)_2][CpW(CO)]\{\mu_3\text{-}\eta^2\text{-}C_2(C_6H_4Me)_2\}$.

of the acetylene (even C_2H_2) to $Mo(S_2CNR_2)_2(O)$ (125,126). For tungsten the starting complexes $W(S_2CNR_2)_2(O)$ are not available, and the oxoacetylene complexes are obtained by an unusual exchange of O for CO using $[Mo\{S_2P(OEt)_2\}_2(O)]_2(\mu\text{-}O)$ as the transfer reagent [Eq. (27) above] (46). As with the ethylene complex discussed above, the acetylene ligand in all of these compounds is cis to the oxo ligand and lies perpendicular to the M–O axis, a configuration that maximizes π donation of electrons from O to M (giving a formal M≡O triple bond) and π back donation from M to the acetylene ligand.

Recently, diacetylene oxo compounds with the metal in a lower oxidation state have been prepared. These are the Re(III) compounds $[ReL(\eta^2\text{-}RC\equiv CR^1)_2(O)]^{n+}$ (L = Cl, Br, I, MeCOO for $n = 0$; L = py, Ph_3P, bipy for $n = 1$; R = R^1 = Me or Et; R = Me, Me_3C for R^1 = H; R = Me for R^1 = Ph) (30,127,127a), which are effectively low-spin d^4 tetrahedral complexes. The C–C axes of the two acetylene ligands are parallel to one another and are almost perpendicular to the Re–O bond. The acetylenes show no tendency to dimerize or to oxidize. Extended Hückel calculations suggest that the Re≡O bond is again close to triple. Two metal orbitals derived from the e set of a tetrahedral ML_4 molecule are nonbonding with respect to the Re–O π interactions but are involved in π-backbonding to the acetylenes. Thus the M–O bond is not destabilized by electrons in π-antibonding orbitals and the acetylenes cannot rotate or couple because of the necessity of π back bonding, which is available only perpendicular to the M–O bond (30,127a).

The most important oxo compounds having multiple bonds between the metal and the organic group are those with alkylidene ligands. There is now clear evidence that compounds with the $M(=CR_2)(=O)$ grouping are the active catalysts in many olefin metathesis reactions. The $M(=O)$ moiety is essential to catalytic activity, although it is only a spectator in the actual reaction. The problem of the assignment of an oxidation state to the metal, oxygen, and alkylidene ligands, already discussed briefly in the Introduction to this review becomes acute when considering these metathesis catalysts. This problem has been treated theoretically in detail by Rappé and Goddard for the model compounds $MCl_2(=CH_2)(=O)$ (M = Cr, Mo, W) (128). They conclude that in such compounds the M–Cl bond is very ionic; that is, an M^+Cl^- description is acceptable. The $M=CH_2$ bond is very covalent and corresponds best to a metal–alkylidene double bond, that is, $M^0=C^0H_2$. However, the same is true for the M=O bond, which gives the metal in the $MCl_2(CH_2)(O)$ complexes a formal oxidation state of $+2$. However, the nature of the M=O bond changes as the other ligands are varied (for instance, when Cl is replaced by Me the M=O bond becomes more ionic). There is therefore no one correct assignment of oxidation states in these compounds. To conform with the other sections of this article we will regard $M{=}O$ as involving M^{2+} and O^{2-} and $M{=}CR_2$ as involving a covalent bond with neither M nor CR_2 charged.

Recipes for metathesis catalysts based on WCl_6 combined with an alkylating agent such as $AlCl_2Et$ were described first in 1967 (*129*). It is now clear that the catalyst precursor is in fact $WCl_4(O)$ and that the active catalyst is very probably the alkylidene derivative $WCl_2(CR_2)(O)$ (*26,130*). Extremely active metathesis catalysts have been obtained from $MX(CH_2CMe_3)_3(O)$ or $W(OCMe_3)_2(CH_2CMe_3)_2(O)$ and a Lewis acid such as $AlBr_3$ (M = Mo, W; X = Cl, Br, or Me_3CO) (*94,95*). The structure of the adduct that results is $[MX(CH_2CMe_3)_3(\mu\text{-}\overline{O})][\overline{A}lBr_3]$, and this is photochemically activated with loss of Me_4C to give the active catalyst, which is presumed to be $[MX(CH_2CMe_3)(CHCMe_3)(\mu\text{-}\overline{O})][\overline{A}lBr_3]$; $[W(OCMe_3)(CH_2CMe_3)_3(\mu\text{-}\overline{O})][\overline{A}lBr_3]$ can be thermally activated (*95*). It should be noted that the Lewis acid must be present and the loss of Me_4C must take place to obtain an active catalyst.

A rather less active catalyst is formed from $WCl_2(PEt_3)_2(CHCMe_3)(O)$ and a Lewis acid such as $AlCl_3$ (*28,131–134*). In this case at least one function of the Lewis acid is to create a vacant coordination site by removing a phosphine ligand, since $WCl_2(PEt_3)(CHCMe_3)(O)$ acts as a metathesis catalyst even in the absence of a Lewis acid. The monophosphine complex has a trigonal pyramidal structure with one Cl and the PEt_3 ligands in the axial position (as determined by X-ray diffraction) (*27,28*). These complexes, though less active than the simple oxohalides of tungsten, are the only isolated and crystallographically characterized oxoalkylidene complexes that catalyze the metathesis of olefins. Neither the active catalyst $WCl_2(PEt_3)(CHCMe_3)(O)$ nor the precursor $WCl_2(PEt_3)_2(CHCMe_3)(O)$ can be obtained directly from a metal complex and the ligands involved, nor from the parent alkyl. They are prepared by the hydrolysis of an alkyne, reaction (37) above, or by the high-yield but obviously complicated exchange reaction (49) (*133*).

$$2[TaCl_3(PEt_3)_2(CHCMe_3)] + 2[W(OCMe_3)_4(O)] \rightleftharpoons$$

$$2[WCl_2(PEt_3)_2(CHCMe_3)(O)] + [TaCl(OCMe_3)_4]_2 \quad (49)$$

The oxo ligand plays an essential role in determining the electronic structure of the metathesis catalyst, but it is a spectator in the actual reactions (50) and (51) (*128,135*).

$$Cl_2M{\overset{O}{\underset{CR_2}{\diagdown}}} + CH_2{=}CH_2 \rightleftharpoons Cl_2M\overset{O\ \ CH_2}{\diamond}{\underset{CR_2}{CH_2}} \quad (50)$$

$$\text{Cl}_2\text{M} \overset{\overset{\text{O}}{\|}}{\underset{\text{CR}_2}{\diagdown}} \overset{\text{CH}_2}{\underset{}{\diagup \diagdown}} \text{CH}_2 \quad \rightleftharpoons \quad \text{Cl}_2\text{M} \overset{\text{O}}{\underset{\text{CH}_2}{\diagup\diagup}} \; + \; \text{CH}_2{=}\text{CR}_2 \qquad (51)$$

The oxo ligand stabilizes the metallocycle and by donation of π electrons (again giving formally an $M{\equiv}O$ triple bond) increases the electron density at the metal, thus allowing initial coordination of the olefin to what is effectively a d^0 center. The Lewis acid usually present with metathesis catalysts may assist in the push–pull of electrons between the oxygen and the metal.

The analogy between $[M{=}O]$ and $[M{=}CR_2]$ suggests that reaction (50) could proceed in a different manner, forming an oxometallocyclobutane via reaction (52). This pathway is of higher energy and is not found.

$$\text{Cl}_2\text{M} \overset{\text{O}}{\underset{\text{CR}_2}{\diagup\diagup}} \; + \; \text{CH}_2{=}\text{CH}_2 \quad \rightleftharpoons \quad \text{Cl}_2\text{M} \overset{\text{O}}{\underset{\overset{\|}{\text{CR}_2}\text{CH}_2}{\diagup\diagdown}} \text{CH}_2 \qquad (52)$$

However, if a dioxo and not an oxoalkylidene complex is initially present, then the analog of reaction (52), namely (53), is the only route available. The product of reaction (53) may then decompose in several ways, reactions (54)–(56) (*128,135*).

$$\text{Cl}_2\text{M} \overset{\text{O}}{\underset{\text{O}}{\diagup\diagup}} \; + \; \text{CH}_2{=}\text{CH}_2 \quad \rightleftharpoons \quad \text{Cl}_2\text{M} \overset{\overset{\text{O}}{\|}\diagup\text{O}}{\underset{\text{CH}_2}{\diagdown}} \text{CH}_2 \qquad (53)$$

$$\text{Cl}_2\text{M} \overset{\overset{\text{O}}{\|}\diagup\text{O}}{\underset{\text{CH}_2}{\diagdown}} \text{CH}_2 \;\; \begin{cases} \nearrow \; \text{Cl}_2\text{M(O)} + \overline{\text{CH}_2\text{CH}_2\text{O}} & (54) \\ \longrightarrow \; \text{ClM(O)(OC}_2\text{H}_4\text{Cl)} & (55) \\ \searrow \; \text{Cl}_2\text{M(CH}_2)\text{(O)} + \text{HCHO} & (56) \end{cases}$$

This mechanism is the same as that proposed by Sharpless on the basis of experimental data for the oxidation of olefins by $CrCl_2(O)_2$ (*22*).

Oxoalkylidene complexes are believed to be the active intermediates in the conversion of carbonyls into olefins via the reaction sequence (57)–(59) (*136,137*). However, none of the proposed intermediates has been isolated, and carbonyls are in fact converted into olefins by $Ta(CH_2CMe_3)_3$-$CHCMe_3$) with concomitant production of $[Ta(CH_2CMe_3)_3O]_n$ (*80*).

Finally, one extremely peculiar reaction of an oxoalkylidene is worth noting. This is reaction (60), in which a $[W{\equiv}CR]$ replaces a $[W{\equiv}O]$ moiety (*138*).

$$WCl_3(thf)_2(O) + MeLi \rightarrow WCl(Me)_2(O) \qquad (57)$$

$$WCl(Me)_2(O) \rightarrow WCl(CH_2)(O) + CH_4 \qquad (58)$$

$$WCl(CH_2)(O) + PhC(O)Me \rightarrow PhMeC{=}CH_2 \qquad (59)$$

$$WCl_2(PEt_3)_2(CHCMe_3)(O) + C_2Cl_6 \rightarrow$$

$$C_2Cl_4 + Et_3PH^+ + Cl^- + WCl_3(Et_3P{=}O)(CCMe_3) \qquad (60)$$

V

METAL CYCLOPENTADIENYL COMPLEXES CONTAINING OXYGEN ATOMS

Complexes containing cyclopentadienyl and oxygen as coligands are of two basic types: those containing a terminal double bond between a metal and oxygen, [M=O], and those containing one or more doubly bridging oxygen atoms, $[M(\mu_2\text{-}O)_nM]$. A few complexes contain both types of groups. Tables II–VI list the known examples classified according to type, and several generalizations can be drawn from the information in these tables.

There are no complexes containing the [M=O] unit for the Group 3 and 4 metals. In this respect cyclopentadienyl oxo complexes are no different from purely inorganic ones; the only known complex of a Group 3 or 4 metal containing an [M=O] group is (OEP)Ti(O), where OEP is octaethylporphinato (206). The number of [M=O] units per metal and the frequency of occurrence increase as one proceeds to the right in the periodic table, culminating in $Cp^5Re(O)_3$.

By far the largest group of cyclopentadienyl oxo complexes are those of general formula $[Cp_2'MX]_2(\mu\text{-}O)$, which are mainly of the Group 4 metals but with a few Group 5 examples as well. The structures of many of

TABLE II

CYCLOPENTADIENYL OXO COMPLEXES OF THE
GROUP 3 METALS

Complex	Ref.
$[Cp^5{}_2Sm]_2(\mu\text{-}O)$	*32*
$[Cp_3U]_2(\mu\text{-}O)$	*139*
$\{[(\eta^5\text{-}C_9H_7)UX(MeCN)_4]_2O\}UX_6$	*140*
X = Cl, Br	
$\{[(\eta^5\text{-}C_9H_6Et)UCl(MeCN)_4]_2(\mu\text{-}O)\}UCl_6$	*140*
$\{[(\eta^5\text{-}C_9H_4Me_3)UCl(MeCN)_4]_2(\mu\text{-}O)\}UCl_6$	*140*

TABLE III

CYCLOPENTADIENYL OXO COMPLEXES OF THE GROUP 4 METALS

Complex	Ref.
$[Cp_2Ti]_2(\mu\text{-}O)$	*141–43*
$[Cp_2T_iX]_2(\mu\text{-}O)$	
X = Cl	*33,34,41,51,51a,142–45*
X = Br, I, NO$_3$	*11,51,145*
X = NCS	*146*
X = η^1-CR^1=CHR^2	
R^1 = R^2 = CF$_3$	*65*
R^1 = R^2 = CO$_2$Me, CO$_2$Et	*52*
R^1 = Ph, R^2 = H	*52*
R^1 = R^2 = Ph	*52,64*
X = μ-OC[Co(CO)$_3$]$_3$	*147*
$\{[Cp_2Ti(H_2O)]_2(\mu\text{-}O)\}^{2+}$	*13,147a*
$[Cp_2ZrX]_2(\mu\text{-}O)$	
X = Br, I, OCN	*145*
X = SPh	*148*
X = Cl	*145,149–51*
X = Me	*67*
$[(\eta\text{-}C_5H_4Bu^t)_2ZrMe]_2(\mu\text{-}O)$	*152*
$[Cp_2^5MH]_2(\mu\text{-}O)$	
M = Zr, Hf	*54*
$[(\text{thInd})_2ZrCl]_2(\mu\text{-}O)$	*145*
thInd = tetrahydroindenyl	
$[Cp_2^5ZrX][Cp_2^5ZrH](\mu\text{-}O)$	*54*
X = OH, Cl	
$[Cp_2ZrCl][CpW(CO)_2L](\mu\text{-}O)$	*145a*
L = CO, PMe$_3$, PhC≡CPh	
$[CpHfMe]_2(\mu\text{-}O)$	*55*
$[CpTiCl_2]_2(\mu\text{-}O)$	*5,47,153*
$[Cp^5Ti]_2(\mu\text{-}\eta^5:\eta^1\text{-}C_5Me_4CH_2)(\mu\text{-}O)_2$	*35,66*
$[CpMCl]_2(\mu\text{-}\eta^5:\eta^5\text{-}C_{10}H_8)(\mu\text{-}O)$	*154,197,197a*
M = Ti, Zr	
$[(\eta\text{-}C_5H_4Bu^t)_2Zr]_2(\mu\text{-}O)(\mu\text{-}E)$	*152*
E = S, Se	
$[Cp_2Zr(\mu\text{-}O)]_3$	*41, 197*
$[Cp_2Hf(\mu\text{-}O)]_3$	*29*
$[CpTiCl(\mu\text{-}O)]_3$	*155*
$[Cp_2TiCl(\mu\text{-}O)]_2[CpTiCl]$	*12*
$[CpTi]_3(\mu\text{-}O)_4(NO)$	*74*
$[Cp^5TiMe(\mu\text{-}O)]_3$	*66a*
$[CpTiCl(\mu\text{-}O)]_4$	*47,50,155,155a*
$[Cp^1TiCl(\mu\text{-}O)]_4$	*49*
$[CpTi(O\text{-}C_6H_2Me_3)(\mu\text{-}O)]_4$	*156*
$[(\eta^5\text{-}C_9H_{11})TiCl(\mu\text{-}O)]_4$	*157*

TABLE IV

Cyclopentadienyl Oxo Complexes of the Group 5 Metals

A. Complexes Containing Terminal [M=O] Groups

Complex	ν(V–O) (cm^{-1})	r(M=O) (Å)	Ref.
$Cp_2VCl(O)$	—	—	158
$Cp_2NbCl(O)$	867	—	79,163,163a
$Cp_2NbR(O)$			
R = n-C_4H_9, 2-$CH_2C_5H_4N$, 2-C_4H_3S	865–870	—	79
R = η^1-$C_7H_5(CF_3)_2$	980	1.63	76
R = Me			74
$[Cp_2Nb(ONMeNO)(O)]_n$			74
$CpVX_2(O)$	950–1000	—	1,159,160
X = Cl, Br			
$Cp^1VCl_2(O)$	965	—	161
$Cp^5VCl_2(O)$	965	1.58	162
$[Cp^5VI(O)]_2(\mu\text{-}O)$	955	—	162
$[Cp^5V(O)(\mu\text{-}O)]_3$	920, 935	—	161
$Cp^5_2TaX(O)$	—	—	83
X = H, Me			

B. Complexes Containing Bridging [M(μ-O)M] Groups

$[Cp^5VI_2]_2(\mu\text{-}O)$	162
$[Cp^1_2NbCl]_2(\mu\text{-}O)$	164
$\{[Cp_2NbCl]_2(\mu\text{-}O)\}(BF_4)_2$	165,166
$[Cp_2Nb(n\text{-}C_4H_9)]_2(\mu\text{-}O)$	78
$[Cp^5VI(O)]_2(\mu\text{-}O)$	162
$[Cp_2VCl][CpVCl_2](\mu\text{-}O)$	167
$[Cp_2VCl][CpVCl](\mu\text{-}O)_2$	168
$[Cp^1NbCl_3(H_2O)]_2(\mu\text{-}O)$	169,170
$[(\eta^5\text{-}C_5Me_4Et)TaCl_2]_2(\mu\text{-}CHPMe_3)(\mu\text{-}O)$	81,82
$[Cp^5VCl(\mu\text{-}O)]_4$	162
$[CpVI]_2[CpV(NO)]_2(\mu\text{-}O)_4$	171,172

TABLE V

CYCLOPENTADIENYL OXO COMPLEXES OF THE GROUP 6 METALS

A. Complexes Containing Terminal [M=O] Groups

Complex	$\nu(M=O)$ (cm^{-1})	$r(M=O)$ (Å)	Ref.
Cp$_2$Mo(O)	793–868		57,173
Cp$_2^1$Mo(O)	793–863	1.72	58
Cp$_2$W(O)	799, 897		57,173
CpMoX(O)$_2$	920		4,59,174,175
X = Cl, Br			
(η^5-C$_5$Et$_5$)W(OCMe$_3$)(O)$_2$			176
(η^5-C$_5$Me$_4$CMe$_3$)W(OCMe$_3$)(O)$_2$			176
(η^5-C$_5$Me$_2$Et$_3$)W(OCMe$_3$)(O)$_2$			176
CpW(CH$_2$SiMe$_3$)(O)$_2$			39
CpW(Fc)(O)$_2$			263
Fc = ferrocenyl, CpFeC$_5$H$_4$			
CpMoCl$_2$(O)	949		4,175,176a,177
CpMoBr$_2$(O)	945		59
Cp^{1}MoCl$_2$(O)			176a
[CpMo(dmpe)(O)]PF$_6$		1.76	178
dmpe = 1,2-bis(dimethylphosphino)ethane			
[CpMo(O)][CpMo(CO)]$_2$(μ-CO)$_2$(μ_3-N)			179
[CpMo(O)][CpMo(CO)](μ-η^1:η^3-CHCHCMe$_2$)(μ-CO)	910	1.71	92
CpMo(SPh)(η^2-C$_2$(CF$_3$)$_2$)(O)	961	1.68	121,122
CpMCl(η^2-C$_2$Ph$_2$)(O)			120
M = Mo, W			
CpWPh(η_2-C$_2$Ph$_2$)(O)	945	1.69	100
Cp'WMe(η^2-C$_2$H$_2$)(O)	940		40,40a
Cp' = Cp, Cp^1, Cp^5			
Cp^{1}W(C(O)Me)(η^2-C$_2$H$_2$)(O)	950	1.66	40,40a
CpW(CH$_2$SiMe$_3$)$_3$(O)			39
CpW(CH$_2$SiMe$_3$)(CHSiMe$_3$)(O)			39
[CpW(O)][CpW(NSiMe$_3$)](μ-S)$_2$	925		180
[CpW(O)][CpW(S)](μ-S)$_2$	928		180

(continued)

TABLE V (continued)

B. Complexes Containing Terminal [M=O] and Bridging M(μ-O)$_n$M Groups

Complex	ν(M=O) (cm^{-1})	r(M=O) (Å)	Ref.
[Cp^{1}Mo(μ-O)$_2$Mo(O)$_2$]$_2$	917, 885, 854	1.72	176a,177,181,182a,183
[Cp$_2$Mo(μ-O)$_2$Mo(O)$_2$]$_2$	918, 900, 880, 843		59,176a,177,182,182a
[(η^5-C$_5$H$_4$Bu^n)$_2$Mo(μ-O)$_2$Mo(O)$_2$]$_2$	912, 890		182a
[Cp^{1}Mo(μ-O)$_2$W(O)$_2$]$_2$	937, 885		182a
[Cp$_2$W(μ-O)$_2$W(O)$_2$]$_2$	940, 890		181,182a
[CpMo(O)$_2$]$_2$(μ-O)	930, 920		4,42,59,182,184
[Cp^{5}Mo(O)$_2$]$_2$(μ-O)	908, 879		42,182
[Cp^{5}Cr(O)(μ-O)]$_2$	910	1.59	42
[CpMo(O)(μ-O)]$_2$	925, 901	1.70	4,59,176a,177,182,186–187
[Cp'Mo(O)(μ-O)]$_2$			176a,177,184
[Cp^{5}Mo(O)(μ-O)]$_2$	914	1.69	182,185
[CpMo(O)(μ-S)]$_2$	920, 895	1.68	188–190
[Cp^{5}Mo(O)(μ-S)]$_2$	907, 900	1.69	191,192
[CpMoI(O)]$_2$(μ-O)		1.68	184,193
[Cp'M(O)]$_2$(μ-NR)(μ-O)	895	1.71	38
Cp' = Cp, Cp^1; M = Mo; R = p-C$_6$H$_4$Me			
Cp' = Cp; M = Mo; R = 1-naphthyl, 1-fluorenyl			38
[Cp^{5}W(O)(μ-O)]$_2$			7
[CpW(O)(μ-S)]$_2$	920, 932		180
[CpW(O)]$_2$(μ-NC$_6$H$_4$Me)(μ-O)			38

C. Complexes Containing Bridging [M(μ-O)$_n$M] Groups

Complex	Ref.
[Cp^{5}W(CO)]$_2$(μ-O)$_2$	7
{[CpMo]$_2$(μ-O)(μ-H)(μ-η^5:η^5-C$_{10}$H$_8$)}$^+$	194
{[CpMo]$_2$(μ-OMo(O)$_3$)(μ-η^5:η^5-C$_{10}$H$_8$)}$^+$	194
[CpW(CO)$_2$L][Cp$_2$ZrCl](μ-O)	145a
L = CO, PMe$_3$, PhC≡CPh	

TABLE VI
Cyclopentadienyl Oxo Complexes of Rhenium

A. Complexes Containing Terminal [Re=O] Groups

Complex	$\nu(M=O)$ (cm^{-1})	$r(M=O)$ (Å)	Ref
$Cp^*Re(O)_3$	909, 878		7,43–45,195–198
$(\eta^5\text{-}C_5Me_4Et)Re(O)_3$	920	1.71	198
$Cp^*ReX_2(O)$			
X = Cl	964, 948		7,198–200
Br	965, 950		7,198,201
Et, OMe, CH(Bu^t)$_2$			7
Ph, CH$_2$CMe$_3$, F			198
I		1.69	198
CH$_2$Ph		1.69	198
Me		1.68	7,198,201
X$_2$ = η^2-O$_2$CCPh$_2$	940, 975		7,200–202
η^2-O$_2$C$_6$Cl$_4$			7,203
η^2-OC(O)CPh$_2$O		1.68	7,196,202
η^2-O$_2$C$_{14}$H$_8$			7
η^2-OC(O)NPh	916, 964	1.69	200,202
η^2-(CH$_2$)$_2$CMe$_2$, η^2-CH$_2$C$_6$H$_4$			198
$(\eta^5\text{-}C_5Me_4Et)ReX_2(O)$			
X = Cl, Br			198
$\{[Cp^*Re(\mu\text{-}OH)(O)]_2\}^{2+}$			7

(continued)

TABLE VI (continued)

B. Complexes Containing Terminal [Re=O] and Bridging [Re(μ-O)$_n$Re] Groups

Complex	ν(M=O) (cm^{-1})	r(M=O) (Å)	Ref.
[Cp^{5}Re(O)(μ-O)]$_2$	930	1.72	195,197–199,201–203
[Cp^{5}Re(O)][Cp^{5}Re{ORe(O)$_3$}$_2$](μ-O)$_2$		1.72	195,201
[Cp^{5}Re(O)][Cp^{5}ReCl$_2$](μ-O)$_2$		1.68	198
[Cp^{5}Re(O)][Cp^{5}Re(CO)](μ-CO)(μ-O)			7

C. Complexes Containing Bridging [Re(μ-O)$_n$Re] Groups

Complex	Ref.
[Cp^{5}Re(CO)$_2$]$_2$(μ-O)	7,196–201
[Cp^{5}ReCl$_2$]$_2$(μ-O)$_2$	198
L$_n$M(μ-O)Re(O)$_3$	205,205a
L$_n$M = Cp(Ph$_3$P)$_2$Ru, Re(CO)$_5$,	
(Ph$_3$P)$_2$Rh(CO), (Ph$_3$P)$_2$Ir(CO),	
(Ph$_3$P)$_3$Os(CO)(H), (Ph$_3$P)$_2$IrCl(H)(CO)	

TABLE VII

Important Parameters in the M–O–M Linkage of $[Cp_2MX]_2(\mu\text{-}O)$ Complexes

Complex	r(M–O) (Å)	θ (deg)[a]	α (deg)[a]	Ref.
$[Cp_2^5Sm]_2(\mu\text{-}O)$	2.09	180	90	32
$[Cp_2Ti]_2(\mu\text{-}O)$	1.84	170.9	92.3	141
$[Cp_2TiCl]_2(\mu\text{-}O)$	1.84	173.8	75.4	51a
$[Cp_2Ti(NO_3)]_2(\mu\text{-}O)$	1.84	171.8	85.9	11
$[Cp_2Ti(CCF_3{=}CHCF_3)]_2(\mu\text{-}O)$	1.86	170	54	65
$[Cp_2Ti(PhC{=}CHPh)]_2(\mu\text{-}O)$	1.86	168.8	68.1	64
$\{[Cp_2Ti(H_2O)]_2(\mu\text{-}O)\}(ClO_4)_2$	1.83	176	—	13
$\{[Cp_2Ti(H_2O)]_2(\mu\text{-}O)\}S_2O_6$	1.83	177	74.1	147a
$[Cp_2Zr(SC_6H_5)]_2(\mu\text{-}O)$	1.96	165.8	61.7	148
$[Cp_2ZrCl]_2(\mu\text{-}O)$	1.94	168.9	74	149
$[Cp_2ZrMe]_2(\mu\text{-}O)$	1.95	174.1	—	67
$[Cp_2HfMe]_2(\mu\text{-}O)$	1.94	173.9	—	55
$[Cp_2^1NbCl]_2(\mu\text{-}O)$	1.92	173	0	164
$[Cp_2Nb(C_4H_9)]_2(\mu\text{-}O)$	1.93	180	4	78
$\{[Cp_2NbCl]_2(\mu\text{-}O)\}(BF_4)_2$	1.88	169	72.5	166

[a] θ is the M–O–M angle and α the Cp_2XM–O–MCp_2X dihedral angle.

these have been determined by X-ray diffraction and the important parameters are listed in Table VII. It is seen that all have essentially linear M–O–M bridges [the largest deviation is to 166° in $[Cp_2Zr(SC_6H_5)]_2(\mu\text{-}O)$ (148)] and most have staggered $Cp_2M(X)$ groups [i.e., the $Cp_2M(X)$–O–M(X)Cp_2 dihedral angle approaches 90° unless the X group hinders this rotation sterically]. This conformation allows maximum overlap of the single empty orbital of each of the $Cp_2M(X)$ fragments (M = Group 4 metal) with a lone pair of electrons an oxygen (207). As a consequence of this overlap the M–O distances are shorter than expected for an M–O single bond. Exceptions to the staggered conformation are the diamagnetic Group 5 derivatives $[Cp_2^1NbCl]_2(\mu\text{-}O)$ (164) and $[Cp_2Nb(C_4H_9)]_2(\mu\text{-}O)$ (78), which are eclipsed ($Cp^1 = \eta^5\text{-}C_5H_4CH_3$). In these cases each of the metals has the d^1 configuration with the frontier orbital singly occupied. The odd electrons are paired via a single occupied π orbital on the bridging oxygen atom. This is possible only in an eclipsed conformation. However, the d^1 complex $[Cp_2Ti]_2(\mu\text{-}O)$ has staggered Cp_2Ti groups and is paramagnetic (141–143). Steric factors may force the dimer to adopt the staggered conformation. In such a case, there can be π bonding between orthogonal lone pairs on oxygen and an empty orbital on each titanium (thus shortening the Ti–O bond distance) but the unpaired electrons localized on each titanium in mutually orthogonal orbitals cannot pair up. The only biscyclopentadienyl oxo complex of Group 3, $[Cp_2^5Sm]_2$-$(\mu\text{-}O)$, is also staggered for steric reasons (32).

There is only one example of a dimer, or higher oligomer, of the metals of Groups 3–5 that has a double oxygen bridge, namely $[Cp^5Ti]_2(\mu\text{-}\eta^1:\eta^5\text{-}C_5Me_4CH_2)(\mu\text{-}O)_2$ (Fig. 2) (35,66). In contrast, there are only a few examples of single oxygen bridges in complexes of the metals of Groups 6 and 7, but double bridges are common.

The questions of the formation of complexes with terminal versus bridging oxygen atoms or of dimers versus trimers, tetramers, or higher oligomers (which undoubtedly exist but have been intractable so far) are very difficult ones. Simple consideration of π versus σ bond energies suggests that an $[M(\mu\text{-}O)_2M]$ bridge will be favored over two isolated $M{=}O$ units in any oxo complex (steric constraints being ignored). This preference for bridging oxygen atoms will be even more marked with the highly oxophilic metals of Groups 3 and 4 and will also be reinforced when a proposed $Cp_mMX({=}O)$ complex is coordinatively unsaturated, e.g., the hypothetical $CpTiCl({=}O)$ or $CpCr({=}O)$ versus the tetramers $[CpTiCl(\mu\text{-}O)]_4$ (50) or $[CpCr(\mu_3\text{-}O)]_4$ which are actually observed. Conversely, the coordination number may restrict dimerization or the number of bridging oxygen atoms, e.g., in the observed $[CpMo(O)_2]_2(\mu\text{-}O)$ (4,59,182) versus an alternative formulation $[CpMo(O)]_2(\mu\text{-}O)_3$. Both formulations have the same number of electrons, 16, around the metal. However, these arguments do not wholly explain why, for instance, $Cp^5V(O)_2$ occurs as a trimer $[Cp^5V(O)(\mu\text{-}O)]_3$ (161) or $Cp^5Cr(O)_2$ occurs as $[Cp^5Cr(O)(\mu\text{-}O)]_2$ (42). The question of structure becomes almost impossible to answer when considering the trimers or tetramers such as $[Cp_2Zr(\mu\text{-}O)]_3$ (41) or $[Cp'MCl(\mu\text{-}O)]_4$ [M = Ti for Cp' = Cp, Cp^1 (50); M = V for Cp = Cp^5] (162), which, apart from having reasonable, albeit coordinatively unsaturated formulations as monomers, can be formulated with any ring size. Neither the coordination number nor the electronic configuration is affected by the number of units in the ring. The energy differences determining the structure actually observed are clearly small and the influences on them subtle. Crystal packing forces may be the deciding factor. Since there is almost no information on the structures in solution the problem cannot be pursued further.

The prototypes of cyclopentadienyl oxo complexes containing the $[M{=}O]$ group are $Cp'_2M(O)$ (M = Mo, W) (57,58) and $Cp^5Re(O)_3$ (43,44). It is distinctly possible that these prototypes will remain unique. Of the Group 3 metals only Ce has a chance to form $Cp_2Ce(O)$; $Cp_2M(O)$ (M = Ti, Zr, Hf) are likely to oligomerize if formed because of the high M–O bond strength; $Cp_2V(O)$ was claimed in the early days of metal–cyclopentadienyl chemistry (208), but later work on $[CpV]_m(\mu_3\text{-}O)_n$ clusters (see Section VI) suggests that the existence of $Cp_2V(O)$ is fleeting at best (*8,209*); nothing is known about possible Nb or Ta analogs. To the right of Group 6, any $Cp_2M(O)$ complexes would have more than 18 electrons. In the case of the $Cp'M(O)_n$ analogs of $Cp^5Re(O)_3$, two obvious candidates, $Cp'V(O)_2$ and $Cp'M(O)_2$ (M = Cr, Mo), are known, but the first is found as a trimer $[Cp^5V(O)(\mu\text{-O})]_3$ (*161*) and the second as a dimer $[Cp'M(O)(\mu\text{-O})]_2$ [M = Cr (*42*) or Mo (*4,59,185,186*)]. Even $Cp^5Re(O)_2$ is known only as a dimer, $[Cp^5Re(O)(\mu\text{-O})]_2$ (*195,199*). Nothing is known about compounds of the heavier members of Group 5 or the lighter members of Group 7. Once again, the caveat of possible differences in structure in the solid state and in solution must be made. There is in fact some evidence that $[Cp^5Re(O)(\mu\text{-O})]_2$ reacts as if it were $Cp^5Re(O)_2$ (*203*). Monomeric complexes such as $Cp^5Ti(O)$ are unlikely because they are coordinatively unsaturated and prone to oligomerization; one might hope to prepare species such as $CpM(O)$ (M = Fe, Co), but even these can be expected to oligomerize to the clusters $[CpM(\mu_3\text{-O})]_4$, as do their sulfur analogs (*210–212*).

If one follows the maximum oxidation state of the metals, then the series of compounds $Cp'M^3O$, $[Cp'M^4(O)]_2(\mu\text{-O})$, $Cp'M^5(O)_2$, $[Cp'M^6(O)_2](\mu\text{-O})$, and $Cp'M^7(O)_3$ may be written, where M^3, M^4, etc. refer to the periodic group. Of these, $Cp^5Re(O)_3$ has been discussed above, as has $Cp^5V(O)_2$ as the trimer $[Cp^5V(O)(\mu\text{-O})]_3$. The Group 6 member of the series is known in $[Cp'Mo(O)_2]_2(\mu\text{-O})$ [Cp' = Cp (*4,59,182*) or Cp^5(*42,182*)]. The Groups 3 and 4 members of the series are probably too coordinatively unsaturated to exist in the form written.

Listed in Tables IV–VI are the M=O distances and ν(M=O) frequencies in cyclopentadienyl oxo complexes containing the [M=O] group, where this information is known. The distances and frequencies are similar to those found in purely inorganic complexes containing the [M=O] moiety (*19*). A crystal structure determination on $(\eta^5\text{-}C_5Me_4Et)Re(O)_3$ shows it to have the expected piano-stool geometry (*198*). A variety of such piano-stool complexes of general formulas $Cp'MX(O)_2$ (for the Group 6 metals) and $CpMX_2(O)$ (Groups 5, 6, and 7) exist, as do derivatives of $Cp_2MX(O)$ (see Tables III–VI). Together, these three types constitute the second largest group of cyclopentadienyl oxo complexes.

An extensive chemistry of cyclopentadienyl oxo complexes has been developed only for $Cp^5Re(O)_3$ and its reduction product *trans*-$[Cp^5Re(O)(\mu\text{-O})]_2$ (*7,200a*). The latter is obtained from the former by oxygen abstraction with Ph_3P in the absence of O_2 (*201*). The same reagent gives, in the presence of O_2, the cluster $[Cp^5Re(\mu\text{-O})_2]_3^{2+}$ (Fig. 7) (*195,204*). This cluster is apparently diamagnetic, a result which contradicts theoretical predictions (*204a*).

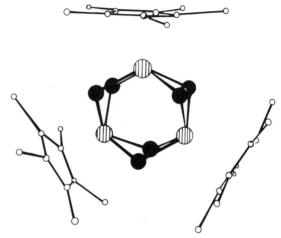

FIG. 7. Structure of $[Cp^5Re(\mu_2-O)_2]_3^{2+}$.

Reductive aggregation of organometallic oxo complexes is a promising route to organometallic clusters containing oxygen atoms; another example is the reduction of $[Cp^1TiCl(\mu-O)]_4$ or $[CpTiCl_2]_2(\mu-O)$ with Al producing $[Cp^1Ti]_6(\mu_3-Cl)_n(\mu_3-O)_{8-n}$ ($n = 0$, 2, or 4) (213) (Figs. 5, 8 and 9). The reactions of $Cp^5Re(O)_3$ and $[Cp^5Re(O)(\mu-O)]_2$ are shown in Schemes 1 and 2. Several points of interest may be noted. First, the ease with which an oxo ligand can be replaced by a halide, or simply removed by Ph_3P, is surprising. Second, there is a tendency for one [Re=O] unit to be

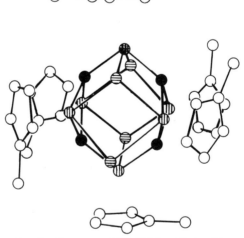

FIG. 8. Structure of $[Cp^1Ti]_6(\mu_3-Cl)_4(\mu_3-O)_4$.

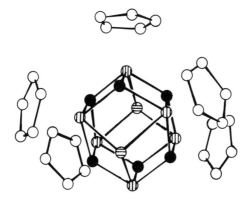

FIG. 9. Structure of $[CpTi]_6(\mu_3\text{-}Cl)_2(\mu_3\text{-}O)_6$.

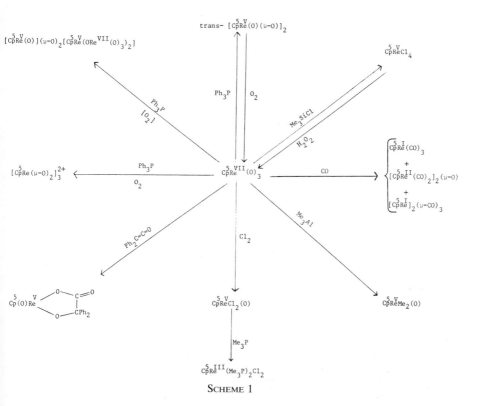

SCHEME 1

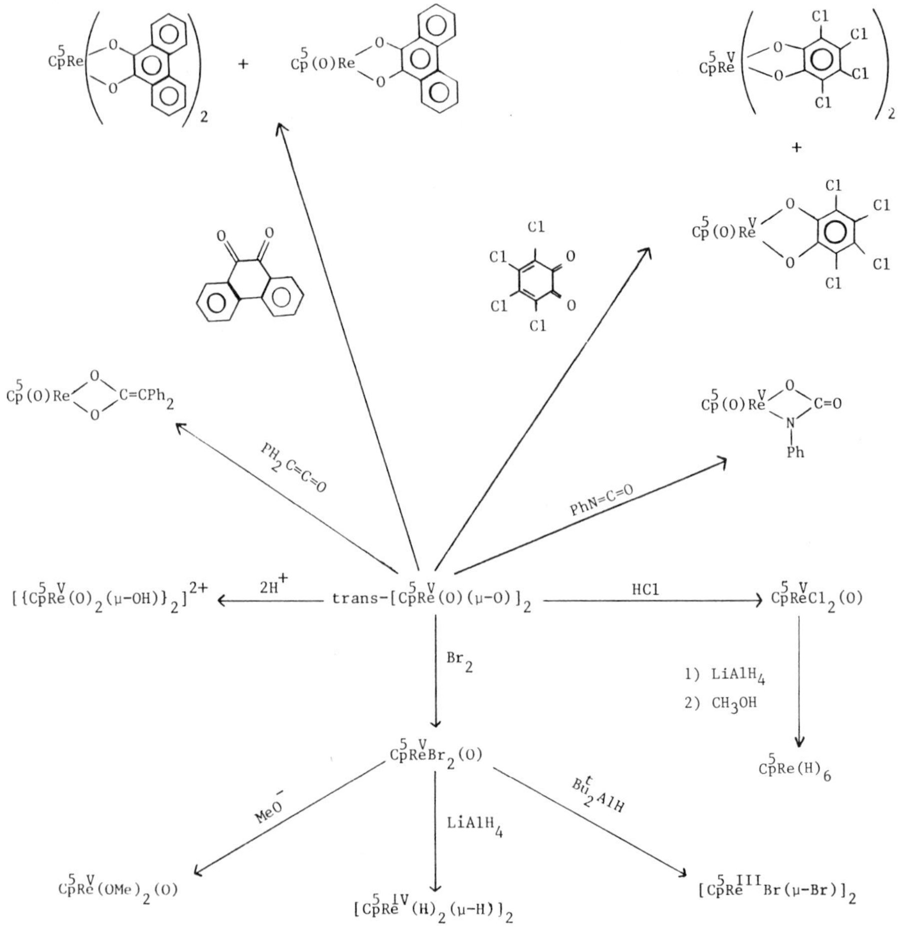

SCHEME 2

retained throughout a reaction unless forcing conditions are used. This may be related to the coordination number of the oxo complex and its reaction products, $[Cp^5Re(O)(\mu\text{-}O)]_2$ appearing to function as the five-coordinate $Cp^5Re(O)_2$ in many reactions. Third, the reaction of $Cp^5Re(O)_3$ with the ketene $Ph_2C=C=O$ should be noted since two oxygen atoms are added to the ketene. This can be compared to the reactions of $Os(O)_4$ (see Section I). Fourth, the reaction between PhNCO and $[Cp^5Re(O)(\mu\text{-}O)]_2$, giving $Cp^5(O)Re\overline{N(Ph)C(O)O}$ (200,202), should be contrasted with that between PhNCO and $[CpMo(O)(\mu\text{-}O)]_2$, in which all the oxygen ligands are replaced by NPh giving $[CpMo(NPh)(\mu\text{-}NPh)]_2$ (187). It is not clear why the presence of one less electron in the molybdenum case alters the course of the reaction so drastically. Fifth, in the reaction between $[Cp^5Re(O)(\mu\text{-}O)]_2$ and the orthoquinone (I) to give

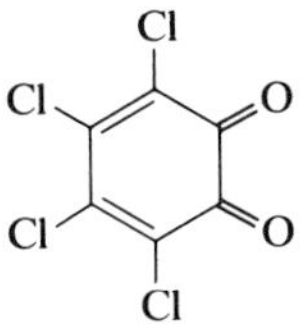

(I)

orthocachetol complexes, the other product is $Cp^5Re(O)_3$ (*203*), so an equation such as (61) may be written. Which oxygen atoms are transferred is not known.

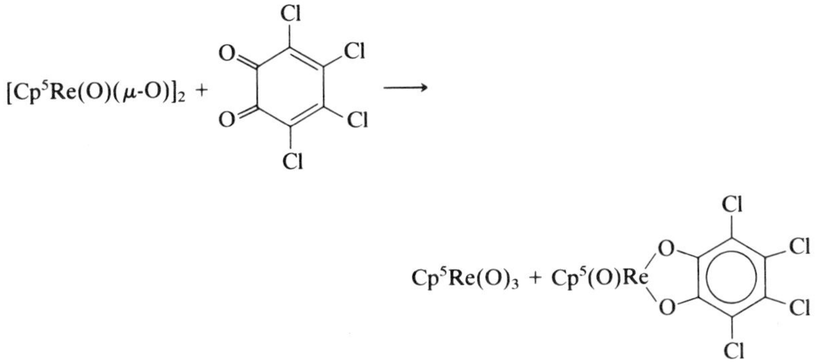

$[Cp^5Re(O)(\mu\text{-}O)]_2$ +

$Cp^5Re(O)_3 + Cp^5(O)Re$ (61)

VI

ORGANOMETALLIC CLUSTERS CONTAINING OXYGEN ATOMS

We discuss under this heading species with a metal–carbon bond, one or more oxygen atoms, and three or more metal atoms arranged in a closed fashion. Compounds containing only terminal [M=O] groups are not considered as oxo clusters, and compounds obtained by adding orga-nometallic fragments to polyoxometallates are discussed in Section VII below. The clusters falling within the restrictions given are listed in Table VIII, together with their structures and magnetic properties, where known. Organometallic clusters containing oxygen atoms are a new phenomenon and little is known about their chemistry, other than their sensitivity to air and water.

The large majority of the clusters in Table VIII contain a μ_3-oxygen atom, that is, an oxygen capping a triangle of metal atoms. Of such clusters

TABLE VIII
Organometallic Clusters Containing Oxygen Atoms

Cluster	Figure no.	Total no. of electrons	No. of cluster electrons	Unpaired electrons	Ref.
[CpY]$_5$(μ_2-OMe)$_4$(μ_3-OMe)$_4$(μ_5-O)	10	78	0	0	214
[Cp'Ti]$_4$(μ_3-S$_2$)$_4$(μ_4-O)(μ_2-O)	11	68	0	0	215
[Cp'Ti]$_4$(μ_3-S$_2$)$_3$(μ_3-S)(μ_2-S)(μ_4-O)	—	66	0	0	215
[CpTi]$_6$(μ_3-O)$_8$	5	86	2	0	53,213,216
[Cp'Ti]$_6$(μ_3-Cl)$_4$(μ_3-O)$_4$ (Cp' = Cp, Cp^l)	8	90	6	2	213
[Cp'Ti]$_6$(μ_3-Cl)$_2$(μ_3-O)$_6$ (Cp' = Cp, Cp^l)	9	88	4	0	213
{[CpTi(OMe)(μ_2-OMe)]$_3$(μ_3-O)}$^+$	12	42	0	0	217
{[CpTi(μ_2-OH)(μ_2-HCOO)]$_3$(μ_3-O)}$^+$	13	42	0	0	218
{[CpZr(μ_2-OH)(μ_2-PhCOO)]$_3$(μ_3-O)}$^+$	—	42	0	0	219
[CpV(μ_3-O)]$_4$	—	56	8	1	10
[CpV]$_5$(μ_3-O)$_6$	14	74	8	0.5	8,209
[CpV]$_6$(μ_3-O)$_8$	—	92	8	1	9,10
Cp$_5$(O)V$_6$(μ_3-O)$_8$	3	89	7	2	10
[Cp$_5$V$_6$(μ_3-O)$_8$]$_2$(μ_2-O)	15	88 × 2	8 × 2^a	1.5	10
[Cp$_5$V$_6$(μ_3-O)$_8$]$_2$[(μ-O)$_2$VCp(NMe$_3$)$_2$]	16	88 × 2 + 2	8 × 2^a	—	9
[Cp$_5$V$_6$(μ_3-O)$_8$]$_2$[(μ_2-O$_8$)V$_4$Cp$_4$]	17	89 × 2 + 0	9 × 2^a	2	9
[CpCr(μ_3-O)]$_4$	1	60	12	Afm[b,c]	2,3,8
[Cp'Cr(μ_3-O)]$_4$	—	60	12	Afm	220
[CpCr]$_4$(μ_3-η^2-C$_5$H$_4$)(μ_3-O)$_3$	4	60	12	Afm	37
{[Cp'Cr]$_4$(μ_3-S)$_3$(μ_3-O)}$^+$	—	59	11	1	221

Complex	No.	Cluster electrons[a]		Afm[d]	Ref.
$[Cp'Cr]_4(\mu_3\text{-}S)_2(\mu_4\text{-}SCuBr_2)(\mu_3\text{-}O)$	18	60	12	Afm	220,221
$[CpCr]_3[Co(CO)_3](\mu_3\text{-}S)_3\{\mu_4\text{-}O(HOOCMe_3)\}$	19	64	16	0	222
$[CpMo(O)(\mu_3\text{-}O)]_4$	—	68	4	—	59
$\{[CpMo(CO)(\mu_2\text{-}CO)]_3(\mu_3\text{-}O)\}$[a]	20	48	12	—	56
$[CpMo]_3(\mu_2\text{-}Cl)(\mu_3\text{-}\eta^2\text{-}C_4(CF_3)_4)(\mu_3\text{-}O)$	21	46	10	0	223
$Cp^5Mo(CO)_2]_2[Fe(CO)_3](\mu_3\text{-}O)$	22	48	16	0	224
$\{[Mo(\mu_2\text{-}O_2CMe)_2(H_2O)]_3(\mu_3\text{-}CMe)(\mu_3\text{-}O)\}^+$	23	48	6	1	225
$[W(\mu_2\text{-}O_2CMe)_2(H_2O)]_3(\mu_3\text{-}CMe)(\mu_3\text{-}O)\}^{2+}$[e]	—	47	5	0	226
$[CpW][Os_3(CO)_9](\mu_3\text{-}CCH_2C_6H_4Me)(\mu_2\text{-}O)$	24	58	24	0	227
$[Cp^5Re(\mu_2\text{-}O)_2]_3^{2+}$	7	46	4	0	7,195,204
$\{[Re(CO)_3(\mu_2\text{-}H)]_3(\mu_3\text{-}O)\}^{2-}$	25	48	24	0	228,229
$\{[Fe(CO)_3]_3[Mn(CO)_3](\mu_4\text{-}O)\}^+$	—	62	30	0	229a
$[Fe(CO)_3]_3(\mu_3\text{-}O)\}^{2-}$	26	48	24	—	230
$Ru(CO)_2]_2[Ru(CO)]\{(\mu_2\text{-}Ph_2AsCH_2AsPh_2)_2\ (\mu_3\text{-}CO)(\mu_3\text{-}O)$	27	48	22	0	231
$[Ru(CO)_2]_2[Ru(CO)]\{(\mu_2\text{-}Ph_2PCH_2PPh_2)_2\ (\mu_2\text{-}H)_2(\mu_3\text{-}O)$	28	48	20	0	232
$\{[Ru(CO)_2]_2[Ru(CO)]\{(\mu_2\text{-}Ph_2PCH_2PPh_2)_2\ (\mu_2\text{-}H)_2(\mu_2\text{-}I)(\mu_3\text{-}O)\}^+$	29	48	18	0	232
$[Os(CO)_3(\mu_3\text{-}O)]_4$	—	72	24	0	14,14a,233
$[Os(CO)_4Os(CO)_2]_3(\mu_3\text{-}CO)(\mu_3\text{-}O)$	30	90	46	0	234
$[CpCo]_3(\mu_3\text{-}CO)(\mu_3\text{-}O)$	31	48	22	—	235
$[(\eta^4\text{-}C_8H_{12})Ir]_3(\mu_2\text{-}I)_3(\mu_3\text{-}O)_2$	32	50	22	—	236

[a] The figures given are the number of cluster electrons in each of the $Cp_5V_6(\mu_3\text{-}O)_8$ units multiplied by the number of units plus any electrons on the metal atoms in the bridge.

[b] Afm, Antiferromagnetic.

[c] Diamagnetic at 20 K.

[d] $\mu_{eff} = 3.14\ \mu B$ at 79 K; Cu^{II}, one unpaired electron.

[e] Note also the existence of $\{[W_2(\mu_2\text{-}O_2CMe)_2(H_2O)]_3(\mu_3\text{-}O)_2\}^{2+}$.

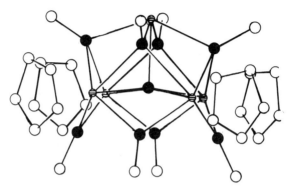

FIG. 10.　Structure of $[CpY]_5(\mu_2\text{-OMe})_4(\mu_3\text{-OMe})_4(\mu_5\text{-O})$.

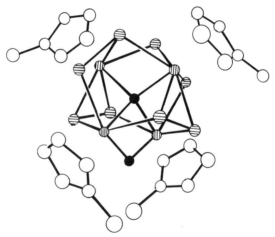

FIG. 11.　Structure of $[Cp^1Ti]_4(\mu_3\text{-S}_2)_4(\mu_4\text{-O})(\mu_2\text{-O})$.

a significant number are of the type $[L_3M]_m(\mu_3\text{-O})_n$, where L_3 is a tridentate ligand (usually $\eta^5\text{-C}_5H_5$) or some combination of uni- and bidentate ligands. The integers m and n may be 3 and 2, 4 and 4, 5 and 6, or 6 and 8. The basic structures of these clusters are a triangle of metal atoms with oxygen above and below the plane for $[L_3M]_3(\mu_3\text{-O})_2$, interpenetrating tetrahedra of metal and oxygen atoms for $[L_3M(\mu_3\text{-O})]_4$ (a cubane structure), a trigonal bipyramid of metal atoms and a trigonal prism of

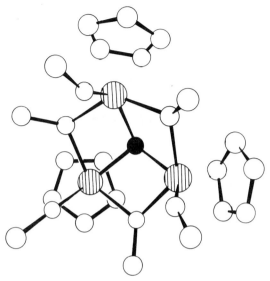

Fig. 12. Structure of $\{[CpTi(\mu_2\text{-}OMe)(OMe)]_3(\mu_3\text{-}O)\}^+$.

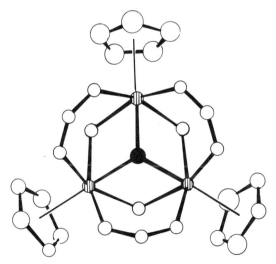

Fig. 13. Structure of $\{CpTi(\mu_2\text{-}OH)(\mu_2\text{-}HCOO)]_3(\mu_3\text{-}O)\}^+$.

oxygen atoms for $[L_3M]_5(\mu_3\text{-}O)_6$, and an octahedron of metal atoms with a cube of oxygen atoms for $[L_3M]_6(\mu_3\text{-}O)_8$. Note that the oxygen atoms are face-bridging and the metal-to-oxygen ratio is much higher than in the familiar edge-bridged polyoxometallates. The relatively low oxygen content is a consequence of the preparative routes to these clusters, only a

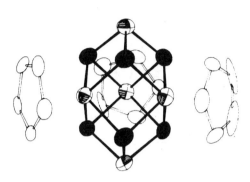

Fig. 14.   Structure of $[CpV]_5(\mu_3\text{-}O)_6$.

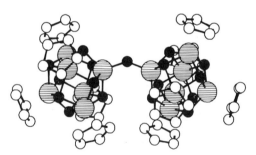

Fig. 15.   Structure of $[Cp_5V_6(\mu_3\text{-}O)_8]_2(\mu\text{-}O)$.

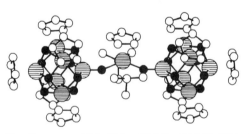

Fig. 16.   Structure of $[Cp_5V_6(\mu_3\text{-}O)_8]_2[(\mu\text{-}O)_2VCp(NMe_3)]_2$.

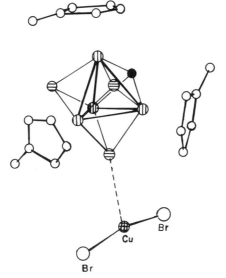

FIG. 17. Structure of $[Cp_5V_6(\mu_3\text{-}O)_8]_2[(\mu_2\text{-}O_8)V_4Cp_4]$.

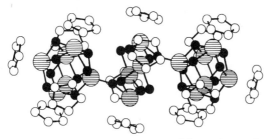

FIG. 18. Structure of $[Cp^1Cr]_4(\mu_3\text{-}S)_2(\mu_4\text{-}SCuBr_2)(\mu_3\text{-}O)$.

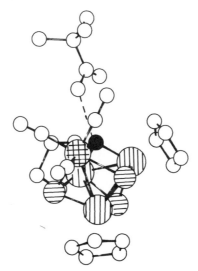

FIG. 19. Structure of $[CpCr]_3[Co(CO)_3](\mu_3\text{-}S)_3(\mu_4\text{-}OHOOCMe)$.

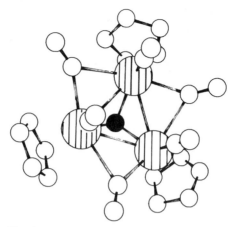

Fig. 20.   Structure of $\{[CpMo(CO)(\mu_2\text{-}CO)]_3(\mu_3\text{-}O)\}^+$.

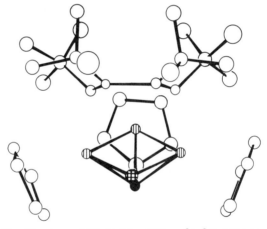

Fig. 21.   Structure of $[CpMo]_3(\mu_2\text{-}Cl)(\mu_3\text{-}\eta^2 \text{:} \eta^2\text{-}C_4(CF_3)_4)(\mu_3\text{-}O)$.

restricted supply of the oxygen-containing reagent being available. This also explains the relatively low formal oxidation states of the metal atoms (III and IV being common, and even zero occurring) and the air and water sensitivity of these clusters.

The $[CpM]_m(\mu_3\text{-}O)_n$ clusters are part of a series of $[CpM]_m(\mu_3\text{-}A)_n$ clusters, where A is an atom from Group 16. The best known example is $[CpFe(\mu_3\text{-}S)]_4$ (210,211,237). Clusters having the $M_m(\mu_3\text{-}S)_n$ core are under intensive investigation because of their presence in ferredoxins and nitrogenase (238). A discussion and comparison of these clusters is beyond the scope of this article but the relationship should be borne in mind when

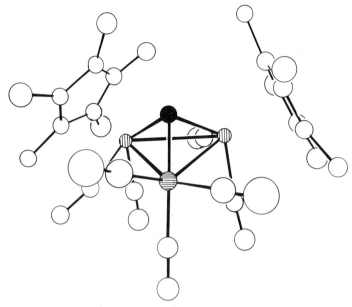

FIG. 22. Structure of $[Cp^5Mo(CO)_2]_2[Fe(CO)_3](\mu_3\text{-}O)$.

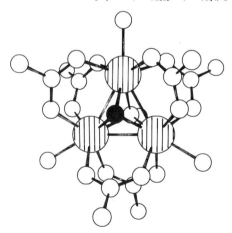

FIG. 23. Structure of $\{[Mo(\mu_2\text{-}O_2CMe)_2(H_2O)]_3(\mu_3\text{-}CMe)(\mu_3\text{-}O)\}^+$.

considering the properties of the clusters containing oxygen. Although the high electronegativity of oxygen compared to sulfur will make oxo clusters different from their sulfur analogs, there will be many useful similarities and contrasts.

The basic structures of the $[CpM(\mu_3\text{-}O)]_4$, $[CpM]_5(\mu_3\text{-}O)_6$, and $[CpM]_6(\mu_3\text{-}O)_8$ clusters are more or less distorted, and the distortion

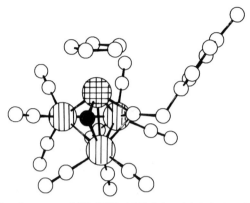

FIG. 24.   Structure of $[CpW][Os(CO)_3]_3(\mu_3\text{-}CCH_2C_6H_4Me)(\mu_2\text{-}O)$.

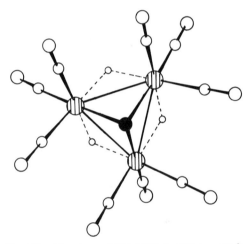

FIG. 25.   Structure of $\{[Re(CO)_3(\mu\text{-}H)]_3(\mu_3\text{-}O)\}.^{2-}$

depends on the electronic arrangement in the cluster (211,237,239). The electronic arrangement is also reflected in the magnetic properties of the clusters. The clusters are held together by M–O bonds and the M–(η^5-C_5H_5) bonding is of the usual type. The $[CpM(\mu_3\text{-}O)]_4$ clusters of idealized T_d symmetry require 48 electrons in 24 orbitals for the M–O and M–Cp bonds, the $[CpM]_5(\mu_3\text{-}O)_6$ clusters (D_{3h}) require 66 electrons in 33 orbitals, and $[CpM]_6(\mu_3\text{-}O)_8$ (O_h) require 84 in 42. In each type of cluster there remain 12 metal orbitals to accommodate any electrons in excess of the required number (these can be considered as cluster electrons) (239). Jahn–Teller distortion of the basic cluster geometry occurs when degenerate orbitals of the highly symmetric idealized structures are partially

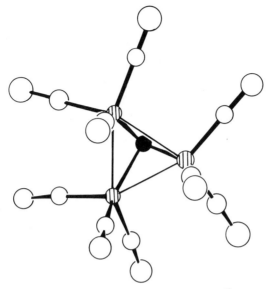

FIG. 26. Structure of $\{[Fe(CO)_3]_3(\mu_3\text{-O})\}^{2-}$.

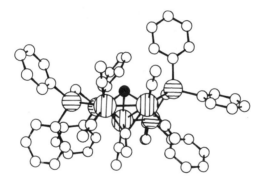

FIG. 27. Structure of $[Ru(CO)_2]_2[Ru(CO)](\mu_2\text{-Ph}_2AsCH_2AsPh_2)_2(\mu_3\text{-CO})(\mu_3\text{-O})$.

occupied. The 12 orbitals are essentially pure metal d in character and of similar energy. Hence unusual magnetic properties are found, for instance, the antiferromagnetism of $[CpCr(\mu_3\text{-O})]_4$. The exact ordering of the energies of the 12 cluster orbitals is still controversial (*237,239,240*) and may vary from metal to metal. Hence the distortions are not readily explicable.

A similar molecular orbital analysis can be made for the $[CpM]_3(\mu_3\text{-O})_2$ or $[L_3M]_3(\mu_3\text{-O})_2$ clusters, and such an analysis has been made for the sulfur analogs such as $[CpCo]_3(\mu_3\text{-S})_2$ (*235,241,242*). At present no simple

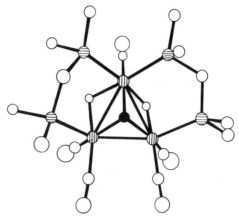

Fig. 28.   Structure of $[Ru(CO)_2]_2[Ru(CO)](\mu_2\text{-}Ph_2PCH_2PPh_2)_2(\mu_2\text{-}H)_2(\mu_3\text{-}O)$. (Only one carbon of the phenyl rings is shown for clarity.)

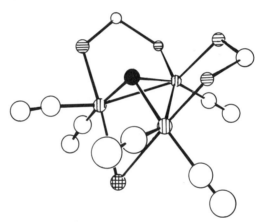

Fig. 29.   Structure of $\{[Ru(CO)_2]_2[Ru(CO)](\mu_2\text{-}Ph_2PCH_2PPh_2)_2(\mu_2\text{-}H)_2(\mu_2\text{-}I)(\mu_3\text{-}O)\}.^+$ (The phenyl rings are not shown for clarity).

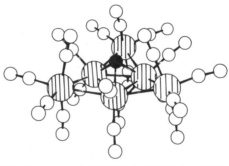

Fig. 30.   Structure of $[Os(CO)_4Os(CO)_2]_3(\mu_3\text{-}CO)(\mu_3\text{-}O)$.

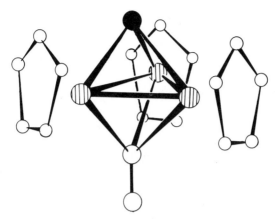

FIG. 31. Structure of $[CpCo]_3(\mu_3\text{-CO})(\mu_3\text{-O})$.

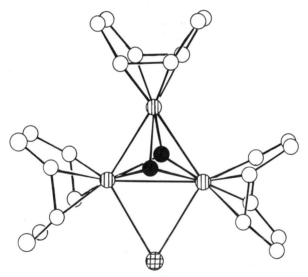

FIG. 32. Structure of $[Ir(\eta^4\text{-}C_8H_{12})]_3(\mu_2\text{-I})(\mu_3\text{-O})_2$.

$[CpM]_3(\mu_3\text{-O})_2$ or $[L_3M]_3(\mu_3\text{-O})_2$ cluster of high symmetry has been obtained; the closest is $[CpCo]_3(\mu_3\text{-CO})(\mu_3\text{-O})$ (Fig. 31) (*235*). Hence a detailed discussion is not warranted.

From the above discussion it is clear that a wide variety of $[L_3M]_m$-$(\mu_3\text{-O})_n$ clusters should exist as long as the number of cluster electrons is between 0 and 24. Clusters with mixtures of metals, e.g., $[CpCr]_3[Co-(CO)_3](\mu_3\text{-S})_3(\mu_3\text{-O})$ (*222*), and of ligands, e.g., $\{[Cp^1Cr]_4(\mu_3\text{-S})_3(\mu_3\text{-O})\}^+$ (*221*), $[CpCo]_3(\mu_3\text{-CO})(\mu_3\text{-O})$ (*235*), and $[Cp^1Ti]_6(\mu_3\text{-Cl})_4(\mu_3\text{-O})_4$ (*213*), exist, and other cations and anions should be obtainable by oxidation or reduction of the parent $[CpM]_m(\mu_3\text{-O})_n$ cluster. It is also possible to link the basic units together to produce arrays. The largest so far has the formula $Cp_{14}V_{16}O_{24}$ and consists of two $Cp_5V_6(\mu_3\text{-}O)_8$ cores linked by a $Cp_4V_4(\mu_2\text{-}O)_8$ rectangle (Fig. 17) (9). These arrays are true organometallic clusters, being soluble in nonpolar solvents, but at the same time have the magnetic properties of metal oxides.

There are a number of points of interest about the other clusters in Table VIII. Note that the carbonyl ligand is relatively common for clusters containing metals of groups 8 and 9, and $\{[Fe(CO)_3]_3(\mu_3\text{-}O)\}^{2-}$ (230) (Fig. 26) contains only CO and O.* This is in contrast to the non-cluster organometallic oxides, for which only one carbonyl is known (see Section IV). Note also that in the clusters the Group 8 metals have five representatives and Group 9 two, with the first-row elements represented in both groups. The only noncluster organometallic oxide of a Group 8 metal is $[Os(CH_2SiMe_3)_4(O)]$ (18); there are none of the Group 9 metals. All of the above facts can be accounted for by the low oxidation states in the clusters compared to the noncluster compounds. The low oxidation states mean that electrons are available for backbonding to CO and also that the less oxophilic metals to the right of the transition series are stabilized sufficiently to form a metal–oxygen bond. Note also the cluster $[Cp^5Re(\mu_2\text{-}O)_2]_3^{2+}$, which contains a triangle of Re atoms with oxygen bridging the edges (204). Such edge-bridging oxygens are the foundation of purely inorganic polyoxometallates, whereas the majority of the clusters in Table VIII have face-bridging oxygen atoms. It remains to be seen whether $[Cp^5Re(\mu_2\text{-}O)_2]_3^{2+}$ is an isolated example, occurring because of the steric bulk of the Cp^5 ligand, or is the first of a new series of organometallic polyoxometallates. The four cluster electrons in $[Cp^5Re(\mu_2\text{-}O)_2]_3^{2+}$ may provide an entry to development of this type of cluster. Finally, the cluster $[CpCr]_4(\mu_3\text{-}\eta^2\text{-}C_5H_4)(\mu_3\text{-}O)$ (37) should be noted. It is a derivative of $[CpCr(\mu_3\text{-}O)]_4$, in which the $\eta^2\text{-}C_5H_4$ fragment (II) replaces a μ_3 oxygen. This cluster is therefore a model for a hydrocarbon attached to a metal oxide surface.

(II)

VII

ORGANOMETALLIC POLYOXOMETALLATES

Parallel to the development of organometallic clusters containing oxygen atoms has been the preparation of organometallic polyoxometallates (243). These are obtained by adding a known organometallic complex to

*As do $[Os(CO)_3(\mu\text{-}O)]_4$ (14), and $[Os(CO)_4Os(CO)_2]_3(\mu\text{-}CO)(\mu_3\text{-}O)$ (Fig. 30) (234).

an existing or preformed polyoxometallate. Two types of product are obtained. In the first an organometallic fragment replaces a peripheral metal–oxygen group, and the organometallic group becomes part of (i.e., is "in") the polyoxometallate. In the second type of product the organometallic fragment becomes attached to the peripheral oxygen atoms of (i.e., is "on") the polyoxometallate.

An example of the first type of reaction is (62):

$$2Cp_2TiCl_2 + 5[Mo_2O_7]^{2-} + H_2O \rightarrow$$

$$2[CpTi(Mo_5O_{18})]^{3-} + 4Cl^- + 2C_5H_6 \quad (244\text{--}246) \quad (62)$$

$[CpTi(Mo_5O_{18})]^{3-}$ (Fig. 33a) is related to $[VMo_5O_{19}]^{3-}$ (Fig. 33b) by replacement of $[V{=}O]^{3+}$ by $[CpTi]^{3+}$. By similar methods $M_{12}PO_{40}^{3-}$ or $M_{11}SiO_{39}^{8-}$ (M = Mo, W) can be converted into $CpTiW_{11}PO_{39}^{4-}$ (247,248), $CpTiW_{11}SiO_{39}^{5-}$ (247), or $CpFe(CO)_2SnW_{11}PO_{39}^{4-}$, (249–251) and a variety of other derivatives containing organotin fragments, $[RSnM_{11}AO_{39}]^{n-}$ (M = Mo, W; A = P for n = 4; A = Si for n = 5) (247,252). A related replacement is $[Cp^5Zr]^{3+}$ for $[C_6H_{11}Si]^{3+}$ to give the cubic $[Cp^5Zr](C_6H_{11}Si)_7O_{12}$ (Fig. 34) (253). The substitution of the organometallic fragment does not markedly alter the molecular structure of the polyoxometallate, and since all the metal centers have zero d electrons the electronic structure also remains the same. However, the organometallic fragment is not as strongly bound as the [M=O] group that it replaces. For instance, whereas $[VMo_5O_{19}]^{3-}$ is stable towards water at 80°, $[CpTiMo_5O_{18}]^{3-}$ is rapidly decomposed under these conditions (245). This instability may be due more to the lower oxidation state of the substituted metal (in this case Ti^{IV} substitutes for V^V) and a consequent increase in the charge density of the polyoxometallate than to any inherent instability of organometallic fragments bound in polyoxometallates.

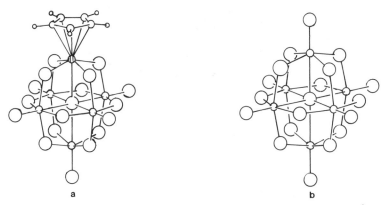

a   b

FIG. 33. Structure of $[CpTi(Mo_5O_{18})]^{3-}$ (a) compared to that of $[VMo_5O_{19}]^{3-}$ (b).

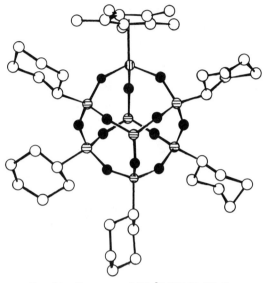

FIG. 34.　Structure of $[Cp^5Zr](C_6H_{11}Si)_7O_{12}$.

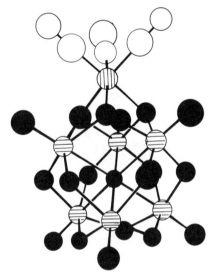

FIG. 35.　Structure of $\{[Mn(CO)_3](Nb_2W_4O_{19})\}^{3-}$.

An example of the second type of reaction is the preparation of $\{[M(CO)_3]$-$(Nb_2W_4O_{19})\}^{3-}$ (M = Mn, Re; Fig. 35) [Eq. (64)]. In these reactions the surface oxygen atoms of the parent polyoxometallate, $W_6O_{19}{}^{2-}$, must be activated by increasing the charge density. To this end Nb^V is substituted

$[(CH_3CN)_3M(CO)_3]^+ + Nb_2W_4O_{19}{}^{4-} \rightarrow$

$$\{[M(CO)_3](Nb_2W_4O_{19})\}^{3-} + 3CH_3CN \qquad (254,255) \quad (63)$$

for W^{VI}, giving $Nb_2W_4O_{19}{}^{2-}$. A combination of X-ray crystallography and ^{17}O NMR evidence indicates that the $[M(CO)_3]$ fragment is bound to three adjacent bridging oxygens (255). Similar attachment of the organometallic fragment to three bridging oxygens is found in $[Cp^5Rh(cis\text{-}Nb_2W_4O_{19})]^{2-}$ [Fig. 36 (256)], $[CpTi(W_9V_3SiO_{40})]^{4-}$ (257), and $[Cp^5Rh(W_9Nb_3SiO_{40})]^{5-}$ (258). In the remarkable $\{[(\eta^4\text{-}C_7H_8)Rh]_5(cis\text{-}Nb_2W_4O_{19})_2\}^{3-}$ the organometallic Rh^I fragment is attached to two terminal $[Nb{=}O]$ oxygen atoms [Fig. 37 (259)]; in $[Cp_3U(NbW_5O_{19})_2]^{5-}$ the attachment is to one terminal $(Nb{=}O)$ oxygen [Fig. 38 (260)]; in $[(Cp_2U)_2(\mu\text{-}\kappa^2\text{-}O\text{-}TiW_5O_{19})_2]^{4-}$ each Cp_2U is attached to a bridging oxygen of one $OTiW_5O_{19}$ unit and the terminal oxygen of both units [Fig. 39 (261)].

Although the organometallic groups are on the surface of the polyoxometallate, they are strongly and covalently bonded to the peripheral oxygen atoms (257). The importance of the compatibility of the organic group with the peripheral oxygen atoms in these anions has been stressed (256,260,261). A packing diagram for $[Cp^5Rh(cis\text{-}Nb_2W_4O_{19})]^{2-}$ is shown in Fig. 40. The polyoxometallate consists of close-packed oxygen atoms with Nb or W in the octahedral holes; the $C_5(CH_3)_5$ group adds another layer and the "intruder" metal atom is encapsulated in a hole between the organic layer and the oxygen atoms.

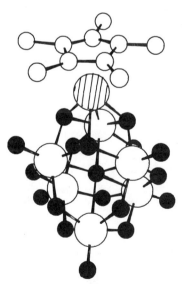

FIG. 36. Structure of $[Cp^5Rh(cis\text{-}Nb_2W_4O_{19})]^{2-}$.

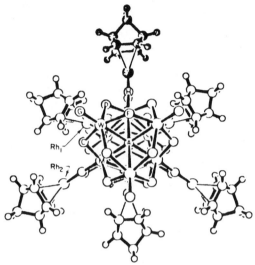

FIG. 37. Structure of $\{[\eta^4\text{-}C_7H_8)Rh]_5(cis\text{-}Nb_2W_4O_{19})_2\}^{3-}$. [Reproduced with permission from C. J. Besecker, W. G. Klemperer, and V. W. Day, *J. Am. Chem. Soc.* **104,** 6158 (1982)].

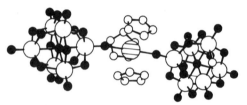

FIG. 38. Structure of $[Cp_3U(NbW_5O_{19})_2]^{5-}$.

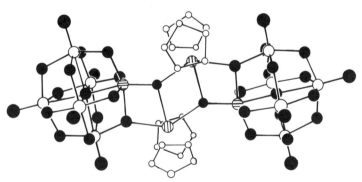

FIG. 39. Structure of $[(Cp_2U)_2(\mu\text{-}\kappa^2O\text{-}TiW_5O_{19})_2]^{4-}$.

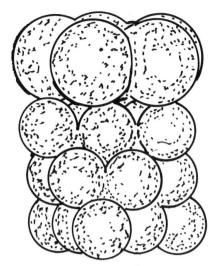

FIG. 40.   Packing diagram of $[Cp^5Rh(cis\text{-}Nb_2W_4O_{19})]^{2-}$.

These organometallic derivatives of polyoxometallates require the breaking of new ground in experimental techniques. Because of the unit cell size and monumental disorder problems, X-ray crystallography (normally the main weapon in a cluster chemist's arsenal) is of limited value in completely characterising the anions. Hence old-fashioned microanalysis, ^{17}O NMR, and most recently fabs mass spectrometry [e.g., for $(Bu_4N)_5[Cp^5Rh(W_9Nb_3SiO_{40})]$ (M = 4053)] have been employed as well (257,262).

These fascinating materials have obvious importance in the study of supported metal catalysts, in ceramics, and in the understanding of metal–carbon bonds in metal oxides. A potential use for them has already been identified: attachment of reactive organic functional groups to the Cp ring in $CpTiW_{11}PO_{39}^{4-}$ allows the Keggin anion to be covalently bonded to macromolecules and thus used as a label in electron microscopy (264,265). A complete review requires a discussion of polyoxometallates and is therefore beyond the scope or length of this article.

ACKNOWLEDGMENTS

We wish to thank Dr. Peter S. White for preparing some of the figures in this article, and Cathy Underhill for assistance in the preparation of the manuscript.

References

1. E. O. Fisher and S. Vigoureux, *Chem. Ber.* **91**, 1342 (1958).
2. E. O. Fischer, K. Ulm, and H. P. Fritz, *Chem. Ber.* **93**, 2167 (1960).
3. F. Bottomley, D. E. Paez, and P. S. White, *J. Am. Chem. Soc.* **103**, 5581 (1981).
4. M. Cousins and M. L. H. Green, *J. Chem. Soc.* 1567 (1964).
5. P. Corradini and G. Allegra, *J. Am. Chem. Soc.* **81**, 5510 (1959).
6. W. L. Carrick, W. T. Reichle, F. Pennella, and J. J. Smith, *J. Am. Chem. Soc.* **82**, 3887 (1960).
7. W. A. Herrmann, *J. Organomet. Chem.* **300**, 111 (1986).
8. F. Bottomley, D. E. Paez, and P. S. White, *J. Am. Chem. Soc.* **104**, 5651 (1982).
9. F. Bottomley, D. E. Paez and P. S. White, *J. Am. Chem. Soc.* **107**, 7226 (1985).
10. F. Bottomley, D. F. Drummond, D. E. Paez, and P. S. White, *J. Chem. Soc. Chem. Commun.*, 1752 (1986).
11. U. Thewalt and H.-P. Klein, *Z. Anorg. Allgem. Chem.* **479**, 113 (1981).
12. H.-P. Klein, U. Thewalt, K. Döppert, and R. Sanchez-Delgado, *J. Organomet. Chem.* **236**, 189 (1982).
13. U. Thewalt and B. Kebbel, *J. Organomet. Chem.* **150**, 59 (1978).
14. B. F. G. Johnson, J. Lewis, I. G. Williams, and J. Wilson, *Chem. Commun.* 391 (1966).
14a. D. A. Bright, *J. Chem. Soc. Chem. Commun.* 1169 (1970).
15. F.-M. Su, C. Cooper, S. J. Geib, A. L. Rheingold, and J. M. Mayer, *J. Am. Chem. Soc.* **108**, 3545 (1986).
16. K. Lane, L. Sallans, and R. R. Squires, *J. Am. Chem. Soc.* **106**, 2719 (1984).
17. J. A. Crayston, M. J. Almond, A. J. Downs, M. Poliakoff, and J. J. Turner, *Inorg. Chem.* **23**, 3051 (1984).
18. A. S. Alves, D. S. Moore, R. A. Andersen, and G. Wilkinson, *Polyhedron* **1**, 83 (1982).
19. W. P. Griffith, *Coord. Chem. Rev.* **5**, 459 (1970).
20. R. A. Sheldon and J. A. Kochi, "Metal-Catalyzed Oxidation of Organic Compounds." Academic Press, New York, 1981.
21. K. B. Wiberg, "Oxidation in Organic Chemistry," Part A. Academic Press, New York, 1965.
22. K. B. Sharpless, A. Y. Teranishi, and J.-E. Bäckvall, *J. Am. Chem. Soc.* **99**, 3120 (1977).
23. K. A. Jorgensen and R. Hoffmann, *J. Am. Chem. Soc.* **108**, 1867 (1986).
24. H. Mimoun, *Angew. Chem. Int. Ed. Engl.* **21**, 734 (1982).
25. C. P. Casey, *J. Chem. Soc. Chem. Commun.* 126 (1983).
26. E. L. Muetterties and E. Band, *J. Am. Chem. Soc.* **102**, 6572 (1980).
27. M. R. Churchill, J. R. Missert, and W. J. Youngs, *Inorg. Chem.* **20**, 3388 (1981).
28. J. H. Wengrovius, R. R. Schrock, M. R. Churchill, J. R. Missert, and W. J. Youngs, *J. Am. Chem. Soc.* **102**, 4515 (1980).
29. R. D. Rogers, R. Vann Bynum, and J. L. Atwood, *J. Crystallogr. Spectrosc. Res.* **12**, 239 (1982).
30. J. M. Mayer, D. L. Thorn, and T. H. Tulip, *J. Am. Chem. Soc.* **107**, 7454 (1985).
31. P. G. Edwards, G. Wilkinson, M. B. Hursthouse, and K. M. A. Malik, *J. Chem. Soc. Dalton Trans.* 2467 (1980).
32. W. J. Evans, J. W. Grate, I. Bloom, W. E. Hunter, and J. L. Atwood, *J. Am. Chem. Soc.* **107**, 405 (1985).
33. F. Bottomley and I. J. B. Lin, *J. Chem. Soc. Dalton Trans.* 271 (1981).
34. G. Fochi and C. Floriani, *J. Chem. Soc. Dalton Trans.* 2577 (1984).

35. F. Bottomley, G. O. Egharevba, I. J. B. Lin, and P. S. White, *Organometallics* **3**, 550 (1985).
36. W. Mowat, A. Shortland, G. Yagupsky, N. J. Hill, M. Yagupsky, and G. Wilkinson, *J. Chem. Soc. Dalton Trans.* 533 (1972).
37. F. Bottomley, D. E. Paez, L. Sutin, and P. S. White, *J. Chem. Soc. Chem. Commun.* 597 (1985).
38. H. Alper, J.-F. Petrignani, F. W. B. Einstein, and A. C. Willis, *J. Am. Chem. Soc.* **105**, 1701 (1983).
39. P. Legzdins, S. J. Rettig, and L. Sanchez, *Organometallics* **4**, 1470 (1985).
40. H. G. Alt and H. I. Hayen, *Angew. Chem. Int. Ed. Engl.* **24**, 497 (1985).
40a. H. G. Alt and H. I. Hayen, *J. Organomet. Chem.* **316**, 105 (1986).
41. G. Fachinetti, C. Floriani, A. Chiesi-Villa, and C. Guastini, *J. Am. Chem. Soc.* **101**, 1767 (1979).
42. M. Herberhold, W. Kremnitz, A. Razavi, H. Schöllhorn, and U. Thewalt, *Angew Chem. Int. Ed. Engl.* **24**, 601 (1985).
43. W. A. Hermann, R. Serrano, and H. Bock, *Angew. Chem. Int. Ed. Engl.* **23**, 383 (1984).
44. A. H. Klahn-Oliva and D. Sutton, *Organometallics* **3**, 1313 (1984).
45. W. A. Herrmann, E. Voss, and M. Flöel, *J. Organomet. Chem.* **297**, C5 (1985).
46. J. L. Templeton, B. C. Ward, G. J.-J. Chen, J. W. McDonald, and W. E. Newton, *Inorg. Chem.* **20**, 1248 (1981).
47. R. D. Gorsich, *J. Am. Chem. Soc.* **82**, 4211 (1960).
48. G. Allegra, P. Ganis, L. Porri, and P. Corradini, *Lincei-Rend. Sci. Fis. Mat. Nat.* **30**, 44 (1961).
49. J. L. Petersen, *Inorg. Chem.* **19**, 181 (1980).
50. A. C. Skapski and P. G. H. Troughton, *Acta Crystallogr.* **B26**, 716 (1970).
51. K. Döppert, *J. Organomet. Chem.* **178**, C3 (1979).
51a. Y. LePage, J. D. McCowan, B. K. Hunter, and R. D. Heyding, *J. Organomet. Chem.* **193**, 201 (1980).
52. B. Demerseman and P. H. Dixneuf, *J. Chem. Soc. Chem. Commun.* 665 (1981).
53. F. Bottomley, D. F. Drummond, G. O. Egharevba, and P. S. White, *Organometallics* **5**, 1620 (1986).
54. G. L. Hillhouse and J. E. Bercaw, *J. Am Chem. Soc.* **106**, 5472 (1984).
55. F. R. Fronczek, E. C. Baker, P. R. Sharp, K. N. Raymond, H. G. Alt, and M. D. Rausch, *Inorg. Chem.* **15**, 2284 (1976).
56. K. Schloter, U. Nagel, and W. Beck, *Chem. Ber.* **113**, 3775 (1980).
57. M. L. H. Green, A. H. Lynch, and M. G. Swanwick, *J. Chem. Soc. Dalton Trans.* 1445 (1972).
58. N. D. Silavwe, M. Y. Chiang, and D.R. Tyler, *Inorg. Chem.* **24**, 4219 (1985).
59. M. Cousins and M. L. H. Green, *J. Chem. Soc. A* 16 (1969).
60. S. M. Rocklage, R. R. Schrock, M. R. Churchill, and H. J. Wasserman, *Organometallics* **1**, 1332 (1982).
61. I. Feinstein-Jaffe, D. Gibson, S. J. Lippard, R. R. Schrock, and A. Spool, *J. Am. Chem. Soc.* **106**, 6305 (1984).
62. I. Feinstein-Jaffe, J. C. Dewan, and R. R. Schrock, *Organometallics* **4**, 1189 (1985).
63. H. Stoeckli-Evans, *Helv. Chim. Acta* **57**, 684 (1974).
64. V. B. Shur, S. Z. Berndyuk, V. V. Burlakov, V. G. Andrianov, A. I. Yanovsky, Yu. T. Struchkov, and M. E. Vol'pin, *J. Organomet. Chem.* **243**, 157 (1983).
65. M. D. Rausch, D. J. Sikora, D. C. Hrincir, W. E. Hunter, and J. L. Atwood, *Inorg. Chem.* **19**, 3871 (1980).

66. F. Bottomley, I. J. B. Lin, and P. S. White, *J. Am. Chem. Soc.* **103**, 703 (1981).
66a. S. G. Blanco, M. P. G. Sal, S. M. Carreras, M. Mena, P. Royo, and R. Serrano, *J. Chem. Soc. Chem. Commun.* 1572 (1986).
67. W. E. Hunter, D. C. Hrincir, R. Vann Bynum, R. A. Penttila, and J. L. Atwood, *Organometallics* **2**, 750 (1983).
68. R. P. Planalp and R. A. Andersen, *J. Am. Chem. Soc.* **105**, 7774 (1983).
69. K.-H. Thiele, W. Schumann, S. Wagner, and W. Brüser, *Z. Anorg. Allgem. Chem.* **390**, 280 (1972).
70. W. T. Reichle and W. L. Carrick, *J. Organomet. Chem.* **24**, 419 (1970).
71. R. Choukroun and S. Sabo, *J. Organomet. Chem.* **182**, 221 (1979).
72. K.-H. Thiele, B. Adler, H. Grahlert, and A. Lachowicz, *Z. Anorg. Allgem. Chem.* **403**, 279 (1974).
73. A. Lachowicz and K.-H. Thiele, *Z. Anorg. Allgem. Chem.* **431**, 88 (1977).
74. A. R. Middleton and G. Wilkinson, *J. Chem. Soc. Dalton Trans.* 1888 (1980).
75. G. Yagupsky, W. Mowat, A. Shortland, and G. Wilkinson, *J. Chem. Soc. Chem. Commun.* 1369 (1970).
76. R. Mercier, J. Douglade, J. Amaudrut, J. Sala-Pala, and J. E. Guerchais, *J. Organomet. Chem.* **244**, 145 (1983).
77. C. Santini-Scampucci and J. G. Riess, *J. Chem. Soc. Dalton Trans* 1433 (1974).
78. N. I. Kirillova, D. A. Lemenovskii, T. V. Baukova, and Yu T. Struchkov, *Sov. J. Coord. Chem.* **3**, 1254 (1977).
79. D. A. Lemenovskii, T. V. Baukova, V. A. Knizhnikov, É. G. Perevalova, and A. N. Nesmeyanov, *Dokl. Akad, Nauk SSSR (Engl. Trans.)* **226**, 65 (1976).
80. R. R. Schrock, *J. Am. Chem. Soc.* **98**, 5399 (1976).
81. M. R. Churchill and W. J. Youngs, *Inorg. Chem.* **20**, 382 (1981).
82. P. Belmonte, R. R. Schrock, M. R. Churchill, and W. J. Youngs, *J. Am. Chem. Soc.* **102**, 2858 (1980).
83. A. van Asselt, B. J. Burger, V. C. Gibson, and J. E. Bercaw, *J. Am. Chem. Soc.* **108**, 5347 (1986).
84. B. Heyn and R. Hoffmann, *Z. Chem.* **16**, 407 (1976).
85. B. Heyn and R. Hoffmann, *Z. Chem.* **16**, 195 (1976).
86. H. Arzoumanian, A. Baldy, R. Lai, J. Metzger, M.-L.N. Peh, and M. Pierrot, *J. Chem. Soc. Chem. Commun.* 1151 (1985).
86a. R. Lai, S. Le Bot, A. Baldy, M. Pierrot, and H. Arzoumanian, *J. Chem. Soc. Chem. Commun.* 1208 (1986).
87. G. N. Schrauzer, L. A. Hughes, N. Strampach, F. Ross, D. Ross, and E. O. Schlemper, *Organometallics* **2**, 481 (1983).
88. G. N. Schrauzer, E. L. Moorehead, J. H. Grate, and L. Hughes, *J. Am. Chem. Soc.* **100**, 4760 (1978).
89. G. N. Schrauzer, L. A. Hughes, and N. Strampach, *Z. Naturforsch.* **37b**, 380 (1982).
90. G. N. Schrauzer, L. A. Hughes, N. Strampach, P. R. Robinson, and E. O. Schlemper, *Organometallics* **1**, 44 (1982).
91. G. N. Schrauzer, L. A. Hughes, E. O. Schlemper, F. Ross, and D. Ross, *Organometallics* **2**, 1163 (1983).
91a. G. N. Schrauzer, E. O. Schlemper, N. H. Liu, Q. Wang, K. Rubin, X. Zhang, X. Long, and C. S. Chin, *Organometallics* **5**, 2452 (1986).
92. W. E. Carroll, M. Green, A. G. Orpen, C. J. Schaverien, I. D. Williams, and A. J. Welch, *J. Chem. Soc. Dalton Trans.* 1021 (1986).
93. E. Carmona, A. Galindo, E. Gutierrez-Puebla, A. Monge, and C. Puerta, *Inorg. Chem.* **25**, 3804 (1986).

94. J. R. M. Kress, M. J. M. Russell, M. G. Wesolek, and J. A. Osborn, *J. Chem. Soc. Chem. Commun.* 431 (1980).
95. J. Kress, M. Wesolek, J.-P. Le Ny, and J.A. Osborn, *J. Chem. Soc. Chem. Commun.* 1039 (1981).
96. I. Feinstein-Jaffe, S. F. Pedersen, and R. R. Schrock, *J. Am. Chem. Soc.* **105**, 7176 (1983).
96a. K. Jacob and K.-H. Thiele, *Z. Anorg. Allgem. Chem.* **508**, 50 (1984).
97. C. Santini-Scampucci and J. G. Riess, *J. Organnomet. Chem.* **73**, C13 (1974).
98. C. Santini-Scampucci and J. G. Riess, *J. Chem. Soc. Dalton Trans.* 195 (1976).
99. J. H. Freudenberger and R. R. Schrock, *Organometallics* **5**, 398 (1986).
100. N. G. Bokiy, Yu. V. Gatilov, Yu. T. Struchkov, and N. A. Ustynyuk, *J. Organomet. Chem.* **54**, 213 (1973).
101. L. M. Engelhardt, R. I. Papasergio, C. L. Raston, G. Salem, and A. H. White, *J. Chem. Soc. Dalton Trans.* 789 (1986).
102. M. F. Lappert, C. L. Raston, G. L. Rowbottom, B. W. Skeleton, and A. H. White, *J. Chem. Soc. Dalton Trans.* 883 (1984).
102a. M. Herberhold, H. Kniesel, L. Haumaier, and U. Thewalt, *J. Organomet. Chem.* **301**, 355 (1986).
103. K. Mertis, D. H. Williamson, and G. Wilkinson, *J. Chem. Soc. Dalton Trans.* 607 (1975).
104. K. Mertis and G. Wilkinson, *J. Chem. Soc. Dalton Trans.* 1488 (1976).
105. K. Mertis, J. F. Gibson, and G. Wilkinson, *J. Chem. Soc. Chem. Commun.* 93 (1974).
106. J. C. Green, D. R. Lloyd, L. Galyer, K. Mertis, and G. Wilkinson, *J. Chem. Soc Dalton Trans.* 1403 (1978)
107. L. Galyer, K. Mertis, and G. Wilkinson, *J. Organomet. Chem.* **85**, C37 (1975).
108. J. F. Gibson, K. Mertis, and G. Wilkinson, *J. Chem. Soc. Dalton Trans.* 1093 (1975).
109. I. R. Beattie and P. J. Jones, *Inorg. Chem.* **18**, 2318 (1979).
110. J. M. Huggins, D. R. Whitt, and L. Lebioda, *J. Organomet. Chem.* **312**, C15 (1986).
111. P. Stavropoulos, P. G. Edwards, G. Wilkinson, M. Motevalli, K. M. A. Malik, and M. B. Hursthouse, *J. Chem. Soc. Dalton Trans.* 2167 (1985).
111a. R. P. Tooze, G. Wilkinson, M. Motevalli, and M. B. Hursthouse, *J. Chem. Soc., Dalton Trans.*, 2711 (1986).
112. A. R. Middleton and G. Wilkinson, *J. Chem. Soc. Dalton Trans.* 1898 (1981).
112a. P. Stravropoulos, P. G. Edwards, T. Behling, G. Wilkinson, M. Motevalli, and M. B. Hursthouse, *J. Chem. Soc., Dalton Trans.*, 169 (1987).
113. W. A. Herrmann, U. Küsthardt, M. Flöel, J. Kulpe, E. Herdtweck, and E. Voss. *J. Organomet. Chem.* **314**, 151 (1986).
114. W. A. Herrmann, R. Serrano, A. Schäfer, U. Küsthardt, M. L. Ziegler, and E. Guggolz, *J. Organomet. Chem.* **272**, 55 (1984).
115. M. J. Almond, J. A. Crayston, A. J. Downs, M. Poliakoff, and J. J. Turner, *Inorg. Chem.* **25**, 19 (1986).
116. M. Poliakoff, K. P. Smith, J. J. Turner, and A. J. Wilkinson, *J. Chem. Soc. Dalton Trans.* 651 (1982).
117. H. Huber, D. McIntosh, and G. A. Ozin, *Inorg. Chem.* **16**, 975 (1977).
118. F. A. Cotton, P. Lahuerta, M. Sanao, and W. Schwotzer, *Inorg. Chim. Acta* **120**, 153 (1986).
119. L. Ricard and R. Weiss, *Inorg. Nucl. Chem. Lett.* **10**, 217 (1974).
120. W. E. Newton, J. W. McDonald, J. L. Corbin, L. Ricard, and R. Weiss, *Inorg. Chem.* **19**, 1997 (1980).
121. J. L. Davidson, M. Green, D. W. A. Sharp, F. G. A. Stone, and A. J. Welch, *J. Chem. Soc. Chem. Commun.* 706 (1974).

122. P. S. Braterman, J. L. Davidson, and D. W. A. Sharp, *J. Chem. Soc. Dalton Trans.* 246 (1976).

122a. L. Busetto, J. C. Jeffery, R. M. Mills, F. G. A. Stone, M. J. Went, and P. Woodward, *J. Chem. Soc. Dalton Trans.* 101 (1983).

123. J. A. K. Howard, R. F. D. Stansfield, and P. Woodward, *J. Chem. Soc. Dalton Trans.* 246 (1976).

124. P. W. Schneider, D. C. Bravard, J. W. McDonald, and W. E. Newton, *J. Am. Chem. Soc.* **94**, 8640 (1972).

125. E. A. Maatta and R. A. D. Wentworth, *Inorg. Chem.* **18**, 524 (1979).

126. E. A. Maatta, R. A. D. Wentworth, W. E. Newton, J. W. McDonald, and G. D. Watt, *J. Am. Chem. Soc.* **100**, 1320 (1978).

127. J. M. Mayer and T. H. Tulip, *J. Am. Chem. Soc.* **106**, 3878 (1984).

127a. J. M. Mayer and T. H. Tulip, *J. Am. Chem. Soc.* **106**, 3878 (1984).

127a. J. M. Mayer, T. H. Tulip, J. C. Calabrese, and E. Valencia, *J. Am. Chem. Soc.* **109**, 157 (1987).

128. A. K. Rappe and W. A. Goddard, III, *J. Am. Chem. Soc.* **104**, 448 (1982).

129. N. Calderon, H. Y. Chen, and K. W. Scott, *Tetrahedron Lett.* 3327 (1967).

130. M. T. Mocella, R. Rovner and E. L. Muetterties, *J. Am. Chem. Soc.* 98, 4689 (1976).

131. R. Schrock, S. Rocklage, J. Wengrovius, G. Rupprecht, and J. Fellmann, *J. Mol. Catal.* **8**, (1980).

132. M. R. Churchill, A. L. Rheingold, W. J. Youngs, R. R. Schrock, and J. H. Wengrovius, *J. Organomet. Chem.* **204**, C17 (1981).

133. J. H. Wengrovius and R. R. Schrock, *Organometallics* **1**, 148 (1982).

134. M. R. Churchill and A. L. Rheingold, *Inorg. Chem.* **21**, 1357 (1982).

135. A. K. Rappe and W. A. Goddard, III, *J. Am. Chem. Soc.* **102,**, 5115 (1980).

136. T. Kauffmann, R. Abeln, S. Welke, and D. Wingbermühle, *Angew. Chem. Int. Ed Engl.* **25**, 909 (1986).

137. T. Kauffmann, M. Enk, W. Kaschube, E. Toliopoulos, and D. Wingbermühle, *Angew. Chem. Int. Ed. Engl.* **25**, 910 (1986).

138. J. H. Wengrovius, J. Sancho, and R. R. Schrock, *J. Am. Chem. Soc.* **103**, 3932 (1981).

139. T. J. Marks and R. D. Fischer, Eds., "Organometallic Chemistry of the *f*-Elements," pp. 1–35. Reidel, Dordrecht, 1979.

140. W. Beeckman, J. Goffart, J. Rebizant, and M. R. Spirlet, *J. Organomet. Chem.* **307**, 23 (1986).

141. B. Honold, U. Thewalt, M. Herberhold, H. G. Alt, L. B. Kool, and M. D. Rausch, *J. Organomet. Chem.* **314**, 105 (1986).

142. F. Bottomley and H. H. Brintzinger, *J. Chem. Soc. Chem. Commun.* 234 (1978).

143. F. Bottomley, I. J. B. Lin, and M. Mukaida, *J. Chem. Soc.* **102**, 5238 (1980).

144. S. A. Giddings, *Inorg. Chem.* **3**, 684 (1964).

145. E. Samuel, *Soc. Bull. Chim. Fr.* 3548 (1966).

145a. E. N. Jacobsen, M. K. Trost, and R. G. Bergman, *J. Am. Chem. Soc.* **108**, 8092 (1986).

146. S. A. Giddings, *Inorg. Chem.* **6**, 849 (1967).

147. B. Stutte, V. Batzel, R. Boese, and G. Schmid, *Chem. Ber.* **111**, 1603 (1978).

147a. U. Thewalt and G. Schleussner, *Angew. Chem. Int. Ed. Engl.* **17**, 531 (1978).

148. J. L. Petersen, *J. Organomet. Chem.* **166**, 179 (1979).

149. A. F. Reid, J. S. Shannon, J. M. Swan, and P. C. Wailes, *Aust. J. Chem.* **18**, 173 (1965).

150. J. F. Clarke and M. G. B. Drew, *Acta Crystallogr.* B30, 2267 (1974).

151. E. M. Brainina, R. Kh. Freidlina, and A. N. Nesmeyanov, *Dokl. Akad. Nauk SSSR* **154**, 143 (1964).
152. G. Tainturier, B. Gautheron, and M. Fahim, *J. Organomet. Chem.* **290**, C4 (1985).
153. U. Thewalt and D. Schomburg, *J. Organomet. Chem.* **127**, 169 (1977).
154. T. V. Ashworth, T. C. Agreda, E. Herdtweck, and W. A. Herrmann, *Angew. Chem. Int. Ed. Engl.* **25**, 289 (1986).
155. H. Köpf, S. Grabowski, and R. Voigtländer, *J. Organomet. Chem.* **216**, 185 (1981).
155a. L. Saunders and L. Spirer, *Polymer* **6**, 6335 (1965).
156. U. Thewalt and K. Döppert, *J. Organomet. Chem.* **320**, 177 (1987).
157. E. Samuel, R. D. Rogers, and J. L. Atwood, *J. Crystallogr. Spec Frosc. Res.* **14**, 573 (1984).
158. A. K. Holliday, P. H. Makin, and R. J. Puddephatt, *J. Chem. Soc. Dalton Trans.* 228 (1979).
159. H. J. De Liefde Meijer and G. J. M. Van Der Kerk, *Recueil* **84**, 1418 (1965).
160. E. O. Fischer, S. Vigoureux, and P. Kuzel, *Chem. Ber.* **93**, 701 (1960).
161. F. Bottomley and L. Sutin, *J. Chem. Soc., Chem. Commun.*, 1112 (1987).
162. F. Bottomley, J. Darkwa, L. Sutin, and P. S. White, *Organometallics* **5**, 2165 (1986).
163. D. A. Lemenovskii, T. V. Baukova, and V. P. Fedin, *J. Organomet. Chem.* **132**, C14 (1977).
163a. P. M. Treichel and G. P. Werber, *J. Am. Chem. Soc.* **90**, 1753 (1968).
164. Yu. V. Skripkin, I. L. Eremenko, A. V. Pasynskii, O. G. Volkov, S. I. Bakum, M. A. Porai-Koshits, A. S. Antsyshkina, L. M. Dikareva, V. N. Ostrikova, S. G. Sakharov, and Yu. T. Struchkov, *Sov. J. Coord. Chem.* **3**, 570 (1985).
165. W. E. Douglas and M. L. H. Green, *J. Chem. Soc. Dalton Trans.* 1796 (1972).
166. K. Prout, T. S. Cameron, R. A. Forder, S. R. Critchley, B. Denton, and G. V. Rees, *Acta Crystallogr.* **B30**, 2290 (1974).
167. F. Bottomley, L. Sutin, and P. S. White, *J. Organomet. Chem,* in press.
168. F. Bottomley and J. Darkwa, *J. Chem. Soc. Dalton Trans.* 399 (1983).
169. J.-C. Daran, K. Prout, A. DeCian, M. L. H. Green, and N. Siganporia, *J. Organomet. Chem.* **136**, C4 (1977).
170. K. Prout and J.-C. Daran, *Acta Crystallogr.* **B35**, 2882 (1979).
171. F. Bottomley, J. Darkwa, and P. S. White, *J. Chem. Soc. Chem. Commun.* 1039 (1982).
172. F. Bottomley, J. Darkwa, and P. S. White, *J. Chem. Soc. Dalton Trans* 1435 (1985).
173. M. Berry, S. G. Davies, and M. L. H. Green, *J. Chem. Soc. Chem. Commun.* 99 (1978).
174. M. Cousins and M. L. H. Green, *J. Chem. Soc.* 889 (1963).
175. M. L. H. Green, J. Knight, and J. A. Segal, *J. Chem. Soc. Dalton Trans.* 2189 (1977).
176. R. R. Schrock, S. F. Pedersen, M. R. Churchill, and J. W. Ziller, *Organometallics,* **3**, 1574 (1984).
176a. M. J. Bunker, A. De Cian, and M. L. H. Green, *J. Chem. Soc. Chem. Commun.* 59 (1977).
177. M. L. H. Green, *J. Less-Common Met.* **54**, 159 (1977).
178. G. S. B. Adams and M. L. H. Green, *J. Chem. Soc. Dalton Trans.* 353 (1981).
179. N. D. Feasey, S. A. R. Knox, and A. G. Orpen, *J. Chem. Soc. Chem. Commun.* 75 (1982).
180. M. Herberhold, W. Jellen, and M. L. Ziegler, *Inorg. Chim. Acta* **118**, 15 (1986).
181. J.-C. Daran, K. Prout, G. J. S. Adam, M. L. H. Green, and J. Sala-Pala, *J. Organomet. Chem.* **131**, C40 (1977).
182. K. Isobe, S. Kimura, and Y. Nakamura, *J. Chem. Soc. Chem. Commun.* 378 (1985).

182a. G. J. S. Adam and M. L. H. Green, *J. Organomet. Chem.* **208**, 299 (1981).
183. K. Prout and J.-C. Daran, *Acta Crystallogr.* **B34**, 3586 (1978).
184. M. J. Bunker and M. L. H. Green, *J. Chem. Soc. Dalton Trans.* 847 (1981).
185. H. Arzoumanian, A. Baldy, M. Pierrot, and J.-F. Petrignani, *J. Organomet. Chem.* **294**, 327 (1985).
186. C. Couldwell and K. Prout, *Acta Crystallogr.* **B34**, 933 (1978).
187. M. L. H. Green and K. J. Moynihan, *Polyhedron* **5**, 921 (1986).
188. D. L. Stevenson and L. F. Dahl, *J. Am. Chem. Soc.* **89**, 3721 (1967).
189. P. M. Treichel and G. R. Wilkes, *Inorg. Chem.* **5**, 1182 (1966).
190. C. A. Poffenberger, N. H. Tennent, and A. Wojcicki, *J. Organomet. Chem.* **191**, 107 (1980).
191. X. You, Z. Zhu, J. Huang, R. F. Fenske, and L. F. Dahl, *Proc. China–Jpn–USA Trilateral Semino, 2nd* 257 (1982); *Chem. Abst.* **102**; 204073k (1985).
192. M. R. DuBois, D. L. DuBois, M. C. VanDerveer, and R. C. Haltiwanger, *Inorg. Chem.* **20**, 3064 (1981).
193. K. Prout and C. Couldwell, *Acta Crystallogr.* **B36**, 1481 (1980).
194. J. Bashkin, M. L. H. Green, M. L. Poveda, and K. Prout, *J. Chem. Soc. Dalton Trans.* 2485 (1982).
195. W. A. Herrmann, R. Serrano, U. Küsthardt, E. Guggolz, B. Nuber, and M. L. Ziegler, *J. Organomet. Chem.* **287**, 329 (1985).
196. W. A. Herrmann, U. Küsthardt, M. L. Ziegler, and T. Zahn, *Angew. Chem. Int. Ed. Engl.* **24**, 860 (1985).
197. W. A. Herrmann, T. Cuenca, and U. Küsthardt, *J. Organomet. Chem.* **309**, C15 (1986).
197a. T. Cuenca, W. A. Herrman, and T. V. Ashworth, *Organometallics* **5**, 2514 (1986).
198. W. A. Herrmann, E. Herdtweck, M. Flöel, J. Kulpe, U. Küsthardt, and J. Okuda, *Polyhedron*, **6**, 1165 (1987).
199. W. A. Herrmann, E. Voss, U. Küsthardt, and E. Herdtweck. *J. Organomet. Chem.* **294**, C37 (1985).
200. W. A. Herrmann, U. Küsthardt, A. Schäfer, and E. Herdtweck, *Angew. Chem. Int. Ed. Engl.* **25**, 817 (1986).
200a. W. A. Herrmann and J. Okuda, *Angew. Chem.* **98**, 1109 (1986).
201. W. A. Herrmann, R. Serrano, U. Küsthardt, M. L. Ziegler, E. Guggolz, and T. Zahn, *Angew. Chem. Int. Ed. Engl.* **23**, 515 (1984).
202. U. Küsthardt, W. A. Herrmann, M. L. Ziegler, T. Zahn, and B. Nuber, *J. Organomet. Chem.* **311**, 163 (1986).
203. W. A. Herrmann, U. Küsthardt, and E. Herdtweck, *J. Organomet. Chem.* **294**, C33 (1985).
204. W. A. Herrmann, R. Serrano, M. L. Ziegler, H. Pfisterer, and B. Nuber, *Angew. Chem. Int. Ed. Engl.* **24**, 50 (1985).
204a. P. Hofmann, N. Rösch, and H. R. Schmidt, *Inorg. Chem.* **25**, 4470 (1986).
205. J. Heidrich, D. Loderer, and W. Beck, *J. Organomet. Chem.* **312**, 329 (1986).
205a. R. Brady, B. R. Flynn, G. L. Geoffrey, H. B. Gray, J. Peone, and L. Vaska, *Inorg. Chem.* **15**, 1485 (1976).
206. R. Guilard, J.-M. Latour, C. Lecomte, J.-C. Marchon, J. Protas, and D. Ripoll, *Inorg. Chem.* **17**, 1228 (1978).
207. J. W. Lauher and R. Hoffmann, *J. Am. Chem. Soc.* **98**, 1729 (1976).
208. H. M. McConnell, W. W. Porterfield, and R. E. Robertson, *J. Chem. Phys.* **30**, 442 (1959).
209. F. Bottomley and P. S. White, *J. Chem. Soc. Chem. Commun.* 28 (1981).
210. R. A. Schunn, C. J. Fritchie, and C. T. Prewitt, *Inorg. Chem.* **5**, 892 (1966).

211. C. H. Wei, G. R. Wilkes, P. M. Treichel, and L. F. Dahl, *Inorg. Chem.* **5**, 900 (1966).
212. G. L. Simon and L. F. Dahl, *J. Am. Chem. Soc.* **95**, 2164 (1973).
213. A. Roth, C. Floriani, A. Chiesi-Villa, and C. Guastini, *J. Am. Chem. Soc.* **108**, 6823 (1986).
214. W. J. Evans and M. S. Sollberger, *J. Am. Chem. Soc.* **108**, 6095 (1986).
215. G. A. Zank, C. A. Jones, T. B. Rauchfuss, and A. L. Rheingold, *Inorg. Chem.* **25**, 1886 (1986).
216. J. C. Huffman, J. G. Stone, W. C. Krusell, and K. G. Caulton, *J. Am. Chem. Soc.* **99**, 5829 (1977).
217. H. Aslan, T. Sielisch, and R. D. Fischer, *J. Organomet. Chem.* **315**, C69 (1986).
218. K. Döppert and U. Thewalt, *J. Organomet. Chem.* **301**, 41 (1986).
219. U. Thewalt, K. Döppert, and W. Lasser, *J. Organomet. Chem.* **308**, 303 (1986).
220. A. A. Pasynskii, I. L. Eremenko, Yu V. Rakitin, V. M. Novotortsev, O. G. Ellert, V. T. Kalinnikov, V. E. Shklover, Yu T. Struchkov, S. V. Lindeman, T. Kh. Kurbanov, and G. Sh. Gasanov, *J. Organomet. Chem.* **248**, 309 (1983).
221. A. A. Pasynskii, I. L. Eremenko, G. Sh. Gasanov, B. Orazsakhatov, V. T. Kalinnikov, V. E. Shklover, and Yu T. Struchkov, *Sov. J. Coord. Chem.* **10**, 347 (1984).
222. I. L. Eremenko, A. A. Pasynskii, G. Sh. Gasanov, B. Orazsakhatov, Yu. T. Struchkov, and V. E. Shklover, *J. Organomet. Chem.* **275**, 71 (1984).
223. J. L. Davidson, K. Davidson, W. E. Lindsell, N. W. Murall, and A. J. Welch, *J. Chem. Soc. Dalton Trans.* 1677 (1986).
224. C. P. Gibson, J.-S. Huang, and L. F. Dahl, *Organometallics*, **5**, 1676 (1986).
225. A. Bino, F. A. Cotton, Z. Dori, and B. W. S. Kolthammer, *J. Am. Chem. Soc.* **103**, 5779 (1981).
226. F. A. Cotton, Z. Dori, M. Kapon, D. O. Marler, G. M. Reisner, W. Schwotzer, and M. Shaia, *Inorg. Chem.* **24**, 4381 (1985).
227. J. R. Shapley, J. T. Park, M. R. Churchill, J. W. Ziller, and L. R. Beanan, *J. Am. Chem. Soc.* **106**, 1144 (1984).
228. A. Bertolucci, M. Freni, P. Romiti, G. Ciani, A. Sironi, and V. G. Albano, *J. Organomet. Chem.* **113**, C61 (1976).
229. G. Ciani, A. Sironi, and V. G. Albano, *J. Chem. Soc. Dalton Trans.* 1667 (1977).
229a. C. K. Schaver and D. F. Shriver, *Angew. Chem.* **99**, 275 (1987).
230. A. Ceriotti, L. Resconi, F. Demartin, G. Longoni, M. Manassero, and M. Sansoni, *J. Organomet. Chem.* **249**, C35 (1983).
231. G. Lavigne, N. Lugan, and J.-J. Bonnet, *Nouv. J. Chim.* **5**, 423 (1981).
232. A. Colombié, J.-J. Bonnet, P. Fompeyrine, G. Lavigne, and S. Sunshine, *Organometallics* **5**, 1154 (1986).
233. U. A. Jayasooriya and C. E. Anson, *J. Am. Chem. Soc.* **108**, 2894 (1986).
234. R. J. Goudsmit, B. F. G. Johnson, J. Lewis, P. R. Raithby, and K. H. Whitmire, *J. Chem. Soc. Chem. Commun.* 246 (1983).
235. V. A. Uchtman and L. F. Dahl, *J. Am. Chem. Soc.* **91**, 3763 (1969).
236. F. A. Cotton, P. Lahuerta, M. Sanau, and W. Schwotzer, *J. Am. Chem. Soc.* **107**, 8284 (1985).
237. Trinh-Toan, B. K. Teo, J. A. Ferguson, T. J. Meyer, and L. F. Dahl, *J. Am. Chem. Soc.* **99**, 408 (1977).
238. R. H. Holm, *Acct. Chem. Res.* **10**, 427 (1977).
239. F. Bottomley and F. Grein, *Inorg. Chem.* **21**, 4170 (1982).
240. P. D. Williams and M. D. Curtis, *Inorg. Chem.* **25**, 4562 (1986).
241. L. R. Byers, V. A. Uchtman, and L. F. Dahl, *J. Am. Chem. Soc.* **103**, 1942 (1981).
242. B. E. R. Schilling and R. Hoffmann, *J. Am. Chem. Soc.* **101**, 3456 (1979).

243. V. W. Day and W. G. Klemperer, *Science* **228**, 533 (1985).
244. W. G. Klemperer and W. Shum, *J. Chem. Soc. Chem. Commun.* 60 (1979).
245. V. W. Day, M. F. Fredrich, M. R. Thompson, W. G. Klemperer, R.-S. Liu, and W. Shum, *J. Am. Chem. Soc.* **103**, 3597 (1981).
246. T. M. Che, V. W. Day, L. C. Francesoni, M. F. Fredrich, W. G. Klemperer, and W. Shum, *Inorg. Chem.* **24**, 4055 (1985).
247. W. H. Knoth, *J. Am. Chem. Soc.* **101**, 759 (1979).
248. R. K. C. Ho and W. G. Klemperer, *J. Am. Chem. Soc.* **100**, 6772 (1978).
249. W. H. Knoth, *J. Am. Chem. Soc.* **101**, 2211 (1979).
250. W. H. Knoth, P. J. Domaille, and D. C. Roe, *Inorg. Chem.* **22**, 198 (1983).
251. P. J. Domaille and W. H. Knoth, *Inorg. Chem.* **22**, 818 (1983).
252. F. Zonnevijlle and M. T. Pope, *J. Am. Chem. Soc.* **101**, 2731 (1979).
253. F. J. Feher, *J. Am. Chem. Soc.* **108**, 3850 (1986).
254. C. J. Besecker and W. G. Klemperer, *J. Am. Chem. Soc.* **102**, 7598 (1980).
255. C. J. Besecker, V. W. Day, W. G. Klemperer, and M. R. Thompson, *Inorg. Chem.* **24**, 44 (1985).
256. C. J. Besecker, V. W. Day, W. G. Klemperer, and M. R. Thompson, *J. Am. Chem. Soc.* **106**, 4125 (1984).
257. R. G. Finke, B. Rapko, and P. J. Domaille, *Organometallics* **5**, 175 (1986).
258. R. G. Finke and M. W. Droege, *J. Am. Chem. Soc.* **106**, 7274 (1984).
259. C. J. Besecker, W. G. Klemperer, and V. W. Day, *J. Am. Chem. Soc.* **104**, 6158 (1982).
260. V. W. Day, W. G. Klemperer, and D. J. Maltbie, *Organometallics* **4**, 104 (1985).
261. V. W. Day, C. W. Earley, W. G. Klemperer, and D. J. Maltbie, *J. Am. Chem. Soc* **107**, 8261 (1985).
262. R. G. Finke, M. W. Droege, J. C. Cook, and K. S. Suslick, *J. Am. Chem. Soc.* **106**, 5750 (1984).
263. M. Herberhold, H. Kriesel, L. Haumaier, A. Gieren, and C. Ruiz-Pèrez, *Z. Natur-Forsch.* **41b**, 1431 (1986).
264. J. F. W. Keara and M. O. Ogan, *J. Am. Chem. Soc.* **108**, 7951 (1986).
265. J. F. W. Keara, M. D. Ogan, Y. Lü, M. Beer, and J. Varley, *J. Am. Chem. Soc.* **108**, 7957 (1986).
266. D. C. Bradley, M. B. Hurthouse, K. M. A. Maliln, and A. J. Nielson, *J. Chem. Soc. Chem. Commun.* 103 (1981).

Recent Developments in NMR Spectroscopy of Organometallic Compounds

BRIAN E. MANN

Department of Chemistry
University of Sheffield
Sheffield S3 7HF, England

I

INTRODUCTION

The rapid development of the computer over the past decade has transformed nuclear magnetic resonance (NMR) spectroscopy from being the simple observation of a given nucleus, to a wide range of complex experiments. Until the late 1970s, the NMR spectrometer consisted of an instrument to which a computer was attached to accumulate spectra, thus enhancing the signal-to-noise ratio, and to perform a Fourier transform if required. In a modern spectrometer, the computer controls the spectrometer, following a sequence programed by the operator or manufacturer. In many experiments, the spectrum is not simply observed, but the nuclei are pretreated by a suitable pulse sequence to prepare them to yield more information. This computer control has opened up a wealth of new experiments, and it is the purpose of this review to examine these experiments, with particular reference to organometallic compounds. The advantages and disadvantages of each experiment will be examined. Little emphasis is placed on the physical background as reviews covering this have already appeared.

The availability of these new experiments has considerably increased the skill necessary to operate an NMR spectrometer. In order to get the best out of a spectrometer, it is essential for the operator not only to be highly skilled in the operation but also to have enough knowledge to decide which of the many experiments available is most likely to yield the desired information. It is very easy to waste a great deal of valuable spectrometer time doing the wrong experiments or using the wrong experimental parameters.

In addition to the computer control of spectrometers, advances in spectrometer and probe design have now made a wide range of nuclei available, both in solution and the solid state, to the organometallic chemist.

This review is organized with one-dimensional experiments first and then the two-dimensional experiments. Each section is subdivided into homo- and heteronuclear experiments. The experimental aspects and some results from multinuclear NMR spectroscopy are then discussed, and finally some examples from solid-state NMR spectroscopy are examined.

There are a number of reviews of these techniques already in print, but the majority are written from the standpoint of physical or organic chemists. In general, this is of no disadvantage to the organometallic chemist, as many of the problems are based on ^{1}H and ^{13}C NMR spectroscopy and are essentially the same as those found for organic compounds. A number of excellent books and reviews have appeared (1–6).

Coupling constants are an extremely valuable tool in establishing the connectivity between various nuclei in molecules, hence leading to important information on the proximity of one nucleus to another. In order to use the coupling information it is necessary to establish which nuclei are coupled. In simple cases, this can be established by comparing coupling constants. Modern NMR spectrometers produce accurate coupling constants, frequently removing the need for time-wasting decoupling experiments. In more complicated cases, decoupling is necessary. Modern NMR spectroscopy has greatly increased our ability to derive coupling constant information and hence determine the connectivity between groups in molecules.

Additional information on connectivity comes from the nuclear Overhauser enhancement (NOE). This effect gives direct information on the through-space connectivity between nuclei, and thus gives information complementary to that obtained from coupling constants.

II

ONE-DIMENSIONAL NMR SPECTROSCOPY

A. Resolution Enhancement

Modern NMR spectrometers have powerful routines to modify the free induction decay and hence the resolution and signal/noise ratio. Exponential multiplication has been used for many years to improve the signal/noise ratio at the expense of resolution and is now standard in the measurement of spectra of all nuclei apart from ^{1}H.

Many of the early resolution enhancement routines led to unacceptible losses in the signal/noise ratio. However, this problem was solved by the introduction of a Gaussian function (7):

$$e^{(-at-bt^2)} \tag{1}$$

This function causes some distortion of the signal shape and a small increase in the noise level, but considerably reduces the linewidth, making coupling information far easier to extract. This is illustrated in Fig. 1. Figure 1a shows a multiplet from the ^{1}H NMR spectrum of $[(\eta^3\text{-}1,2,5\text{-}$

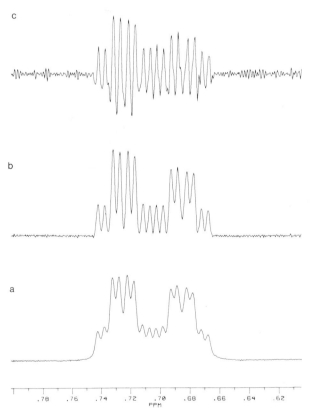

FIG. 1.   Partial ^{1}H NMR spectrum of $[(\eta^3\text{-}1,2,5\text{-}C_8H_{13})PdCl]_2$ in $CDCl_3$. (a) Normal ^{1}H NMR spectrum without any manipulation of the FID apart from Fourier transform and phasing, showing the signal due to H_6. (b) The same FID as in (a) but with limited Gaussian enhancement. (c) The same as (b) but with more extreme Gaussian enhancement.

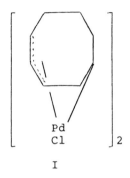

I

$(C_8H_{13})PdCl]_2$, **I**, which has received no resolution enhancement. Figure 1b shows the same multiplet after resolution enhancement using Gaussian multiplication. Clearly, it is far easier to extract coupling information from the enhanced spectrum than from the normal spectrum. Gaussian enhancement is now used routinely as part of the normal signal processing for ^1H NMR spectra in the same way the exponential broadening has routinely been applied to ^{13}C NMR spectra for many years. If too much enhancement is used, then negative components appear on both sides of the signal; see Fig. 1c. This enhancement procedure works well for any nucleus, provided the signal/noise ratio is sufficient. It is commonly used in our laboratory for ^{31}P and occasionally ^{13}C NMR spectra and has even been used with ^{103}Rh NMR spectra to extra small coupling constants. Some caution is necessary in using resolution enhancement. It is quite easy to generate a badly shimmed magnetic field on a superconducting magnet so that all signals are split. This splitting will be present on all signals, including the reference TMS, and is therefore easily detected. However, enhancement can emphasize this splitting, and the splitting can easily be erroneously attributed to coupling.

Recently, a superior enhancement routine has been described, which results in even less loss of signal/noise ratio (8). The free induction decay (FID) is multiplied by

$$e^{(-t/T_2^*)}/[e^{(-t/T_2^*)^2} + e^{(-[T-t]/T_2^*)_2}] \qquad (2)$$

where T is the total acquisition time.

B. *Decoupling Difference NMR Spectroscopy*

Homonuclear decoupling has long been established as a valuable tool in determining which nuclei are coupled. The experiment works well when all the signals are clearly resolved. Unfortunately, in the more complicated molecules, signals can overlap. It is then often difficult to establish which signals have changed on decoupling. This problem is removed by using difference spectroscopy. In its simplest form, two spectra are recorded. In the first speectrum, the decoupling frequency is placed in a region of the spectrum where there is no signal. In the second spectrum, the decoupling frequency is placed on a signal. When one spectrum is subtracted from the other, all the unchanged resonances cancel. All that are left are the decoupled signal and the signals due to the nuclei coupled to the decoupled nucleus and signals showing an NOE (see below). The appearance of signals in the decoupling difference spectrum is illustrated in Fig. 2. Figure 2a shows the normal spectrum without decoupling. Figure 2b shows a difference spectrum, where the 4.5-Hz coupling has been removed by

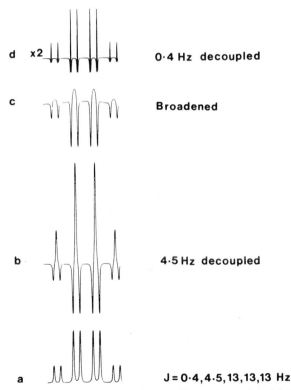

d x2   0·4 Hz decoupled

c   Broadened

b   4·5 Hz decoupled

a   J = 0·4, 4·5, 13, 13, 13 Hz

FIG. 2.   (a) Simulated multiplet, using the coupling constants 0.4, 4.5, 13, and 13 Hz. (b) Difference spectrum after decoupling the 4.5-Hz coupled signal. (c) As (b), but using a lower decoupling power. (d) Difference spectrum after decoupling the 0.4-Hz coupled signal. [Reproduced with permission from J. K. M. Sanders and J. D. Mersh, *Prog. N.M.R. Spectroscopy* **15**, 353 (1983).]

decoupling. The coupled spectrum appears as a negative signal and the decoupled spectrum as a positive signal. When the decoupling is efficient and the coupling removed is large, both the coupled spectrum and the decoupled spectrum are derived from this method of presentation. In practice, it is advantageous to use a lower decoupling power than is usual for normal decoupling (see below), when complete decoupling may not occur. When this happens, the "decoupled" spectrum is broad and only the coupled spectrum is clearly resolved; see Fig. 2c. The experiment is particularly advantageous when the coupling is small, so that is is not resolved in the coupled spectrum. Decoupling then results in a sharpening of the signal, which is very clear in the difference spectrum, Fig. 2d.

This experiment is subject to Bloch–Siegert shifts, which cause difficulties when the signals are close to the irradiation point (9). The presence of the decoupling frequency ν_{irr}, generating an effective magnetic field B_2, causes the other signals to shift from the original position ν_0 to a new frequency ν_n. The shift is given by

$$\nu_n - \nu_0 = \frac{\gamma B_2{}^2}{4\pi(\nu_{irr} - \nu_0)]}\tag{3}$$

The size of B_2 gives the decoupling power used for the experiment. The effect is worst when B_2 is large and $(\nu_{irr} - \nu_0)$ is small. It is therefore advisable to use low decoupling power, which produces effects such as that shown in Fig. 2c. The presence of the Bloch–Siegert shift results in error signals when the spectra are subtracted, and these can mask the true signals. It is possible to differentiate between error signals and true signals from the form of the signals. The true signal appears as shown in Fig. 2, while the error signal appears as an out-of-phase signal. As a result of this shift, the use of simple decoupling difference spectroscopy is best on high-field NMR spectrometers and works extremely well when the coupled signals are separated by at least 500 Hz from the decoupling frequency. The size of the Bloch–Siegert shift can be reduced by using low decoupling power. As long as the power is high enough to perturb the coupled signal, then the coupling connectivity can be established. Alternatively, the spectra can be corrected for the Bloch–Siegert shift, using Eq. (3). This approach works well but requires extensive digitization of the transformed spectrum and consequently considerable computing power (10). The experiment is susceptible to spectrometer instability and is therefore recorded using the procedure

$$\begin{aligned}&{}^1\text{H(observe)} \quad \{\text{read} - \quad -D_1 - (\theta - \text{acquire})_n - \text{store}\}_m\\&{}^1\text{H(decouple)} \{ \qquad \nu_l \qquad\qquad\qquad\qquad\qquad \}_m\end{aligned}\tag{4}$$

The decoupled spectra are recorded in small blocks, typically eight, with different irradiation frequencies ν_1. The separate spectra are read from and stored on disk many times during the experiment. There is a delay D_1, typically 10 s, for the effects of the previous irradiation to decay and any NOE effects due to the new irradiation frequency to build up to a steady level. The θ is the exciting pulse. This procedure helps to reduce any effects due to spectrometer instability. A blank irradiation point is included in the list of frequencies used for ν_1.

This experiment is demonstrated in Fig. 3. Figure 3a shows a partial ^{1}H NMR spectrum of $[(\eta^3\text{-}1,2,5\text{-}C_8H_{13})PdCl]_2$, **I**, in CDCl$_3$ with the decoupling frequency placed between signals at δ 4.50. It is essential to carry out

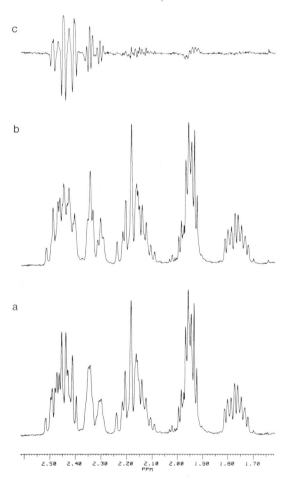

c

b

a

2.50 2.40 2.30 2.20 2.10 2.00 1.90 1.80 1.70
PPM

FIG. 3.   Partial 400-MHz ^{1}H NMR spectrum of $[(\eta^1\text{-}1,2,5\text{-}C_8H_{12})PdCl]_2$ in CDCl$_3$. (a) Spectrum with decoupling, off resonance, at δ 4.50. (b) Spectrum with decoupling H$_1$ at $\delta 5.57$. (c) Difference spectrum produced by subtracting (a) from (b).

this experiment as the blank, rather than a blank with no decoupling. In the decoupling experiment, there is time sharing between the decoupling irradiation and signal detection on the observing channel to prevent the detection circuitry being swamped by the decoupling signal. This leads to a significant change in the size of the observed signal. Figure 3b shows the ^{1}H NMR spectrum of $[(\eta^3\text{-}1,2,5\text{-}C_8H_{13})PdCl]_2$ with the decoupling frequency placed on the signal at δ 5.57. Figure 3c shows the difference between the two signals. The signals unaffected by the decoupling have virtually vanished, and all that remain are the decoupled signals, which are present in only one spectrum and do not vanish on subtraction, and the signals that change on decoupling. This experiment clearly identifies the two H_8 protons at δ 2.43 and δ 2.33. The changes in the signal at δ 2.33 are obvious when spectra in Fig. 3a and 3b are compared, but the changes in the signal at δ 2.43 are masked by the presence of a second overlapping proton signal. The changes are obvious in Fig. 3c, where the original coupled signal is observed as a negative triplet of doublets. The decoupled positive spectrum is far from being perfect, but this reflects the use of insufficient decoupler power. A balance has to be struck between efficient decoupling and inducing Bloch–Siegert shifts in the unchanged signals. The Bloch–Siegert shift is probably the major cause of the residual signals observed at δ 2.15 and δ 1.95.

C. Nuclear Overhauser Enhancement Difference Spectroscopy

The (NOE) is one of the most valuable tools in determining the connectivity between protons and to a lesser extent between protons and other nuclei. It permits the determination of which nuclei are physically close together. The effect relies on the relaxation mechanism of protons, which is predominantly dipole–dipole; i.e., the energy associated with one proton nucleus is passed most efficiently to another proton. The effect acts directly through space and depends on the (distance between the nuclei) (6). Because of the sixth power, the effect falls off very rapidly with distance. It is therefore sensitive to the nearest neighbors. Typical applications include the determination of which ligands are cis on the metal, the configuration about a double bond, and cis ligands in a cyclic system. If another relaxation mechanism is dominant, then no, or a reduced, NOE effect is observed. The most common alternative mechanism is paramagnetic relaxation due to a paramagnetic impurity, e.g., a metal complex, either from decomposition of the compound under study or introduced accidentally from a spatula, washing glassware with chromic acid, or simply dissolved O_2.

The maximum NOE that can be observed is $\gamma_A/2\gamma_B$, where γ_A is the gyromagnetic ratio for the observed nucleus and γ_B the gyromagnetic ratio for the irradiated nucleus. In the usual experiment where ^1H is both observed and irradiated, the maximum NOE is 0.5; i.e., the intensity of the resonances can increase by up to 50%. Usually this enchancement is not achieved, but with modern instrumentation an enhancement of 1% can easily be detected. If the experiment is performed with other nuclei, considerable enhancements can occur. For ^{13}C, with ^1H decoupling, the maximum enhancement is 1.99. Caution is necessary for nuclei with negative γ. For ^{15}N and ^{29}Si, γ is negative, leading to maximum theoretical NOE values of -4.93 and -2.52, but if other relaxation mechanisms are operating the enhancement can be -1, leading to no net signal. These values have been calculated assuming rapid molecular tumbling, i.e., $\omega\tau_c \ll 1$. At 9.4 T, i.e., 400 MHz for ^1H observation, this condition is met by molecules of moderate molecular weight, <1000, at room temperature, but not at low temperatures or in viscous solvents. When $\omega^2\tau_c^2 \gg 1$, then for ^1H the maximum enhancement is -1, i.e., the signal vanishes. This is equivalent to decoupling the second proton, and this then transmits a -1 "enhancement" to the next proton until the whole spectrum vanishes! At present, this is rarely a problem for organometallic compounds, but it will become a problem as higher field strengths become available. For nuclei such as ^{13}C, ^{15}N, and ^{29}Si, slow tumbling results in the maximum NOE approaching zero.

The power of NOE measurements is illustrated here for $(\eta^5\text{-}C_5H_5)$-$Mo(MeN{=}CPhNMeCHPh)(CO)_2$. This compound consists of two isomers, **II** and **III** (11). The data are best accumulated using the pulse sequence

$$^1H(\text{observe}) \quad \{\text{read} - [D_1 - (\nu_1 - D_2)_m - D_3 \; \theta - \text{acquire}]_n - \text{write}\}_p$$

$$^1H(\text{decouple}) \{ \qquad [\text{off} \quad (\text{ on }\;)_m \qquad \text{off} \qquad]_n \qquad \}_p$$

$$(5)$$

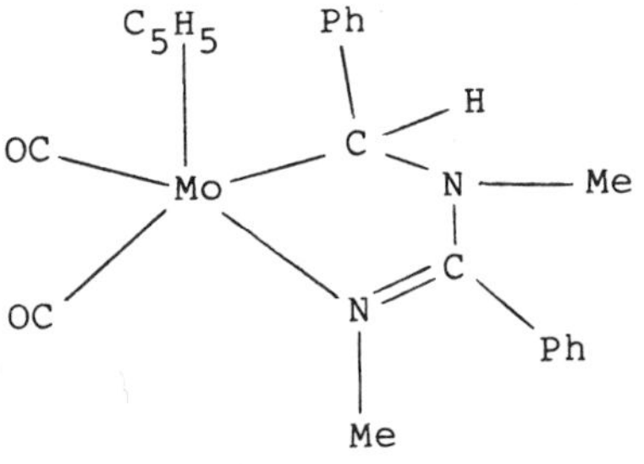

II 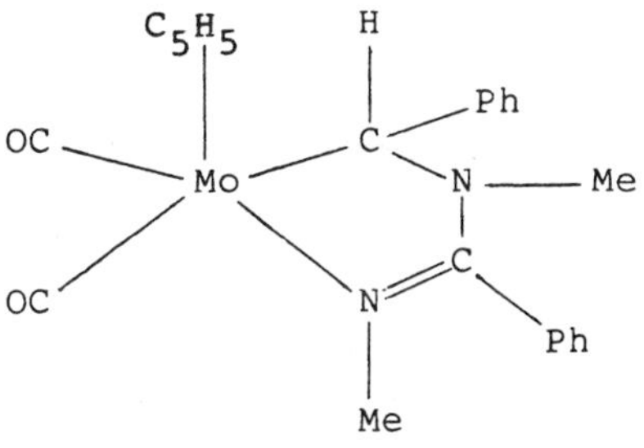   III

After an initial relaxation period D_1, typically 10 s, the ^{1}H decoupler is turned on at very low power, and the decoupler frequency ν_1 is cycled through the multiplet to reduce the effects of selective population transfer. The irradiation frequency is kept on one line of the multiplet for a short time D_2, typically 5 ms, before it is transferred to another line of the multiplet. This is repeated several thousand times, m, to build up an irradiation time of several seconds. There is a short delay D_3, typically 50 ms, to allow the spectrometer to stabilize after switching off the decoupler. An observing pulse is applied and the spectrum is acquired. Typically eight spectra are acquired and then the spectrum is stored on disk. This is repeated for each multiplet. Once all the multiplets have been irradiated, the previously recorded spectra are read back from disk and more data added to them. This procedure tends to average out effects of spectrometerr instabiity. The cycling of the irradiation frequency through the multiplet reduces effects due to selective population transfer (12). In this experiment, the response is important, not the resolution, so the data are normally processed with a large, e.g., 5 Hz, line broadening.

In order to differentiate between the two isomers, **II** and **III**, NOE measurements were carried out. Figure 4a shows the normal ^{1}H NMR spectrum of $(\eta^5\text{-}C_5H_5)Mo(MeN=CPhNMeCHPh)(CO)_2$ in CDCl$_3$. Figures 4b to 4e show the NOE difference ^{1}H NMR spectra. Each spectrum is the difference between the normal spectrum and one where a particular proton has been irradiated. Unaffected protons are unchanged and give no signal. The irradiated signal appears negative, while the signals that show an NOE appear positive. Figure 4b shows the effect of irradiating the cyclopentadienyl group, 12, at δ 4.9. The irradiated cyclopentadienyl group appears as a negative signal. An NOE is observed in the phenyl protons at δ 7.0–7.3, showing that this is isomer **II**, with the phenyl group pointing toward the cyclopentadienyl group. An additional NOE is observed into the methyl group, 1, at δ 3.3. Figure 4c shows the effect of irradiating the proton, 8, at δ 5.6. An NOE is observed into the *o*-protons of the adjacent phenyl group and the methyl group, 7, at δ 3.1. Significantly, there is no NOE into the cyclopentadienyl group, consistent with H^8 pointing away from it. In contrast, when H$^{8'}$ is irradiated, cyclopentadienyl group 12′ shows an NOE, consistent with H$^{8'}$ being cis to this group; see Fig. 4d. The *o*-hydrogen of the adjacent phenyl group and methyl 7′ also show NOE effects. Figure 4e shows the effect of irradiating the cyclopenta dienyl group 12′ at δ 5.4. An NOE is observed into the cis hydrogen H$^{8'}$ and the methyl group 1′.

The technique is also applicable to heteronuclear NOE measurements and has been used to determine the proximity of protons to ^{31}P nulei. This technique has been applied to RhCl(PPh$_3$)(Ph$_2$PCH$_2$PPh$_2$), **IV**, to identify

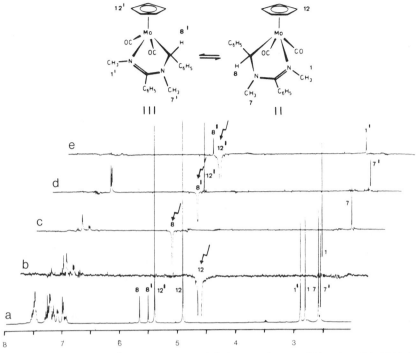

FIG. 4.   The 400-MHz ^{1}H NMR spectrum (η^5-C$_5$H$_5$)Mo(MeN=CPhNMeCHPh)(CO)$_2$ in CDCl$_3$. (a) The normal spectrum and (b to e) the NOE difference spectrum with preirradiation at (b) δ 4.9, (c) δ 5.6, (d) δ 5.5 and (e) δ 5.4. [Reproduced with permission from H. Brunner, J. Wachter, I. Bernal, G. M. Reisner, and R. Benn, *J. Organomet. Chem.* **243**, 179 (1983).]

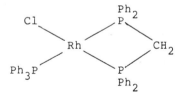

IV

the Ph$_2$PCH$_2$PPh$_2$ ^{31}P signals in the spectrum (*13*). This experiment is shown in Fig. 5. Figure 5a shows the ^{31}P–$\{^1$H$\}$ NMR spectrum. Figure 5b shows the calculated spectrum and Figure 5c shows the NOE difference spectrum with preirradiation at the Ph$_2$PCH$_2$PPh$_2$, CH$_2$, proton resonance. A strong NOE is observed for the ^{31}P nuclei at δ −13.3 and δ −38.5

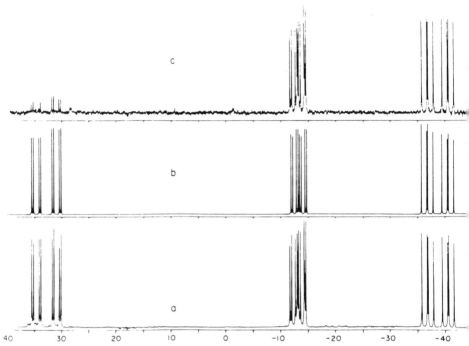

Fig. 5.   The 101.25-MHz ^{31}P NMR spectrum of RhCl(PPh$_3$)(Ph$_2$PCH$_2$PPh$_2$) with complete ^{1}H decoupling. (a) Normal spectrum. (b) Calculated spectrum. (c) NOE difference spectrum obtained by preirradiation of the Ph$_2$PCH$_2$PPh$_2$ CH μ protons and subtraction of a control spectrum obtained with off-resonance ^{1}H preirradiation. [Reproduced with permission from K. W. Chiu, H. S Rzepa, R. N. Sheppard, G. Wilkinson, and W.-K. Wong, *Polyhedron* **1**, 809 (1982).]

which must be from the Ph$_2$PCH$_2$PPh$_2$ ligand, while little or no NOE is observed from the PPh$_3$ group at δ 32.7. This method offers a useful approach to assigning heteronuclear signals.

D. *Heteronuclear Experiments*

A number of experiments use coupling between ^{1}H and the X nucleus to provide an improved signal/noise ratio and signal sorting on the basis of the number attached protons. In principle, this approach can use the coupling between any sensitive nucleus, e.g., ^{19}F, ^{31}P, and the X nucleus, but the availability of equipment has prevented extensive use involving any nucleus other than ^{1}H. The two main experiments are INEPT and DEPT, which have attracted the most attention. However, the easiest test to apply to ^{13}C is to use J modulation to distinguish between C, CH$_2$ and CH, CH$_3$ groups.

Originally, off-resonance decoupling was used to determine the number of protons attached to each carbon atom. In this experiment an intense ^{1}H single-frequency irradiation is applied off-resonance, e.g., at δ 15 or $-$ 5. This experiment works adequately at low frequencies, i.e., at and below 100 MHz for ^{1}H, but even then it is frequently difficult to differentiate between CH and CH$_3$ groups. Because of inhomogeneity in the decoupling field, the outer lines of the $1:3:3:1$ quartet are broadened and can be lost in the noise, giving rise to an apparent doublet. At higher frequencies, the large frequency spread of the ^{1}H signals makes the experiment difficult and of low sensitivity. Even at lower frequencies, the loss in sensitivity due to signal spread because of residual coupling makes off-resonance decoupling insensitive compared with J-modulation, INEPT, and DEPT.

1. *J Modulation Test*

The J modulation test is a *single* experiment that produces a spectrum with the C and CH$_2$ groups 180° out of phase with the CH and CH$_3$ groups; see Fig. 6. The experiment consists of the pulse sequence

$$^{13}\text{C} \qquad \left(D_1 - \frac{\pi}{2} - D_2 - \pi - D_2 - \text{observe} \right)_n \qquad (6)$$

$$^1\text{H decoupler} \leftarrow \text{on} \rightarrow\!\leftarrow \text{off} \longrightarrow\!\leftarrow\!\!\longrightarrow \text{on} \longrightarrow$$

Ideally, D_1 should be set equal to $5T_1$ for the slowest-relaxing carbon atom, but in practice it is set at between T_1 and $2T_1$. For a typical compound, 3 s is commonly used. $\pi/2$ and π represent a 90° and a 180° pulse, respectively. These pulse lengths must be accurate to better than 10%. The decoupler is on during the period D_1 to build up the NOE. It is normally used at low power to avoid sample heating. The decoupler is on at normal decoupling power during acquisition to produce a decoupled spectrum. D_2 is set equal to $1/^1J(^{13}\text{C}-^1\text{H})$. Unfortunately, most compounds contain a variety of different carbon atoms with a range of $^1J(^{13}\text{C}-^1\text{H})$ from 125 for sp^3 CH groups to 165 Hz for sp^2 CH groups. The presence of electron-withdrawing groups or sp CH groups extends the range of $^1J(^{13}\text{C}-^1\text{H})$ even further, e.g., 250 Hz for acetylene. Typically $^1J(^{13}\text{C}-^1\text{H})$ is taken as 140 or 145 Hz, and D_2 is consequently 0.007 s^{-1}. The technique works well if $^1J(^{13}\text{C}-^1\text{H})$ lies between 125 and 165 Hz. The experiment is illustrated in Fig. 6*b* for $(\eta^5\text{-C}_5\text{H}_4\text{-menthyl})\text{RuCl}(\text{PPh}_3)_2$, **V**, in CDCl$_3$, using a value of $^1J(^{13}\text{C}-^1\text{H})$ of 145 Hz. The experiment leads to the unambiguous assignment of the number of protons attached to each carbon atom, by combining the phase of the signal with the chemical shift and intensity.

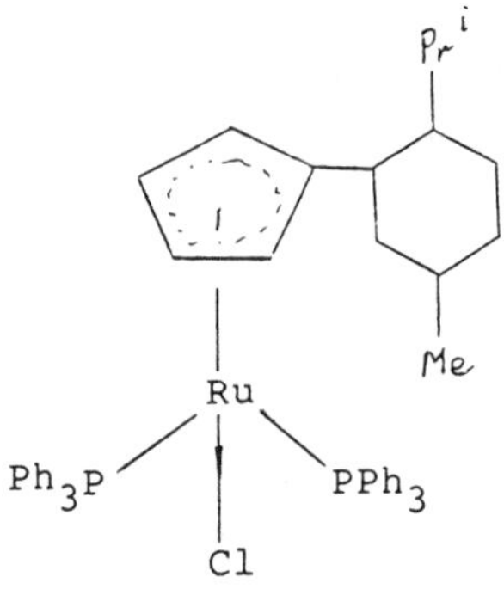

V

For the majority of compounds, ca. 90% in our laboratory, the combination of this one experiment with a knowlede of chemical shifts (*14*) permits the unambiguous sorting of C, CH, CH_2, and CH_3 groups. Confusion between C and CH_2 groups can also be resolved by this experiment or by using INEPT or DEPT; see Sections II,D,2; II,D,3, respectively. Setting D_2 equal to $1/\{2^1J(^{13}C-^1H)\}$ has no effect on the intensity of carbon atoms without attached protons, but ideally should null the signals due to proton-bearing carbon atoms. In practice, the signals due to proton-bearing carbon atoms are rarely nulled, but the intensity is usually reduced; see Fig. 6c. There is little change in intensity for the cyclopentadienyl ^{13}CH groups due to the relatively large values of $^1J_{CH}$. It is preferable to carry out an INEPT or DEPT measurement. In these experiments, only proton-bearing carbon atoms are observed (see Fig. 12a). Confusion between CH and CH_3 groups is best resolved by using DEPT, optimized for CH observation (see Fig. 12b).

2. Insensitive Nuclei Enhanced by Polarization Transfer (INEPT)

This technique is advantageous in increasing the sensitivity of the nucleus and making the ^1H relaxation time dominant, rather than the X nucleus relaxation time, which is usually far longer.

The sensitivity of an NMR experiment depends on the population difference between the energy levels. For an $I = 1/2$ nucleus, e.g., ^1H or ^{13}C, there are two energy levels, $m_I = 1/2$ and $m_I = -1/2$. As the energy gap between the two energy levels is small compared with the thermal energy, kT, the population difference is proportional to the energy gap. For ^1H, the energy gap, and hence NMR frequency, is ca. four times that

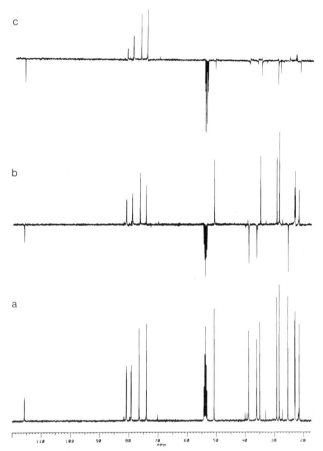

FIG. 6. The 100.62-MHz ^{13}C NMR spectrum of $(\eta^5\text{-}C_5H_4\text{-menthyl})RuCl(PPh_3)_2$ in CDCl$_3$. (a) Normal spectrum without J modulation; (b) with J modulation, using a value of D_2 of 0.007 s; (c) with J modulation, using a value of 0.0035 s.

for ^{13}C. This is one of several factors that leads to the higher sensitivity of ^{1}H compared with ^{13}C NMR spectroscopy. The INEPT experiments take advantage of this greater population difference to enhance the sensitivity of the coupled nucleus (15). In effect, these experiments observe the ^{13}C nuclei attached to the excess population of H nuclei with $m_I = 1/2$. This results in a four-fold intensity enhancement of the ^{13}C NMR signal, and the effective relaxation time is that of ^{1}H rather than that of ^{13}C. Unfortunately, there are also losses. Proton-bearing ^{13}C atoms normally undergo a nuclear Overhauser enhancement of up to 2.99 times the original intensity. Unfortunately, INEPT does not gain intensity from the NOE. As, for most organometallic compounds, there is at least a twofold NOE for proton-bearing carbon atoms, the gain is immediately reduced to, at best, a signal doubling. The technique is, however, particularly advantageous when the NOE is negligible. This occurs for compounds of high molecular weight. At 9.4 T, i.e., 400-MHz ^{1}H observation, the technique becomes advantageous for molecular weights >1000. At 1.9 T, i.e., 80-MHz ^{1}H observation, the corresponding molecular weight is >5000. Alternatively, if the compound is contaminated with paramagnetic impurities, the NOE is quenched, and INEPT or DEPT becomes advantageous. A further disadvantage is that carbon atoms not coupled to ^{1}H are invisible to this technique. It is therefore necessary to carry out a conventional spectrum to detect these carbon atoms.

The real advantage of this technique lies with other $I = 1/2$ nuclei. For ^{103}Rh, no NOE has been observed, but the potential gain of using INEPT or DEPT is 31.8 times, leading to a reduction in the time necessary for the experiment in excess of 1000! The gain is even more dramatic as ^{103}Rh relaxation times are frequently long, requiring slow pulsing for simple observation. Similar large gains can be made for ^{57}Fe and ^{183}W, but the technique is also advantageous for nuclei such as ^{29}Si and ^{195}Pt. In INEPT and DEPT, the ^{1}H relaxation time is dominant, permitting much more frequent pulsing. Using this technique, quantities of $(\eta\text{-}C_5H_5)RhH_2$-$(SiEt_3)_2$ as low as 10 mg yield INEPT ^{103}Rh NMR spectra within 1 hr. There are several variations on the INEPT experiment (16). The simplest version of INEPT uses the pulse sequence

$$^1H \quad \left\{ D_1 - \frac{\pi}{2}(x) - D_2 - \pi - D_2 - \frac{\pi}{2}(y) \quad \right\}_n$$
$$^{103}Rh \quad \left(\qquad\qquad\qquad \pi - D_2 - \frac{\pi}{2} \text{ acquire} \right)_n \qquad (7)$$

D_1 is $5T_1$, where T_1 is the spin–lattic relaxation time for the hydride, not for ^{103}Rh. In practice, this relaxation delay is taken as ca. $2T_1$. D_2 is $1/\{4J(^{103}\text{Rh}, {}^1\text{H})\}$. The presentation of the spectra is not as would be expected at first sight. This is illustrated in Fig. 7. Figure 7a shows the ^{103}Rh NMR spectrum of a monohydride, $(\eta\text{-}C_5Me_5)RhH(SiEt_3)(C_2H_4)$, and consists of a -1, 1 doublet. The separation of the doublet is $J(^{103}\text{Rh}, {}^1\text{H})$. It is fruitless to try decoupling this doublet as a null signal results. Figure 7b shows the ^{103}Rh NMR spectrum of a dihydride, $(\eta\text{-}C_5Me_5)RhH_2(SiEt_3)_2$ and consists of a -1, 0, 1 triplet. This can be differentiated from a -1, 1 doublet by the line separation, which is $2J(^{103}\text{Rh}, {}^1\text{H})$. The true value of $J(^{103}\text{Rh}, {}^1\text{H})$ is determined from the ^{1}H NMR spectrum. Figure 7c shows the ^{103}Rh NMR spectrum of a trihydride, $(\eta\text{-}C_5Me_5)RhH_3(SiEt_3)$, and consists of a -1, -1, 1, 1 quartet. It is

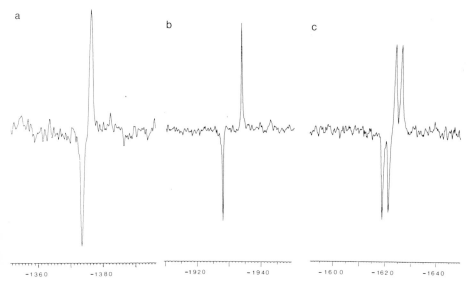

FIG. 7. The 12.62-MHz ^{103}Rh NMR spectra obtained using the INEPT pulse sequence (2) with $D_2 = 0.0074$ s, corresponding to $^1J(^{103}$Rh–^{1}H) of 34 Hz. The spectra are referenced to Ξ 3.16 MHz. (a) (η^5-C$_5$Me$_5$)RhH(η^2-C$_2$H$_4$)(SiEt$_3$); (b) (η-C$_5$MeH$_5$)RhH$_2$(SiEt$_3$)$_2$; (c) (η-C$_5$MeH$_5$)RhH$_3$(SiEt$_3$).

possible to obtain normal spectra, but this pulse sequence has the advantage that only $2D_2$ is waited between the exciting pulse and observation. During this time, relaxation is the minimum possible. This is important if the coupling constant being used is small. The delay D_2 can then be significant compared with T_2 relaxation of the metal. This is why that this approach has been used to detect ^{109}Ag in [Ag$_2${μ-1,2-[(2-C$_5$H$_4$N)CH = N]$_2$-cyclohexane)$_2$]$^{2+}$, **VI**, where $^3J(^{109}$Ag–^{1}H) is only 6.3 and 9.3 Hz for the two different imine protons; see Fig. 8 (*17*).

It is also advantageous to obtain a predictable multiplet to provide a

VI

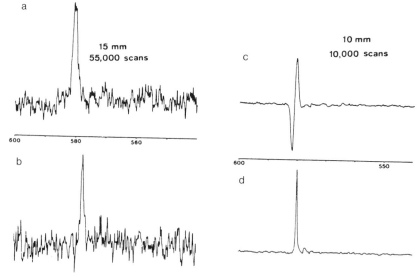

FIG. 8. The ^{109}Ag NMR spectrum of $[Ag_2\{\mu\text{-}1,2\text{-}[(2\text{-}C_5H_4N)CH{=}N]_2\text{-}cyclohexane\}_2]^+$ in CD_3OD. (a) Using direct observation with a 15-mm probe and 55,000 scans. (b) As in (a), but with ^1H decoupling. (c) Using INEPT, with a 10-mm probe and 10,000 scans, ^1H coupled. (d) As in (c), but with ^1H decoupling. Note that decoupling heats the sample and changes the chemical shift. [Reproduced with permission from C. Brevard, G. C. van Stein, and G. van Koten, *J. Am. Chem. Soc.* **103**, 6746 (1981).]

recognizable pattern in the metal NMR spectrum and thus differentiate it from an artifact that could be misassigned as a signal.

Spectra that, show normal multiplets, e.g., 1, 3, 3, 1 quartet (*18*), can be obtained using the longer pulse sequence

$$^1H \left(D_1 - \frac{\pi}{2}(x)\right) - D_2 - \pi - D_2 - \frac{\pi}{2}(y) - D_3 - \pi - D_3 - \frac{\pi}{2}(x) \Bigg)_n$$

$$^{103}Rh \left(\qquad\qquad \pi - D_2 - \frac{\pi}{2} \quad - D_3 - \pi - D_3 - \quad \text{acquire}\right)_n$$

$$(8)$$

Introduction of the additional delay of $2D_3$ can produce additional significant T_2 relaxation if there is only a small coupling constant, and hence a long D_3 is used. The value of the further delay, D_3, depends on the multiplicity of the signal; see Fig. 9, which shows the intensity of the signal as a function of D_3 for any MH, MH_2, and MH_3 system. Clearly, the intensity of the MH doublet is maximum at $D_3 = 1/\{4J(M, {}^1H)\}$, for an MH_2 triplet at $D_3 = 1/\{8J(M, {}^1H)\}$, and for an MH_3 quartet at $D_3 = 1/\{10J(M, {}^1H)\}$. It is therefore impossible to select a perfect value for D_3 unless the number

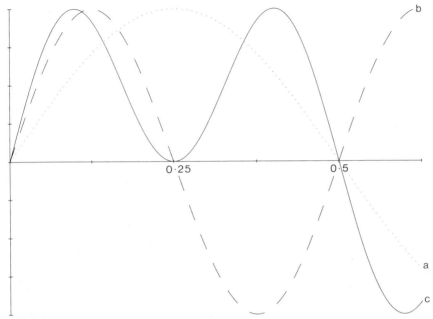

FIG. 9.   Plot of the intensity of the signal produced using pulse sequence (3) as a function of the delay D_3. (a) For an MH Group; (b) for an MH_2 group; (c) for an MH_3 group.

of attached protons is known. As a consequence, an there are potential dangers. If an MH_2 group is believed to be an MH group and D_3 is set as $1/\{4J(M, {}^1H)\}$, then the result is a null signal! Problems can also arise with the value chosen for D_2. It is always safe to set $D_2 = 1/\{4J(M-{}^1H)\}$ for MH_n spin systems, but for A_2X_2 spin systems, such as those containing the $M_2(\mu\text{-}H)_2$ group, e.g. $(\eta^5\text{-}C_5Me_5)_2Rh_2(\mu\text{-}H)_2X_2$, then D_2 must be $1/\{(8J(M-{}^1H)\}$ for optimum signal to noise (19). A value of D_2 of $1/\{4J(M-{}^1H)\}$ leads to a null signal. The same problem applies to $[AX]_2$ spin systems and has been demonstrated for trans-$[O_3PCH\!=\!CHPO_3]^{2-}$ (20).

   Although the simple pulse sequence, (7), cannot be decoupled as a null signal would result, this second pulse sequence, (8), can be decoupled and becomes

$${}^1H \quad \left(5D_1 - \frac{\pi}{2}(x) - D_2 - \pi - D_2 - \frac{\pi}{2}(y) - D_3 - \pi - D_3 - \text{decouple}\right)_n$$

$${}^{103}Rh \quad \left(\phantom{5D_1 - \frac{\pi}{2}(x) - D_2} \pi - D_2 - \frac{\pi}{2} - D_3 - \pi - D_3 - \text{acquire}\right)_n$$

$$(9)$$

This pulse sequence leads to fully ^{1}H-decoupled spectra and generally gives the strongest signals provided the $J(M-{}^1H)$ coupling constant is substantial.

This technique works extremely well for many $I = 1/2$ nuclei and is especially valuable whenever there is *one* substantial $J(M-{}^1H)$. Nuclei for which, the technique is particularly valuable include ^{15}N, ^{29}Si, ^{119}Sn, ^{183}W, and ^{195}Pt (*21*). For these nuclei the nuclear Overhauser enhancement is usually of little or no value. For nuclei with negative gyromagnetic ratios a negative NOE can be observed. This is often a problem for ^{15}N and ^{29}Si, where the signal intensities may undergo "enhancements" of up to -4.94 and -2.52, respectively. An "enhancement" of -1.0 produces a null signal. This case has been demonstrated for ^{29}Si in Me_4Si, where no signal is detected at $-62.5°C$ (*22,23*). No such problem exists for the INEPT experiment, where the maximum enhancements are 9.87 and 5.03, respectively. The heavier nuclei rarely exhibit nuclear Overhauser enhancements, as relaxation mechanisms other than dipole–dipole are normally dominant. Problems can be encountered when there are several $J(M-{}^1H)$. As a consequence, problems are found in recording ^{29}Si INEPT NMR spectra for $SiEt_3$ groups, where the magnitudes of $^2J(^{29}Si-{}^1H)$ and $^3J(^{29}Si-{}^1H)$ are such that it is very difficult to choose a suitable value of D_2 to obtain an enhanced signal.

The INEPT experiment suffers from a major problem for low-frequency nuclei. For ^{103}Rh, the length of the π pulse on the equipment in our laboratory is 160 μs. This produces adequate excitation only up to $1/(4 \times 160 \times 10^{-6})$ Hz, i.e., 1563 Hz from the carrier frequency. This gives only a 248-ppm spectral width for ^{103}Rh at 9.4 T, compared with the complete chemical shift range of ca. 12,000 ppm. It is therefore necessary either to have a good estimate of the ^{103}Rh chemical shift or to take a series of spectra separated by 3000 Hz in carrier frequency. Alternatively, the position of the ^{103}Rh signal can be located approximately by observing ^1H and pulsing ^{103}Rh at various frequencies until there is a response (see below). There is no corresponding problem with the ^1H pulse, as the position of the hydride is accurately known from the ^1H NMR spectrum.

The INEPT experiment is not restricted to using polarization transfer between ^1H and the other nuclei. In principle, any pair of nuclei that show resolved coupling can be used. In practice, the experiment is restricted by having to purchase additional equipment to perform it. For the organometallic chemist, polarization transfer between ^{31}P and another nucleus is particulary promising. The sensitivity gain is 0.405 that found when using ^1H, but many organometallic compounds contain phosphorus, usually as a tertiary phosphine, directly attached to the metal. This provides a large $^1J(M–^{31}P)$ and hence makes INEPT feasible. This approach has been applied to observing ^{57}Fe, ^{103}Rh, and ^{183}W (24).

Although the INEPT experiment gives a considerable improvement in the intensity of low-frequency nuclei, the low-frequency nuclei are still difficult to observe. Thus ^{103}Rh is 3.12×10^{-5} less sensitive than ^1H. Even with the improved sensitivity of the INEPT experiment, it is 10^{-3} less sensitive than ^1H. The ideal way to observe the low-frequency nuclei would be via the INDOR experiment, which was readily available on the continuous-wave spectrometers. Modern spectrometers are rarely equipped for continuous-wave measurement. An INEPT experiment has been devised to carry out such an experiment; see Section III,E,1.

The INEPT experiment is also used extensively to sort between CH, CH_2, and CH_3 carbon atoms, using pulse sequence (9). Figure 9 shows the signal intensity of each of these types of atoms as a function of D_3. Clearly, when $D_3 = 1/\{4^{13}J(C–^1H)\}$, only CH groups are observed; see Fig. 9a. When $D_3 = 3/\{8^1J(^{13}C–^1H)\}$, CH and CH_3 groups are positive and CH_2 groups are negative; see Fig. 9b. When $D_3 = 1/\{8^1J(^{13}C–^1H)\}$, all the ^{13}C signals are positive; see Fig. 9c. It is therefore theoretically possible to obtain a CH-only ^{13}C NMR spectrum by using a D_3 of $1/\{4^1J(^{13}J(^{13}C–^1H)\}$. Subtraction of the spectra with a D_3 of $3/\{8^1J(^{13}C–^1H)\}$ from that with a D_3 of $1/\{8^1J(^{13}C–^1H)\}$ should yield only CH_2 signals. Subtraction of the CH-only spectra from the spectrum with a D_3 of $1/\{(8^1J(^{13}C–^1H)\}$ then yields CH_3-only spectra.

The problem is that in real compounds, there is no single value of $^1J(^{13}C–^1H)$, but it covers a range of values. This is illustrated using $(\eta$-menthylC$_5$H$_4)$RuCl(PPh$_3)_2$, V. The signals shown in Fig. 10 do not show the aromatic carbon atoms, but only the cyclopentadienyl CH atoms at δ 73–82, and the saturated carbon atoms at δ 20–51. TMS is observed at δ 0.00. The sp^3 carbon atoms have $^1J(^{13}C–^1H)$ of ca. 125 Hz, while the cyclopentadienyl CH atoms have $^1J(^{13}C–^1H)$ of ca. 180 Hz. The spectra in Fig. 10 were recorded using values of D_3 varying between 0.0005 s for Fig. 10a and 0.0035 s for Fig. 10g. Examination of the spectra shows that when $D_3 = 0.0025$ s, only the CH carbon atoms are observed. This is because all the CH_2 and CH_3 groups have $^1J(^{13}C–^1H) = $ ca. 125 Hz, and these sorts of carbon atoms are nulled when $D_3 = 1/\{4^1J(^{13}C–^1H)\}$, i.e., 0.0020 s. In Fig. 10f the CH (cyclopentadienyl) groups have become negative because D_3, 0.003 s, is a little larger than $1/\{2^1J(^{13}C–^1H)\}$, 0.0028 s, yet the CH (menthyl) carbon atoms do not become negative until D_3 is ca. 0.004 s. Unfortunately, 0.003 s is $3/\{8^1J(^{13}C–^1H)\}$ for the saturated methyl carbon atoms. This leads to a large error signal when the

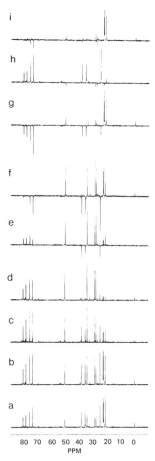

FIG. 10.   Partial INEPT ^{13}C NMR spectrum of $(\eta^5$-methylC$_5$H$_4)$RuCl(PPh$_3)_2$ in CDCl$_3$. The spectra was recorded with D$_3$ set D$_3$ set equal to (a) 0.0005 s, (b) 0.001 s, (c) 0.0015 s, (d) 0.002 s, (e) 0.0025 s, (f) 0.003 s, and (g) 0.0035 s, (h) Spectrum obtained by subtracting spectra to give the CH$_2$ signals preferentially; (i) spectrum optimized for CH$_3$ signals.

CH$_2$-only spectrum is produced; see Fig. 10h. The menthyl CH carbon signals have virtually vanished, but the cyclopentadienyl carbon atoms are still substantial. Figure 10i shows the CH$_3$-only ^{13}C NMR spectrum. Here the subtraction is reasonable but far from being perfect.

The selection is not very good, and this experiment is better performed using the DEPT pulse sequence; see Section II,D,3 (*25,26*). In principle the same procedure could be applied to other nuclei, but it is rare for a suitable occasion to arise on scientific grounds, rather than to produce a pretty picture for a thesis or a paper!

3. Distortionless Enhancement by Polarization Transfer (DEPT)

The DEPT pulse sequence is

$$^{1}\text{H} \left(5D_1 - \frac{\pi}{2}(y) - D_2 - \pi - D_2 - \theta(x) - D_2 - \text{decouple}\right)_n \quad (10)$$

$$^{13}\text{C} \left(\qquad\qquad \frac{\pi}{2} - D_2 - \pi \quad - D_2 - \text{acquire} \right)_n$$

It uses different pulse angles θ to do the editing. For observations of CH, CH_2, and CH_3 groups all to be positive, $\theta = \pi/4$; see Fig. 11a. At $\theta = \pi/2$, only CH groups are observed, Fig. 11b, while for $\theta = 3\pi/4$, the CH and CH_3 groups are positive and the CH_2 groups are negative, Fig. 11c. Once again the spectra can be added and subtracted to produce CH-only (Fig. 11b), CH_2-only (Fig. 11d), and CH_3-only (Fig. 11e) subspectra. Comparison with the INEPT spectra in Fig. 10 clearly demonstrates that the DEPT pulse sequence is far superior in determining the number of attached protons. The major advantage of DEPT is that it is considerably less sensitive to the magnitude of $^{1}J(^{13}\text{C}-^{1}\text{H})$ when used for spectral editing. It is therefore the preferred technique for spectral editing. It does appear to be far more sensitive to spectrometer imperfections than INEPT. It requires accurate phase shifting on both the observation and decoupling channels, and it requires the $\pi/4$, $\pi/2$, $3\pi/4$, and π pulse lengths to be accurately determined at each phase shift. Not all spectrometers permit the setting of the pulse length for each phase shift, but the accuracy is generally good enough to produce tolerable results. The major problem with DEPT is that the delay D_2 is $1/\{2J(^{13}\text{C}-^{1}\text{H})\}$ rather than the shorter $1/\{4J(^{13}\text{C}-^{1}\text{H})\}$. As a consequence, it requires a longer time than the basic INEPT sequence and is therefore less suitable when applied to other nuclei where the coupling constant is small. Signal intensity can be lost during the delays due to T_2 dephasing.

4. Incredible Natural Abundance Double Quantum Transfer Experiment (INADEQUATE)

The INADEQUATE pulse sequence was originally designed to detect $^{13}\text{C}-^{13}\text{C}$ coupling without the confusion of the ^{13}C singlets due to the species without adjacent ^{13}C atoms (27). The basic pulse sequence is

$$^{13}\text{C} \left(D_1 - \frac{\pi}{2(x)} - D_2 - \pi - D_2 - \frac{\pi}{2} - 10 \ \mu\text{s} - \frac{\pi}{2} \text{acquire}\right)_n \quad (11)$$

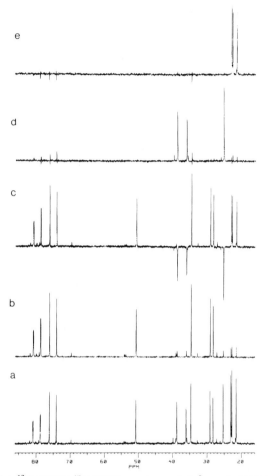

Fig. 11. Partial ^{13}C DEPT ^{13}C NMR spectrum of $(\eta^5\text{-menthylC}_5H_4)RuCl(PPh_3)_2$ in CDCl$_3$. (a to c) Spectrum with the pulse angle θ set to (a) $\pi/4$, (b) $\pi/2$, and (c) $3\pi/4$. (d and e) Difference spectra optimized for CH$_2$ and CH$_3$ only respectively. D_2 was set at 0.0035 s corresponding to $^1J(^{13}C-^1H)$ of 143 Hz.

but it is also necessary to use a phase-cycling routine. D_2 is $1/\{4J(^{13}C-^{13}C)\}$. Once again D_1 is a relaxation delay. This pulse sequence produces out-of-phase doublets. In-phase doublets are produced by the longer sequence

$$^{13}C\left(D_1 - \frac{\pi}{2}(x) - D_2 - \pi - D_2 - \frac{\pi}{2} - 10\ \mu s - \frac{\pi}{2} - D_2 - \pi - D_2 - \text{acquire}\right)_n$$

(12)

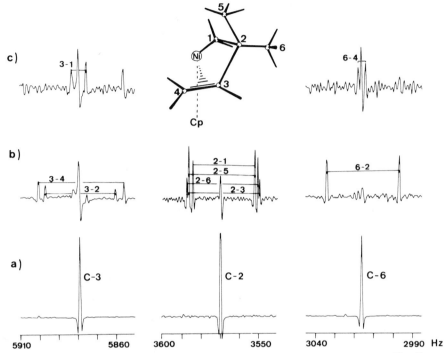

FIG. 12. Use of the INADEQUATE pulse sequence for the determination of $J(^{13}C-^{13}C)$ in $(\eta\text{-}C_5H_5)Ni(\eta^1,\eta^2\text{-}2,2\text{-dimethyl-3-butenyl})$. (a) Signals of carbon atoms 3, 2, and 6; INADEQUATE NMR spectra with $D_2 = 0.0062$ s, corresponding to $J(^{13}C-^{13}C) = 40$ Hz; (c) as in (b), but with $D_2 = 0.08$ s, corresponding to $J(^{13}C-^{13}C) = 3$ Hz. [Reproduced with permission from R. Benn and H. Gunther, *Angew. Chem. Int. Edit. Engl.* **22**, 350 (1983).]

Figure 12 shows the application of this pulse sequence to $(\eta\text{-}C_5H_5)Ni(\eta^1,\eta^2-2,2\text{-dimethyl-3-butenyl})$, **VII** (28). Figure 12a shows the signals due to carbon atoms 2, 3, and 6 recorded normally. Figure 12b shows the spectrum obtained using the INADEQUATE pulse sequence

VII

with $D_2 = 0.0062$ s, corresponding to $^1J(^{13}C, ^{13}C) = 40$ Hz. Figure 12c shows the result of changing D_2 to 0.08 s, corresponding to $^2J(^{13}C, ^{13}C) = 3$ Hz.

INADEQUATE is of particular value in establishing the connectivity between nuclei of low abundance and is not restricted to ^{13}C. It can therefore be applied to nuclei such as ^{29}Si, ^{77}Se, and ^{183}W.

5. Selective Excitation and DANTE

There are two principal methods for selectively exciting nuclei. The first method is to use a low-power signal, which has a narrow bandwidth. This is normally generated through the decoupling channel of the spectrometer. It may be a continuous decoupling signal that changes the Boltzmann population from the equilibrium value to equalize the populations of the two energy levels. Alternatively, the decoupler may be used as a low-power pulse amplifier. The effective bandwidth of a pulse depends on its length, PW. Efficient excitation occurs over a range $1/(4PW)$. Hence, if a pulse power is selected to give a π pulse length of $50,000$ μs, then the pulse is fully effective only within 5 Hz of the chosen frequency, giving a bandwidth of 10 Hz. Nuclei will be somewhat perturbed within 50 Hz of this frequency. This pulse sequence is normally used for nuclei other than ^1H, where few spectrometers are equipped with a low-power pulse or decoupling channel.

a. Delays Alternating with Nutation for Tailored Excitation (DANTE). The normal $\pi/2$ pulse excites nuclei over a wide range of frequencies. For most purposes this is what is required, but it can be convenient to excite one specific nucleus. This can be done in two ways. First, a long low-power pulse can be applied through a low-power amplifier. This amplifier is readily available for ^1H as part of the ^1H decoupler but usually has to be purchased as an additional amplifier for other nuclei. The second method is to use the DANTE pulse sequence, that is,

$$\{D_1 - (\theta - D_2)_m - \theta - \text{acquire}\}_m \tag{13}$$

where θ is a small pulse, applied $m + 1$ times to generate an overall $\pi/2$ pulse. D_1 is a relaxation delay. Each θ pulse is separated by a delay D_2. This produces excitation within $1/mD_2$ of the carrier frequency. Hence if m is 20 and D_1 is 0.001 s, excitation is within 50 Hz of the carrier frequency. In addition to excitation at the carrier frequency ν_0, there is excitation at $\nu_0 \pm p/D_1$, where p is an integer. Thus if D_1 is 0.001 s, there are sidebands at $\nu_0 \pm 1000p$ Hz.

The experiment was originally introduced to assist the analysis of ¹H-coupled ¹³C NMR spectra, where overlapping multiplets cause considerable confusion (29,30). Individual carbon atoms can be excited in the presence of complete decoupling and then observed without decoupling. It is probable that this experiment would now be carried out by two-dimensional methods; see Section III,E. The DANTE pulse sequence is now used for selective excitation in kinetics involving magnetization transfer; see Section V.

6. Broadband Decoupling using Multiple Pulses

Broadband decoupling has been applied for many years to obtain X-nuclei NMR spectra without ¹H coupling. Many techniques are available. Recently a review of the various methods has appeared (31). There are two basic approaches, one using analog and the other pulse methods of decoupling the nuclei.

For ¹³C spectra, broadband decoupling leads to much sharper spectra with NOE (32). At low fields, the modulation of the ¹H single frequency is adequate, but it suffers from the disadvantage of being inefficient. The power is unevenly applied, so that to achieve decoupling over the whole ¹H NMR spectrum high power levels are essential. The problem becomes worse at high fields, where considerable power must be applied to obtain decoupling over the whole ¹H spectral width. As a consequence, there is considerable sample heating. Sample heating is undesirable for a number of reasons. It may cause sample decomposition. In the case of polar solutions, the solution may boil. The sensitivity decreases with increasing temperature due to increased T_1 values, poorer Boltzmann ratios, and increased electronic noise. Usually, temperature gradients are induced in the sample. Most chemical shifts are temperature-dependent, and consequently there is considerable signal broadening with consequent loss of signal/noise (33). This last factor is significant for ¹³C but can be disastrous for the heavier nuclei. Many metal nuclei have a temperature dependence of more than 1 PPM/deg. For ¹⁹⁵Pt at 9.4 T, this is ca. 100 Hz/deg. Hence a temperature gradient of only 0.1 C° over the sample will produce a 10-Hz linewidth. This is particularly a problem when $^1J(X–^1H)$ is large, e.g., $^1J(^{195}Pt–^1H)$ in a platinum hydride, and with the large spectral width when hydrides are present. When noise decoupling is used in such circumstances, it is necessary to use a large decoupling power with the consequent heating and signal broadening.

The usual analog methods of producing the frequency spread in the ¹H decoupling are inefficient and do not produce a uniform power distribution. Recently, pulse sequences have been introduced that are far more efficient and can be adjusted either to produce less heating or to cover a greater spectral width. There are a number of variations on each pulse sequence, depending on the number of pulses that make up the repeating unit. The two main pulse sequences are MLEV and WALTZ. Each method applies a sequence of pulses to the nuclei to be decoupled.

The simplest MLEV sequence (*34*) is MLEV-4. Usually, the more complex pulse sequence, MLEV-16, is used as it is less susceptible to pulse imperfections (*35*). These pulse sequences are applied continuously to the nucleus to be decoupled.

The WALTZ sequence is similar. The basic sequence is WALTZ-4 (*36*), which has been extended to become the WALTZ-16 pulse sequence. The lattu pulse sequence is generally preferred.

The introduction of these pulse sequences to decouple has permitted the observation of much sharper signals due to more efficient decoupling and the reduction of heating effects.

III

TWO-DIMENSIONAL NMR SPECTROSCOPY

Many of the experiments so far examined can be summarized by the sequence preparation–evolution–detection. In two-dimensional NMR spectroscopy the evolution period is an additional variable. Instead of a single experiment being performed, a series of experiments are performed with incresing evolution periods. The FIDs are then transformed twice, once in the normal sense and once as a function of evolution period. This leads to two experimental frequency variables. The basic background to two-dimensional NMR spectroscopy has been described previously (*1,6*), and in this review emphasis is placed on the experiments and the information available. Numerous two-dimensional NMR experiments have been described. In this review, attention is focused on the most commonly used experiments.

The presentation of the results can be confusing. In conventional one-dimensional NMR spectra, there are two variables, frequency and signal intensity. The presentation of such results on a two-dimensional piece of paper presents no problems as the spectra can be plotted using these two variables. The problem with two-dimensional NMR spectra is that there are three variables, requiring a three-dimensional representation on a two-dimensional piece of paper. Originally, stacked plots were used; see Fig 13a. Such spectra can produce attractive decorations for theses, papers, offices, and even homes, but they are difficult to analyze. Stacked

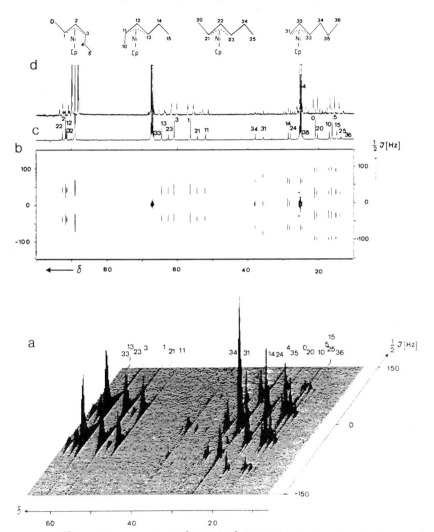

FIG. 13. ^{13}C NMR spectra of Ni(η^3-C$_6$H$_{11}$)(η^5-C$_5$H$_5$). (a) Stacked plot of a J-resolved 100.6-MHz two-dimensional ^{13}C NMR spectrum. (b) Contour map representation of the stacked plot. (c) ^{1}H-decoupled one-dimensional spectrum. (d) ^{1}H-coupled one-dimensional spectrum. [Reproduced with permission from R. Benn, Z. *Naturforsch.* **37b**, 1054 (1982).]

plots are also very time-consuming to plot. It is now usual to present the spectra as projections, cross sections, or contour plots. These approaches are best illustrated by taking some examples. Figure 13a shows a ^{13}C NMR two-dimensional experiment as a stacked plot, where the chemical shift information is in the *x* dimension and coupling constant information is in the y dimension. The stacked plot is difficult to interpret, but the "^{1}H decoupled" ^{13}C NMR spectrum is given by the projection of the stacked plot onto the x axis. Cross sections are of particular value when the purpose of the experiment is to obtain coupling patterns. The most common presentation for a two-dimensional NMR spectrum is as a contour plot. Just as a map represents the three-dimensional structure of the land by contour lines marking places of equal altitude, the three-dimensional nature of a two-dimensional NMR spectrum has contour lines joining places of equal intensity. Figure 13b shows the contour plot representation of the stacked plot in Fig. 13a. Each contour is a cross section through the stacked plot. Comparison of the two figures shows that the contour plot is far simpler to analyze.

A. J-Resolved Two-Dimensional Homonuclear   NMR Spectra

The examples given above are taken from a single J-resolve two-dimensional ^{1}H NMR spectrum of a mixture of four isomers of Ni(η^3-C_6H_{11})(η^5-C_5H_5), **VIII**, **IX**, **X**, and **XI** (37). This experiment (38) uses the pulse sequence

$$\text{Preparation } |\text{evolution}| \text{ detection}$$

$$(D_1 - \pi/2 - D_2 - \pi - D_2 - \text{acquire})_n \tag{14}$$

A series of spectra are measured with different values of D_2 to provide the second dimension. D_1 should be $5T_1$, but normally a value of $2T_1$ is used.

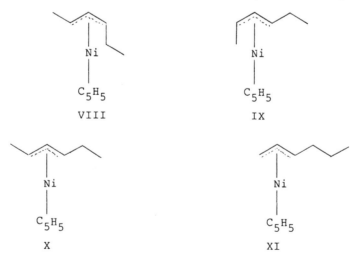

The experiment consists of three parts. There is, first, a preparation time, which consists of a relaxation delay and a $\pi/2$ pulse. There is then the evolution period, which contains the delays for the second dimension and a refocusing π pulse. Finally, there is the detection period, during which the signal is detected. The experiment consists of many separate measurements with different D_2 values. The spectra must be Fourier-transformed twice. The x direction gives the chemical shift, while coupling constant information is separated into the y direction. A projection onto the x axis gives back the normal one-dimensional NMR spectrum. Each multiplet lies along a diagonal of unit gradient. It is therefore possible to tilt each multiplet. The resulting spectrum is shown as a contour plot in Fig. 14a. As a consequence of this tilting, a projection onto the x axis now has no ^{1}H coupling and is an "^{1}H decoupled" ^{1}H NMR spectrum. The coupling pattern is given as a cross section through each peak; see Fig. 14b. Using this technique, it is easy to separate overlapping multiplets. The individual multiplets can be obtained by plotting the cross sections of each peak; see Fig. 14c. Note that the spectrum is symmetric about the center. As a consequence, it is usual to symmetrize the spectrum. The normal symmetrization procedure is to compare points that should have the same intensity and set both equal to the lower intensity. This procedure has the effect of reducing the noise level and minimizing artifacts. Any residual

a

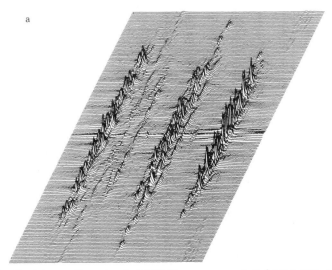

FIG. 14. Partial two-dimensional ^{1}H NMR spectrum of $[(\eta^3\text{-}1,2,5\text{-}C_8H_{13})PdCl]_2$ in CDCl$_3$. (a) Stacked plot; (b) projection of the spectrum onto the x-axis; (c) cross sections of the signals observed in (b).

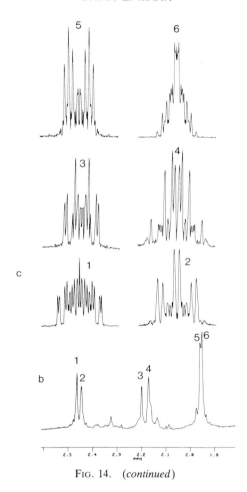

FIG. 14.   (*continued*)

coupling in the projection must be due to a heteronucleus, e.g., ^{31}P. It is therefore possible to distinguish between homo- and heterocoupling.

This experiment is not restricted to ^{1}H NMR spectra, but can be applied to any homonuclear coupled spectra, e.g., complicated ^{19}F or ^{31}P NMR spectra.

B. Homoscalar-Correlated Two-Dimensional NMR Spectroscopy

There are two closely related homoscalar-correlated two-dimensional NMR experiments, *c*orrelated *s*pectroscop*y*, COSY, and *s*pin *e*cho corre-lated *s*pectroscopy, SECSY (*39,40*). COSY gives the most easily interpreted spectra and as a consequence is the most popular. COSY is also a very forgiving two-dimensional experiment, and errors in setting up the experiment are rarely disastrous. It is normally the first pulse sequence to attempt when learning how to perform two-dimensional experiments. COSY has its origins in an idea due to Jeener (41). The pulse sequence is particularly simple:

$$\left(D_1 - \frac{\pi}{2} - D_2 - \theta - \text{acquire}\right)_n \tag{15}$$

where D_1 is ideally $5T_1$ but in practice T_1, D_2 is the variable time used to generate the second dimension, and θ is a pulse between $\pi/4$, for COSY-45, and $\pi/2$, for COSY-90. A contour plot of a COSY two-dimensional NMR spectrum of $[(\eta^2,\eta^1\text{-}1,2,5\text{-}C_8H_{13})PdCl]_2$, I, is given in

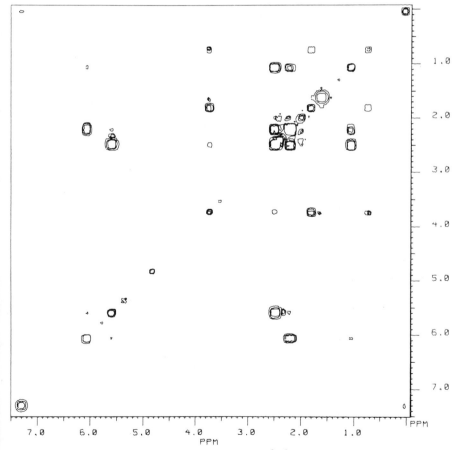

FIG. 15.   Two-dimensional COSY spectrum of $[(\eta^2,\eta^1\text{-}1,2,5\text{-}C_8H_{13})PdCl]_2$ in $CDCl_3$.

Fig. 15. The diagonal represents the normal spectrum. The off-diagonal contours correlate coupled protons. For example, H-5 is at δ 3.73. There is therefore a signal at $x = \delta$ 3.73 and $y = \delta$ 3.73 on the diagonal. It is coupled to signals at δ 2.48, 1.76, 1.02, and 0.71, due to the adjacent CH_2 signals. As a consequence, there are off-diagonal contours at x δ 3.73, $y = \delta$ 2.48, 1.76, 1.02 and 0.71 and at x δ = 2.48, 1.76, 1.02, 0.71, $y = \delta$ 3.73 correlating these two signals. There is no coupling between H-5 and H-1 or H-2, so there are no off-diagonal signals at $x = \delta$ 3.73, $y = \delta$ 6.06 or 5.60 and $x = \delta$ 6.06 or 5.60, $y = \delta$ 3.73. A COSY spectrum must be symmetric about the diagonal. Hence the signal/noise ratio is normally improved by symmetrizing the matrix of data about the diagonal.

The choice of θ affects the signal/noise ratio of the spectrum. If $\theta = \pi/2$, the signal/noise is optimized, but strong responses are found from second-order spectra. The choice of $\theta = \pi/4$ results in poorer signal/noise but second-order coupled responses are suppressed, leaving the diagonal less cluttered.

The choice of the maximum value of D_2 has consequences. D_2 is normally increased from a very small value, e.g., 3 μs to a value comparable with $1/J(^1H-^1H)$. It sets the values of the coupling constants where the response is optimized. It is tempting to choose a long maximum value for D_2 to obtain good digitization, but a maximum value of D_2 of 1 s will produce a response from coupling constants as small as 0.3 Hz, leading to confusing results. It is normal to use a maximum value of D_2 of ca. 0.2 s. If a response from small coupling constants is required, then it can be produced by increasing D_2 from $\{1/J(^1H-^1H) = 0.2$ s$\}$ to $1/J(^1H-^1H)$. This procedure has the advantage that the experimental time is kept short, for both acquisition and transform.

The SECSY experiment uses the closely related pulse sequence

$$\left(D_1 - \frac{\pi}{2} - D_2 - \theta - D_2 - \text{acquire} \right)_n \qquad (16)$$

The presentation of the results is not as convenient as in the COSY experiment, which is normally used.

Both these experiments are suitable to apply to determine the homocoupling between any nuclei, e.g., ^{19}F, ^{31}P, ^{183}W, or ^{195}Pt.

C. INADEQUATE Two-Dimensional NMR Spectra

The major problem with both COSY and SECSY is that the plot close to the diagonal, for COSY, or the x axis, for SECSY, becomes crowded. This is particularly a problem for low-abundance nuclei such as ^{13}C. The problem is avoided by using the double quantum technique. INADE-

QUATE, which suppresses signals without homonuclear coupling (*42,43*). In principle, this technique is potentially very powerful for ^{13}C NMR spectroscopy. It correlates ^{13}C signals via $^1J(^{13}C, {}^{13}C)$. It therefore provides information on which carbon atoms are attached to each other and provides connectivity information. Unfortunately, as it requires two ^{13}C nuclei to be in specific sites in a molecule, only one molecule in 10,000 gives signals, and the sensitivity of the ^{13}C NMR experiment is reduced by a factor of 100 compared to the normal ^{13}C experiment. Consequently ca. 0.5 g of compound is required.

The INADEQUATE experiment uses the pulse sequence

$$\left(D_1 - \frac{\pi}{2}(x) - D_2 - \pi(y) - D_2 - \frac{\pi}{2}(x) - D_3 - \frac{3\pi}{4} - \text{acquire}\right)_n \quad (17)$$

D_1 is once again a relaxation time, ca. $2T_1$, $D_2 = 1/\{4J(^{13}C, {}^{13}C)$, and D_3 is the second variable to give the second dimension. Unfortunately, ^{13}C INADEQUATE only observes one molecule in 10^5 and is very insensitive. Consequently, this technique is of little value. It has been applied to $^{183}W - {}^{183}W$ interactions in polytungstates.

D. *Nuclear Overhauser Effect Two-Dimensional NMR Spectroscopy (NOESY)*

Nuclei can be correlated not only by coupling constants, but also by the nuclear Overhauser effect. This is achieved by the pulse sequence

$$\left(D_1 - \frac{\pi}{2} - D_2 - \frac{\pi}{2} - D_3 - \frac{\pi}{2} - \text{acquire}\right)_n \quad (18)$$

D_1 is the relaxation delay, D_2 the second variable to give the second dimension, and D_3 a mixing time for the NOE to develop. This sequence produces a spectrum analogous to that produced in the COSY experiment. Indeed, it can be combined with the COSY experiment to give both sets of measurements in the same instrument time using the pulse sequence

$$\left\{D_1 - \frac{\pi}{2} - D_1 - \frac{\pi}{2} - \text{acquire(COSY)} - D_3 - \frac{\pi}{2} - \text{acquire(NOESY)}\right\}_n$$
$$(19)$$

Unfortunately, the NOESY experiment is not very successful for small molecules. The problem arises with the relatively long T_1 for protons in relatively low molecular weight molecules. The NOE builds up with a rate T_1^{-1}. As T_1 is long, D_3 must be long. During this time, field inhomogeneity causes signal decay and loss in signal/noise. Consequently, the NOESY experiment works best for high molecular weight compounds. Few organometallic compounds are suitable, and one-dimensional NOE measurements are usually preferable.

The NOESY experiment is of value when exchange is involved. Nuclei can be correlated by exchange in exactly the same way as nuclei can be correlated by the NOE (44). This provides a very valuable method for determining the exchange pathway of nuclei. The experiment has been used to map out the exchange pathway in $Cr(CO)_2(CS)\{P(OMe)_3\}_3$ (45). A variant on the pulse sequence was used:

$$\left(D_1 - \frac{\pi}{2} - D_2 - \frac{\pi}{2} - D_3 - \pi - D_4 - \text{acquire}\right)_n \quad (20)$$

All the three possible isomers are present in solution, **XII**, **XIII**, and **XIV**.

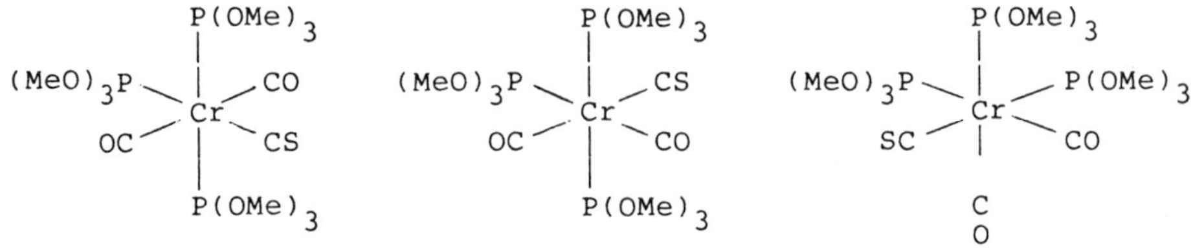

Each isomer gives rise to an AB_2 ^{31}P spin pattern. The NOESY experiment in Fig. 16 clearly demonstrates exchange between isomers. Thus the two equivalent ^{31}P nuclei at δ 191.4 from the mer–cis isomer clearly exchange with both types of phosphorus in the mer–trans isomer, at δ 188.6 and δ 181.2. These observations are consistent with a trigonal twist mechanism. No exchange was definitely detected involving the fac isomer. The response between the two types of phosphorus nuclei in each isomer could be due to a ^{31}P–^{31}P NOE. A quantitative analysis using this type of experiment would clarify this point, as the rate of buildup of the NOE is related to T_1.

A quantitative analysis has been performed for carbonyl scrambling in $Os_3(\mu\text{-}H)_2(CO)_{10}$ (46). This is only a two-site problem.

E. J-Resolved Two-Dimensional Heteronuclear NMR Spectroscopy

The experiments described in Section II,D,1 for one-dimensional investigations can be used to separate $J(X-^1H)$ from the X chemical shift information. This experiment is most commonly applied to ^{13}C. In

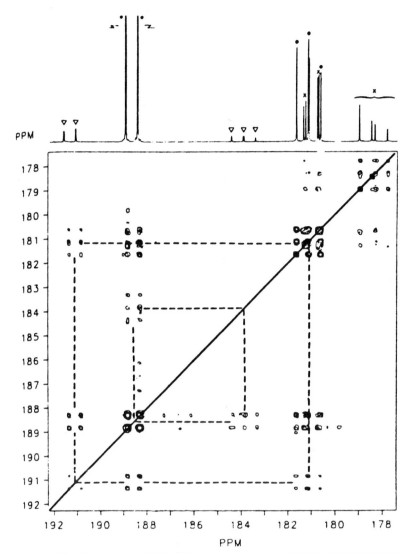

Fig. 16. Two-dimensional NMR ^{31}P contour map for $Cr(CO)_2(CS)\{P(OMe)_3\}_3$ in $CD_3C_6D_5$ at 61°C on a Varian XL-300 NMR spectrometer. All three isomers exhibit an AB_2 coupling pattern. [Reproduced with permission from A. A. Ismail, F. Sauriol, J. Sedman, and I. S. Butler, *Organometallics* **4**, 1914 (1985).]

a

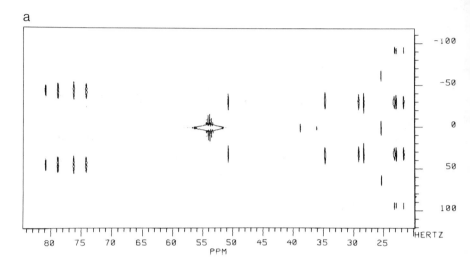

b

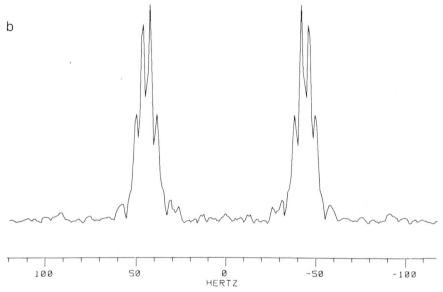

Fig. 17 Partial ^{13}C NMR spectrum of $(\eta\text{-}C_5H_4\text{-methyl})RuCl(PPh_3)_2$ in CD_2Cl_2, (a) Contour plot showing the separation of the ^{13}C chemical shift and ^{1}H coupling. (b) Cross section through the signal at δ 74.1 to show the ^{1}H coupling.

complicated molecules it is difficult to derive $J(^{13}C-^{1}H)$ due to the overlap of the ^{1}H coupled multiplets in the ^{13}C NMR spectrum. This problem is alleviated by using the two-dimensional pulse sequence

$$^{1}H \; ((\quad \text{decouple} \quad) \qquad \text{decouple})_n \qquad (21)$$

$$^{13}C \left(D_1 - \frac{\pi}{2} - D_2 - \pi - D_2 - \text{observe} \right)_n$$

Figure 17a shows the application of this pulse sequence to $(\eta\text{-}C_5H_4\text{-menthyl})RuCl(PPh_3)_2$, **V**, as a contour plot. It is very easy to distinguish between singlet, doublets, triplets, and quartets. Detailed information on the multiplets is obtained by plotting the cross sections; see Fig. 17b, which shows a cross section through the signal at δ 74.1. Note that the pulse sequence has the effect of halving the coupling constant. This experiment is of limited value for the organometallic chemist.

F. Heteroscalar Correlated Two-Dimensional NMR Spectroscopy

There are several versions of this basic experiment. The most common one involves determing the chemical shift of the proton(s) attached to each carbon atom. This experiment uses either INEPT or DEPT to provide the correlation. It assists the assignment of ^{13}C NMR signals, but provides little information on how these building blocks fit together. This vital information can be obtained directly from ^{1}H NMR spectroscopy using $J(^{1}H-^{1}H)$ or NOE measurements. Alternatively, $J(^{1}H-^{1}H)$ information can be obtained from the RELAY two-dimensional experiment. This experiment correlates ^{13}C signals with protons by the two-step process involving $^{1}J(^{13}C-^{1}H)$ and $J(^{1}H-^{1}H)$, e.g., in a $^{1}H-^{13}C-C-^{1}H$ fragment, thus providing $^{13}C-^{1}H$ correlation over two or three bonds. Where there are parts of the molecule that are not linked by $J(^{1}H-^{1}H)$, they can sometimes be linked via $^{2}J(^{13}C-^{1}H)$ and $^{3}J(^{13}C-^{1}H)$, e.g., in the fragment H–C–O–C(O)–C–H. This is done using the two-dimensional experiment COLOC.

1. Heteronuclear Correlation Using $^{1}J(X-^{1}H)$

The definitive assignment of ^{13}C signals usually relies on determining the connectivity between assigned protons and the ^{13}C nuclei to which they are attached. This can be done by selective decoupling if the protons are well separated, or by a series of single-frequency decoupling experiments.

These experiments are preferable when the spectrum contains only a few well-separated signals. In other cases, two-dimensional NMR spectroscopy provides an elegant alternative. The experiment uses the pulse sequence

$$^1\text{H} \left(D_1 - \frac{\pi}{2}(x) - D_2 - D_2 - D_3 - \frac{\pi}{2}(y) - D_4 - \text{decouple} \right)_n \quad (22)$$

$$^{13}\text{C} \left(\qquad\qquad \pi \qquad\qquad \frac{\pi}{2} \qquad\qquad \text{acquire} \right)_n$$

where D_1 is the relaxation time, D_2 is a variable delay to give the second dimension, D_3 is $1/2J(^{13}\text{C}-^1\text{H})$, and D_4 is $1/4J(^{13}\text{C}-^1\text{H})$ to observe all CH_n groups and $1/2J(^{13}\text{C}-^1\text{H})$ to observe only CH groups. This experiment is normally performed in a more sophisticated form. Pulse sequence (22) yields spectra that retain $J(^1\text{H}-^1\text{H})$. Better resolution is obtained by removing $J(^1\text{H}-^1\text{H})$, which is found using the sequence

$$^1\text{H} \left(D_1 - \frac{\pi}{2}(x) - D_2 - \frac{\pi}{2}(y) - D_3 \right.$$

$$\left. - \pi - D_3 - \frac{\pi}{2}(y) - D_2 - D_3 - \frac{\pi}{2}(x) - D_4 - \text{decouple} \right)_n \quad (23)$$

$$^{13}\text{C} \left(\pi \qquad\qquad\qquad \frac{\pi}{2} \qquad\qquad \text{acquire} \right)_n$$

The delays have the same meaning as in pulse sequence (22). A typical contour plot for $(\eta\text{-}C_5H_4\text{-menthyl})\text{RuCl}(\text{PPh}_3)_2$ is given in Fig. 18. It is then easy to read off the plot which protons are attached to which carbon atoms. Thus the ^1H at $\delta\,3.1$ is attached to the ^{13}C at $\delta\,74$. In the case of CH_2 groups with inequivalent ^1H nuclei, two responses are observed for the two different protons. Thus one CH_2 group has the ^{13}C signal at $\delta\,39$ and the ^1H signals at $\delta\,1.2$ and 2.55. There is an artifact in the middle at $\delta\,1.9$. Often this is all that is required to assign a ^{13}C NMR spectrum.

This experiment can equally well be carried out for nuclei other than ^{13}C and certainly could provide useful information for ^{15}N or ^{29}Si. The experiment is rarely of use for metal nuclei, as when more than one inequivalent metal nuclei are present, the $^1J(M, {}^1\text{H})$ are different enough to provide unambiguous correlation without needing the time-consuming two-dimensional NMR measurement.

The reverse of this experiment provides a powerful method for detecting very insensitive nuclei such as ^{187}Os. As noted in Section II,D,2, it is possible to invert the INEPT experiment to carry out the observation of insensitive nuclei with ^1H sensitivity (47). Thus ^{187}Os has 1.14×10^{-3} the sensitivity of ^{13}C. Direct INEPT observation has failed to produce any

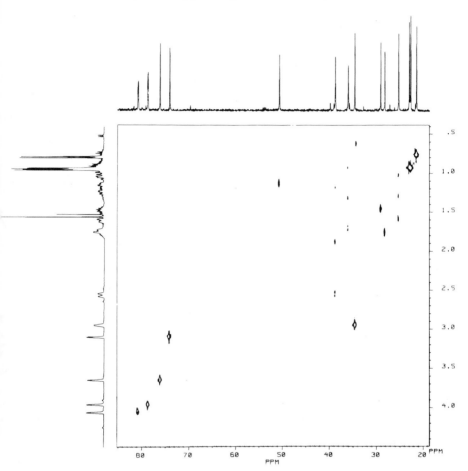

Fig. 18. Partial ^{13}C–^{1}H correlated two-dimensional NMR spectrum of (η-C$_5$H$_4$-menthyl)RuCl(PPh$_3$)$_2$ in CDCl$_3$.

signal, but inverted INEPT, observing ^{1}H, has detected ^{187}Os relatively easily. This experiment must be carried out as a two-dimensional experiment to determine the ^{187}Os chemical shift (48). The ^{187}Os frequency is first estimated by observing ^{1}H and pulsing ^{187}Os using the pulse sequence

$$^1\text{H} \quad \left(D_1 - \frac{\pi}{2} - D_2 - \pi - D_2 - \text{acquire} \right)_n \qquad (24)$$

$$^{187}\text{Os} \, (\qquad\qquad \pi \qquad\qquad)_n$$

D_1 is a relaxation delay, and D_2 is $1/\{2J\,(^{187}\text{Os}–^1\text{H})$. The ^{187}Os frequency is stepped through the expected range. The hydride resonance

SS=9.100   SS=9.105   SS=9.108   SS=9.110   SS=9.112   SS=9.114   SS=9.116   SS=9.118   SS=9.120

Fig. 19.   The 250-MHz^{1}H NMR spectra of the hydride signal of $[\{(\eta^6\text{-}p\text{-cymene})\text{Os}(\mu_3\text{-}H)\}_4]^{2+}$ in $(CD_3)_2CO$ as a function of ^{187}Os frequency using pulse sequence (25). The ^{187}Os frequency was stepped from 9.100 to 9.120 MHz.

shows satellite peaks of intensity 0.84% that of the central signal; see Fig. 19. When a ^{187}Os frequency of 9.100 MHz was used, there was little or no perturbation of the ^{187}Os satellites, but when a frequency of 9.112 MHz was used, the ^{187}Os satellites were inverted. It can be estimated that the best inversion occurs at 9.1126 MHz, which gives approximately the ^{187}Os chemical shift. When the correct $\nu(^{187}\text{Os})$ is used, these satellite peaks are inverted, giving an approximate ^{187}Os chemical shift. The accurate ^{187}Os chemical shift is obtained from the inverted heterocorrelation two-dimensional experiment using the pulse sequence

$$^1\text{H} \quad \left(D_1 - \frac{\pi}{2} - D_2 - \pi - D_2 - \frac{\pi}{2} - D_3 - \frac{\pi}{2} - \text{acquire}\right)_n \quad (25)$$

$$^{187}\text{Os}\left(\qquad\qquad \pi \qquad\qquad \frac{\pi}{2} \qquad\qquad \frac{\pi}{2} - \text{acquire}\right)_n$$

where D_1 is a relaxation delay, D_2 is $1/\{4J(^{187}\text{Os}, {}^1\text{H})\}$ and D_3 is the two-dimensional variable delay. The contour plot for $[(\eta\text{-}p\text{-cymene})_2\text{Os}_2(\mu\text{-H})_3]^+$ is given in Fig. 20a. The ^{187}Os signal appears as a $1:0:0:1$ quartet in the cross section in Fig. 20b. This pattern arises from the method of recording the spectrum. In the data processing, phasing problems are avoided by using magnitude spectra. This removes all sign information from the intensity, making all signals positive.

This method requires a substantial coupling constant between the insensitive nucleus and ^{1}H. Obvious candidates include ^{29}Si, ^{57}Fe, ^{77}Se, ^{103}Rh, and ^{183}W. Recently the method has been applied to ^{57}Fe in $(\eta^5\text{-}C_5H_5)\text{FeH(dppe)}$ using the two-dimensional pulse sequence (49).

$$^1\text{H} \quad \left(5T_1 - \frac{\pi}{2} - D_1 - \quad - D_2 - \pi - D_2 - \quad \text{acquire}\right)_n \quad (26)$$

$$^{57}\text{Fe} \left(\qquad\qquad \frac{\pi}{2} \qquad\qquad \frac{\pi}{2} \qquad\qquad \right)_n$$

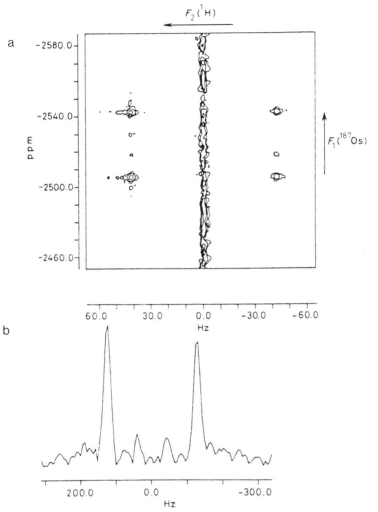

FIG. 20.   Two-dimensional $^1H-^{187}Os$ NMR spectrum of $[(\eta\text{-}p\text{-cymene})_2Os_2(\mu\text{-}H_3)]^+$ using pulse sequence (26). Two-dimensional plot. (b) Cross-section showing lines separated by $3J(^{187}Os, \, ^1H)$. [Reproduced, in part, with permission from J. A. Cabeza, A. Nutton, B. E. Mann, C. Brevard, and P. M. Maitlis, *Inorg. Chim. Acta* **147**, 115 (1986).]

where D_1 is $1/\{2J(^{57}Fe-^1H)$, and D_2 is the two-dimensional delay. This yielded the two-dimensional NMR spectrum given in Fig. 21.

This experiment can equally well be applied to any nucleus that couples to a weak nucleus. Thus ^{31}P can be used as the observing nucleus. This experiment was also applied to $(\eta^5\text{-}C_5H_5)FeH(dppe)$ using pulse sequence (26) with the pulse applied to ^{31}P rather than ^1H; see Fig. 22.

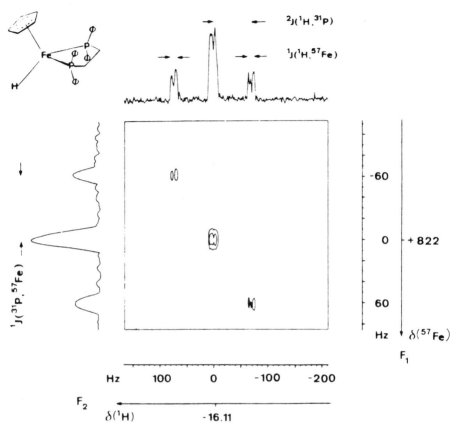

F$_{IG}$. 21.   Reverse two-dimensional ^{1}H–$\{^{57}$Fe$\}$ NMR spectrum of (η^5-C$_5$H$_5$)FeH(dppe) in CD$_3$C$_6$D$_5$. [Reproduced with permission from R. Benn and C. Brevard, *J. Am. Chem. Soc.* **108,** 5622 (1986).]

2. COLOC

The two-dimensional pulse sequence COLOC uses long-range coupling constants to perform X–H correlation (*50*). It is of particular value in determining the correlation between protons and quaternary carbon atoms. It uses the pulse sequence

$$^1\mathrm{H}\left(5T_1 - \frac{\pi}{2} - D_2 - \pi - (D_3 - D_2) - \frac{\pi}{2} - D_4 - \text{decouple}\right)_n \qquad (27)$$

$$\mathrm{X}\left(\qquad\qquad \pi \qquad\qquad \frac{\pi}{2} \qquad\qquad \text{acquire}\right)_n$$

where D_2 is the variable delay to produce the second dimension, while D_3 and D_4 should halve the value determined in an optimized INEPT

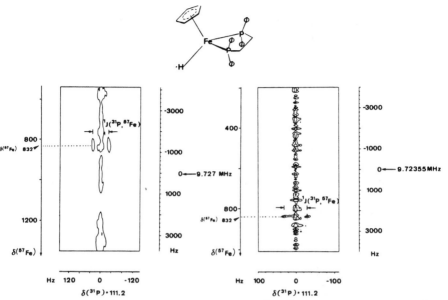

Fig. 22. Reverse ^{1}H-decoupled two-dimensional ^{31}P-{^{57}Fe} NMR spectrum of (η^5-C$_5$H$_5$)FeH(dppe) in CD$_3$C$_6$D$_5$. The ^{57}Fe chemical shift is somewhat different from that in Fig. 21 because of the heating effect of the ^{1}H decoupler. [Reproduced with permission from R. Benn and C. Brevard, *J. Amer. Chem. Soc.* **108**, 5622 (1986).]

experiment. The experiment is extremely valuable in determining the connectivity in molecules.

3. RELAY

The two-dimensional pulse sequence called RELAY enables the connectivity of CH fragments in the molecule to be connected by coupling ^{13}C–^{1}H–^{1}H (*51,52*). It uses the pulse sequence

$$^1H\left(D_1 - \frac{\pi}{2} - D_2 - D_2 - \frac{\pi}{2} - D_3 - \pi - D_3 - D_4 - \frac{\pi}{2} - D_5 - \text{decouple}\right)_n$$

$$^{13}C\left(\qquad \pi \qquad\qquad \frac{\pi}{2} \qquad \text{acquire}\right)_n$$

$$(28)$$

where D_1 is the relaxation delay, D_2 is the delay to produce the second dimension, D_3 is $1/10J(^1$H–^{1}H), D_4 is $1/2J(^{13}$C–^{1}H), and D_5 is $1/4J$-(^{13}C–^{1}H) for all CH$_n$ groups or $1/2J(^{13}$C–^{1}H) for only CH groups. Once again this experiment could be applied in other cases, but suitable molecules are uncommon.

IV

MULTINUCLEAR NMR SPECTROSCOPY

It is now possible to purchase NMR spectrometers that are equipped to observe at any frequency required. Broadband probes exist that cover most nuclei. Thus in our laboratory most work on nuclei other than ^{1}H, ^{13}C, and ^{31}P is performed using a 10-mm probe which covers the wide frequency range of 12–162 MHz, i.e., ^{103}Rh to ^{31}P. It is often advantageous to use these broadband probes. Samples can be prepared in the NMR tube at low temperature, and a wide range of nuclei, e.g., ^{31}P, ^{13}C, and ^{195}Pt, can be recorded without removing the sample from the probe. The ^{1}H NMR spectrum can also be recorded using the decoupling coils of this probe. The ^{1}H NMR spectra may not be quite as good as would be obtained using a 5-mm ^{1}H probe, but they are good enough to characterize many compounds. Figure 23 shows ^{13}C and ^{31}P NMR spectra obtained on [Rh(^{13}COMe)(^{13}CO)(PMe$_2$Ph)I$_3$]$^-$, **XV**, which is stable only at low temperatures. The sample was not removed from the probe between recording the two spectra, but the probe was simply retuned.

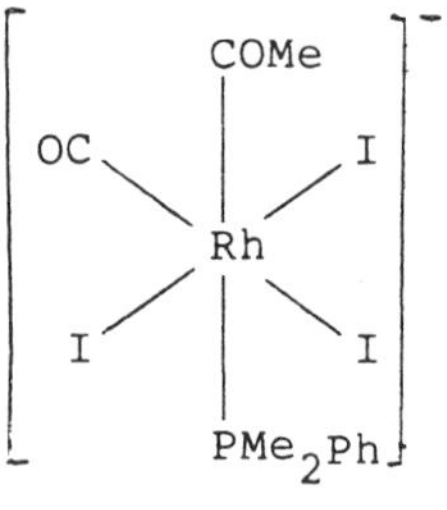

XV

There are a number of problems associated with observing a wide range of nuclei:

(a) The probe is constructed of materials that usually contain NMR-active nuclei, e.g., ^{11}B, ^{17}O, ^{27}Al, ^{29}Si, ^{63,65}Cu. These nuclei in the probe give rise to broad signals, which at least cause a distorted baseline. For ^{29}Si this rarely causes a problem, as the compounds give sharp signals and it is easy to remove the broad signal.

The simplest method for removing broad signals is to remove the first points of the FID. This may be achieved in three ways. First, a delay can be put between the pulse and the beginning of the acquisition of the FID. Second, after acquisition, the data points in the FID can be left-shifted to remove the initial data points and to introduce a zero point at the end of

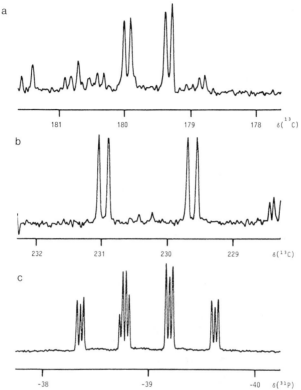

FIG. 23.   (a) Partial ^{13}C NMR spectrum of [Rh(^{13}COMe)(^{13}CO)(PMe$_2$Ph)I$_3$]$^-$ in CD$_2$Cl$_2$ at $-80°$C, showing the ^{13}CO signal. (b) As in (a), but showing the ^{13}COMe signal. (c) ^{31}p NMR spectrum of [Rh(^{13}COMe)(^{13}CO)(PMe$_2$Ph)I$_3$]$^-$ recorded at $-80°$C in CD$_2$Cl$_2$.

the FID. Both these methods introduce major phasing problems if several signals are present. Acceptable signals are then obtained by Gaussian enhancement followed by a magnitude calculation on the transformed spectrum. The Gaussian enhancement attempts to correct for the inherent broadening associated with the magnitude calculation. Third, there are a number of pulse sequences, which are in effect T_2-based. The theory is to apply the pulse sequence

$$\left(D_1 - \frac{\pi}{2} - D_2 - \pi - D_2 - \text{acquire}\right)_n \qquad (29)$$

D_2 is a relaxation delay, which would not normally be included for quadrupolar nuclei, which relax rapidly. The time D_2 is chosen so that the broad signal decays while the sharp signals remain. The various methods have recently been examined, and it was proposed that the pulse sequence

$$\left\{ D_1 - \left(\frac{\pi}{2}\right)_x - D_2 - \pi_x - D_3 - \text{acquire}(-) - D_1 - \left(\frac{\pi}{2}\right)_{-x} - D_2 - \pi_{-x} \right.$$

$$- D_3 - \text{acquire}(+) - D_1 - \pi - \left(\frac{\pi}{2}\right)_x - D_2 - \pi_x - D_3 - \text{acquire}(+)$$

$$\left. - D_1 - \pi - \left(\frac{\pi}{2}\right)_{-x} - D_2 - F_{-x} - D_3 - \text{acquire} \right\}_n \tag{30}$$

yields the best results, where D_1 is a relaxation delay and D_2 and D_3 are delays chosen to minimize the spectral distortions from acoustic ringing (53–56).

This method works when the chemical shift of the quadrupolar nucleus is known approximately. In practice, long pulses are required, causing problems (see below). This procedure is rarely satisfactory on a high-resolution NMR spectrometer.

(b) When a pulse is applied to the probe, it can cause the probe to "ring" and to emit the same frequency for a short period to time, up to 500 μs. The duration of this ringing increases with decreasing frequency. Fourier transform to this ringing produces an undulating baseline. The intensity of the ringing can be much greater than the genuine signal and can cause digitization problems analogous to those when a neat solvent is present in a ^{1}H NMR spectrum.

The methods described above can be applied to probe ringing. An alternative method exists in this case. Spectra are acquired using $\pi/2$ and $3\pi/2$ pulses. The $\pi/2$ pulses are obtained by attenuating the $3\pi/2$ pulse by 10 dB, so that they are the same length but of different power. The $3\pi/2$ produces three times as much probe ringing as the $\pi/2$ pulse, but the true signal is negative compared with a positive signal from the $\pi/2$ pulse. Thus, subtracting three $\pi/2$ FIDs from each one $3\pi/2$ FID removes the probe ringing and leaves the true signal. This method works well but suffers from the narrow spectral widths due to the long pulses required (57).

(c) On high-resolution NMR spectrometers, pulse power is frequently lacking. This is a serious problem, which is exacerabated by further losses associated with the construction of broadband systems. For the best signal/noise, it is often advantageous to use relatively large pulses, e.g., $\pi/2$, for quadrupolar nuclei and to pulse rapidly so that the FID lasts for about half the data acquisition time. This method of operation can cause considerable difficulties. On some spectrometers, the pulse amplifier cannot withstand continuous operation and may only have a 1% duty cycle; i.e., the pulse should last for only 1% of the time between pulses. To adhere to this condition it can be necessary to reduce the pulse width or increase the time between pulses. This results in more time being required to measure a spectrum. For ^{95}Mo NMR spectroscopy, it is often adventitious to apply a $\pi/2$ pulse every 2ms. As a $\pi/2$ pulse for ^{95}Mo can be 50 μs, this can result in a 2.5% duty cycle for the spectrometer, which may endanger the pulse amplifier.

There is an additional problem in using long pulses. A pulse has its maximum effect within 1/[4(pulse width)] and has zero effect at 1/(pulse width). [The pulse is effective as the function $(\sin x)/x$.] Thus a 50-μs pulse has its full effect only within 5 kHz of the carrier frequency, giving a spectral width of 10 kHz with quadrature detection, and the null point is 20 kHz from the carrier frequency. Although a spectrometer may have a spectral width of 150 kHz, it is necessary to use a pulse width of less than about 6 μs to use this spectral width. There is a consequent loss in sensitivity. This is a severe problem at high frequency. A spectral width of 150 kHz is equivalent to only 1750 ppm for ^{195}Pt at 9.4 T (400 MHz for ^{1}H), while the complete chemical shift range for ^{195}Pt is ca. 12,000 ppm. As a consequence of such problems, it is often necessary to carry out the measurement of an unknown at several different carrier frequencies.

(d) Temperature control of the probe is critical for many heteronuclear observations. As the chemical shifts of many nuclei can be very temperature-dependent, e.g., 1 ppm/C° is common, a temperature difference over a sample of 0.01 C° can cause a 1-Hz line broadening for ^{195}Pt at 9.4 T. The air conditioning of the spectrometer room is rarely this good. It is therefore frequently important to acquire heteronuclear signals with temperature control. Even for ^{13}C observation, the quality of the spectrum can be markely improved with accurate temperature control. Thus Allerhand has obtained very sharp ^{13}C NMR signals by using accurate thermostating and composite pulse decoupling (*33*). Unfortunately, the commerically available probes rarely have good enough temperature control, and there is often a major discrepancy between the set temperature and the actual temperature.

V

MAGNETIZATION TRANSFER MEASUREMENTS

In this experiment, one of the exchanging sites is "labeled" with a nonequilibrium Boltzmann population. Chemical exchange transfers this "label" to other exchanging sites in the molecule, where it can be detected in the form of a decrease in the intensity of the signals from these sites. The easiest method to use, especially if the exchange is between two sites, is to decouple at one site. Most NMR spectrometers are equipped to decouple ^1H but not other nuclei. The decoupling method is best when the rate of exchange is comparable with T_1^{-1} for the sites being examined. Both the alternative methods are essentially the same. One uses a low-power transmitter to apply a selective π pulse to one site, and in the other this selective pulse is generated using the DANTE pulse sequence; see Section II,D,5. These selective pulse methods are preferable when the rate of exchange is considerably faster than T_1^{-1}. Three major problems lead to erroneous results from these measurements. These problems are associated with the use of an abundant nucleus, e.g., ^1H, ^{19}F, or ^{31}P. The experiment involves perturbing the Boltzmann population of one nucleus and examing how that perturbation is transferred to the other sites. This perturbation can be transferred by the NOE; see Section II,C. The NOE only occurs significantly intramolecularly in compounds containing more than one ^1H environment, and it builds up at a rate determined by T_1. Hence this problem can be avoided by using low-abundance nuclei, e.g., ^{13}C, where $^{13}C-^{13}C$ NOEs are negligible, by examining intermolecular exchange, or by setting the conditions, e.g., temperature, so that the exchange is fast compared with T_1^{-1}. Decoupling can remove intramolecular coupling, preventing the use of peak heights for the rate measurements and necessitating integration of the spectra. This problem is easily avoided by using the decoupler to perturb the system and then turning if off before observing the effect of the perturbation. The third problem, which can be associated with either decoupling or selective pulsing, involves the use of too narrow excitation, resulting in partial excitation of a multiplet. This leads to selective population transfer causing multiplet distortions, and makes the analysis of the results more difficult. This problem is easily avoided by using strong enough excitation.

This approach to chemical exchange is generally superior to line shape analysis. Line shape analysis simply provides information on the rate of leaving each site. It provides *no* information on the fate of each type of nucleus. As a consequence, mechanisms must be deduced from incomplete information. Magnetization transfer permits attachment of the label of nonequilibrium Boltzmann population to each site, and the movement of each nucleus can be monitored. This provides unambiguous information on the reaction pathway and mechanism.

A. *Magnetization Transfer Using Decoupling*

The original experiment described by Forsén and Hoffman (*58,59*) used a continuous-wave spectrometer and consequently required slow exchange to carry out the measurements within the instrumental limitations. Modern

Fourier transform (FT) NMR spectrometers now make the experiment very easy to perform. A number of methods are available to carry out the experiment, but one of the easiest methods used the sequence

$$^1\text{H(observe)} \quad (D_1 - \leftarrow D_2 \rightarrow - \pi/2 - \text{observe})_n \qquad (31)$$

$$^1\text{H(decouple)} \, (\qquad \text{on} \qquad \text{off} \qquad \qquad)_n$$

where D_1 is a relaxation delay, typically $5T_1$, and D_2 is a variable time to give the exchange rate, with values ranging from 0 to $5T_1$. Figure 24 shows the experiment applied to $\text{Os}(\eta^6\text{-}C_8H_8)(\eta^4\text{-}1,5\text{-cyclooctatetraene})$. For a

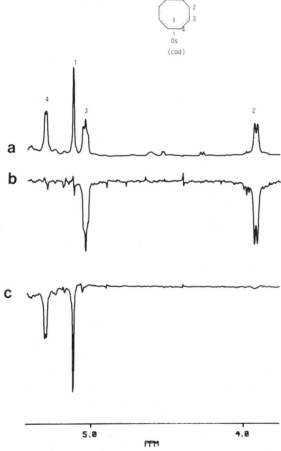

FIG. 24. (a) Partial 400.13-MHz ^{1}H NMR spectrum of $\text{Os}(\eta^6\text{-}C_8H_8)(\eta^4\text{-}1,5\text{-cyclo-}$ octatetraene) in D_8-toluene. (b) Difference spectrum where H_2 is preirradiated and the saturation is transferred exclusively to H_3. (c) Difference spectrum where H_4 is preirradiated and the saturation is transferred exclusively to H_1.

two-site exchange problem, analysis of the data is easy. One signal is irradiated for different times. As the time increases, the observed signal decreases from its initial value of M_0 to its equilibrium value of M_∞. The time dependence of the signal, $M(t)$, follows the equation

$$M(t) = M_0 \left| \frac{\tau}{T_1} + \frac{\tau_1}{\tau} e^{-t/\tau_1} \right| \qquad (32)$$

where

$$\tau_1^{-1} = T_1^{-1} + \tau^{-1} \qquad (33)$$

and τ is the residence time in the observed site. Hence a plot of $\ln\{M(t) - M_\infty\}$ against time yields a straight line with gradient $-\tau_1$. T_1 can be calculated from $\tau_1 M_0 / M_\infty$.

An alternative approach is to measure M_∞ and T_1 separately. The rate of exchange can then be calculated using the equation (60)

$$\tau = \frac{M_0}{M_0 - M_\infty} T_1 \qquad (34)$$

The problem with this approach is that if the T_1 values are different in the two exchanging sites, then exchange tends to average them, leading to erroneous values. Provided that the T_1 is measured in the presence of decoupling at the second site, it is possible to extract a meaningful T_1 value (61).

This equilibrium approach can also be applied to multisite exchange problems (62). For example, it has been applied to $(\eta^6\text{-}C_8H_8)Cr(CO)_3$ to demonstrate that the mechanism of exchange is predominantly [1,3] shifts, with a small contribution from [1,2] shifts (63). Previous work using line shape analysis had failed to distinguish between [1,3] and random shifts (64).

B. Magnetization Transfer using Selective Pulses

There is very little difference between the experiments using a low-power selective pulse and a DANTE pulse. Due to limitations imposed by many NMR spectrometers, the approach using the DANTE pulse is described here, but it can easily be modified for the low-power selective pulse by replacing the DANTE pulse with the selective pulse.

For magnetization transfer measurements, the DANTE pulse sequence given in Section II,D,5 is modified to give the sequence

$$\left\{ D_1 - (\theta - D_2)_m - \theta - D_3 - \frac{\pi}{2} - \text{acquire} \right\}_n \qquad (35)$$

In this case a selective π pulse is applied through the $m + 1$ small θ pulses. During the period D_3, chemical exchange transfers the magnetization from the selectively excited signal to other signals and the degree of transfer is measured by applying the $\pi/2$ general pulse. D_3 is an experimental variable and normally the experiment is repeated ca. 10 times with values of D_3 varying from 1 μs to $5T_1$. The experiment with a D_3 of 1 μs gives the spectrum after the application of the selective pulse. This is essential as the selective pulse may not be exactly a π pulse, and if other resonances are close enough to be perturbed, then this perturbation is also determined. The experiment with a D_3 of $5T_1$ gives the reference equilibrium spectrum.

This technique is most powerful when applied to multisite exchange problems. Such problems are often encountered with polynuclear carbonyl clusters. One of the first examples of the application of this technique was to $Ir_4(CO)_{11}PEt_3$, **XVI**; see Fig. 25. There are seven different carbonyl sites (65). A selective π DANTE pulse was applied to carbonyl b.

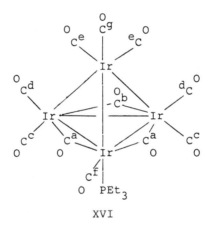

XVI

Examination of the spectra in Fig. 25 shows that the magnetization is first transferred to carbonyl d. There are then two pathways. The magnetization is transferred then to carbonyl a and finally to carbonyl f. Alternatively, the magnetization is transferred to carbonyls c and e. These data are consistent with there being two mechanisms, which could be the merry-go-round mechanism about the bridged face and a merry-go-round mechanism about the unsubstituted face. Unfortunately, the qualitative analysis could not rule out a number of other mechanisms, and at the time of the work the problem of quantitative analysis had not been solved. Recently this problem has been solved and a computer program to fit the data is now available (66). This procedure has been applied to the seven-site exchange problem $[Re_3(\mu-H)_3(CO)_{10}]^{2-}$ and has been used to demonstrate the presence of four separate rate processes (67).

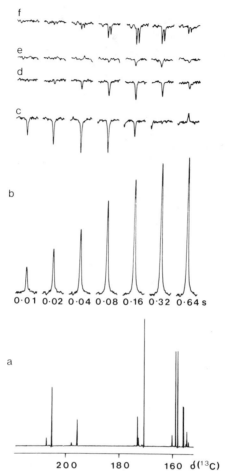

Fig. 25. (a) Partial 100.62-MHz ^{13}C NMR spectrum of $Ir_4(CO)_{11}PEt_3$, enriched ca. 30% in ^{13}CO. (b) Difference spectra showing the recovery of carbonyl b after applying a selective π pulse using the DANTE plus sequence. The times refer to the delay between applying the selective π pulse and the general $\pi/2$ pulse. (c to f) Difference spectra showing magnetization transfer to (c) carbonyls d, (d) carbonyls a, (e) carbonyls f and (f) carbonyls c and e. [Reproduced with permission from B. E. Mann, A. K. Smith, and C. M. Spencer, *J. Organomet. Chem.* **244**, C17 (1983).]

VI

RELAXATION MEASUREMENTS

Relaxation measurements can provide valuable additional information on organometallic compounds. Recently, Crabtree has observed that coordinated H_2 molecules have very short T_1 values (68). This arises because the ^{1}H relaxation arises predominantly from the dipole–dipole mechanism. The value of T_{1dd} is determined by two major factors, the H–H distance r_{HH} and the correlation time τ_c; see the equation

$$T_{1dd}^{-1}(\text{intra}) = \frac{\mu_0^2 \gamma_I h^2 \, I \, (I + 1)}{40 \, \pi^2 r_{HH}^6} \left| \frac{\tau_c}{1 + \omega_I^2 \tau_c^2} + \frac{4\tau_c}{1 + 4\omega_I^2 \tau_c^2} \right| \qquad (36)$$

where ω_I is the NMR frequency in radians per second. A coordinated H_2 molecule retains a very short r_{HH} of 0.80 to 0.90 Å leading to a very short T_1 of ca. 0.03 s; this can be compared with 0.5 to 1.5 s for a terminal hydride, where the H–H distances are much larger. Recently this approach has been extended to accurately determine the H–H distance. T_1 is determined as a function of temperature. At high temperatures $\omega_I^2 \tau_c^2 \ll 1$, but on cooling molecular tumbling slows down, and T_1 passes through a minimum. Analysis of the curve yields τ_c and r_{HH}. The r_{HH} appears to be accurate to 0.01 Å.

Relaxation measurements can also give information on molecular motion. This approach has been used to determine the activation energy for cyclopentadienyl rotation in some ferrocenes (69). For example, in Fe(η^5-C$_5$H$_5$(η^5-C$_5$H$_4$Bu^n), the n-butyl side chain acts as an anchor into the solvent. As the relaxation time depends inversely on the correlation time, the relaxation times for the substituted ring ^{13}CH groups are shorter than for the unsubstituted cyclopentadienyl ring. This is because, in addition to molecular tumbling, the cyclopentadienyl ring is rotating. Analysis of the results yielded the rate of rotation and hence, from the temperature dependence, the activation energy for ring rotation.

VII

SOLID-STATE NMR SPECTROSCOPY

Over the past few years it has become possible to obtain spectra on solid samples. Most of the work in this area has been on commercially interesting problems such as polymers and zeolites. The little work on organometallic compounds has clearly illustrated the potential of the method.

A. Room Temperature Solid-State NMR Spectroscopy

The X-ray structure of (Ph$_3$P)$_2$PtP(mesityl)=CPh$_2$ shows the P(mesityl)=CPh$_2$ to be σ-bonded to the platinum through the phosphorus, but the ^{31}P NMR spectrum in solution at $-55°$C shows the PPh$_3$

groups to be inequivalent, with a low $J(^{195}\text{Pt}-^{31}\text{P})$ of 505 Hz (70,71). This solution spectrum is inconsistent with a σ-bonded structure, **XVII**, and it was suggested that the solution structure has a π-bonded $P(\text{mesityl})=\text{CPh}_2$ group, **XVIII**. It was possible, but very unlikely, that the

XVII   XVIII

structure M could give rise to the solution spectrum. The problem was resolved by determining the ^{31}P NMR spectrum in the solid state; see Fig. 26. These spectra illustrate the problems and the potential of solid-state NMR spectroscopy. At first sight the spectra are a mess. There are many signals, most of which are spinning sidebands. A simple solid-state NMR spectrum would be very broad due to both coupling to other nuclei and chemical shift anisotrophy. The coupling is not only the normal small coupling that occurs through bonds; there is a large coupling through space that is lost in solution due to molecular motion. This through-space coupling leads to a broad signal. The major problem arises from chemical shift anisotropy. In solution, an average chemical shift is observed. For a nucleus in a static environment, e.g., in the solid state, the chemical shift is dependent on the orientation of the molecule with respect to the applied magnetic field. Thus the ^{31}P chemical shift is different for the ligand $(\text{mesityl})\text{P}=\text{CPh}_2$ when the magnetic field is orthogonal to the $\text{C}—\text{P}=\text{C}$ plane than when the field is along the $\text{P}=\text{C}$ bond. A polycrystalline sample contains molecules in all possible orientations. As a consequence, a wide range of chemical shifts are found, leading to a broad signal. Spinning at the magic angle, 54° 44′, with respect to the applied magnetic field and decoupling can remove this broadening, but the spinning must be fast compared with the width of the signal in hertz. Spin rates of 3000 Hz are readily available, and spin rates of 10,000 are possible. At high spin rates the forces applied to the sample are enormous, and disintegration of the sample can occur. Magic angle spinning produces sharp signals when the chemical shift anisotropy is small. When the chemical shift anisotrophy is large, many spinning sidebands are produced, and the central signal may not be the strongest one. This is the case for P_x in Fig. 26. The chemical shift anisotropy for P_x is expected to be substantial, and consequently an

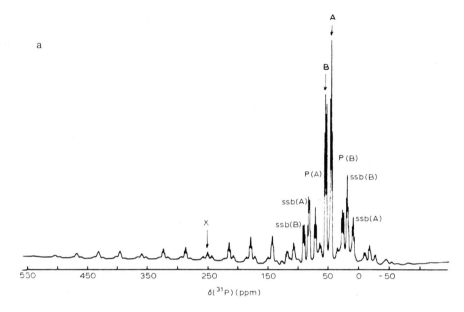

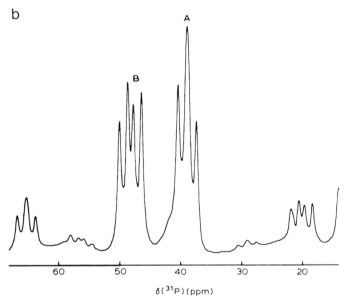

FIG. 26. An 81.013-MHz ^{31}P NMR spectrum using cross polarization and magic angle rotation of $Pt(PPh_3)_2\{P(mesityl)=CPh_2\}$ in the solid state. (a) Spinning speed 2940 Hz. (b) Expansion of (a) in the P_A and P_B region. (c) Spinning speed 2345 Hz. [Reproduced with permission from H. W. Kroto, S. I. Klein, M. F. Meidine, J. F. Nixon, R. K. Harris, K. J. Packer, and P. Reams, *J. Organomet. Chem.* **280,** 281 (1985).]

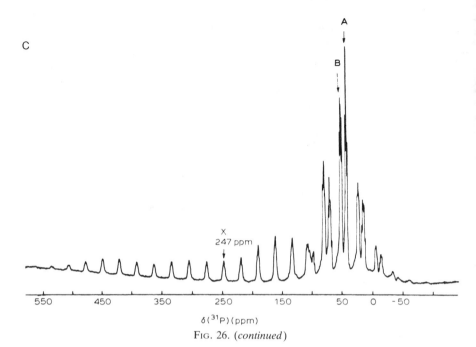

FIG. 26. (*continued*)

extensive set of spinning sidebands is observed. The signals can be separated from spinning sidebands by repeating the measurements at different spin rates. As the extent of spinning sidebands depends on signal width in hertz, the sidebands can be reduced by working at lower fields. The PPh$_3$ ligands are in an approximately tetrahedral environment, leading to a small chemical shift anisotropy and only a few spinning sidebands, making the signals due to P$_A$ and P$_B$ readily identifiable. In solution, an averaged signal for P$_A$ and P$_B$ would be expected. The two phosphorus nuclei can be interconverted by rotation about the Pt–P bond. This is a facile process in solution, but in the solid state considerable lattice disruption would be necessary, and two signals are observed. Figure 26b shows an expansion of the signals due to P$_A$ and P$_B$. The fine structure yields $^2J(^{31}P-^{31}P)$. $^1J(^{195}Pt-^{31}P)$ is derived as 4720 Hz. This is consistent with the σ-bonded structure.

This work illustrates two of the potential uses of solid-state NMR spectroscopy. First, it provides a method for obtaining chemical shift and coupling constant information for a compound with a structure defined by X-ray crystallography. This information is valuable when there is a rapid equilibrium in solution between this and other compounds. A concentration study can lead to equilibrium constants. This method does not appear to have been applied to organometallic compounds, but it has been applied to the complexes $[CdCl_n]_{2-n}$, by determining their ^{113}Cd chemical shifts in the solid state and for the exchange-averaged mixtures is solution. Second, it provides a possible method for determining coupling constants between equivalent groups in order to determine stereochemistry. Extensive use has been made of $^2J(^{31}P, {}^{31}P)$ to determine the stereochemistry of tertiary phosphine complexes of second- and third- row Group 8 transition metals. Frequently, in solution, rapid rotation about the M–P bond results in chemical equivalence of the two tertiary phosphines. In the solid state the tertiary phosphines are frequently inequivalent and should yield $^2J(^{31}P, {}^{31}P)$ in the solid-state ^{31}P NMR spectrum.

B. Variable-Temperature Solid-State NMR Spectroscopy

Until very recently, variable-temperature solid-state NMR spectroscopy with magic angle spinning has been restricted close to room temperature because of the great difficulty of spinning the sample at high speed. This problem is now partially solved and variable-temperature NMR spectra are appearing. Solution-state NMR spectroscopy becomes very difficult below $-100°C$ due to limited solubility in the restricted range of solvents available. When the technique is fully developed, there is the possibility of reaching very low temperatures and freezing out very low energy processes. At present the greatest potential is offered by processes that are too fast to freeze out in solution, yet, because they require considerable crystal rearrangement to occur, are higher energy in the crystal. This has already been noted for $Pt(PPh_3)_2\{(mesityl)P{=}CPh_2\}$ in Section VI,A. Although in solution rotation of the $Pt–P(mesityl){=}CPh_2$ bond would be very fast, it is frozen out in the solid state by crystal packing forces providing a high barrier to the rotation of such a large and nonconical ligand. Unfortunately, most examples chosen for study do not require much spatial rearrangement, and the barrier in the solid state is similar to that in solution. For example, in the $Co_2(CO)_8$ the barrier for carbonyl scrambling between terminal and bridging positions is 11.7 ± 0.6 kcal/mol in the solid state (72). Similarly, the reorientational movement of $[(\eta^5C_6H_7)\text{-}Fe(CO)_3]^+$ is approximately the same in the solid state as in solution (73).

REFERENCES

1. A. D. Bax, "Two-Dimensional Nuclear Magnetic Resonance in Liquids." Delft Univ. Press, Delft and Reidel, Dordrecht, 1982.
2. J. K. M. Sanders and B. K. Hunter, "Modern N.M.R. Spectroscopy, A Guide for Chemists." Oxford Univ. Press, Oxford, 1987.

3. A. E. Derome, "Modern N.M.R. Techniques for Chemistry Research." Pergamon, Oxford, 1987.
4. J. K. M. Sanders and J. D. Mersh, *Prog. N.M.R. Spectrosc.* **15**, 353 (1983).
5. R. R. Ernst, G. Bodenhausen and A. Wokaun, "Principles of Nuclear Magnetic Resonance in One and Two Dimensions." Clarendon Press, Oxford, 1987.
6. G. A. Morris, *Magn. Reson. Chem.* **24**, 371 (1986).
7. A. G. Ferrige and J. C. Lindon, *J. Magn. Reson.* **31**, 337 (1978).
8. D. D. Traficante and D. Ziessow, *J. Magn. Reson.* **66**, 182 (1986).
9. R. Freeman and W. A. Anderson, *J. Chem. Phys.* **37**, 85 (1962) and references therein.
10. J. D. Mersh and J. K. M. Sanders, *J. Magn. Reson.* **50**, 289 (1982).
11. H. Brunner, J. Wachter, I. Bernal, G. M. Reisner, and R. Benn, *J. Organomet. Chem.* **243**, 179 (1983).
12. M. Kinns and J. Sanders, *J. Magn. Reson.* **56**, 518 (1984).
13. K. W. Chiu, H. S. Rzepa, R. N. Sheppard, G. Wilkinson, and W.-K. Wong, *Polyhedron* **1**, 809 (1982).
14. B. E. Mann and B. F. Taylor, "^{13}C N.M.R. Data for Organometallic Compounds." Academic Press, New York, 1981.
15. G. A. Morris and R. Freeman, *J. Am. Chem. Soc.* **101**, 760 (1979).
16. J. Ruiz, B. E. Mann, C. M. Spencer, B. F. Taylor, and P. M. Maitlis, *J. Chem. Soc. Dalton Trans.*, in press (1986).
17. C. Brevard, G. C. van Stein, and G. van Koten, *J. Am. Chem. Soc.* **103**, 6746 (1981).
18. O. W. Sorensen and R. R. Ernst, *J. Magn. Reson.* **51**, 477 (1983).
19. B. E. Mann, P. M. Maitlis, and C. M. Spencer, unpublished results.
20. G. M. Blackburn, B. E. Mann, C. M. Spencer, unpublished results.
21. D. M. Doddrell, D. T. Pegg, W. Brooks, and M. R. Bendall, *J. Am. Chem. Soc.* **103**, 727 (1981).
22. G. C. Levy, *J. Am. Chem. Soc.* **94**, 4793 (1972).
23. G. C. Levy, J. D. Cargioli, P. C. Juliano, and T. D. Mitchell, *J. Am. Chem. Soc.* **95**, 3445 (1973).
24. C. Brevard and R. Schimpf, *J. Magn. Reson.* **47**, 528 (1982).
25. D. M. Doddrell, D. T. Pegg, and M. R. Bendall, *J. Magn. Reson.* **48**, 323 (1982).
26. D. T. Pegg, D. M. Doddrell, and M. R. Bendall, *J. Chem. Phys.* **77**, 2745 (1982).
27. A. Bax, R. Freeman, and S. P. Kempsell, *J. Am. Chem. Soc.* **102**, 4849 (1980).
28. R. Benn and A. Rufinska, *J. Organomet. Chem.* **238**, C27 (1982).
29. G. Bodenhausen, R. Freeman, and G. A. Morris, *J. Magn. Reson.* **23**, 171 (1976).
30. G. A. Morris and R. Freeman, *J. Magn. Reson.* **29**, 433 (1978).
31. A. J. Shaka and J. Keeler, *Org. Magn. Reson.* **19**, 47 (1987).
32. R. R. Ernst, *J. Chem. Phys.* **45**, 3845 (1966).
33. A. Allerhand and C. H. Bradley, *J. Magn. Reson.* **67**, 173 (1986).
34. M. H. Levitt and R. Freeman, *J. Magn. Reson.* **43**, 502 (1981).
35. M. H. Levitt, R. Freeman, and T. Frenkiel, *J. Magn. Reson.* **47**, 328 (1982).
36. A. J. Shaka, J. Keeler, T. Frenkiel, and R. Freeman, *J. Magn. Reson.* **52**, 335 (1983).
37. R. Benn, *Z. Naturforsch.* **837**, 1054 (1982).
38. W. P. Aue, J. Karhan, and R. R. Ernst, *J. Chem. Phys.* **64**, 4226 (1975).
39. A. Bax, and R. Freeman, *J. Magn. Reson.* **42**, 164 (1981); ibid. **44**, 542 (1981).
40. K. Nagayama, A. Kumar, K. Wuthrich, and R. R. Ernst, *J. Magn. Reson.* **40**, 321 (1980).
41. J. Jeener, Lecture, Ampere International Summer School, Basko Polje, Yugoslavia, 1971.
42. T. H. Mareci and R. Freeman, *J. Magn. Reson.* **48**, 158 (1982).
43. D. L. Turner, *J. Magn. Reson.* **49**, 175 (1982).

44. G. Bodenhausen and R. R. Ernst, *J. Am. Chem. Soc.* **104**, 1304 (1982).
45. A. A. Ismail, F. Sauriol, J. Sedman, and I. S. Butler, *Organometallics* **4**, 1914 (1985).
46. G. Hawkes, L. Y. Lian, E. W. Randall, and K. D. Sales, *J. Chem. Soc. Dalton Trans.* 225 (1985).
47. D. A. Vidusek, M. F. Roberts, and G. Bodenhausen, *J. Am. Chem. Soc.* **104**, 5452 (1982).
48. J. A. Cabeza, B. E. Mann, C. Brevard, and P. M. Maitlis, *J. Chem. Soc. Chem. Commun.* 65 (1985).
49. R. Benn and C. Brevard, *J. Am. Chem. Soc.* **108**, 5624 (1986).
50. H. Kessler, C. Grlesinger, J. Zarbock, and H. R. Loosli, *J. Magn. Reson.* **57**, 331 (1984).
51. P. H. Bolton, *J. Magn. Reson.* **48**, 336 (1982).
52. A. Bax, *J. Magn. Reson.* **53**, 149 (1983).
53. S. L. Pratt, *J. Magn. Reson.* **49**, 161 (1982).
54. D. Canet, J. Brondeau, J. P. Marchal, and B. Robin-Lherbier, *Org. Magn. Reson.* **20**, 51 (1982).
55. P. S. Belton, I. J. Cox, and R. K. Harris, *J. Chem. Soc. Faraday Trans.* 2 **81**, 63 (1985).
56. I. P. Gerothanassis and J. Lauterwein, *J. Magn. Reson.* **66**, 32 (1986).
57. G. A. Morris and M. J. Toohey, *J. Magn. Reson.* **63**, 629 (1985).
58. S. Forsen and R. A. Hoffman, *J. Chem. Phys.* **39**, 2892 (1963); *ibid* **40**, 1189 (1964).
59. S. Forsen and R. A. Hoffman, *Prog. N.M.R. Spectroscopy* **1**, 173 (1966).
60. B. E. Mann, *J. Magn. Reson.* **21**, 17 (1976).
61. B. E. Mann, *J. Magn. Reson.* **25**, 91 (1977).
62. M. M. Hunt, W. G. Kita, B. E. Mann, and J. A. McCleverty, *J. Chem. Soc. Dalton Trans.* 467 (1978).
63. J. A. Gibson and B. E. Mann, *J. Chem. Soc. Dalton Trans.* 1021 (1979).
64. F. A. Cotton, D. L. Hunter, and P. Lahuerta, *J. Am. Chem. Soc.* **96**, 4723, 7926 (1974).
65. B. E. Mann, A. K. Smith, and C. M. Spencer, *J. Organomet. Chem.* **244**, C17 (1983).
66. M. Grassi, B. E. Mann, B. T. Pickup, and C. M. Spencer, *J. Magn. Reson.* **69**, 92 (1986).
67. T. Beringhelli, G. D'Alfonso, H. Molinari, B. E. Mann, B. T. Pickup, and C. M. Spencer, *J. Chem. Soc. Chem. Commun.* 796 (1986).
68. R. H. Crabtree and D. G. Hamilton, *J. Am. Chem. Soc.* **108**, 3125 (1986).
69. B. E. Mann, C. M. Spencer, B. F. Taylor, and P. Yavari, *J. Chem. Soc. Dalton. Trans.* 2027 (1984).
70. H. W. Kroto, S. I. Klein, M. F. Meidine, J. F. Nixon, R. K. Harris, K. J. Packer, and P. Reams, *J. Organomet. Chem.* **280**, 281 (1985).
71. Th. A. van der Knaap, F. Bickelhaupt, J. G. Kraaijkamp, G. Van Koten, J. P. C. Bernards, H. T. Edzes, W. S. Veeman, E. de Boer, and E. J. Baerends, *Organometallics* **3**, 1804 (1985).
72. B. E. Hansen, M. J. Sullivan, and R. J. Davis, *J. Am. Chem. Soc.* **106**, 251 (1984).
73. M. J. Buckingham, B. W. Fitzsimmons, and I. Sayer, *J. Chem. Soc. Chem. Commun.* 339 (1984).

Index

Cumulative List of Contributors